WINCH AND CABLE SYSTEMS

FORESTRY SCIENCES

Baas P, ed: New Perspectives in Wood Anatomy. 1982. ISBN 90-247-2526-7
Prins CFL, ed: Production, Marketing and Use of Finger-Jointed Sawnwood. 1982. ISBN 90-247-2569-0
Oldeman RAA, et al., eds: Tropical Hardwood Utilization: Practice and Prospects. 1982. ISBN 90-247-2581-X
Den Ouden P and Boom BK: Manual of Cultivated Conifers: Hardy in Cold and Warm-Temperate Zone. 1982. ISBN 90-247-2148-2 paperback; ISBN 90-247-2644-1 hardbound.
Bonga JM and Durzan DJ, eds: Tissue Culture in Forestry. 1982. ISBN 90-247-2660-3
Satoo T and Magwick HAI: Forest Biomass. 1982. ISBN 90-247-2710-3
Van Nao T, ed: Forest Fire Prevention and Control. 1982. ISBN 90-247-3050-3
Douglas J: A Re-appraisal of Forestry Development in Developing Countries. 1983. ISBN 90-247-2830-4
Gordon JC and Wheeler CT, eds: Biological Nitrogen Fixation in Forest Ecosystems: Foundations and Applications. 1983. ISBN 90-247-2849-5
Hummel FC, ed: Forest Policy: A Contribution to Resource Development. 1984. ISBN 90-247-2883-5
Duryea ML and Landis TD, eds: Forest Nursery Manual: Production of Bareroot Seedlings. 1984. ISBN 90-247-2913-0
Manion PD, ed: Scleroderris Canker of Conifers. 1984. ISBN 90-247-2912-2
Staaf KAG and Wiksten NA: Tree Harvesting Techniques. 1984. ISBN 90-247-2994-7
Duryea ML and Brown GN, eds: Seedling Physiology and Reforestation Success. 1984.
ISBN 90-247-2949-1
Boyd JD: Biophysical Control of Microfibril Orientation in Plant Cell Walls. 1985. ISBN 90-247-3101-1
Findlay WPK, ed: Preservation of Timber in the Tropics. 1985. ISBN 90-247-3112-7
Samset I: Winch and Cable Systems. 1985. ISBN 90-247-3205-0
Leary RA: Interaction Theory in Forest Ecology and Management. 1985. ISBN 90-247-3220-4

Winch and cable systems

by

IVAR SAMSET
Norwegian Forest Research Institute
N-1432 Ås-NLH
Norway

1985 **MARTINUS NIJHOFF/DR W. JUNK PUBLISHERS**
a member of the KLUWER ACADEMIC PUBLISHERS GROUP
DORDRECHT / BOSTON / LANCASTER

Distributors

for the United States and Canada: Kluwer Academic Publishers, 190 Old Derby Street, Hingham, MA 02043, USA
for the UK and Ireland: Kluwer Academic Publishers, MTP Press Limited, Falcon House, Queen Square, Lancaster LA1 1RN, UK
for all other countries: Kluwer Academic Publishers Group, Distribution Center, P.O. Box 322, 3300 AH Dordrecht, The Netherlands

Library of Congress Cataloging in Publication Data

Samset, Ivar.
Winch and cable systems.

(Forestry sciences)
Includes index.
1. Winches. I. Title. II. Series.
TJ1362.S25 1985 621.8'64 85-15239
ISBN 90-247-3205-0

ISBN 90-247-3205-0 (this volume)

PRINTED IN THE NETHERLANDS

CONTENTS

		Page
	Preface	11
	Introduction	13
	List of symbols	17
1.0.	The history of rope transportation	23
1.1.	From antiquity to the present	23
1.1.1.	The oldest equipment	23
1.1.2.	The evolution	25
1.2.	Fiber ropes and wire ropes	28
1.3.	The development of cable systems in Europe	28
1.4.	The development of winch and cable systems in the United States and British Columbia	36
1.5.	The development of winch and cable systems in Norway	46
2.0.	The factor of safety and lifetime of wire ropes	54
2.1.	The factor of safety	54
2.2.	The factor of safety and lifetime	55
2.3.	Bending of wire ropes	57
3.0.	The winch and the wire rope	59
3.1.	The pulling force and the speed of the winch	59
3.2.	Drum capacity	64
3.3.	Capstan and grooved pulley drives	66
3.4.	Braking the winch drum or the pulley	73
4.0.	Resistance from the load	78
4.1.	The weight of trees and stems	78
4.2.	Pull requirement during skidding	80
5.0.	Main, haul-back and hoist lines	85
5.1.	Salvage winch on a vehicle	85
5.2.	Ground skidding	86
5.2.1.	Uphill skidding	86
5.2.2.	Downhill skidding	88
5.2.3.	Skidding on horizontal surface	88
5.3.	Highlead skidding	88
5.3.1.	High skidding without haul-back line	89
5.3.2.	Highlead skidding with brakeable haul-back line	90
5.3.3.	Highlead skidding with a carrying wheel on the brakeable haul-back line	90
5.4.	Braking forces in the haul-back line	92
5.4.1.	Lateral skidding with single hoist line	93
5.4.2.	Lateral skidding with double hoist line and terrain block	94
5.4.3.	Skyline and carriage with double hoist line or hoist drum	95
6.0.	Anchors, spars and intermediate supports	97
6.1.	Anchors	97
6.1.1.	Green trees or stumps as anchor	97
6.1 2.	Earth anchor	102
6.1.3.	Rock bolt anchor	102
6.2.	Head spar and tail spar	104
6.2.1.	Guyline tension	105
6.2.2.	Spar tree	106
6.2.3.	Steel spar	108

Page

6.3.	Intermediate supports	112
6.3.1.	Four strut supports	112
6.3.2.	Two strut supports	112
6.3.3.	Single strut support	114
7.0	Lines in long span	117
7.1	Horizontal and vertical components of the tension	117
7.2.	Cable tension weight	118
7.3.	Elastic elongation of the cable	119
7.4.	Distributed load on the cable	122
7.4.1.	Unevenly distributed load	122
7.4.2.	Evenly distributed load from the weight of the cable	123
7.4.3.	The construction of a cable curve	128
7.5.	Cable tension during load transport	128
7.5.1.	Concentrated load on the cable	128
7.5.2.	Cable tension and deflection due to both the cable weight and the concentrated load	130
7.5.3.	Simplified calculation of the cable tension	132
7.5.4.	Multispan skyline	136
7.6.	The length of the skyline, elastic elongation and pre-tension	140
7.6.1.	Equations for calculating the cable length	140
7.6.2.	The pre-tension of the cable	142
7.7.	Running skyline	146
7.7.1.	Line tensions due to the rope weight (q) and the load (Q)	146
7.7.2.	Line tension due to load resistance	149
7.7.3.	Total line tension due to Q, q in S_T and μ (μ = efficiency)	152
7.7.4.	Power in the lines and transmissions	156
7.7.5.	The torque and the variation of the drum speed with an interlocked winch	159
7.8.	Measurement of cable tension	161
7.8.1.	Measurement by means of extensometer	161
7.8.2.	Tension measurement by means of watch and cable oscilliations	161
8.0	The yarding crew	166
8.1.	Organization of the yarding crew	167
8.2.	Ergonomic work load	168
8.3.	Signal systems	169
8.3.1.	Field telephone	169
8.3.2.	Short distance communication with radio	171
8.3.3.	Acoustic signals	172
8.3.4.	Arm signals	175
9.0.	The function of cable systems	175
9.1.	Cable skidding or cable transport	176
9.2.	Underway and terminal function of cable systems	176
9.3.	Access roads and feeder roads	178
10.0.	Cable ways as access roads	181
10.1.	Wire or wire rope for gravity transport	181
10.2.	Stationary pendulum cable way	181
10.2.1.	The equipment	182
10.2.2.	The operational method	186
10.2.3.	The production	191

Page

10.3. Movable pendulum cable way .191
10.3.1. The equipment .192
10.3.2. The operational method .195
10.3.3. The production .196
10.4. Twin cable systems with circulating carriages .198
10.4.1. The equipment .200
10.4.2. The operational method .203
10.4.3. The production .206
10.5. Single cable system with circulating main line and carriages .208
11.0. Cable cranes as feeder roads .209
11.1. Yarding production cycle with cable cranes .212
11.1.1. The turn cycle .212
11.1.2. The chokers and the rigging .213
11.1.3. Chokers for automatic unhooking .215
11.1.4. Collecting loads from small timber .222
11.1.5. Collecting loads from big timber .226
11.1.6. The chasing .229
11.1.7. The landing .229
11.1.8. Yarding and loading in one operation .235
11.1.9. Yarding and conversion in one operation .237
11.2. Continuous main line system. The mono cable .240
11.2.1. Lasso cable conveyors .240
11.2.1.1. The equipment .241
11.2.1.2. The operational method .243
11.2.2. Nansei mono cable .245
11.2.2.1. The equipment .245
11.2.2.2. The operational method .246
11.2.3. Steyr thinning-paternoster .246
11.2.3.1. The equipment .246
11.2.3.2. The operational method .247
11.3. Cable cranes with top mounted yarders .249
11.3.1. Downhill yarding. Gravity inhaul systems .250
11.3.1.1. The equipment .252
11.3.1.2. The operational method .260
11.3.2. Uphill yarding. Gravity outhaul systems .264
11.3.2.1. Mobile yarders with tower .266
11.3.2.2. Yarders with swing boom for uphill yarding .272
11.3.2.3. Uphill cable crane yarding combined with tractor skidding .275
11.3.2.4. Double haul-back line as skyline .276
11.3.2.5. The production .279
11.4. Cable cranes with yarders at valley station .280
11.4.1. The yarders .282
11.4.2. The operational methods .286
11.4.2.1. Carriages with hoist tackle .287
11.4.2.2. Carriages with single hoist line .301
11.4.3. The production .313
11.5. Cable cranes with all round mobile yarders .315
11.5.1. The chassis .319
11.5.2. Tight skyline systems .319
11.5.2.1. Carriages with hoist winch .320

Page

11.5.2.2. The production .342
11.5.3. Mobile towers and yarders .344
11.5.3.1. The yarders .347
11.5.3.2. The carriages .352
11.5.3.3. The production .359
11.5.4. Running skyline systems .362
11.5.4.1. Tractor mounted double-drum winch with clutches and brakes .362
11.5.4.2. Endless main and haul-back lines .365
11.5.4.3. Interlocked winch with slipping clutch variation .367
11.5.4.4. Interlocked winches with planetary drives .373
11.5.4.5. Interlocked winches with hydrostatic drive .377
11.5.4.6. Interlocked winches with belt variators .386
11.5.4.7. Supports for cable cranes with running skylines .402
11.5.4.8. Yarding cranes .404
11.5.4.9. The yarding production .410
12.0. The rig up work of cable cranes .413
12.1. The planning .414
12.1.1. Layout of the road network .414
12.1.1.1. The access roads .415
12.1.1.2. The feeder roads .415
12.1.1.3. The terminals .416
12.1.2. The location of the cable cranes .416
12.1.3. The planning method .417
12.2. Rig up tools .418
12.2.1. Yarder mounted auxiliary winches .418
12.2.1.1. Guyline drums .418
12.2.1.2. Straw line drums .419
12.2.1.3. Rig up drums - rig up spools .421
12.2.1.4. The rig up line method .423
12.2.1.5. Separate rig up winches .423
12.2.1.6. The straw line method .428
12.2.1.7. The spar tree .434
12.3. Pre-rigging .436
12.4. The rig up work .439
12.4.1. Light cable cranes .440
12.4.1.1. Light cable cranes for ground lead yarding .440
12.4.1.2. Light cable cranes for downhill yarding with top mounted winch .445
12.4.1.3. Light farm-tractor mounted cable cranes for downhill or uphill yarding .448
12.4.1.4. Light farm-tractor mounted cable cranes for uphill yarding and gravity outhaul .451
12.4.2. Medium sized cable cranes .452
12.4.2.1. Traditional standing skyline cable crane .452
12.4.2.2. Radio controlled standing skyline cable cranes .455
12.4.2.3. Running skyline cable cranes .458
12.4.2.4. Steep uphill yarding with running skyline .462
12.4.2.5. Steep uphill yarding with standing skyline (gravity outhaul) .464
12.4.3. Highlead yarding with large yarders .466
12.5. Time consumption for rig up work .473
12.5.1. Light cable ways and cable cranes .474
12.5.2. Medium sized cable cranes .475

Page

12.5.3. Large cable cranes with mobile towers 477
13.0. Balloon logging 478
13.1. Examples on aerostatic and aerodynamic lift 478
13.1.1. Helicopter 478
13.1.2. Airships 481
13.1.3. Hybrid vehicles 483
13.1.3.1. The Piasecki Helistat 484
13.1.3.2. The Cyclo-Crane 485
13.2. The history of balloon logging 486
13.3. The gas and the balloon envelope 489
13.4. The equipment 491
13.4.1. The natural shape balloon 491
13.4.2. The yarder 493
13.4.3. Transfer vehicle 494
13.4.4. Landing machines 494
13.4.5. The bedding area 495
13.4.6. Balloon size for yarding 2 m^3/load 495
13.5. The operational method 495
13.5.1. The turn cycle 496
13.5.2. Main and haul-back lines with double-drum yarder 497
13.5.3. Inverted live skyline and main line with double-drum yarder 498
13.5.4. Balloon running skyline with 3-drum yarder 498
13.5.5. Main and haul-back lines with two single-drum yarders 499
13.5.6. One main line and two haul-back lines with 3-drum winch (Japanese system) 502
14.0. The dimensions, productivity and some economic aspects of cable crane yarding 503
14.1. The cable crane dimensions 506
14.2. The productivity 510
14.3. Some economic aspects 512
14.4. A simple method for the calculation of machine costs 514
14.5. The energy consumption 516
14.6. The law of discontinuous evolution 518
Litterature 521
Index 535

EXAMPLES OF CALCULATIONS

Example 1. Factor of safety in a wire rope 55
Example 2. Factor of safety of a rope which is bent around a sheave 58
Example 3. The pull and the power of a winch 62
Example 4. The torque, speed and pull of a winch 62
Example 5. The acceleration and retardation distances of a cable crane 63
Example 6. The drum capacity 65
Example 7. The pull and the torque of a capstan 68
Example 8. The pull of a groove wheel pulley 71
Example 9. The load on the bearings of a groove wheel pulley due to the tackle effect 72
Example 10. The pull of and the load on the bearings in a groove wheel pulley 72
Example 11. Heat generation caused by braking a winch or a capstan 73
Example 12. The braking length of the cable way 75

Example 13. The braking power of a fan brake 75
Example 14. The braking power of a fan brake on a cable crane 76
Example 15. The braking force of a band brake 77
Example 16. The reduction of the braking force in a band brake with opposite direction of rotation 78
Example 17. The pull of a salvage winch 86
Example 18. Maximum pull on sloping terrain 88
Example 19. The slope angle when the load sliding downhill 88
Example 20. The main line pull of one and two drum winches with or without carrying wheel on the haul-back line 91
Example 21. The braking force in the haul-back line during lateral skidding 95
Example 22. The drawbar pull when ground skidding or high skidding the load with the butt end or the top end forward 95
Example 23. The maximum tension when anchoring in a green stump 97
Example 24. The bending force on a spar tree caused by an angle in the skyline 105
Example 25. Guyline tension 106
Example 26. The elastic elongation of a cable 122
Example 27. Simplified calculation of cable tension 136
Example 28. The length of a cable shoe 140
Example 29. The calculation of cable tension, length of the cable, elastic elongation and pre-tension of a skyline cable 143
Example 30. The tension in a running skyline 148
Example 31. The maximum tension in a running skyline during uphill transport 151
Example 32. The maximum pull during uphill running skyline transport when the load approaches the landing 152
Example 33. The tension in a running skyline caused by load resistance and weight of the load and cables 153
Example 34. The measurement of cable tension by means of stop watch 163

PREFACE

The book is a civil engineering handbook on winch and cable systems.

The handbook may be used as textbook for university studies in civil engineering and forestry and as the basis for studies in schools on a technical level. It should be a useful reference book for construction engineers, civil engineers, logging engineers, foresters and leaders of operational activities under difficult terrain conditions.

The content in the book is based on more than 35 years experience with practical winch and cable operations. As a leader of the Norwegian Institute of Forest Operations, the author has carried out research work in this field since 1947. The Institute is the owner of yarders, winches, cable cranes etc., and with its own cable crews the Institute operates as a contractor in its own research forests as well as in other state or privately owned forests throughout Norway. The research work also includes other cable crane operations in Norway and other countries.

As the leader of the Joint FAO/ECE/ILO Study Group on Mechanized Forest Operations the author studied cable operations in most of the Eastern and Western European countries. As president of the International Union of Forestry Research Organizations the author visited most forest countries in the world. Information from research and practical cable crane operations were collected. The handbook is based on material on winch and cable systems used in Japan, New Zealand, Soviet Union, Central Europe, Northern Europe, Eastern United States, Western United States and British Columbia.

The author has substantial experience as a cable crane worker, leader of cable crews and planner of cable crane operations. The combination of theoretical knowledge and global experience on cable operations has been of great help in the preparation of this book.

The handbook comprises the statics and dynamics in cable systems. The main part of winch and cable systems and methods available in the world have been discussed.

Most of the pictures, drawings, calculations etc. in the book have been done by the author. Many colleagues, research organizations, companies and manufacturers have contributed with pictures or figures and with other background material. They are credited in each case.

Karin Westereng has typed the text. *Solveig Heggen* and *Harriet Ask Kihle* have made most of the drawings. *Magnus Gartland* has helped with the editing and corrections of the text. *Charles N. Mann*, US North West Forest and Range Experiment Station, Seattle, has checked the terminology, *Margaret and Ola Sørhagen* have corrected and adjusted the English text.

Herewith I am thanking all the organizations, institutions and persons who have given valuable help in connection with the preparation of this handbook. Special thanks to *Martinus Nijhoff/Dr W. Junk Publishers* and to *A.C. Plaizier* for their help in connection with the publishing of this book.

Ivar Samset
Professor Dr.
Norwegian Forest Research Institute
1432 Ås-NLH
Norway
December 1984

INTRODUCTION

It has been estimated that the growth of the productive forestry resources in the world is between 2,500 and 3,000 million cubic meters annually. Dr. King, the previous assistant director general of FAO, declared during his speech to the XVI IUFRO World Congress in Oslo, Norway 1976, that it should be possible to manage the forestry resources of the world in such a way that the growing potential would be twice as high. Some authors are even more optimistic. In his lecture October 1977, at the University of Maine, USA, Dr. Egon Gläsinger, previous Director General of FAO, suggested that almost 4,000 million hectares may be suited for forest growth in the world which with proper management are capable of a considerable increase in the annual yield.

The situation today is that we are annually cutting a little over 2,000 million cubic meters timber from the world's forests, half of which is being used as fuel. This harvesting and the corresponding forestry activities take place in the accessible part of the world's forests.

Some forest areas of the world are inaccessible due to remote location or due to difficult terrain conditions. It is assumed that approximately 25 percent of the world's forests are inaccessible for forest operations and other forest activities. This represents some 350 million hectares of forest land (IUFRO, Symposium on mountain logging, Seattle 1979). The main part of these areas is not utilized at all, and some areas are operated by rough logging methods in an exploitative type of forestry.

If these areas could be managed using good silviculture and economically operational methods, the annual cut in the world may be increased by some 700 million cubic meters annually. *A good solution to the problems involved in the operation of the world's mountain forests is one of the most important measures to improve the world's forests today.*

In spite of the fact that mountain logging has a long tradition, the exploitation, and later on the management, started first in the forests on easy terrain. The exploitation on difficult terrain in mountainous regions came later, and even under these conditions the easiest part was taken first. Very often we may see second growth at the foothills while the upper part of the mountain area is covered with over-mature untouched stands.

Modern operations in the mountain forests are not limited to exploitation. The operational methods must also take care of the silvicultural needs. Logging has gradually become an accepted part of the total maintenance and management of the forests and the environment.

Silviculture handles the primary part of forestry production. It involves the input from nature and the adjustments which are being done in order to optimize this part of the forest production.

Forest operations handle the secondary part of forest production and involves the input of men and machines into forestry.

»Logging» is only one part of forest operations. »Forest engineering» represents another part of the total discipline: *Forest operations.* This development of the discipline from logging through forest work science and forest engineering to forest operations and techniques reflects the changes in forest management today. Instead of the previous *biological forestry* on one side, and *exploitative forestry* on the other side, modern forest management takes into account silviculturel as well as operations in a *dynamic forestry production model.*

During the last 30 years a greater change has taken place in the forestry activities than ever before. Modern technology has led to evolutionary changes in the harvesting and transport of forest products. This evolution has first and foremost taken place under easy terrain conditions and under large scale operations. New harvesting machines and better transport systems have been introduced and some parts of the work with the trees in the field have been transferred to the mill. These developments have led to a considerable reduction of labour in forestry. In northern Europe, for example, today's labour consumption is only 20 percent of what it was 30 years ago under easy terrain conditions.

The changes in the operational efficiency were necessary in order to keep up with increases in labour costs, as a consequence of the ever lasting inflation. In Norway, for example, the increase of labour costs during the two periods 1950–60 and 1960–70 was 10–11 per cent annually. The increase of the machine costs (costs per kilowatt hours) was approximately 3 per cent annually during the same two decades. After 1970 the annual labour costs increased by 21 per cent annually, while the machine costs increased by 5.5 per cent annually. *It has never before been more needed to transfer operational activities from labour-intensive to mechanized methods.*

The first introduction of mechanized operations had an exploitative nature and was concentrated on clearfellings of mature stands. Later on more flexible, forestry-oriented machines and methods have been introduced, and gradually, we are reaching a stage where highly mechanized operational methods can be combined with an intensive silvicultural program. In the future it will be of importance to adjust and harmonize the silviculture and the operation according to each others limits and possibilities. This increase in operational efficiency under easy terrain conditions has made it possible to harvest and operate the forests with the same relative economy as 30 years ago.

Under steep and difficult mountainous conditions the situation is different. During the past 30 years the mountain logging operations have changed from the first labour-intensive cable ways and cable cranes to remote controlled cable cranes, interlock winches, etc. In spite of the considerable development in winch and cable systems, the increase in operational efficiency has not kept up with inflation during the same period. The differences between the logging costs in difficult terrain and in easy terrain have steadily increased. Only in forests with dense stands, high volume per hectare and timber of high value is mountain logging being carried out with reasonable economy. Often mountain logging has been an exploitative harvest of clearfellings in over-mature stands.

In many regions of the world the final cut in second growth stands has involved trees of approximately the same dimensions. This has led to more worldwide standardized equipment sizes than was the case in the past.

In forests with low volume per hectare and small tree dimensions such as in the Nordic countries, the costs of mountain logging have increased more rapidly than the inflation. Thinning operations in such stands are most difficult to carry out with reasonable economy.

In some places wheel skidders or crawler tractors are being used for the harvest of timber on steep terrain. Many investigations have shown that tractors can only be used safely in terrain with slopes less than 40 per cent. In steeper terrain tractor roads should be built parallel to the terrain contours, usually with a spacing of 40 – 50 meters.

The timber may be rolled downhill to the tractor road where forwarders transport the timber to a truck road. The combination of lateral ground skidding by means of tractor mounted winch and tractor skidding along the tractor roads may be used.

A dense network of tractor roads may be used on slopes between 40 and 70 per cent. These roads, however, very often lead to serious erosion problems especially where the soil has a high proportion of clay. The negative environmental effect of a such dense road network should be avoided. Instead, an open spacing of feeder truck roads combined with short distance (200–300 m) cable crane logging gives fewer erosion problems and should be preferred in many cases.

Special efforts should be initiated to develop methods suited for forest operations under mountainous conditions. It is hoped that the information available in this book may help with this task.

LIST OF SYMBOLS

The symbol list is divided into the following sub-chapters:

Lengths and slopes
Dimensions and volum units
Elasticity and changes in the length of a cable
Weights and loads
Forces
Powers, torques and frictions
Time concepts
Time units
Availability factors

Lengths and slopes

L	= Horizontal distance of a cable span between the supports	(m)
L_1	= Length along the cable in a span	(m)
$L_1(q)$	= Length of the empty (pre-tensioned) cable (formula 108)	(m)
$L_1(qQ)$	= Length of the cable when a concentrated load is located in the middle of the span (formula 109)	(m)
L_2	= The span distance, which is the length of the straight line between two supports in a span	(m)
L_a	= The acceleration or retardation distance along the cable	(m)
L_v	= The distance along the cable when driving with »constant» speed ($L_1 = 2 \cdot L_a + L_v$)	(m)
L_{tot}	= Theoretical length of cable way related to transport distance for evaluation of operational costs ($L_{tot} = 2 \cdot L_T$)	
sk	= Lateral skidding distance towards the cable	(m)
B	= The width of the area covered by one cable installation	(m)
h	= Height difference between the supports in a span	(m)
h_1	= Height difference between the skyline and the terrain	(m)
y	= The vertical distance between the horizontal line through the lower support and the cable line at a distance x from the left hand support	(m)
f_x	= Cable deflection, which is the vertical distance from the straight line between the supports to the cable line at a distance x from the left hand support	(m)
f_m	= Rope sag, which is the maximum deflection at the centre of the span	(m)
α	= 1) The angle of the slope between the straight line between the supports (L_2) and the horizontal line	(degrees)
	2) Also used as the angle that the steel rope is surrounding a pulley	(radians)
β	= The angle between the horizontal line and the tangent to the cable curve	(degrees)
β_1	= The angle between the horizontal line and the cable tangent at the *uphill* support in a span	(degrees)
β_2	= The angle between the horizontal line and the cable tangent at the *downhill* support in a span	(degrees)
β_1 and β_2	= Also used as the angle of the cable tangent of both sides of a cable shoe (support)	(degrees)
γ	= 1) The difference between the angle of the cable tangent and the slope angle (α), ($\gamma = \beta - \alpha$)	(degrees)
	2) Also used as the difference between the cable tangent angles on both sides of a cable shoe (support) ($\gamma = \beta_2 - \beta_1$)	(degrees)

Dimensions and volume units

R	= The radius of a cable drum or a braking drum	(m)
D	= The diameter of a cable drum or a block sheave	(mm)
d	= The diameter of the steel rope	(mm)
r_1	= The radius of the steel rope	(mm)
r_2	= The radius of the groove in a sheave or a pulley	(mm)
δ	= The diameter of the wire in a rope	(mm)
$D_{1.3}$	= The breast height diameter of the trees including bark	(cm)
h_r	= The relative height of the trees (SAMSET et al. 1969, page 302)	

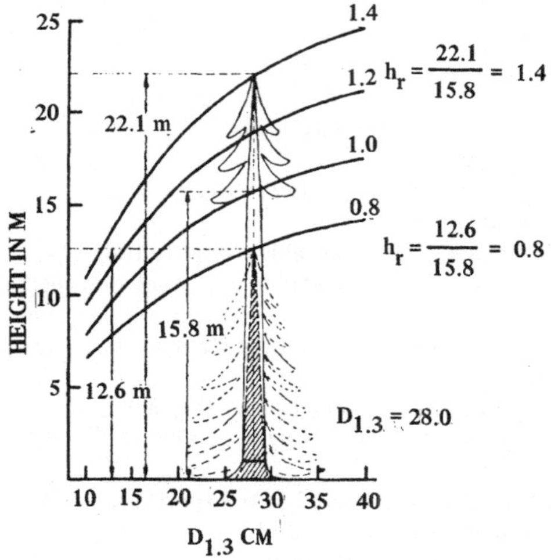

v	= 1) Cubic meter per log outside bark	(m^3)
	2) The designation of speed	(m/s)
	3) The frequency for a forward- and back-running wave on the skyline ($v = \frac{1}{t}$).	(Hz)
v_1	= Cubic meter per log inside bark	(m^3)
$\overline{v}$	= Cubic meter per stem outside bark	(m^3)
$\overline{v}_1$	= Cubic meter per stem inside bark	(m^3)
V	= Cubic meter per winch load outside bark	(m^3)
V_1	= Cubic meter per winch load inside bark	(m^3)
n	= 1) Number of logs, stems or trees per winch load	
	2) Also the designation for number of years after the purchase of a machine (winch or cable crane)	
$\overline{V}$	= Cubic meter per decare outside bark (merchantable timber)	(m^3/daa)
$\overline{V}_1$	= Cubic meter per decare inside bark (merchantable timber)	(m^3/daa)
daa	= Decare = 1,000 m^2	
ha	= Hectare = 10 daa	

Elasticity and changes in the length of a cable

ΔL_1	= Elastic elongation of the cable	(m)
ΔL_1 (Q)	= The part of the elastic elongation caused by the concentrated load (Q)	(m)
ϵ	= The specific elongation of the cable	
A	= Net cross sectional area of cable or steel rope	(mm^2)

E	= Modulus of elasticity. For steel in tension $E = 220\,000\ N/mm^2$	(N/mm^2)
E_r	= Modulus of elasticity for the steel rope	(N/mm^2)
P_E	= Changes in total length of cable in *meters per N of the tension*	(m)
k_1	$= L_1\ (qQ) - (P_E \cdot H + \frac{L}{\cos\alpha})$	
k_2	$= \frac{q^2 L^3 \cos^3 \alpha}{24}$	

Weights and loads

Q	= Weight of the carriage with load	(N)
Q_m	= The mass of the carriage and the load $(\frac{Q}{9.81})$	(kg)
Q_{st}	= Weight of the stem	(N)
Q_{tr}	= Weight of the tree (above the stump)	(N)
Q_t	= The weight of the top part for trees or stems during skidding	(N)
Q_r	= The part of the weight of the butt-end of stems or trees during skidding. $(Q = Q_t + Q_r + \text{possible carriage})$	(N)
Q_{Terr}	= Weight of that part of the load which is in contact with the ground during skidding	(N)
Q_L	= That part of the load, including the carriage, which is carried by the cable during skidding $(Q = Q_L + Q_{Terr})$	(N)
$\bar{q}$	= Weight of the steel rope per meter of the rope length (L_1)	(N/m)
q	$= \frac{\bar{q}}{\cos\alpha}$ = weight of the steel rope per meter of the horizontal distance (L)	(N/m)
q_1	= Weight of the main line per meter of the horizontal distance	(N/m)
q_2	= Weight of the haul-back line per meter of the horizontal distance	(N/m)
q_x	= The function for the load distribution along the cable per meter of the horizontal distance	(N/m)
$\bar{m}$	= The mass of the steel rope $(\frac{\bar{q}}{9.81})$ per meter of the rope (L_1)	(kg/m)
m	= 1) $\frac{\bar{m}}{\cos\alpha}$ = the mass of the steel rope per meter of the horizontal distance (L) 2) m is also the designation for meter	(kg/m)

Forces

S	= The tension	(N)
S_{Br}	= The breaking load	(N)
s	= 1) Factor of safety 2) The average slope of the cable terrain between the head spar and the tail spar in per cent 3) Also the designation of seconds	 (%) (s)
σ	= Stress due to tension	(N/mm^2)
σ_{Br}	= Breaking stress due to tension	(N/mm^2)
S_T	= Pulling force	(N)
S_B	= Braking force	(N)
S_a	= That part of the pulling force which is caused by the slope when skidding	(N)
S_b	= Tension in the guyline	(N)
S_1	= Tension in the main line	(N)
S_2	= Tension in the haul-back line	(N)

S_r	= Tension in the haul-back line caused by braking	(N)
S'	= Cable tension at the *uphill* support in a span	(N)
S''	= Cable tension at the *downhill* support in a span	(N)
H	= Horizontal force (The horizontal component of the cable tension)	(N)
H_f	= Horizontal component of the tension of the unloaded cable (empty cable)	(N)
V	= Vertical force. The vertical component of the cable tension	(N)

Powers, torques and frictions

P	= Engine power	(kW)
P_1	= Power in the main line	(kW)
P_2	= Power in the haul-back line	(kW)
P_T	= Pulling power. (Power needed to move the load)	(kW)
P_B	= Braking power	(kW)
W_B	= Braking work or braking torque	(Nm)
M	= Driving torque	(Nm)
M_B	= Torque for driving double-drum winch when braking the haul-back drum	(Nm)
M_1	= Torque for driving an interlocked double-drum winch	(Nm)
mW	= Milliwatt	(mW)
MHz	= Megahertz	(MHz)
kHz	= Kilohertz	(kHz)
h/m^3	= Manpower requirement in manhours per cubic meter	(h/m^3)
E	= 1) Designation of energy consumption expressed as kilowatthours (kWh), which is the same as 3.6 megajoule 2) Also designates effective time in connection with the time concepts	(kWh)
1 kWh	= 3.6 MJ	
μ	= The efficiency of the winch or of the blocks and ropes	
t	= 1) Time in seconds 2) Also used for the time in seconds for a returning wave on the skyline	(s) (s)
a	= Acceleration	(m/s^2)
n	= 1) Revolutions per minute of the power take-off 2) Also used about the number of wires in a rope	(r/min.)
U	= Ratio for the revolution per minute between the power take-off and the winch drum	
tl	= The ratio between the pulling force and the load for multiple sheave blocks $(\frac{S}{Q})$.	
K	= The pulling force on a handle	(N)
T_2	= The force on the handle-side of the brake lining	(N)
T_1	= The force on the other side of the brake lining ($T_1 > T_2$)	(N)
f	= The coefficient of friction	
f_r	= Rolling resistance between the wheels of the carriage and the cable ($f_r \sim 0.02 - 0.05$)	
f_{gu}	= Friction coefficient when skidding loads on forest floor $(\frac{H}{Q})$.	
f_{go}	= Friction coefficient when skidding loads with uplifted front end on forest floor $(\frac{H}{Q_{Terr}})$.	
CT	= Transport coefficient ($CT = \frac{H}{Q}$)	

CL = Load resistance coefficient (the pull H in relation to the timber weight of the load both for stem skidding and for full-tree skidding), $(CL = \frac{H}{Q_{st}})$.

Time concepts

Ca = All available time in a period (One week = 7 · 24 h = 168 hours)
To = Total time required directly or indirectly to carry out a certain task
Wo = The time spent in performing of task at a working place
E_0 = The time required to perform a specified work element which directly or indirectly changes or transports the work object
D = Delay time, which is an interruption that interfers with the continuity of a performance.
D(n) = Unavoidable delay time, which is an inevitable interruption due to the nature of the work and its continuity along rational lines
D(w) = Delay times caused by the organization of the work
D(m) = Time lost due to faults with the machines, the repair and maintenance which take place on the work place
D(p) = Personal delay time caused by the worker which is necessary for him when carrying out the work in a rational manner
M = Main time. The part of the effective time which directly changes or transports the working object
B = Fixed [B(fix)] or variable [B(var)] by-times which is the part of the effective time which indirectly changes the work object
BW (rep)= Repair time. Time for repair and maintenance which is carried out at other times than during the normal work place time. The repair time may take place at the work place or in the work shop. The time for transportation of the machine to and from the work shop is included.

Calendar time.

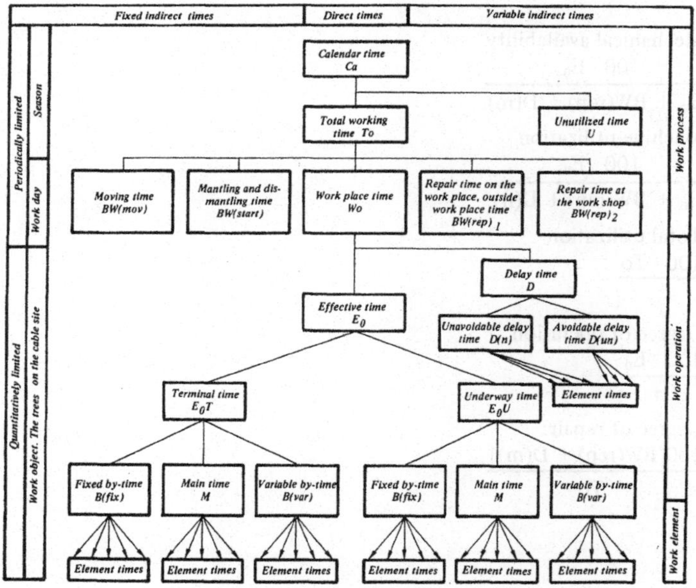

Delay time.

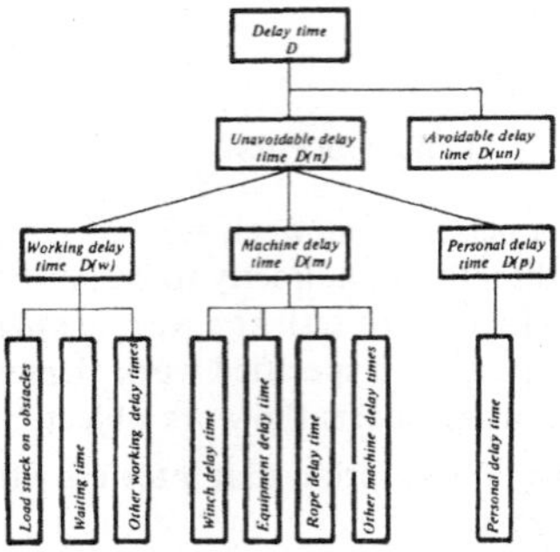

Time units

d	= Number of days at work for the machine or the working crew	(d)
dv	= Mandays	(dv)
h	= Number of hours at work for the machines or the working crew	(h)
hv	= Manhours	(hv)
min	= Minutes	(min)
cmin	= 1/100 of a minute	(cmin)
s	= Seconds	(s)
	The time units are placed in front of the time concepts in combined abreviations, i.e.:	
hWo	= Work place time in hours	(h)
dvTo	= Total time in number of mandays	(dv)

Availability factors

MA = Mechanical availability

$$MA = \frac{100 \cdot E_0}{E_0 + BW(rep) + D(m)} \qquad (\%)$$

MU = Machine utilization

$$MU = \frac{100 \cdot E_0}{E_0 + BW(rep) + D(n)} \qquad (\%)$$

TU = Total utilization

$$TU = \frac{100 \cdot To}{Ca} \qquad (\%)$$

OP = Degree of operation

$$OP = \frac{100 \cdot E_0}{Ca} \qquad (\%)$$

REP = Degree of repair:

$$REP = \frac{100[BW(rep) + D(m)]}{To} \qquad (\%)$$

1.0. THE HISTORY OF ROPE TRANSPORTATION

1.1. From antiquity to the present

1.1.1. The oldest equipment

Before other transport means were available man found it tempting to move through the trees like animals aided by hanging rines. Men hunting or being hunted could swing themselves quickly over obstacles by means of these hanging rines which simplified the journey and eased the load. From time in memorial the men of the primeval forest took advantaged of the line technique. Liana and other hanging rines were used as suspension bridges or to transport material and people over difficult terrain.

The wheel is one of mankind's basic inventions. It originates from Asia during the early stone-age, some 4,000 to 5,000 years ago. The line technique is probably older than the wheel and is thus the oldest method of transport for mankind (STEPHAN 1926).

The rope and line technique was not known in the history of Greek civilization. In ancient civilized nations of East Asia, such as Japan and India with their rugged and difficult mountain terrain, the technique of plaiting ropes from fibres or strips of skin was developed early on. A Japanese sketch from the 14th century showed a cable system which was in use already in ancient Japan (figure 1). It was called »Yaen» which means wild monkey and is still a name of primitive cable ways in Japan (SCHNEIGERT 1966). This method has been in use up to our time, e.g. for crossing rivers or narrow creeks in the jungles in Brazil or New Guinea.

Figure 1. A Japanese drawing from the 14th century of a cable system which have been used in ancient time in Japan. It is called »Yaen» which means a primitive cable way. Similar cable systems have been in use until recently, f.ex. in India and Brazil (STEPHAN 1926, SCHNEIGERT 1966).

Figure 2. Old Japanese cable way with main and haul-back lines (STEPHAN 1926).

Figure 3. Primitive Indian cable way crossing a river in Kaschmir in 1905 (STEPHAN 1926).

In Japan the rope systems were further developed and in the 14th century one had rope-ways for the transport of goods with skylines, main and haul-back lines (figure 2). In other parts of the Orient little material development took place. The Indian mountain population used the same rope ways up until modern times as the Japanese used in ancient times. In Kashmir the same method was used as late as 1925. The ropes were plaited from strips of skins (figure 3). The Indian people may have been more occupied with philosophical thinking than with materialistic developments. STEPHAN (1926) said laconically that a concious development from the simpleancient transport means was not likely to take place among the Oriental population in India due to their peculiar and philosophic character.

1.1.2. The evolution

When Aristotle wrote his book »*Historia Animalium*» some 330 years B.C., he initiated a natural science. Aristotle distinguished between nature and non-nature and announced that human beings should live and work in good relationship with nature. His ideas are still valid. The old Hellenic people had good knowledge in mathematics and physics, which is demonstrated by their impressive buildings. Still their knowledge did not lead to a development of working machines and work saving methods. (WRIGHT 1967).

During the darkness of the Middle-ages some of these ideas were overpowered by religious speculations with philosophical belief and superstition. During the late Middle-ages around the 13th century new ideas developed and the scholastical philosophy tried to build bridges between Aristotlean natural science and religion.
New winds started to blow through European cultural lives, especially after the English philosopher, Sir Francis Bacon, wrote his book »*Novum erganum*» in 1620 (BACON 1620). He was opposed to some of the nature-bound thoughts of Aristotle as well as philosophical superstition and he composed the proverb »Knowledge is power». That is, using our knowledge of nature in order to manage our natural resources. The industrial revolution had started. James Watt invented the steam engine in 1769 and man was no longer limited to dependence on muscle power, the water wheel, the windmill or gravity.

Production of steel was known since the Iron age but this manual process was costly and gave poor quality. The invention of the *Bessemer process* in 1855 and the *Thomas process* in 1879 was fundamental also for the development of cable systems. For the first time in history a cheap mass-production of steel was possible and the development of steel ropes, engine powered winches etc. started.

The development of cable systems took place in the following eras:

1. Antiquity: The Oriental civilizations.
2. Late Middle Age (1300-1600): Japan and Europe.
3. The New industrial Age (1850-1930): Europe, Japan and America.
4. Present time (1930-): All countries with difficult terrain.

The various eras were to some extent partly isolated in time and place in relation to each other, especially during the initial part of the development. Still there are great similarities in the development of cable systems, also in regions where influence from outside was hardly possible. There must be many similarities in the behaviour of mankind and in the way he is tackling problems. In spite of differences in living conditions and environment, the way in which he solved his day-to-day problems had many common features. This also influenced the development of rope systems, especially during the first eras.

There exist, however, differences in the development due to variation in local conditions. There are differences in the original development of rope systems in Japan, Central-Europe, North-Europe, Sovjet Union, United States and Canada. This is especially the case after 1900.

In Central Europe the cable systems should overcome the mountainous conditions and the height differences. The cable systems were developed to tackle special terrain conditons (twin cable with circulating carriages, pendulum cable ways, cable cranes etc.)

In West America the winch systems were first and foremost developed to handle the big tree dimensions from their mature forests. Donkey yarders and winches were developed. Cable equipment was designed in such a way that it could be put together to make up different cable ways according to local requirements. While the European development first concentrated on the development of cable systems the American development concentrated on cable equipment.

In modern times mankind's way of living is becoming more alike, due to easier communication between the various nations and regions of the world. This has influenced the design of the cable systems as well.

Figure 4. A steel wire for gravity transport was built in Fai in the Alps by Hohenstein 1859 (STEPHAN 1926).

There are two main reasons for this global uniformation process.

A. The communication between people dealing with development and research has been easy due to radio, television and other means of communication. Organized research co-operation gives research workers better contact with each other. It has become easier to take new ideas home from anywhere in the world, and put them to use under local conditions.

B. The transfer from labour intensive to capital intensive operational methods replaces manpower by machine power. Mechanized methods are less influenced by the manual input than manual or motormanual methods and are therefore less influenced by individual or local conditions.

Modern technical development has taken advantage of the better motordesigns as well as hydraulic, pneumatic and electronic components. The invention of the silicone chip and the microprocessor may lead to revolutionary developments in all operational activities. The Finish philosopher George Henrik von Wright (WRIGHT 1967) said that atomic energy is

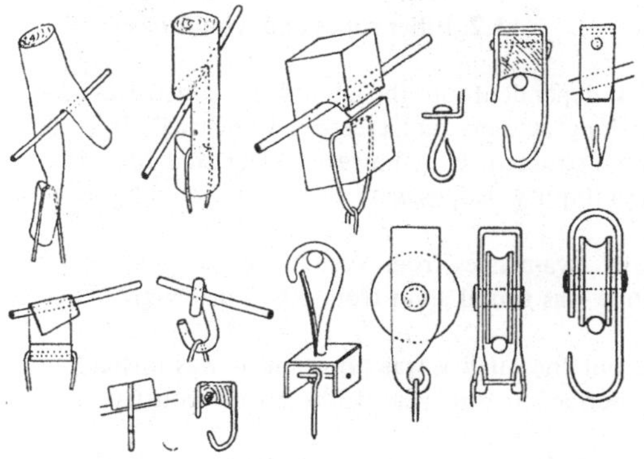

Figure 5. On the steel wire for gravity transport various types of wooden hooks, iron hooks or blocks have been used to carry the load.

not basically different from other sources of power which can replace manual work. The microprocessors which have appeared lately are much more influencial because of their use in automation of work sequences and remote control. While machine power replaces man power, the microprocessors are taking over part of the thinking and can reduce the number of workers in an operation. It is to be expected that these developments will also influence the operational methods concerning winch and cable systems in time to come, partly by reducing the cable crew and partly by the use of remote control.

Figure 6. The first donkey yarder was constructed by John Dolbeer and built by Murry Bros, in San Francisco in 1885. Residues and slash from the logging operation were used as fuel for the steam donkey (ANDREWS 1954).

1.2. Fiber ropes and wire ropes

Line transportation is dependent on the ability to construct ropes of long lengths. The length was limited as long as liana and creepers were used. It is not known when the first long length rope was constructed. It is, however, clear that fibre ropes were used by oriental civilizations already in antiquity. Ropes were also consturcted by using strips of skin.

Pompeii was covered by lava-masses from Vesuvius in the year 79 A.D. During the excavation a 4.5 m long metal rope was found spun from 57 bronze wires. Each wire was 0.7 mm thick.

Iron was also known, but the quality was poor and it was difficult to manufacture the wires. Leonardo da Vinci described in the year 1500 a rope constructed of iron or bronze wires.

The first experiments with wire drawing took place in Nurnberg in Germany in the 14th century, and this method was adopted by French engineers in the 16th century. In the 18th century many European countries made experiments with wire drawing without obtaining any practical results.

Professor Purkinje at the University in Vienna constructed the first endless wire rope in 1925. This rope was used to pull wagons for transporting goods along a steep mountain railway.

After many years of experiments steel wire ropes were put to use in mining in Klausdahl in Hartz 1834. These ropes were 20 and 40 m long. The braking strength was 392 N/mm^2. At first each rope had 3 parts with 4 wires in each part and later, 6 parts were used in the rope (6x4). The wires and the parts were laid or twisted the same way and the construction was called ›*Lang's lay*›.

The Swedish engineer C.P. Bergsten studied the wire rope in Klausdahl and in 1835, Stora Kopparbergs Bergslags A.B. in Falun, Sweden, made a wire rope.

In 1835 Bergsten started his own wire rope factory in Falun. During the following years many countries such as France and England started drawing wires and making wire ropes. After the invention of the ***Bessemer process*** in 1855 the production of steel wires and wire ropes speeded up. In Norway the production of fibre ropes started in Mandal Reperbane in 1774 in order to supply sailships with ropes (DROTTNINGHAUG 1975). Tønsberg Reperbane started to manufacture fibre ropes in 1796 and wire ropes in 1885 (STAVSETH 1948).

1.3. The development of cable systems in Europe

After the stagnation during the dark middle-ages, a more general interest in finding technical solutions for work processes developed. During the 15th century the development of cable systems, for the harvesting of forests in mountainous regions in central Europe, took place.

Keysers von Einstädt showed in 1405 how men on horseback could be transported in carriages over cliffs and narrow valleys using skylines. The carriage was pull across by a main line and returned by a haul-back line (STEPHAN 1926). Johan Hartlieb described a type of cable way in 1411. A basket was moved forward and back over the cliff by a hand driven grooved pulley (figure 7). It was a mono cable basically of the same type as to-days running skyline systems.

Figure 7. Cable way constructed by Johan Hartlieb 1411 (STEPHAN 1926).

Marinius Jacobus Taccola described another type of cable way in 1440 (figure 8). A skyline over a narrow valley was used to transport guns from one mountain to another. The gun was pulled along the span by means of the main line and a haul-back line returned the carriage. Oxen were used to pull it.

Typical for most of these constructions were that some of the lines were used to carry the load while others were used to move it. Horizontal spans were mostly used and the skylines had the same function as roads. The 17th century's cable systems in Western Europe were principally the same as those used in Japan during the 14th century. They replaced roads over difficult terrain. In the book »*Machinae Novae*» written by Faustus Verantius in 1617 a cable way was described that had skylines for carrying the load and an endless line to transport the carriage along the skyline (figure 9). This cable way was used for transportation across gorges and rivers. A similar type of cable way was built in 1536 by the Spanish emperor of South Africa. It was built between Santanda and Merida and it was in operation until around 1800 (STEPHAN 1926).

Figure 8. A cable way poposed by Marinius Jacobus Taccola 1440 (STEPHAN 1926).

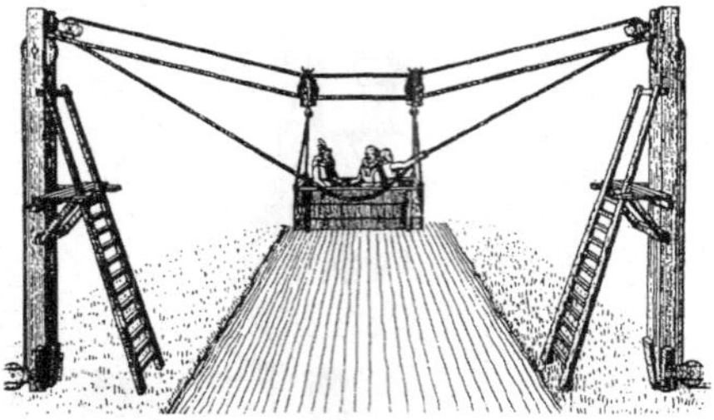

Figure 9. A cable system according to Faustus Verantius 1617 (STEPHAN 1926).

The Dutchman Adam Wybe built a twin cable system with circulating carriages near Gdansk in 1644. Small baskets for earth transportation were attached at short distances from each other along the lines. The system was used for the construction of the fortification of Danzig. Similar constructions were described by Robert Southwell in 1692 (figure 10) and Jacob Leopold in 1714. These developments showed that the cable way was not limited to having a connecting function between two points over difficult terrain (underway functions). The constructions led to higher production throught the simplification of loading and unloading (terminal function). The cable way was transformed from being a road connection only, to being a means of transport.

So far, the cable way had been man or animal powered. The force of gravity was also used although no reliable kind of brakes had been invented. Most of the cable ways therefore, were built on small uphill or downhill slopes.

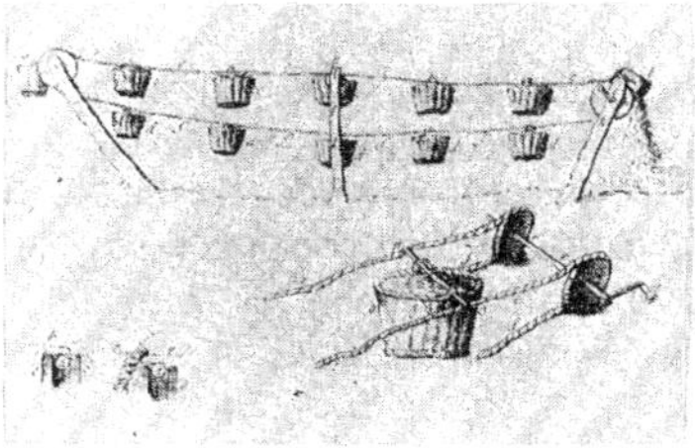

Figure 10. A cable system constructed by Robert Southwell, England 1692 (STEPHAN 1926).

The steam engine was available as power unit in 1769, the gasoline engine in 1861 and the diesel engine in 1893. The production of cheap steel for construction was initiated in 1855. As a result of these inventions the number of development increased. The mining engineer Freiherr von Dücker constructed a trial cable way in Bad Oegenhausen in 1861, where a fixed skyline and supports were used (figure 11). The skyline was a single wire (round-steel) with a diameter of 12.5 mm. The new construction allowed the carriage wheels to pass the cable shoes at the supports. It took another 10 years before this type of cable way was used in practice under the name of the »*German system*». In 1867 the English engineer Hodgeson developed a type of cable way partly inspired by the Adam Wybes construction (figure 12). In some cases the Hodgeson cable way had only one endless line which carried and pulled the circulating carriages. In others there were two skylines (twin cable) carrying the carriages and a special pulling line pulled the carriages. This type of cable way became known as »*the English system*». Such a cable way was built for mining purposes in Colorado in 1868.

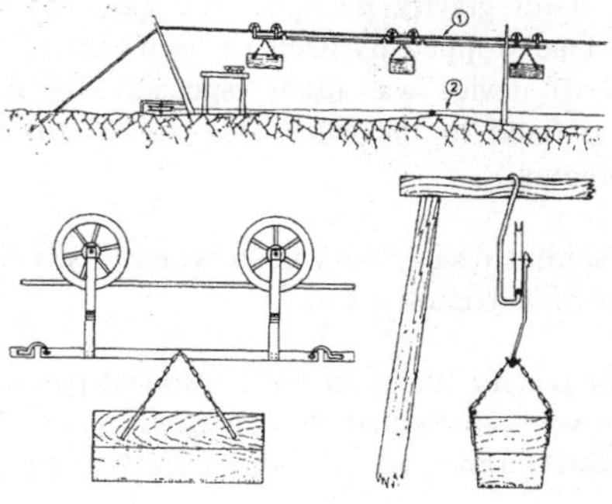

Figure 11. Twin cable system with circulating carriages after von Dücker, Germany 1871 (»The German system»). (STEPHAN 1926).
1. Skyline made of single wire.
2. Main line pulling the carriages.

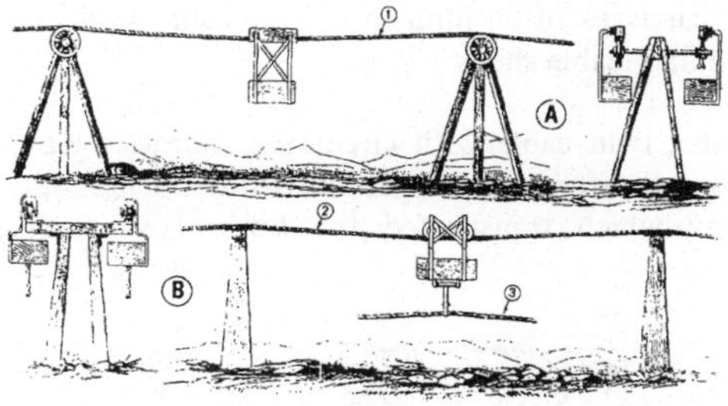

Figure 12. Twin cable system constructed by Hodgson, England 1869 (»The English system»). (STEPHAN 1926).
A.1. Running skyline.
B.2. Fixed skyline.
B.3. Main line pulling the carriages.

The first cable systems for timber transport were aerial transport wires. They were used for transporting the timber down from high locations in the Central European mountainous regions.

At first, hemp rope was used to carry the load and one such unit was in operation in Sorrent in Germany in 1825 (HAFNER 1941). On the mountain slopes of the volcano Vesuvius in Italy, a 500 meter long hemp rope for gravity transport was in operation during the same period. The small logs were hooked to the hemp rope skyline by means of wooden hooks (GIORDANO 1959). Later on a steel wire with a diameter between 5 and 8 mm was used as skyline.

In Cansiglio in Northern Italy a steel wire for gravity transport of timber was used in 1836. The hooks were made partly of wood and partly of steel. The wooden hooks gave the best result because the frictional heat carbonized the contact surface. This in turn resulted in low friction and less wear. The steel hooks did not last long because the heat and friction cut through the hook. Later small wheels were used, especially for transport over long distances.

At the lower end of a rope for gravity transport the speed of the load decreased since the skyline flattened out. The load stopped by itself or by means of a buffer device fixed to the rope. Sometimes a throw-off device was used, especially for fuel wood transportation. A triangular shaped knife was fixed to the wire. This cut off the wooden hook and the fuel wood fell down onto the landing.

Aerial transport wires of similar design have been widely used in Central Europe as well as in Norway up to the present time (figure 4 and 5).

Around 1870 two wires for gravity transport were installed parallel to each other. Two small carriages, one on each wire were connected to each other by a braking line. This was tried out about the same time in Switzerland and Southern France and was the beginning of the pendulum cable ways.

In 1873 the Swiss engineer König built a 2,100 meter long, twin cable with circulating carriages for timber transport in Schlierental in the canton Unterwalden in Switzerland. It was a gravity transport system with an average slope of 33%. The cable systems developed by the German Dücker and the Englishman Hodgeson were used as a model and there were many supports along the line. At first the skyline was supported by wheels on the supports. Since the carriages often jumped off the cable at the supports the carrying wheels were replaced by ordinary cable shoes.

In Russia a 900 meter twin cable with circulating carriages was built in 1870, in order to transport timber from the forests to the Pokrowskiv Nizhgorod railway station. This was a horizontal cable way which transported the timber over soft ground with low bearing capacity.

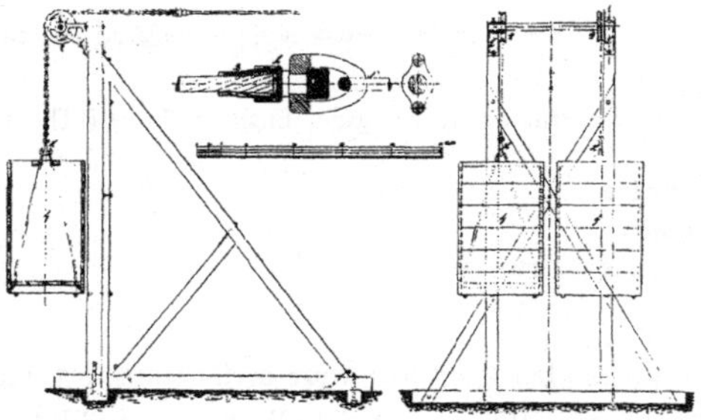

Figure 13. Gravity weights for tightening the skylines, constructed (and patented) by A. Bleichert 1870 (STEPHAN 1926).

These cable ways became more useful after some improvements made by the mechanical engineers Bleichert in Germany and Obach in Austria. Bleichert invented and patented a weight for tensioning the skyline in 1870, in order to avoid overloading the skyline (figure 13). Obach designed a screw connection between the carriage and the endless pulling line, which made it possible to disconnect and stop the carriage at the loading or unloading station without stopping the pulling line and the grooved pulley which powered the line (figure 14). The twin cable system with circulating carriages was built at Minden in 1875. The total length was 920 meters and 10 carriages with equal spacing between them moved along the skyline. The constructors first built a model of the cable way in 1:50 scale in order to determine the size of the weight and the cable tension. This cable way had a production of 40 tons per day and was operated by 15 men. STEPHAN wrote in 1926: »Die heute so wichtige Ersparniss an Bedienungspersonal hatte demnach damals noch keine besondere Bedeutung». It was of importance to save manpower at that time and today this has become a challenge.

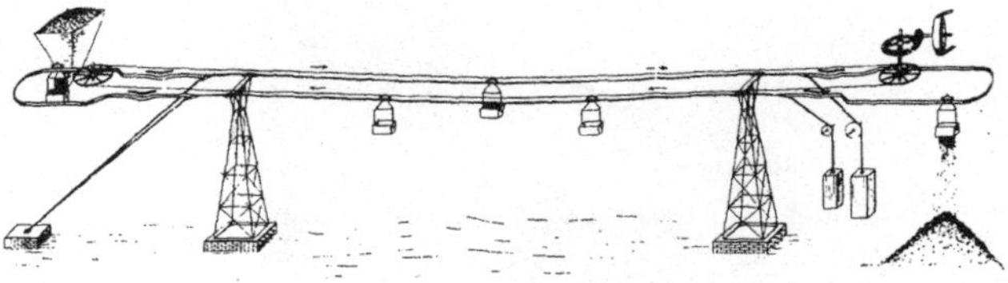

Figure 14. Twin cable system with circulating carriages according to »The German system» from 1871 (STEPHAN 1926).

Figure 15. A. Swiss agricultural winch with transport wheels (WESETH 1952)
B. German self-propelled agricultural and forestry winch (Bergkuli).

The cable way in Minden 1875 had a production of 2.5 tons per man per day. A twin cable with circulating carriages was studied at Stengestad in Telemark, Norway in 1955 and had a production of 15-20 tons per man per day.

Many companies started to manufacture cable equipment during the period 1860-1870. Germany had a leading position and many twin cables with circulating carriages were manufactured for industrial purposes. Around 1900 many twin cables with circulating carriages and pendulum cable ways were built for timber transport. They were very often used to supply sawmills with timber from the forests in the mountainous regions of Central Europe.

During the period 1920-1940 winch equipment was developed for pulling agricultural equipment on steep farmland - such as in the Tyrol and in the Alps (PESTAL 1961). To

Figure 16. On steep farm land the agricultural winch is used for pulling ploughs, harrows, etc. and for pulling transport carriages (WESETH 1952).

begin with these were single-drum farm winches and later on double-drum winches were developed. The winches were mounted on runners and the main line was used to pull the winch along the ground to the winch station. Some winches were equipped with wheels for transportations and could be transported as a trailer. In other cases the winch was self-propelled. The winch motor was used to drive the transport wheels (figure 15). Steering systems were developed. The winch was used to pull agricultural equipment (figure 16). Such agricultural winches were manufactured by small local workshops, often on order from the agricultural administration.

Around 1920 the agricultural winches were used to power simple and light cable cranes for timber transport in the farm forests in the Alps (HAFNER 1964, MEYR 1980). Fuelwood and timber were transported and these cable crane systems were developed mainly during the period 1930-1935.

The first cable crane designed spesifically for timber transport in Central Europe was constructed in 1939 by the Swiss cable crane pioneer J. Wyssen. The development of these cable cranes was influenced by experiences with agricultural winches and gravity powered cable ways (steel wires for gravity transport, pendulum cable ways etc.).

The cable cranes had a fixed skyline with automatic carriage and carriage stops. The winch was used for lateral skidding of the logs to the skyline and to hoist the load from the ground and up to the carriage where the hook was automatically locked to the carriage. Gravity was used for downhill transport and the winch brake was used to control the speed along the skyline down to the lower landing. The big height differences necessitated powerful brakes which generated a lot of heat. The winches therefore, were later equipped with fan brakes (SAMSET 1979B).

Figure 17. The Austrian Koller truck mounted K-800 yarder with steel spar which can be lowered for transport.

The winch had two speeds. The high ratio was used to return the carriages uphill along the skyline with a speed of 3-6 meters per second. The low ratio was used for lateral skidding with a speed of 0.5 - 1.0 meters per second.

During the period 1940 - 1960 many different cable cranes were developed. Some were big, stationary cable cranes and some were light, movable cable cranes for skidding (MEYR 1980). (Switzerland: WYSSEN and KÜPFER; Germany: HUBER; Austria: HINTEREGGER, GOSCH, GANTNER, KOLLER, GIRAK, de PRELIS, BACO and others).

WYSSEN made a further step forward with the development of a cable crane carriage which could be stopped automatically anywhere along the skyline without the use of carriage stops. Later other manufacturers designed other carriages which could be automatically stopped. This was an advantage, since it made it possible to carry out lateral skidding and loading from several places along the skyline and the load could be released onto different piles at the landing.

After 1960, mobile cable cranes with steel towers were built. These cable cranes were mounted on truck or tractor chassis (figure 17). (GÖSSER 1961; HINTEREGGER 1964; KOLLER 1965; STEYR 1979).

The development of Central European winch and cable equipment increased after 1940. It was influenced by research activities, and new methods and machines appeared as a result of the cooperation between manufacturers and researches (STEINLIN, Switzerland;

KAUFMANN, Germany; GIORDANO, Italy; MEYR and PESTAL, Austria and others). The development was also influenced by the new methods and machines which appeared in Canada, The United States and Northern Europe. ISACHSEN, Norway, built a tractor mounted double-drum winch with steel tower (tip tower) in 1951 (the first cable yarder on farm-tractor) (figure 41). A global exchange of ideas began to take place. The winches were mounted on mobile chassis and were equipped with steel towers. This allowed a certain freedom when positioning the winch in the terrain.

1.4. The development of winch and cable systems in The United States and British Columbia

The technical revolution which took place in Europe from 1870 reached »the new world», first in the Eastern United States and Eastern Canada.

An English engineer Joseph GLYNN wrote in 1873 a handbook on cranes where he described a capstan with a planetary drive of similar type as used in today's interlocked winches (GLYNN 1873). (Figure 18).

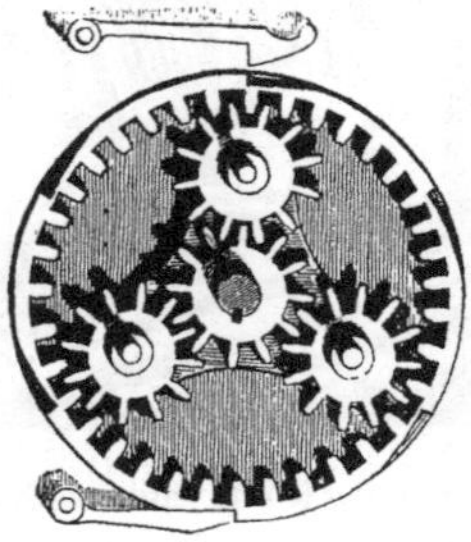

Figure 18. English planetary drive for a capstan from 1873 of a type similar to to the planetary drive in the Moxy winch from 1980 (GLYNN 1873).

Figure 19. Double-drum steam driven winch. Trenton, New Jersey, USA 1892 (ANONYMOUS 1892).

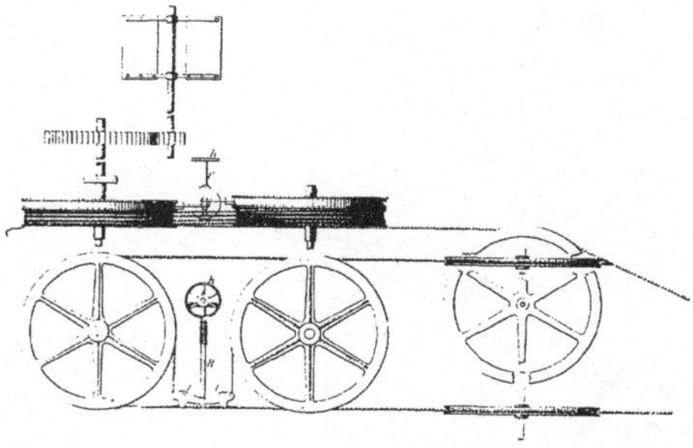

Figure 20. Groove wheel brake with fan brake for pendulum cable way. Trenton, New Jersey, USA 1892 (ANONYMOUS 1892).

Figure 21. Twin cable system with circulating carriages. The southern states of USA 1892 (ANONYMOUS 1892).

The Trenton Iron Company, New Jersey, published a handbook in 1892 where 2 and 4 drum steam driven winches were shown (figure 19) as well as grooved pulleys, capstans with air-brakes (figure 20) and well developed carriages for loading and unloading cable ways (figure 21). ANON (1892).

The Trenton Iron Company built cable ways on licence from the company A. Bleichert & Co. in Leipzig. They built in 1880 a 500 meter long twin cable system with circulating carriages for transporting cellulose from the mill to the railway for the Lorentide Pulp Company. A study of the production in 1889, showed that this cable way transported 50 tons of cellulose on a 10 hour day and the line speed and the speed of the carriages was 1.4 meters per second.

The colonists and settlers in The United States were driven by a pioneering spirit and many inventions appeared. The Trenton Iron Company built winch and cable equipment for many industrial purposes. In 1892 the first cable crane for skidding and transporting timber was

Figure 22. Cable crane at Avondale, USA 1892. Payload 3 tons (ANONYMOUS 1892).

built (figure 22). This was an important creation since it combined lateral skidding, loading and transport in one transport sequence. The cable crane was powered by a 3-drum steam-driven winch (main drum, haul-back drum and hoist drum). It could operate on any slope. This design was ahead of its time, and the cable cranes were not generally used until after 1939 in European forestry.

The forest operations on the American West Coast and in British Columbia started later than in the Eastern part of United States and Eastern Canada. It was difficult to handle the big trees and to begin with, the operations were concentrated on the easy terrain in the valleys. During 1885 - 1900 they started using mobile sawmills. They were put up on the felling site and moved from site to site as the operation went on. Later, forest railways were built in the valleys. They were built as steep as possible in order to reach the timber on the hillsides. After 1919 trucks and truck roads were put to use, and the operation could include more steep terrain.

In the beginning the main problem was to skid the heavy timber over smooth ground in the valleys to the mobile sawmills or to the forest railroads. These skid roads were covered by small logs in order to reduce the friction. Draught oxen were used and usually a team of 12 oxen were used to pull the load (figure 23). Teams of 12 horses were also used. They were, however, less popular since they were not so good at steady pulling as the oxen. The oxen walked slower than the horses and the skidding speed was less. According to woodlands manager DRAKE at Simpson Logging Company, Oregon, in 1949, a team of 12 horses could maintain a steady pulling force of approximately 1.5 metric ton and a maximum pulling force when starting the load of approximately 3.5 metric ton. A team of 12 oxen could maintain a steady pulling force of approximately 2.5 metric ton while the maximum force when starting the load was approximately the same as for the team of horses. With a coefficient of friction of 0.3, the horse team could pull logs of 5 ton while the oxen team could skid logs of up to 8 metric ton. By laying out the log covered skid road with a small down-

Figure 23. A team of 20 oxen was necessary to pull the big logs along log covered skid roads. Oregon, USA 1893 (ANDREWS 1954).

Figure 24. Single-drum steam driven winch for skidding logs along skid roads. The line horse pulls the main line back to the felling site. Oregon, USA 1900 (ANDREWS 1954).

hill gradient and greasing the logs, the team of oxen or horses could pull much bigger logs (figure 23).

In 1885 John DOLBEER from Cresent City, California invented a donkey yarder (figure 6). A vertical capstan was powered by a steam engine. The donkey yarder replaced the team of oxen or horses for pulling the heavy timber along the log covered skid roads. Immediately after 1890 the capstan was replaced by a single drum winch which reduced the stress on the labourers.

A small sledge or a short log (the pig) was attached to the end of the main line. During skidding a man sat on the »pig» with a handle for steering the log as it was pulled along the skid road. He was also greasing the square logs on the skid road, especially along heavy parts of the road, to enable the heavy timber to glide easily and smoothly. A line horse pulled out the main line and the »pig» in order to fetch a new log (figure 24).

During the period 1905-1910 double-drum winches were put into operation instead of the previous single-drum winches. The haul-back line from the haul-back drum was used to pull out the main line. Each of the two drums were controlled by clutch and brake.

During this period petrol engines and later diesel engines gradually replaced the previous steam engines. They started to hang the steering blocks for the main and haul-back lines at a certain height over the terrain or in spar trees. This helped to lift the front end of the loads during skidding, it reduced the friction and it was easier to pass obstacles and less work had to be put into the construction of skid roads. The highlead system came into use during the period 1910 - 1915 (figure 25).

Figure 25. During the period 1905–1915 the double-drum winch and the high lead system was put into operation in the American west coast logging (ANDREWS 1954).

Figure 26. Logging railways were the original access roads. The winches were used to skid the logs to the railroad. They were also used to pull the empty wagons uphill or to brake the loaded wagons downhill. The steep railroads opened up higher forest areas (ANDREWS 1954).

Due to the winch technique it was possible to place the forest railroads on steep slopes and it was possible to reach higher elevations with forest operations. Normally the railroads were built with a maximum slope of 14% in the load direction. Since the winch could be used to pull up the empty waggons and the locomotives along the railroad and since the winch with brakes could be used to brake the loaded waggons down the steepest railroads, the side roads branching out from the main roads could be built on very steep slopes. Slopes between 35 and 45% were often used, and the steepest known railroad was 62% at Lyman Timber Company in the state of Washington. At the lower end of such steep railroads a side track branched out and a big pile of brush was placed at the end of the side track. In the case of a failure, for example if the braking line from the winch broke, the waggons could be directed over onto the side track in order to avoid an accident.

From 1920 the network of forest truck roads was laid out and truck transportation gradually replaced the timber transport on forest railroads. The high lead system and other cable systems with slack skyline or fixed skyline necessitated the use of spar trees and tail trees in order to lift the load above the ground. In the virgin forest of heavy timber it was easy to find big tree dimensions for spar trees. They often had a height of 50 - 60 m and a top diameter of 40 - 50 cm. A specialist group among the forest workers appeared, the so-called high climbers and high riggers. By using climbing irons on their feet and a climbing rope around the stem they could climb a tree. If the tree had a DBH of 2 m the climbing rope had to be 7 meter long. The high climber slung the climbing rope upwards one step at a time and climbed after. In this way he entered the tree step by step. On the uphill tour he cut off those branches which were in the way.

As soon as the high climber reached a height of approximately 50 m, where the top diameter was approx. 45 cm, he tied himself to the tree by means of the climbing rope and cut off the top of the tree with his axe. The moment the top fell off the top end of the spar tree swayed from side to side often 2 - 3 meter each way and the high climber had to hang firmly on to the tree to avoid falling down.

The high climbers were a much admired group of workers and the work was full of exciting moments. Accidents seldom happened.

During a competition the famous high climber Philip GRABINSKI used 1 minute and 3 seconds to climb a spar tree of 50 m height. He used 18 seconds to climb down again. (ascending speed 2.8 km per hour, descending speed 9.9. km per hour). This is probably a world record.

Figure 27. The preparation of a spar tree (WILLAMETTE 1925).
A. The high climber climbs the tree.
B. The top of the tree falls.
C. The spar tree is ready.

Figure 28. Winch with spar that can be lowered for transport (LIDGERWOOD MANUFACTURING CO. 1915).

Some places no good spar trees could be found and a stem was raised and fixed with guy-lines.

In 1915 Lidger Wood Manufacturing Company constructed a 20 m steel spar on the winch (figure 28). Later the height was increased to 35 m. To begin with not many steel spars were used. In today's operation the steel spar is the only practical solution.

During the period 1885 - 1915 winches were used to pull the heavy timber over smooth terrain in the valleys. From 1905 tractors were put into operation and they took over the skidding in the smoothest part of the terrain (figure 29).

Benjamin HOLT constructed the first crawler tractor in 1904. It was powered by a 40 hp steam engine and the tractor had a special wheel for steering. In the beginning special winches were used to skid the logs up to the tractor road and the tractor pulled the logs along skid-roads to the forest railroads. In 1905 - 1912 belt steering instead of front wheel steering was used (CATERPILLAR 1939). An arch or a sulky behind the tractor lifted the front end of the logs. The crawler tractors took over all the skidding work in easy and semi-difficult terrain. The bulldozer blade at the front end of the tractor was used to construct skid roads.

Gradually the purpose of the winch and cable technique changed to become the means for taking out the timber from the most difficult part of the terrain and this has especially influenced development after 1945.

The first donkey yarders and winches were invented and manufactured in small workshops. Bigger companies like Washington Steel & Iron Works, Skagit Steel & Iron Works, Williamette Iron & Steel Works, Madill and others with a high engineering standard, took over the development and manufacture of winch and cable equipment. During the period between the two world wars a wealth of cable systems with double or multiple drum winches, blocks, cable

Figure 29. One of the first crawler tractors for logging was built in Stockton, USA 1912 (ANDREWS 1954).

shoes, riggs and other equipment was developed. The manufacturers published handbooks which showed how to put cable equipment together into a cable system. It was a great variety of inventions and many variations of cable systems with and without skylines appeared. The universities in British Columbia and on the West Coast of The United States systematized the methods and the cable systems. Professor John Kenneth PEARSE contributed much to this work (PEARSE 1947).

During the pioneering period 1890 - 1915 the equipment and methods were developed by workers with little theoretical education: »Go to the job and do it». The need for a professional background became obvious.

During the Pacific Coast Logging Congress in Seattle 1909 the need for a university course in logging was strongly emphasized. Representatives from forestry and forest industries felt that logging engineering was a fundamental science which should be included in the university programme. This scientific field should include knowledge about the working methods, the fundamental engineering principles for logging machines and the operational methods. Planning and layout of transport network should also be dealt with (CORNWALL 1909).

The Faculty of Forestry at the University made a plan for forest engineering in 1910 and in 1911 the University employed professor E.T. CLARK to lead the education in Logging Engineering at the University of Washington (SCMITZ 1973).

Thus, in 1909 the initiative was taken to provide an education in forest engineering at a university level. It was therefore the logging problems in difficult terrain which created the need for this.

The engineering and economic part of the discipline was first taken care of. The part dealing with work science came later, in 1927, and was initiated by professor Hugo HILF at the University in Eberswalde in Germany.

Because of the literature that was written about the West American winch and cable methods these were made known in other parts of the world. American manufacturers introduced the West American winch and cable technique in tropical forest areas such as in the Philippines and in Africa.

Japan had its own development with gravity powered cable ways and cable cranes. Professor Seihei KATO at the Tokyo University systematized the Japanese cable cranes. He also translated handbooks on West American development. During the later years new trends have again appeared in the Japanese development of cable technique, for example concerning remote control of winches and carriages.

The complicated cable systems which appeared on the American West Coast 1925 - 1940 could only be used on large scale operations (figure 30). The installation of such cable cranes was labour consuming and was heavy work (PEARSE 1947, ANDREWS 1954, BROWN 1949). In the 1960's interlock winches and running skyline systems as well as swing booms and grapple yarders were introduced (figure 31). The cable cranes were mounted on mobile chassis and it was easy to move them from site to site and to rig them in different terrain (SAMSET 1973).

During the later years the problems of thinning in young second growths have appeared. Central European and Norwegian equipment have been tried out in British Columbia and the West Coast of The United States. At the same time American methods have been tried out in Japan, Central Europe and Norway. At the moment (in the 1980's) a similarity in development in various parts of the mountainous regions in the world exists. This may influence and harmonize the world wide winch and cable techniques in the years to come (IUFRO 1980).

Figure 30. Modern West American cable ways have telescope spars which can be lowered and laid horizontally over the self-propelled winch during the movement from one logging site to another.

Figure 31. The grapple yarder with swing boom (instead of steel spar) allows the logs to be piled to the side of the yarder.

1.5. The development of winch and cable systems in Norway

One of the first Norwegian foresters J.B. BARTH reported in 1857-59 to the Ministry of Agriculture about the forestry conditions in Norway. He said that already in 1860 rigorous and heedless cutting took place on easy terrain and close to the rivers while the forests on remote areas and difficult terrain remained untouched. The same phenomenon can still be observed today and has been shown many times, for example in connection with the terrain classification of the Norwegian forests made by the national survey of Norway (SAMSET 1975).

Originally the downhill transport from steep mountainous hillsides was carried out along natural shoots using gravity. The timber was cut in full lengths, limbed and barked, while the tops remained on the trees. In the late winter time with hard snow and ice these stems were sent down the natural shoots with the top end forward. It was dangerous work and much damage was caused to the remaining stands. Stones were forced into the surface of the stems which damaged the machinery at the sawmills and papermills located at the lower end of the rivers. In 1909 timber shoots were introduced (figure 33). Some of them were dry shoots and some were wet shoots. One of these wet shoots was built already in 1873 in Tinn in Telemark. The length was 2,400 m and the hight difference 407 m. The shoot was 1 m wide and 77 cm deep. The slope varied from 2% at the intake and was steepest at the middle with a slope at 35%. At the lower end the slope was 22%. The shoot connected a lake in the mountains to the main river at Lake Tinnsjø. A 4-man crew pushed the logs into the upper end of the shoot, averaging 420 logs per hour.

Figure 32. Shoots for downhill transport of tree-length wood at the Bandak Lake, Telemark.

Figure 33. Timber shoot (dry) of the same type as used in Brakandalen, Telemark 1909.

Figure 34. Steel wire for gravity transport of hay or fuel wood from the mountain to the farm (REINTON 1961).

Figure 35. The construction of the twin cable system with circulating carriages started in Svelgen started in 1917 (LUND 1918, Norwegian Technical Museum).

Figure 36. The steel wire for gravity transport was used to carry milk down from the alps to the farms. The method usually relied on gravity with a single steel wire. In some cases a brake was placed at the top (B). The first brakes of this type came into use around 1900 (REINTON 1961).

Figure 37. Oskar Svendsen with a brake built of wood. Research operation with movable pendulum cable way in Valebø, Telemark 1951.

Figure 38. Research operation with steel rope (8 mm 6x7+1) for gravity transport of fuelwood in Valebø, Telemark 1951.

The first cable way in Norway was used in Fyksesund, Hardanger, on the Norwegian West-coast in 1865 (BYRKJELAND 1974). It was a steel wire for gravity transport. Such installations were used by the farmers to transport hay and fuelwood from the alps downhill to the farms (REINTON 1961). Such steel wires for gravity transport became very popular and were frequently used by farmers in the Norwegian mountainous regions during the period 1870 - 1900. Some of them are still in use.

The development of industrial cable ways in Norway followed the traditions of Central Europe. The first industrial cable way in Norway was installed in 1891 on Osterøy near Bergen. It was constructed according to similar principles as the German Bleichert's cable way. During the years 1900 - 1920 many stationary cable ways were put into operation, most of them similar to the Bleichert system (figure 14) and some of them constructed for downhill gravity transport like a pendulum cable way.

The steel wires used by the many farmers for gravity transport initiated new ideas, and many local designs for simplifying the work evolved. In 1900 the farmers started to use a brake at the upper end of the cable way in order to control the speed of the carriage along the skyline. The skyline was a 5 - 8 mm thick steel wire and the braking line was a fibre rope made of hemp. The farmer could return the empty carriage, pulling up the braking line manually (figure 36).

During the period 1900 - 1910 the farmers started to use 2 skylines with a grooved pulley brake made of wood. The pulley was built in such a way that the direction of the wood fibres was perpendicular to the surface of the pulley (figure 37). A log pressing against one of the wheels was used to brake the wheel and to control the speed.

The first cable transportation of wood products took place in Lysekloster near Bergen in 1883. This consisted of a steel wire for gravity transport of logs.

Figure 39. A. Light carriage for movable pendulum cable way.
B. and C. Håland pendulum cable way with hoist in the carriages to simplify the loading and unloading, Årdal, Sogn 1952.

During the period 1930-1940 local blacksmiths constructed a large variety of systems in order to overcome practical problems (figure 39 and 42).

The development of engine powered winches started immediately before 1940. These were single-drum winches powered by electrical motors or combustion engines and were used to pull agricultural equipment on steep farmland. In 1942 Lars RØRLIEN from Voss produced a single-drum agricultural winch and this was further developed in 1952 for cable crane transport over distances up to 400 m.

In 1951 Ludvik ISACHSEN invented a double-drum winch for farm-tractors. It was mounted at the rear end of the farm-tractor and was equipped with a 6 m high steel tower. This winch was used for highlead yarding up to 100 m and both winch drums (main drum and haul-back drum) were controlled by clutch and brake (figure 41).

The development of medium and long distance cable cranes started in 1950 (A/S ANLEGGS-TRANSPORT, HORDALAND MEKANISKE VERKSTAD, OSKAR SVENDSEN, RAGNVALD NESTESTOG, PER IGLAND, VOSS SVEIS OG MEKANIKK, JOHS. SEIM etc.).

There is a difference between the Central European and Northern European mountain conditions. In Central Europe there were many high altitude roads for example those leading to the Alps. The first Central European cable cranes, therefore, had a single-drum motor-driven winch which was located at the upper end of the cable crane.

Figure 40. The Vossa-winch (type V-800) was constructed by Lars Rørlien, Voss 1942. It was used to pull ploughs, harrows and transport wagons on steep farm land. Motor power, 6 kW. Speed v = 0.75 m/sec. Pull S = 6.4 kN.

Figure 41. The first double-drum winch for an agricultural tractor was built by Ludvik Isachsen, 1951. The winch was equipped with a 6 m high tiltable steel spar which could be lowered for transport (Mofossen, Telemark 1953).

Figure 42. The Løland cable crane was constructed in 1952. The cable crane was propelled by a double groove wheel pulley, one for pulling the carriage and the other to power the hoist drum in the carriage. The JoBu Combi-Cat built in 1967 uses the same system (Løland, Sogn , 1953).

In Norway the access roads to the steep mountain forests were located in the valleys at the bottom of the hillsides. The first cable cranes, therefore, had a valley mounted winch located at the lower end of the cable crane. The winch was equipped with 2 drums, a capstan and a drum, or grooved pulley. It was a heavy work to pull out the hoist line (main line) high up on the hillside. Many inventions tried to lighten this work load for the terrain crew: Lølands cable crane, Olav Sveen cable crane, Vitra-winch constructed by Johs. Seim and Lunnek constructed by Hordaland Mechanical Workshop.

From a work scientific point of view a revolutionary innovation was made in 1965 when Ragnvald NESTESTOG invented the radio controlled winch. The main and haul-back lines were driven by a grooved pulley. The winch operator could lower and pull in the hoist line from the carriage by radio remote control. This simplified the work during the yarding, and also during the installation of the cable crane on the logging site. This cable crane was independent of terrain conditions and the winch could be mounted at the lower or the upper end of the cable crane. The crane could also be used for horizontal transport, for example over clefts and valleys. The Alp winch built by Igland Mechanical Workshop had similar capabilities. It had to be controlled by a winch operator since it was not equipped with remote control. The Alp winch became rather popular since it was cheap and reliable. The Isachsen Mechanical Workshop constructed another cable crane (Combi-Cat) with similar advantages.

After 1970 many new constructions appeared and the cable cranes have been mounted on mobile chassis for fast movement from one logging site to another. They are equipped with steel towers (tip tower, telescope tower or tip/telescope towers), they have auxiliary equipment to enable fast rigging and dismantling of the cable crane on the felling site. Running skylines or live skylines have replaced the previous fixed skyline systems. The long distance cable cranes are too labour intensive for the workers in the terrain. Most cranes are therefore constructed for distances of 300 - 500 m and the cable crane operations are combined with a dense network of feeder roads. Wherever possible, it is preferable to use very light cable cranes for distances of 200 - 300 m which can be mounted on skidders or farm tractors. They can only be used where the terrain and soil conditions allows a dense road network with a maximum spacing of 300 - 500 m between the feeder roads. This type of equipment has become rather popular since it may be used by the farmers in the farm forests. Another advantage is that the operations are no longer limited to clearfellings only. With 300 m spacing between the feeder roads it is even possible to use some of the light cable cranes in thinnings.

2.0. THE FACTOR OF SAFETY AND LIFETIME OF WIRE ROPES

The wire rope is a machine component which must be chosen to suit the winch, the grooved pulley or the capstan. The goal is to find a winch or cable system which leads to an appropriate operational method. In order to achieve a high availability, the winch and rope components should be dimensioned so as to be as strong as possible, while the need for flexibility and movability leads to a reduction in the dimensions. It is, therefore, necessary to achieve a good compromise between the two considerations. Normally the lifetime of the wire rope increases in proportion to an increased factor of safety.

2.1. The factor of safety

The factor of safety is the relationship between the tension in the wire rope and the braking load.

$$s = \frac{S_{Br}}{S} \qquad (1)$$

The elastic limit in a wire rope is often half the breaking load. The factor of safety, therefore, should normally be greater than s = 2. Stationary ropes such as fixed skylines or guylines often have as low a factor of safety as possible in order to reduce weight. Skylines can often have a factor of safety s = 2.5–3.

Inside the ropes running on sheaves the wires move a little in relation to each other. The outer wires are subjected to additional loads when the rope is bent. Main and haul-back lines are subjected to additional wear caused by friction against stones, gravel etc. and acidic humus may cause corrosion. In order to allow for these conditions, the factor of safety should be increased on main and haul-back lines and other running lines. The following factors of safety are recommended for winch and cable systems in forestry:

Fixed and live skylines: s = 2.5 – 3.0
Running lines: s = 4 – 5

Only in those cases where the need for light equipment plays an important role and wear is of little importance, such as short main lines on single-drum tractor mounted winches, a lower factor of safety may be used, often s = 3.0.

Example 1: **The main line on a winch has a diametre D = 12 mm and a construction 6x19+1 Filler. The breaking stress is σ_{Br} = 1770 N/mm². The maximum pull in the main line is 25 kN. The breaking load may be found in the wire rope catalogues. The resulting factor of safety is:**

$$s = \frac{S_{Br}}{S} = \frac{86.1 \text{ kN}}{25 \text{ kN}} = \underline{3.4}$$

2.2. The factor of safety and lifetime

USHER & SPROULE (1951) carried out a laboratory test at the Ontario Research Foundation in Canada (figure 43). The tested ropes had the following specifications:

Rope diameter:	d = 16 mm
Construction:	6x19+1 Filler
Breaking load:	S_{Br} = 153 kN
Diameter of the thickest wires:	δ = 1.14 mm

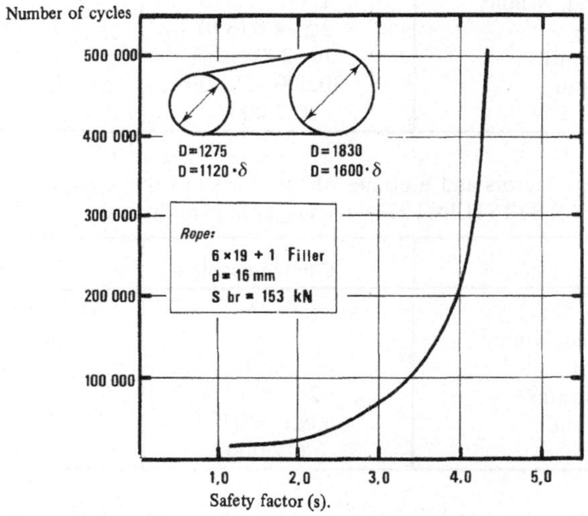

Figure 43. Measurement of the life time of an endless rope. Number of cycles before breakage with various tensions in the line (USHER and SPROULE 1955).

The test rope was driven around two grooved pulleys. The driving wheel had a diameter of D = 1,830 mm. The ratio between the pulley and the thickest wires in the rope was D/δ = 1,600. This pulley rotated one revolution forward and back again so that one cycle consisted of one clockwise and one anticlockwise revolution.

The wear was measured at the smallest pulley which had a diameter of D = 1,275 mm and the ratio of the diameters was D/δ = 1,120. Half the cycle caused a movement of the rope from »straight» to »bent» and back to »straight».

The rope tension could be varied continously by adjusting the distance between the pulleys. The maximum tension in the wire rope was S = 90 kN. At maximum tension the factor of safety was s = 153/90 = 1.7.

The lifetime of the rope was defined as the total number of cycles undergone by the rope before it had to be scrapped due to excessive wear.

When the factor of safety was less than s = 3 the rope lifetime was very low (figure 43). According to these investigations, running wire ropes should at least have a factor of safety of s = 4.5 or more. Joint FAO/ECE/ILO/ECE Committee on Forest Working Technique and Training of Forest Workers made a survey of safety factors and lifetimes of ropes used in winch and cable systems in forestry. The investigation considered results from 12 countries (WETTSTEIN 1971).

Table 1. Safety factors and lifetimes of the ropes in cable ways (after WETTSTEIN 1971). (Averages in parantheses).

	Carrying cable	Hauling cable
Type	6x7+1	6x19+1
Breaking load, N/mm²	1,470–1,570	1,570
Safety factor	2.5–4.0 (3.0)	3.0–5.5 (4.2)
Lifetime, months	36–240 (126)	10–24 (16)
Lifetime, cu.m.	20,000–25,000 (22,000)	8,000–25,000 (14,000)

Table 2. Safety factors and lifetimes of the ropes in cable cranes (300–700 meters). (after WETTSTEIN 1971). (Averages in parantheses).

	Carrying cable	Hauling cable
Type	6x7+1	6x19+1
Breaking load, N/mm²	–	1,570–1,860
Safety factor	3.0	3.2
Lifetime, months	24–60 (35)	6–17 (10.5)
Lifetime, cu.m.	6,000–20,000 (12,000)	2,000–8,000 (4,900)

Table 3. Safety factors and lifetimes of the hauling ropes for skidding winches (after WETTSTEIN 1971).

Vehicle	Safety factor	Lifetime	
		Months	cu.m.
Agricultural tractor	2.4	3.3	1,460
Wheeled skidder	2.5	3.4	2,120
Bulldozer	2.1	2.3	1,750
Truck	3.0	5.0	1,900
Highlead system	2.7	10.0	4,000

Large stationary cable cranes had a higher factor of safety than the small movable cranes. The lifetime of the ropes was 25 - 35 % longer for cable cranes for distances of more than 700 meters than for short distance cable cranes.

Table 1 - 3 gives the results from these investigations. Table 1 deals with cable ways for distances longer than 700 meters, such as pendulum cable ways and twin cables with circulating carriages. Table 2 deals with cable cranes for distances of 300 to 700 meters. Table 3 shows the factor of safety and the lifetime of the main line on tractor mounted winches. The main line had a construction: 6x19+1 in most cases.

2.3. Bending of wire ropes

The tension in a rope is carried by the single wires in the rope.

If the rope is hanging in a free span the stress due to tension is:

$$\boxed{\sigma = \frac{S}{\Pi/4 \cdot n \cdot \delta^2}} \qquad (2)$$

If there are several wire dimensions in the rope such as in seal constructions, warrington constructions, filler constructions etc. the stress due to tension is:

$$\sigma = \frac{S}{\Pi/4\ (n_1 \cdot \delta_1^2 + n_2 \cdot \delta_2^2 + \ldots)} \qquad (3)$$

In these formulae it is assumed that the wires lie parallel to each other in the rope. Each wire, however, has a spiral form in the rope and in reality the conditions are somewhat more unfavourable than that shown in formula 2 and 3. This can be taken care of by increasing the factor of safety.

When the rope passes sheaves, capstans, grooved pulleys or drums the tension of the wires in the rope increases due to bending. Each of the wires is being bent and the wires in the outer part of the rope are subjected to a higher tension than those in the inner part.

The wires move a little in relation to each other as the rope is being bent when passing a sheave. Research has shown that a rope which was lubricated frequently had a lifetime that was three times greater than a dry rope which had not been lubricated (SØLSNES 1941).

There should be a minimum ratio between the sheave or the drum diameter and the diameter of the wire in the rope to ensure safe operation:

Passenger transport:	$D \geqq 1{,}000 \cdot \delta$
Winches for forestry and cable systems:	$D \geqq 400 \cdot \delta$
Minimum:	$D \geqq 200 \cdot \delta$

There should also be a minimum ratio between the sheave diameter and the rope diameter. This ratio varies with the rope construction and the following ratio is recommended (ANON 1975).

Table 4. The ratio between the drum and rope diameter.

Rope construction	Min. Ratio D/d
6x 7	42
6x12	26
6x19	26
6x19 Seale	32
6x24 Seale	26
6x19 Filler	26
6x31 W.S.	26
6x36 W.S.	21
6x41 W.S.	18
8x19 Seale	26
18x 7	35

Table 4 refers to wire ropes with the following braking stresses due to tension: 1 570 N/mm² and 1770 N/mm². The ratio is valid for speeds up to 1.1 m/sec. At greater speeds the sheave or drum diameter should be increased by 5 % for each increase in speed of 0.5 m/sec.

A wire rope which is bent round a sheave or a drum recieves additional stresses which can be allowed for by using the ROULAEUX formula (SØLSNES 1941):

$$\sigma = \frac{S}{\Pi/4 \cdot n \cdot \delta^2} + \frac{\delta}{D} \cdot E \qquad (4)$$

BACH (1924) was of the opinion that the additional stress due to the bending of the rope should be $\frac{3}{8} \cdot \frac{\delta}{D} \cdot E$ since the wires are not parallel to each other in the rope. The wires have a spiral path in the rope. Bach reduced the additional stresses because the moduls of elasticity for the wire rope is less than the moduls of elasticity for the wires. As a result the stress due to tension and bending may be calculated according to the following formula:

$$\boxed{\sigma = \frac{S}{\Pi/4 \cdot n \cdot \delta^2} + \frac{\delta}{D} \cdot E_r} \qquad (5)$$

Example 2: **A 10 mm wire rope (6x19+1) with a breaking stress σ_{Br} = 1770 N/mm² has a tension of S = 12 kN. The rope is being bent around a sheave. The wire diameter in the rope is δ = 0.6 mm and the sheave diameter is D = 300 mm. The modulus of elasticity of the rope is E_r = 86.3 kN/mm².**

$$\sigma = \frac{12\ 000}{\Pi/4 \cdot 114 \cdot 0.6^2} + \frac{0.6}{300} \cdot 86\ 300$$

$$\sigma = 372 + 173 = \underline{545\ \mathrm{N/mm^2}}$$

The factor of safety is as follows:

$$s = \frac{\sigma_{Br}}{\sigma_s} = \frac{1\ 770}{545} = \underline{3.25}$$

Example 2 shows that the stress due to tension only is σ_s = 372 N/mm² and the resulting factor of safety is $s = \frac{1770}{372} = 4.8$. When the bending of the rope is taken into account the factor of safety is reduced to 3.25.

3.0. THE WINCH AND THE WIRE ROPE

3.1. The pulling force and the speed of the winch

The power of the motor must be suited to the pulling force and speed required from the winch. Tractor mounted winches often have excessive available power with respect to the work the winch shall carry out. If a tractor mounted winch is used in thinnings, the line speed may be approximately v = 1 m/sec. In this case the available pulling force would often be unnecessarily high (formula 7). If the wire rope is dimensioned according to this force it would be unnecessarily heavy to work with. It is more convenient to use a thinner and lighter wire rope. The winch should be equipped with a safety device, for example a torque limiter, in order to avoid overloading the wire rope.

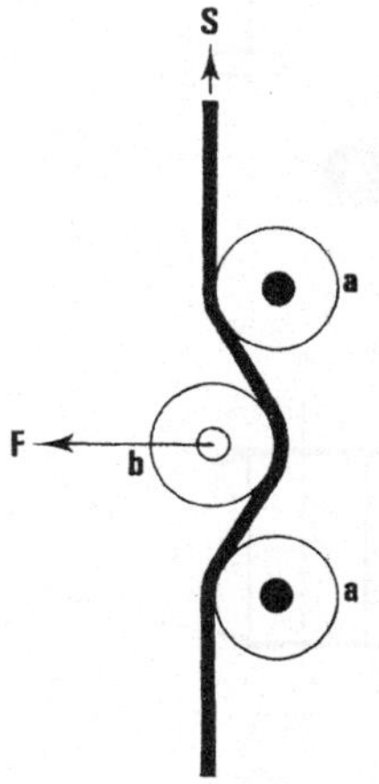

Figure 44. Measurement of the tension in a running wire rope.

Modern winch and cable constructions are often driven by powerful engines (100 - 200 kW) resulting in high line speeds and high line tensions. Such winches are mostly equipped with steel spars. The winch operator may easily overload the spar, the guy lines and the main and haul-back lines. The winch, therefore, should be equipped with a safety device. The main line or the haul-back line may pass two fixed wheels (figure 44a) and a movable wheel (figure 44b). The line tension (S) causes the rope to press against the movable wheel with a force (F). This force may be transferred to an instrument which enables the winch operator to see what tension there is in the running lines. The force (F) may also control an electrical signal, which can operate a relief valve in the hydraulic system or a slipping device connected to the winch drum. This method is being used by interlock winches for running skyline systems where the load is lifted off the ground by the carriage which moves along the haul-back line.

When operating in small sized timber stands the size of the loads are limited. It is then an advantage to work with high speeds. If the winch is used for ground skidding with low speeds the load may hit stumps and stones. Using high speed the load may pass the obstacles more easily due to the elasticity in the lines.

When lines are used in long spans large amplitude oscillations may develope in the wire ropes due to acceleration, retardation or caused by the load colliding with terrain obstacles. If such lines are running directly onto the drum these oscillations may cause poor spooling. The wire rope must pass guide blocks, for example in the steel spar or in a spar tree. The winch should be placed or mounted so that the center line between the guide block and the winch is perpendicular to the winch drum axis. The distance between the the guide block and the winch drum should be at least 20 times the length of the drum. The maximum fleet angle should be less than 1.5° - 2° (α in figure 45). There also exist some constructions which secure good spooling on the drum irrespective of the distance between the guide block and the winch drum.

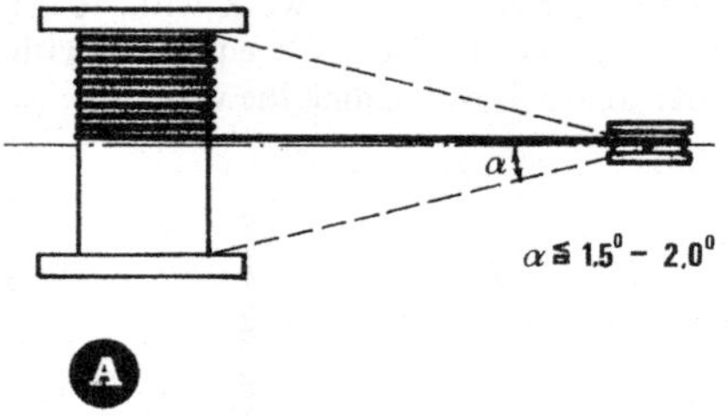

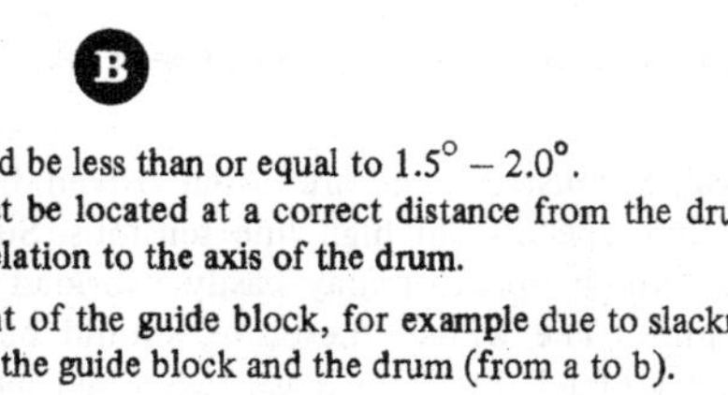

Figure 45. The fleet angle (α) should be less than or equal to 1.5° – 2.0°.
A. The guide block must be located at a correct distance from the drum and the centre line at right (90°) angle in relation to the axis of the drum.
B. A sideways movement of the guide block, for example due to slackness in the tower, reduces the distance between the guide block and the drum (from a to b).

The efficiency of the winch depends on the transmissions and the winch construction. That is, the efficiency between the power in the drive shaft from the engine and the power in the running lines. Standard winch constructions often have an efficiency of:

$$\mu = 75\% - 85\%$$

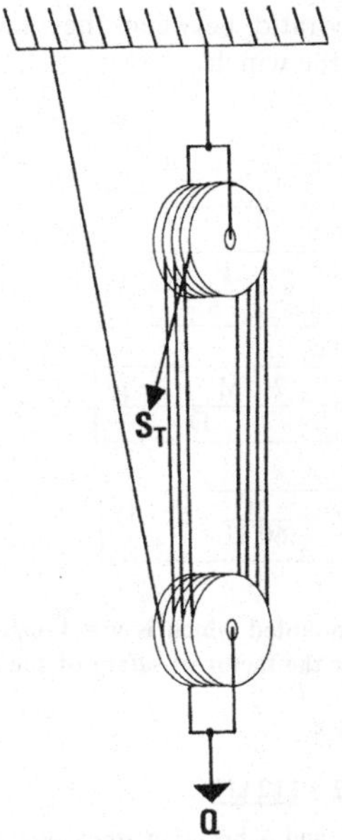

Figure 46. Multiple sheave block.

In addition there is a loss of power in the winch and cable system partly caused by the running lines passing sheaves and partly caused by the friction when part of the lines are sweeping against the ground. The loss of power may be 2 - 4 % each time the wire rope passes a block, depending on the ratio D/d and the angle between the ingoing and outgoing part of the rope which passes the block. The loss of power when the rope passes blocks causes a reduction of the hoisting power when using multiple sheave blocks (Q in figure 46). Table 5 gives a survey of the loss of power in various multiple sheave blocks according to the formula:

$$\boxed{S_T = tl \cdot Q} \tag{6}$$

Table 5. The efficiency and the relationship S_T/Q with multiple sheave blocks (After Wire rope engineering handbook, ANNON. 1943).

Number sheaves	Effect, %	$tl = \frac{S_T}{Q}$
2	96.2	0.520
3	92.5	0.361
4	88.9	0.281
5	86.5	0.234
6	82.2	0.203
7	79.0	0.181
8	76.0	0.165
9	73.1	0.152
10	70.2	0.142

The following formulae describe the ratio between the engine power or torque in the driving shaft and the speed and pull of the winch:

$$S_T = \frac{P \cdot \mu}{v} \tag{7}$$

$$v = \frac{P \cdot \mu}{S_T} \tag{8}$$

$$S_T = \frac{2 \cdot M \cdot U \cdot \mu}{D} \tag{9}$$

$$v = \frac{n}{60\ U} \cdot \Pi \cdot D \tag{10}$$

Example 3: **The line speed of a tractor mounted winch is v = 1 m/sec. The maximum engine power at the driving shaft is 40 kW. It is assumed that the factor of safety of the wire rope should be at least s = 3.5.**

$$S_T = \frac{40 \cdot 0.80}{1} = \underline{32\ kN}$$

$$S_{Br} = s \cdot S_T = 3.5 \cdot 32 = \underline{112\ kN}$$

A wire rope with the construction 6x37+1 and a breaking stress due to tension of 1770 N/mm² is being used. According to the wire rope catalogue a rope diameter of D = 13 mm has a breaking load of 107 kN which give a factor of safety of s = 3.3. When choosing a rope diameter of D = 14 mm the breaking load is 124 kN and the resulting factor of safety is s = 3.9. Both ropes may be used. If it is an advantage to work with a light rope the thinnest of the two may be chosen. If the thickness of the rope makes no difference regarding the work the thickest line should be chosen because it would have a longer life on condition that the diameter of the drum and the diameter of the block sheave allow it.

Example 4: **The torque on the driving shaft of a tractor mounted winch is M = 500 Nm at n = 540 RPM. The efficiency is μ = 0.8. The gear ratio between the driving shaft and the winch drum is U = 10. The empty drum diameter is D_t = 0.14 m and the full drum diameter D_f = 0.40 m.**

Line speed with the drum empty:

$$v_t = \frac{540}{60 \cdot 10} \cdot \Pi \cdot 0.14 = \underline{0.40\ m/sec.}$$

Line speed with the drum full:

$$v_f = \frac{540}{60 \cdot 10} \cdot \Pi \cdot 0.40 = \underline{1.13\ m/sec.}$$

Line pull with the drum empty:

$$S_t = \frac{2 \cdot 500 \cdot 10 \cdot 0.8}{0.14} = \underline{57\ kN}$$

Line pull with the drum full:

$$S_f = \frac{2 \cdot 500 \cdot 10 \cdot 0.8}{0.40} = \underline{20\ kN}$$

An acceleration would add an extra pull. On short spans (figure 47, 1), the carriage accelerates up to a maximum speed whereafter the carriage retards again to zero.On long spans (figure 47, 2), the carriage accelerates up to the maximum speed and keeps this speed until retarded. The total transport distance along the skyline in this case is La + Lv + La.

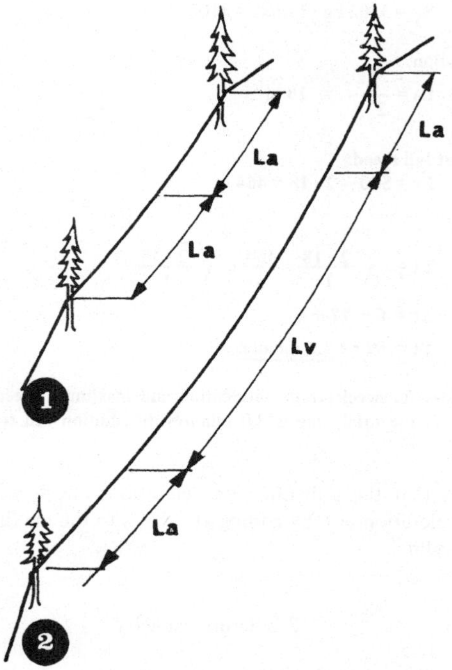

Figure 47. Acceleration and retardation distances along the cable.

The relationship between speed, acceleration, the moving mass (the carriage, the lines and the load) and the transport time is given by the following formulae:

Line pull in the wire rope: $$S_T = (Q_m + 2L_l \cdot \bar{m}) \cdot a \tag{11}$$

Distance of acceleration: $$La = \frac{v^2}{2a} \tag{12}$$

The time for acceleration or retardation: $$t = \sqrt{\frac{2 \cdot La}{a}} \tag{13}$$

The time for full speed transportation: $$t = \frac{L_v}{v} \tag{14}$$

Eksample 5: **The total length of a cable way is L_1 = 500 m. The mass of the carriage and the load is Q_m = 2500 kg. The length of the used main and haul-back lines are 1000 m, 12 mm, 6x19+1 (Filler). According to the wire rope catalogue the mass of the rope is $\bar{m}$ = 0.602 kg/m. The acceleration is a = 1 m/s² and the maximum speed is v = 6 m/s.**

The line pull due to acceleration:

$$S_T = 3102\ \text{kg} \cdot 1\ \text{m/s}^2 = 3102\ \text{N}$$

The distance of acceleration:

$$La = \frac{6^2}{2 \cdot 1} = 18\ \text{m}$$

The transport distance at full speed:

$$Lv = 500 - 2 \cdot 18 = 464\ \text{m}$$

The transport time:

$$\Sigma t = \sqrt{\frac{2 \cdot 18}{1}} + \frac{464}{6} + \sqrt{\frac{2 \cdot 18}{1}}$$

$$\Sigma t = 6 + 77 + 6$$

$$\Sigma t = 89\ \text{s} = \underline{1.5\ \text{minutes.}}$$

If we use the same times for acceleration, retardation and maximum speed during the transport operation and the return operation, the total time is 3.0 minutes. In addition will come the terminal times from the loading and unloading.

This example shows that the pull due to acceleration is S_T = 3.1 kN. It is usually sufficient to allow for the acceleration pull by adding 10 - 15 % to the maximum pull when transporting the load along the skyline.

3.2. Drum capacity

The angle between the barrel of the drum and the flange should be 90° in order to ensure good spooling. Some times grooves for the wire rope are made on the barrel surface in order to ensure good spooling of the first layer. The device for attaching the end of the rope to the drum should be constructed so that it does not spoil the spooling.

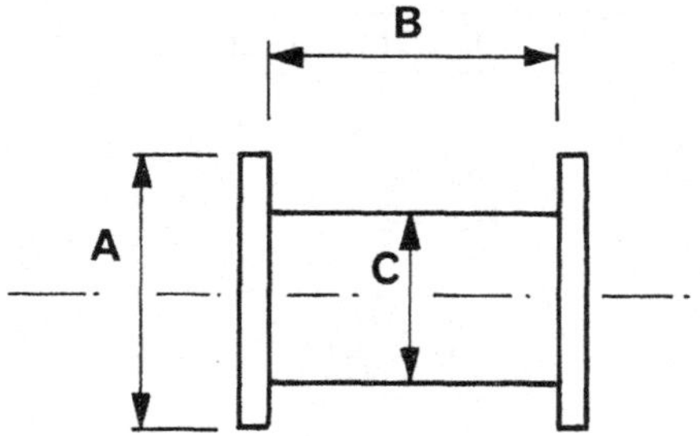

Figure 48. The winch drum.

Using the designations in figure 48, the length of wire rope on the drum may be calculated as follows:

Length of one coil at the middle drum diameter:

$$\Pi \cdot (C + \frac{A - C}{2})$$

Number of coils: $\frac{B}{d}$

Number of layers: $\frac{A - C}{2d}$

The length of wire rope on the drum:

$$L = \Pi \cdot (C + \frac{A - C}{2}) \cdot \frac{B}{d} \cdot \frac{A - C}{2d}$$

$$\boxed{L = \frac{\Pi \cdot B \cdot (A + C) \cdot (A - C)}{4d^2} = \frac{\Pi \cdot B}{4d^2}(A^2 - C^2)} \quad (15)$$

The diameter of the flange should be somewhat bigger than A. According to I.S.O. Standard (ANNON 1978) the diameter of the full drum should be 2 · d less than the flange diameter. The flange diameter should according to this be A + 4d. (d = rope diameter).

The drum capacity according to this calculation only applies when the spooling is good. This is not always the case, especially for single drum tractor mounted winches. In this case the flange should be somewhat higher, often 25% higher than assumed in formula 15.

Example 6: **The flange diameter of a drum is 0.4 m. The barrel length is 0.4 m and the diameter of the barrel is 0.2 m. The diameter of wire rope is 10 mm:**

$$L = \frac{\Pi \cdot 0.4\,(0.4 + 0.2) \cdot (0.4 - 0.2)}{4 \cdot 0.01^2}$$

$$L = \underline{377\ m}$$

A suitable length for this drum would be: $L = \frac{377}{1.25} \sim 300$ **m.**

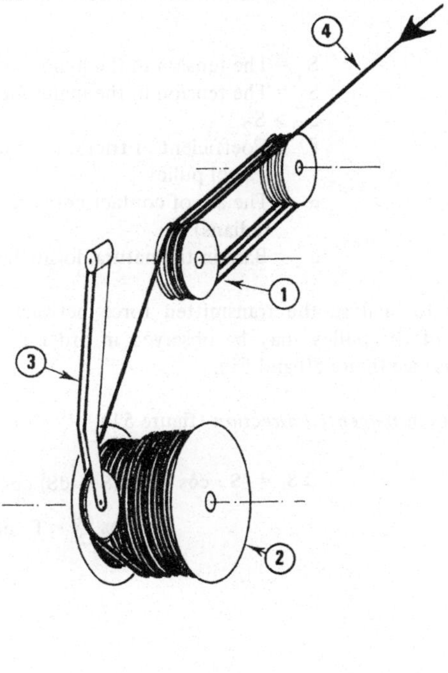

Figure 49. Pulley with storage drum.
1. Pulley.
2. Storage drum.
3. Coiling arm.
4. Main line.

3.3. Capstan and grooved pulley drives

Some cable system use an endless line to pull or brake the loaded and unloaded carriages. This is the case with pendulum cable ways, twin cables with circulating carriages, or some Japanese and Norwegian cable cranes. A capstan or a grooved pulley is used to pull or brake the endless line (figure 52 and 54). The grooved pulley or the capstan may also be used to pull or brake a single main line or haul-back line (figure 49). (Used in Austrian and Swiss winches). The grooved pulley or the capstan works together with a storage drum. In figure 49 the pulley (1) operates on the main line (4). A spooling arm (3) leads the main line onto the storage drum (2). The storage drum has a pre-tension with respect to the grooved pulley and 10% of the power is generally used for pre-tension. This may be done hydraulically or by means of a slipping device at the storage drum.

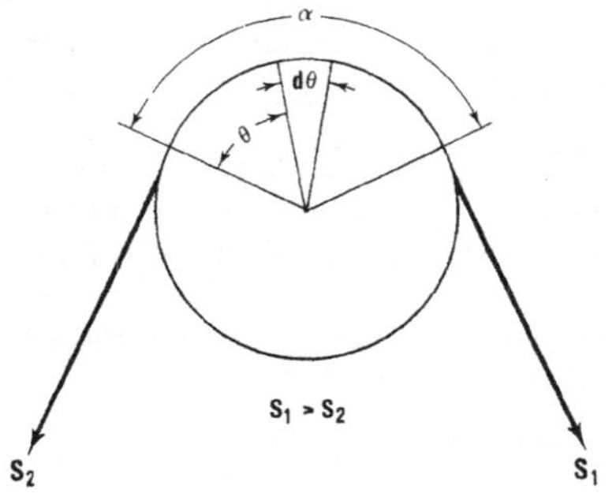

Figure 50. The angle between the in- and outgoing rope on a capstan or a groove wheel.

In order to analyse the tension in the line transmitted by the pulley the following symbols are used:

S_1 = The tension in the main line into the pulley.
S_2 = The tension in the main line out from the pulley.
$S_1 > S_2$
f = Coefficient of friction. f = 0.2 for a wire rope against a steel pulley.
α = The arc of contact between the wire rope and the pulley (in radians).
e = Base in the natural logarithm.

In order to analyse the transmitted force between the pulley and the wire rope a small segment of the pulley may be observed in order to find the forces in tangential and radial directions (see figure 50 and 51).

The forces in tangential direction (figure 51)

$$\Sigma S_t = -S \cdot \cos\frac{d\theta}{2} + (S + dS)\cos\frac{d\theta}{2} - f \cdot dP = 0$$

$$dS \cdot \cos\frac{d\theta}{2} = f \cdot dP$$

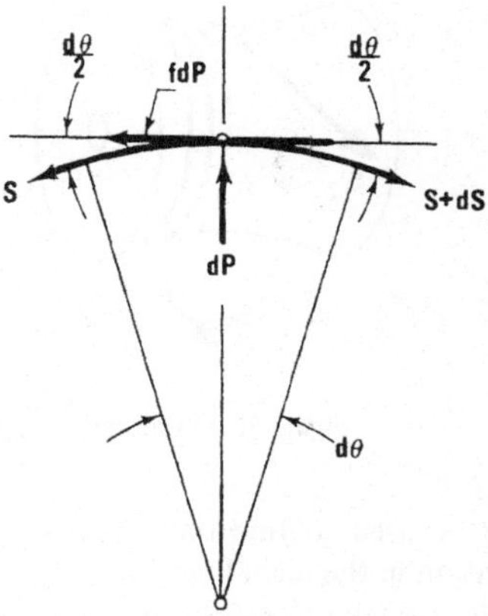

Figure 51. An element of the rope on the capstan or groove wheel.

The angle dθ is very small so that the cosine to the angle is approximately 1.

$$\cos \frac{\theta}{2} \sim 1$$

$$dS = f \cdot dP \tag{16}$$

The forces in radial direction:

$$\Sigma S_r = -S \cdot \sin \frac{d\theta}{2} - (S + dS) \sin \frac{d\theta}{2} + dP = 0$$

$$-2S \cdot \sin \frac{d\theta}{2} - dS \cdot \sin \frac{d\theta}{2} + dP = 0$$

The sine to small angles θ is approximately equal to the angle in radians:

$$\Sigma S_r = -S \cdot d\theta - dS \cdot \frac{d\theta}{2} + dP = 0$$

The expression $dS \cdot \frac{d\theta}{2}$ is a product of very small units and approximately equal to 0.

$$-S \cdot d\theta + dP = 0$$

$$dP = S \cdot d\theta$$

By inserting this equation into formula (16) we get:

$$\frac{dS}{S} = f \cdot d\theta$$

By integrating both sides of this equation we get:

$$\int_{S_2}^{S_1} \frac{dS}{S} = \int_0^{\alpha} f \cdot d\theta$$

$$\ln \frac{S_1}{S_2} = f \cdot \alpha$$

$$\boxed{\frac{S_1}{S_2} = e^{f \cdot \alpha}} \tag{17}$$

In order to achieve sufficient frictional force (figure 52) the wire rope is taken n turns around the capstan. This leads to the following formula:

$$\boxed{\frac{S_1}{S_2} = e^{f(2\Pi n + \alpha)}} \tag{18}$$

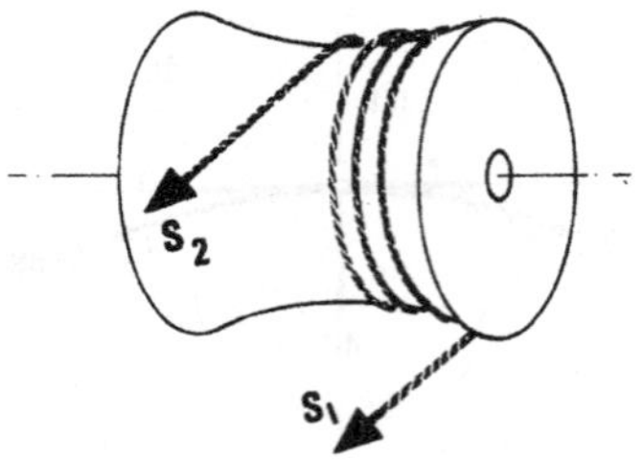

Figure 52. Capstan.

The following formula may be used to find the necessary pre-tension in the slackest line in order to obtain sufficient tension in the main line (S_1):

$$S_2 = \frac{S_1}{e^{f(2\Pi n + \alpha)}} \tag{19}$$

Example 7: **It is assumed that a capstan shall achieve a pull of S_1 = 10 kN. The line runs 3 full turns around the capstan (n = 3) with an addition of $\alpha = 180° = 2\Pi \frac{180}{360} = 3.14$ radians. The coefficient of friction for steel against steel is taken as f = 0.2:**

$$S_2 = \frac{10\ 000}{e^{0.2\,(2\Pi 3\, +\, 3.14)}}$$

The result of this calculation shows that the tension in the line on the »slack» side of the drum should be S_2 = 123 N in order to achieve a pull in the main line of S_1 = 10 000 N.

With a radius of the capstan of R = 15 cm the driving torque (M) is:

$$M = (S_1 - S_2) \cdot R = (10 - 0{,}123) \cdot 0.15$$

$$M = 1.48 \text{ kNm}$$

When a capstan is used there should be a small angle between the in and outgoing ropes. If not, there will be shock loads caused by the ropes interferring with each other (see figure 53).

The rope around a capstan slips side-ways during operation. The friction leads to wear.

This problem is avoided when using grooved pulley which in general causes less wear than a capstan. A grooved pulley system consists of a driving wheel and a leading wheel. The axis of the leading wheel forms a small angle with the axis of the driving wheel in order to steer the wire rope into the next groove on the driving wheel. The driving wheel has usually a bigger diameter than the leading wheel (figure 54).

The grooves in the wheel should have a profile like a saddle key with a 10% bigger radius than the radius of the wire rope. In industrial constructions the depth of the grooves is equal to the diameter of the wire rope. In cable ways or cable cranes for forestry it is necessary that short splices or quick connectors can pass the wheels. The shape of the grooves and the distance between them should allow this ($r_2 = 1.10 \cdot r_1$ and $\beta = 120°$ in figure 56).

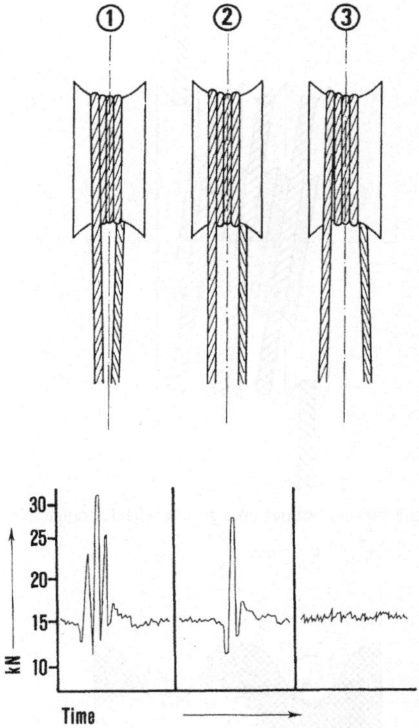

Figure 53. The angle between the in- and outgoing rope on a capstan is of importance (PROCKTER 1977).
1. Acute-angled lines give high climbing forces in the lines.
2. With parallel lines the extra forces are somewhat less.
3. Correct solution.

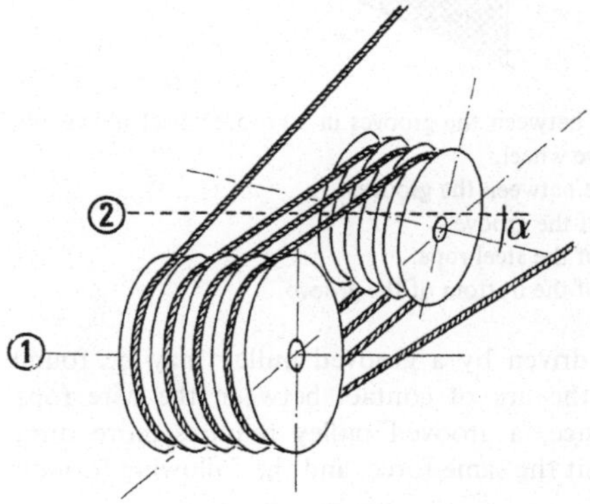

Figure 54. The groove wheel No. 2 should have a certain angle (α) in relation to groove wheel No. 1. The groove wheel No. 2 will then lead the line from the underside of the first groove to the upper side of the second groove in groove wheel No. 1.

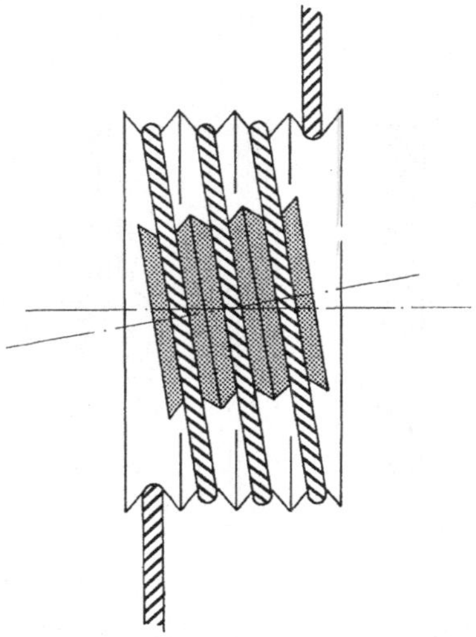

Figure 55. The angle between the two centerlines on a groove wheel arrangement.

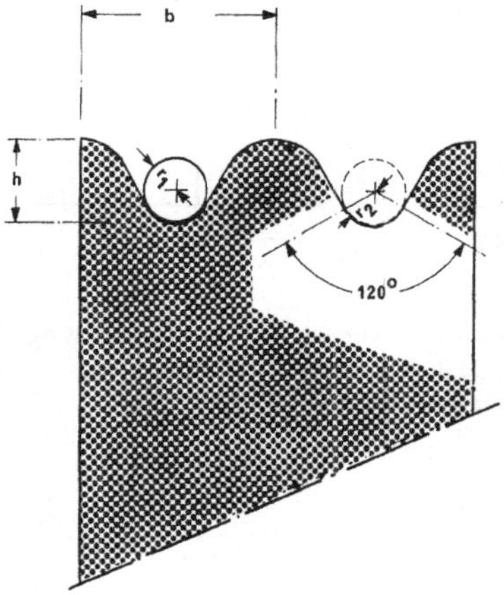

Figure 56. A large distance between the grooves in a groove wheel allows splicings or couplings on the rope to pass the groove wheel.
b. The distance between the grooves.
h. The depth of the groove.
r_1. The radius of the steel rope.
r_2. The radius of the bottom of the groove.

The tension in the lines driven by a grooved pulley may be found in a similar way as for a capstan, however, since the arc of contact between the wire rope and the driving wheel is less than the circumference, a grooved pulley requires more turns of the wire rope than a capstan in order to transmit the same force, and the following formula should be used:

$$\frac{S_1}{S_2} = e^{f \cdot n \cdot \alpha} \tag{20}$$

n is the number of grooves on the driving pulley and α is the arc of contact (figure 57).

Example 8: **A grooved pulley has the following specifications (figure 58).**

Main line pull:	S_1 **= 10 kN**
Coefficient of friction:	**f = 0.2**
Number of grooves on the driving pulley:	**n = 4**
Arc of contact on the driving pulley:	α **= 217° = 3.79 radians**

$$S_2 = \frac{10\ 000}{e^{0.2 \cdot 4 \cdot 3.79}} = 482 \text{ N}$$

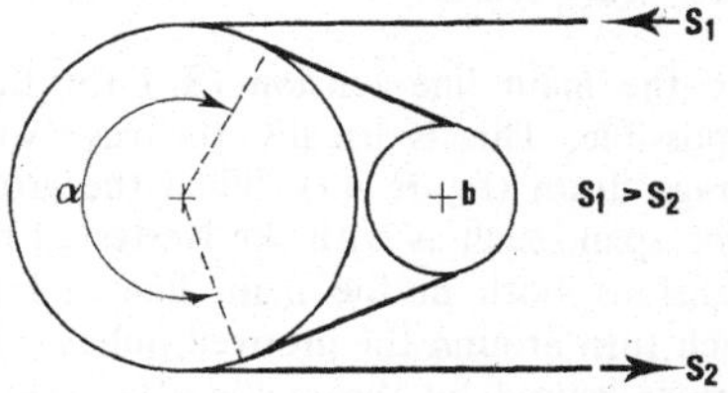

Figure 57. The wire rope passing a groove wheel arrangement.

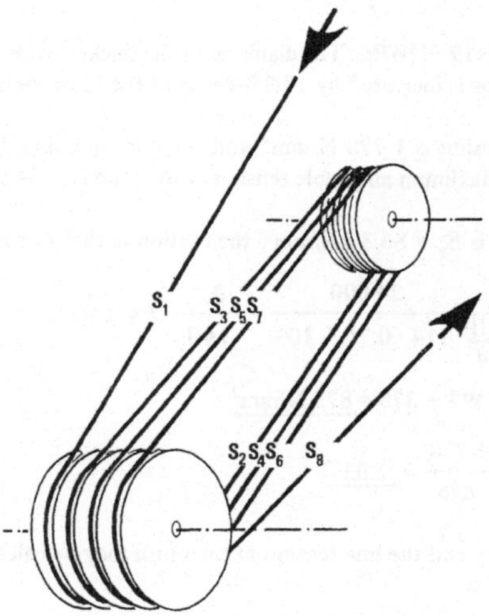

Figure 58. A groove wheel arrangement.

There exists a tackle effect between the driving wheel and the leading wheel in a grooved pulley system. This is shown in example 9.

Example 9: **The main line tension in a grooved pulley system is $S_1 = 10$ kN. The coefficient of friction is $f = 0.2$ and the arc of contact on the driving wheel is $\alpha = 3.79$ radians. When using the symbols in figure 58 the line tension in each of the turns may be calculated by means of formula (20) which leads to the following result:**

$$S_2 = 4\,686 \text{ N} = S_3$$
$$S_4 = 2\,196 \text{ N} = S_5$$
$$S_6 = 1\,029 \text{ N} = S_7$$

The total load on the bearing of the leading wheel is:

$$2 \cdot 4\,686 + 2 \cdot 2\,196 + 2 \cdot 1\,029 = 15.8 \text{ kN}$$

To obtain the load on the bearing of the driving wheel the main line tension and the tension in the slack-line should be added (see example 8).

$$15.8 + 10 + 0.5 = 26.3 \text{ kN}$$

The examples 8 and 9 give the main line tension (S_1) for the condition where the pre-tension (S_2) is as low as possible. This is usually the case when using a grooved pulley system together with a storage drum (figure 49). When the grooved pulleys are pulling an endless line hanging in a free span, such as with the Nestestog radio controlled cable crane, there may be high line tensions both in the main line and in the haul-back line. This influences the tension in each turn around the grooved pulleys. In such a case there may be very high loads on the bearings caused by the tackle effect between the driving wheel and the leading wheel. This is shown in example 10.

Example 10: **A cable crane carriage is being pulled along the skyline by means of an endless line. The endless line is driven by grooved pulleys and the diameter of the smallest pulley (the leading wheel) is D = 160 mm.**

The endless line is a 12 mm 6x19+1IWRC. The diameter of the thickest wires in the rope is $\delta = 0.7$ mm. The metallic cross-section of the rope is increased by 10% because of the filler wires and steel core.

The breaking stress due to tension is 1 770 N/mm² and the corresponding breaking load is 86.1 kN. With a factor of safety of s = 3.6 the maximum allowable tension in the rope is $S_1 = 24$ kN.

Since the modulus of elasticity is $E_r = 86.3$ kN/mm² the tension in the rope may be calculated by formula (5):

$$\sigma = \frac{24\,000}{\frac{\Pi}{4} \cdot 114 \cdot 0.7^2 + 10\%} + \frac{0.7}{160} \cdot 86\,300$$

$$\sigma = 497 + 378 = \underline{875 \text{ N/mm}^2}$$

$$s = \frac{1\,770}{875} = \underline{2.02}$$

There are 4 grooves in the pulley and the line tension in each turn may be calculated by using formula (20) in a similar way as in example 9.

$$S_1 = 24 \text{ kN}$$
$$S_2 = 11.24 \text{ kN} = S_3$$
$$S_4 = 5.27 \text{ kN} = S_5$$
$$S_6 = 2.47 \text{ kN} = S_7$$
$$S_8 = 1.15 \text{ kN} = S_9$$
$$S_{10} = 0.54 \text{ kN}$$

The load on the bearing on the driving wheel is:

$$\Sigma S = 24 + 2 \cdot 11.24 + 2 \cdot 5.27 + 2 \cdot 2.47 + 2 \cdot 1.15 + 0.54$$
$$\Sigma S = \underline{\underline{64.8\ kN}}$$

Additional shock loads may cause even greater stresses on the bearing. The bearings therefore, must be designed with a good safety margin.

3.4. Braking the winch drum or the pulley

The engine produces the necessary power to drive the winch or the pulley. They are connected to the driving shaft by a coupling. The brake gives the necessary retarding power to brake and stop the winch drum or the pulley. The available braking torque should always be higher than the maximum driving torque in order to stop the winch or the pulley in critical situations even when the engine is running. The wire rope should have a sufficiently large factor of safety in order to sustain the tension from the highest braking or driving torque.

On simpler winches a hydraulic motor may be used to drive the winch as well as to brake it.

On gravity powered cable ways the most important task for the brake is to control the speed of the carriage along the skyline and to stop the carriage (figure 59 A). If the winch is located at the lower end of a cable way (figure 59 B and C) the braking torque must be sufficient to overcome the maximum driving torque from the engine plus the forces caused by gravity. On a running skyline system the braking torque should also produce the tension necessary to lift the load from the ground (figure 59 D and E).

In gravity powered cable ways heat is generated during braking. The generated heat should be dissipated through a cooling system.

Example 11: **The height difference of a gravity powered cable way is 400 m and the load is 14.7 kN. The total braking torque needed is:**

$$W_B = 14.7 \cdot 400 = 5\,880\ \text{kNm}$$

The heat generated by the braking is:

$$W_B = 5\,880\ \text{kJ}$$

$$W_B = \frac{5\,880}{4.19} = 1\,404\ \text{kcal}$$

The 5 880 kilojoule or 1 404 kilocalories of generated heat have to be dissipated during the time it takes to run the load along the skyline.

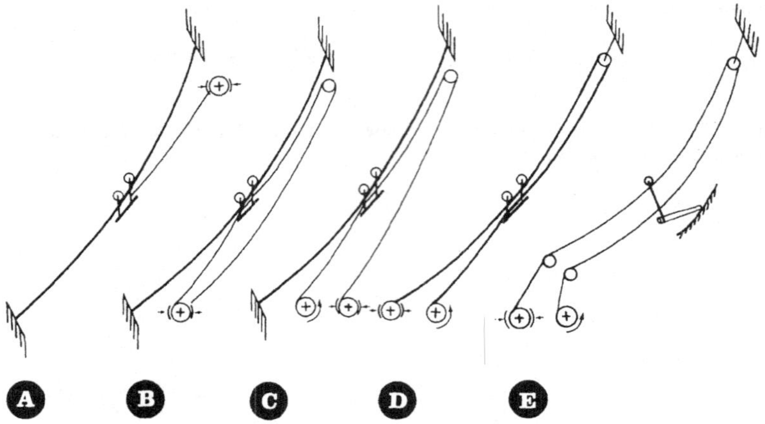

Figure 59. A. The brake mounted at the top of the cable way.
B., C. and D. The brake at the lower end of the cable way.
E. Brake on the haul-back drum of a double-drum winch.

The braking torque must be sufficient to brake the load at the steepest part along the skyline. The load path is a hyperbolic parabola and the steepest part is located immediately below the upper cable shoe in a span. There, the tension in the braking line is:

$$\boxed{S_2 = Q \cdot \sin\beta} \quad (21)$$

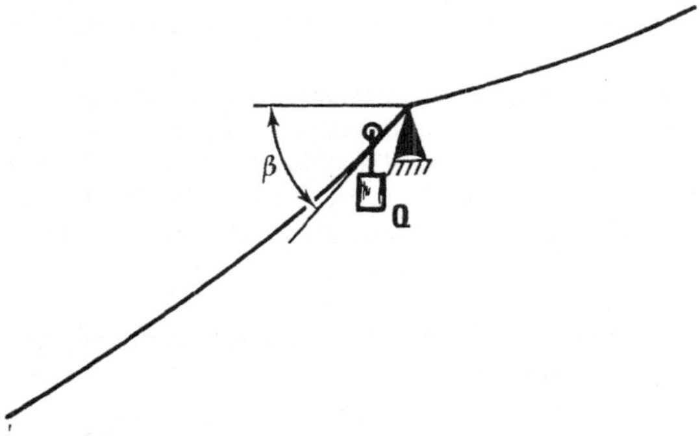

Figure 60. The cable is steepest immediately below the support (the cable shoe).

In addition to the weight of the carriage and the load, it is necessary to add the vertical component of the tension in the main and haul-back lines at the carriage.

If the angle β is 90° the skyline is vertical. In this case $\sin\beta$ is 1 and $S = Q$. Ordinarily the brake is dimensioned to cope with the weight of the carriage with the load in addition to the possible maximum torque from the engine.

The braking drum or pulley on gravity driven cable ways are often equipped with an extra brake for safety in addition to the ordinary brake. This is especially the case with big stationary cable ways.

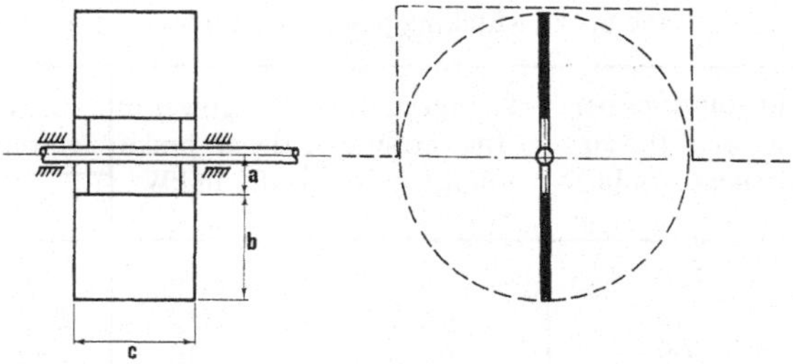

Figure 61. Fan brake.

The distance needed to stop the load during braking (La) may be calculated as follows:

$$S_B = m \cdot a = m \cdot \frac{v^2}{2 \cdot La}$$

$$La = m \cdot \frac{v^2}{2 \cdot S_B} \qquad (22)$$

In this formula m is the mass of the carriage, load and braking line in kg, and a is the deceleration in m/s^2.

Example 12: **The speed of the carriage with the load along the skyline of a cable way is 15 m/sec (54 km/ hour). The mass of the carriage, load and braking line is 2 500 kg and the braking force is 10 kN. Consequently the braking distance is:**

$$La = 2\,500 \cdot \frac{15^2}{2 \cdot 10\,000} = \underline{28 \text{ m}}$$

The necessary braking power may be high in a gravity powered cable way (see example 11). The ordinary brake may be assisted by a fan brake or a wind brake. The fan brake may be equipped with two wings as shown in figure 61. According to CZITARY (1962) the braking power may be calculated by the following formula, using the dimensions in figure 61, and n is the revolution per minute of the fan:

$$P_B = \frac{c[(a+b)^4 - a^4]\, n^3}{3\,264\,000} \qquad (23)$$

Example 13: **A gravity powered pendulum cable way is being braked by a capstan with a diameter D = 40 cm. In order to relieve the mechanical brake the shaft of the capstan is directly connected to a 2 winged fan with the following specifications (see figure 61): a = 0.3 m, b = 0.6 m, c = 0.8 m.**

$$P_B = \frac{0.8[(0.3 + 0.6)^4 - 0.3^4]\, n^3}{3\,264\,000}$$

$$P_B = \frac{n^3}{6\,296\,296}$$

If n = 400 revolutions per minute, the speed is v = 8.4 m/s and

$$P_B = \frac{400^3}{6\,296\,296} = 10.2 \text{ kW}$$

The braking force in the wire rope is:

$$S_B = \frac{10.2 \text{ kN} \cdot \text{m/s}}{8.4 \text{ m/s}} = 1.214 \text{ kN}$$

The braking force and the braking power may be calculated for various revolutions per minute and the results are shown in table 6.

Table 6. Braking power and force of a fan brake.

Revolutions per min. of the air brake n(r/min.)	The speed of the carrier in m/s	The power of the air brake P_B in kW	Braking force in the rope S_B	
			N	kp
100	2.1	0.16	76	7
200	4.2	1.27	303	31
300	6.3	4.30	682	70
400	8.4	10.2	1 213	124
500	10.5	19.9	1 896	193
600	12.6	34.3	2 730	278

If there are no openings between the wings and the axle of a fan brake the CZITARY formula may be simplified:

$$P_B = \frac{c \cdot b^4 \cdot n^3}{3\,264\,000} \tag{24}$$

Example 14: **The winch drum on a cable crane is equipped with a fan brake with a gear ratio of 1:10. The fan brake is rotating 10 times faster than the winch drum. The width of the wing is c = 35 cm and the height of the wing is b = 35 cm. At a speed of 3.000 r/min. the braking power is:**

$$P_B = \frac{0.35^5 \cdot 3000^3}{3264000} = \underline{43.4 \text{ kW}}$$

These examples show that the fan brake has little influence at low speeds and that the braking power increases with increasing speeds. The advantage of using a fan brake is that a high additional braking power is available at the high running speeds where it is most important to relieve the mechanical brakes. Professor SEIHEI KATO at Tokyo University did some research which showed that the braking power of a fan brake is higher when half of the circumference is covered by an airtight lid.

The most commonly used winch brake, especially on smaller winches is the band brake. It is therefore of interest to study the ratio between the forces applied to the front end (T_2) and on the rear end (T_1) of the band. (Figure 62).

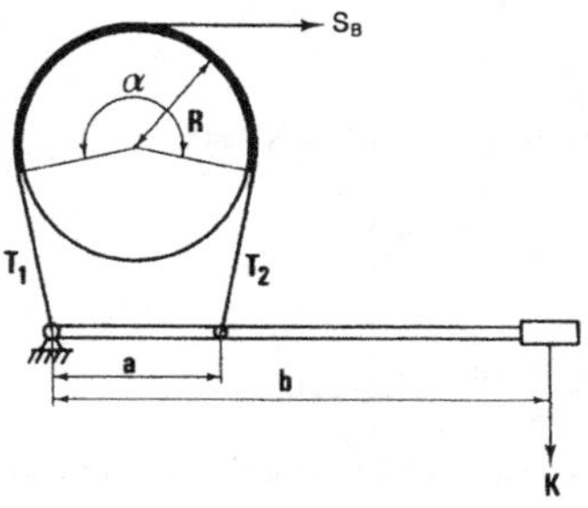

Figure 62. Band brake.

The designations according to figure 62 is the following:

R = Radius of braking drum in m
K = Applied force on the handle in N
T_2 = Applied force at the front end of the band brake in N
T_1 = Force at the rear end of the band brake in N
f = Friction coefficient, f = 0.3
S_B = The braking force at the surface of the braking drum in N
e = The base of the natural logarithm
W_B = The braking torque in Nm

According to formula (17) the ratio between the forces at the two ends of the band brake is:

$$\frac{T_1}{T_2} = e^{f \cdot \alpha}, \text{ when } T_1 > T_2$$
$$T_1 = T_2 \cdot e^{f \cdot \alpha}$$
$$W_B = T_1 \cdot R - T_2 \cdot R = (T_1 - T_2) \cdot R$$
$$W_B = (T_2 \cdot e^{f \cdot \alpha} - T_2) \cdot R$$
$$W_B = T_2 \cdot R \cdot (e^{f \cdot \alpha} - 1)$$
$$T_2 = K \cdot \frac{b}{a}$$

$$\boxed{W_B = K \cdot \frac{b}{a} \cdot R \cdot (e^{f \cdot \alpha} - 1)} \qquad (25)$$

If the winch or the pulley has the same radius as the braking drum the braking force (S_B) may be calculated as follows:

$$S_B = \frac{W_B}{R}$$

Example 15: **The arc of contact between a braking drum and a band brake is 250° or 4.36 radians. The force applied to the handle is 0.3 kN. The band brake has the following specifications: (see figure 62).**

$$\mathbf{a = 10\ cm}$$
$$\mathbf{b = 100\ cm}$$
$$\mathbf{R = 0.45\ m}$$
$$W_B = 0.3 \cdot \frac{100}{10} \cdot 0.45 \cdot (e^{0.3 \cdot 4.36} - 1)$$
$$W_B = 3.64 \text{ kNm}$$

If the pulley or the winch drum has the same radius as the braking drum the braking force in the line is:

$$S_B = \frac{3.64}{0.45} = 8.1 \text{ kN}$$

When using a band brake it is important that the direction of rotation of the braking drum and the applied force (K) are the same. If the braking drum rotates in the opposite direction the braking torque is as follows:

$$\boxed{W_B = K \cdot \frac{b}{a} \cdot R \cdot (1 - \frac{1}{e^{f \cdot \alpha}})} \qquad (26)$$

Example 16: **A band brake has the same specifications as in example 15, but the direction of rotation of the braking drum is opposite that of the applied force (formula 26):**

$$W_B = 0.3 \cdot \frac{100}{10} \cdot 0.45\,(1 - \frac{1}{e^{0.3 \cdot 4.36}}) = \underline{0.99\ kNm}$$

$$S_B = \frac{0.99}{0.45} = 2.2\ kN$$

The example 16 shows that there will be a considerable reduction in the braking force if a band brake is wrongly constructed.

Modern winch and cable systems are often equipped with the same type of brakes as in trucks or tractors (shoe-brakes, disk-brakes, servo-brakes etc.). These constructions are more safe and have the same braking force in both rotating directions. It is referred to truck or tractor litterature where such brake constructions are described (see i.e. P.M. HELT (1952) the automative chassis page 358 - 428).

4.0. RESISTANCE FROM THE LOAD

The cable cranes in forestry are used for uphill or downhill transport of logs, stems or full trees. The total weight to be carried by the cable crane consist of the payload, the carriage as well as the vertical components of the main line, the haul-back line, hoist line etc. The rolling resistance between the carriage wheels and the skyline is approximately $f_r = 0.02 - 0.05$ while the frictions coefficient between the load and the forest floor is approximately $f_{gu} = 0.5 - 0.8$. It is an advantage to lift the load free from the terrain during the transport. Often the height difference between the skyline and the terrain is too little to allow the load to be lifted from the ground. Sometimes intermediate supports may be used. They lead to longer rig up time for the cable crane. In addidition, the transport speed is heigher on a single span than when intermediate supports have to be passed.

4.1. The weight of trees and stems

Many investigations have been carried out throughout the world in order to find the biomass of the tree as well as the distribution of the weight on limbs, top, bark and the stem. As an example some results from measurements carried out by the Norwegian Forest Research Institute, may be given (SAMSET 1962, WESTERN 1970). The results are valid for Norway Spruce, *Picea Abies*. The resulting coefficient of friction and coefficients of transport when pulling the trees or stems on the ground should, however, be useful. Figure 63 shows the weighing of the trees and the stems.

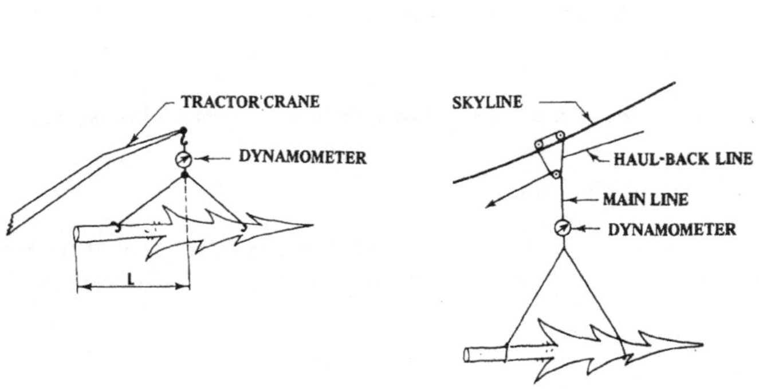

Figure 63. Measurement of the weight and centre of gravity of trees.

There is a high correlation between the amount of branches on the tree and the shape or taper of the stem. This was measured by means of the relative height of the tree (h_r) (KLEM 1934, SAMSET 1950, SAMSET et. al. 1969). The relative height of the tree is the actual height of each individual tree divided by the height according an average height curve. This height curve shows the height of *Picea Abies* in Norway according to results from the Norwegian National Survey. This average or »normal» height curve is called $h_r = 1.0$. Any tree which is shorter has a relative height less than 1.0 ($h_r < 1.0$). If the actual tree is longer than the height according to the average curve $h_r > 1.0$.

Figure 64 shows the relationship between the weight of the tree or the stem, the breast height diameter ($D_{1.3}$) and the relative height (h_r). Table 7 and 8 give some typical data for Norway Spruce.

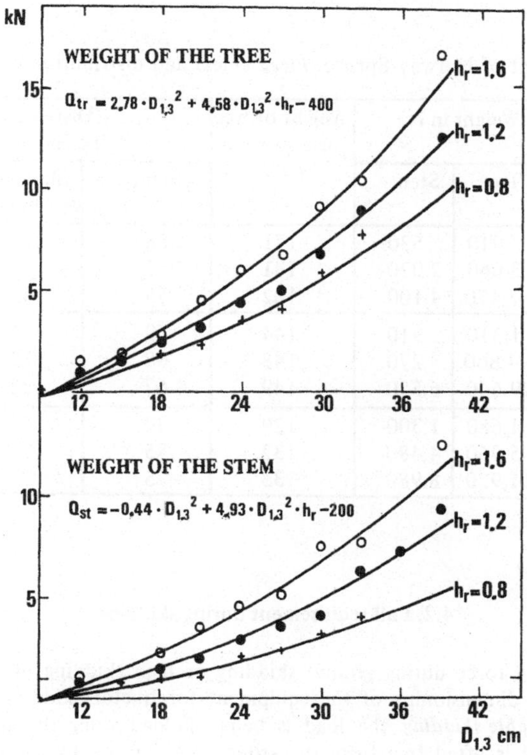

Figure 64. The smoothed functions of the weight of the tree with and without branches and tops (Norway Spruce, *Picea abies*).

Table 7. Tree and stem volume and length of Norway Spruce *(Picea abies)*.

h_r	DBH	m³ per stem	Number of stems per cu.m.	Length in meters		Length to centre of gravity in % of the height	
				Tree	Stem	Tree	Stem
0.8	15	0.066	15.2	8.7	6.2	37	38
	25	0.252	4.0	11.9	9.8	36	34
	35	0.513	1.9	13.7	11.9	35	31
1.2	15	0.113	8.8	13.0	9.1	39	38
	25	0.409	2.4	17.9	14.6	38	36
	35	0.819	1.2	20.4	17.7	37	33
1.6	15	0.163	6.1	17.4	12.0	41	40
	25	0.560	1.8	23.8	19.3	40	40
	35	1.123	0.9	27.3	23.6	40	37

Table 8. Weight of Norway Spruce *(Picea abies)* and distribution of stem and slash.

h_r	DBH	Weight in N		Weight of tree in % of stem	Distribution of the tree weight, %		
		Tree	Stem		Stem	Slash (top and branch)	Total
0.8	15	910	530	171	58	42	100
	25	3,660	2,020	181	55	45	100
	35	7,470	4,100	182	55	45	100
1.2	15	1,310	910	144	69	31	100
	25	4,860	3,270	148	68	32	100
	35	9,640	6,550	149	67	33	100
1.6	15	1,680	1,300	129	77	23	100
	25	5,960	4,480	132	75	25	100
	35	11,920	8,980	133	75	25	100

4.2. Pull requirement during skidding

The required pulling force during ground skidding or high skidding of trees or stems is of importance for the dimensioning of the equipment for tractor skidding as well as for cable skidding. During *cable skidding* the load is being skidded along the ground. During *cable transport* the load is lifted free from the ground and carried by the carriage (figure 65).

Ground skidding. If the direction of the pull is parallel to the ground the whole length of the load touches the ground during the skidding (figure 66, 1).

Highlead skidding. If the front end of the load is being lifted above the ground during skidding while the rear end touches the ground the load is being *highlead skidded*. The pulling force

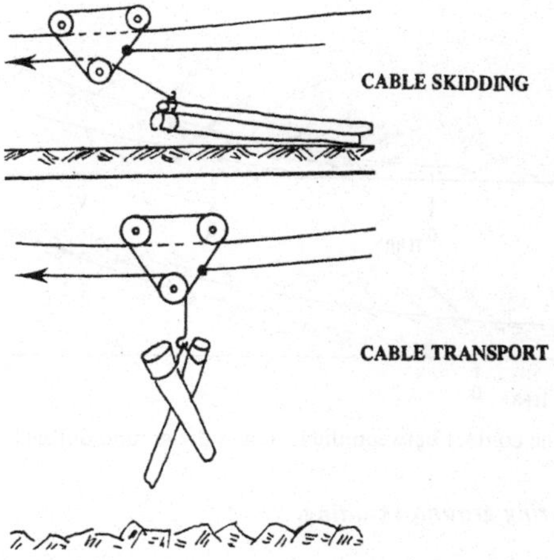

Figure 65. The difference between cable skidding and cable transport.

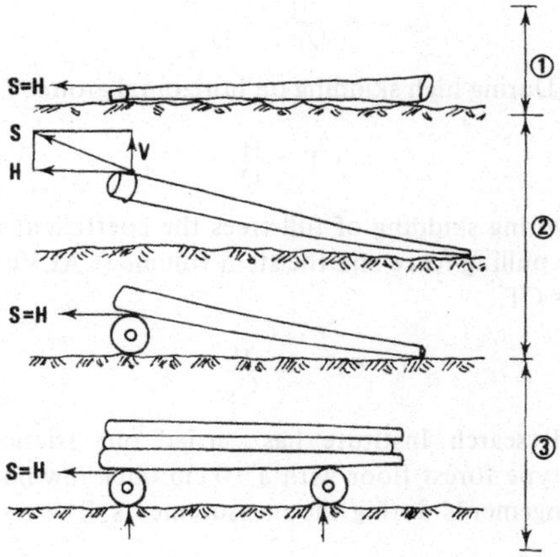

Figure 66. 1. Ground skidding of a load.
2. High skidding or half carried load.
3. Fully carried load.

may be decomposed into a vertical force (V) and a horizontal force (H). The front end of the load may be lifted by the hoist line or the rig on a cable crane or by a sulkey on a wheeled skidder. A highlead skidded load is therefore the same as a half carried load (figure 66, 2).

Fully carried load. The load is being carried by the vehicle or lifted free from the ground by the carriage on a skyline (figure 65 and figure 66, 3).

The coefficient of friction as well as the coefficients of transport and skidding may be defined by the formulae 27 – 30 using the symbols of figure 67 (CALVERT & GAVLIEKI 1967).

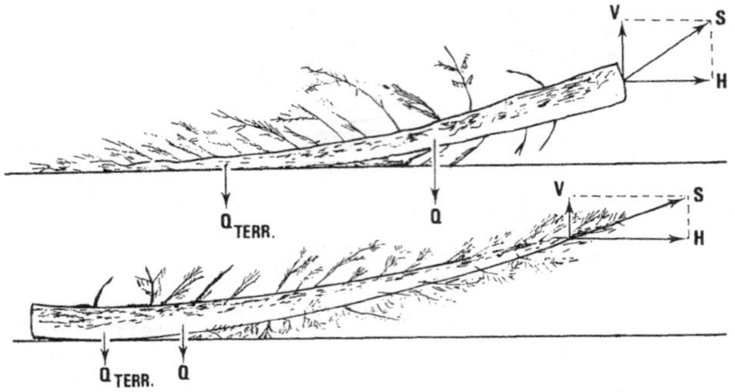

Figure 67. The contact between the load and the ground during high skidding.

Coefficient of friction during ground skidding.

$$f_{gu} = \frac{H}{Q} \tag{27}$$

Coefficient of friction during high skidding.

$$f_{go} = \frac{H}{Q_{Terr}} = \frac{H}{Q - V} \tag{28}$$

Coefficient of transport. (During high skidding on horizontal ground CT = f_{gu}).

$$CT = \frac{H}{Q} \tag{29}$$

Coefficient of skidding. During skidding of full trees the coefficient of skidding describes the relationship between the pulling force and the stem volume (CALVERT & GAVLIEKI 1967). When skidding stems CL = CT.

$$CL = \frac{H}{Q_{st}} \tag{30}$$

The Norwegian Forest Research Institute has carried out friction measurements during skidding on a vaccinium type forest floor with a 10 cm thick raw humus layer (DAHL 1973). Figure 68 shows the arrangements during the measurements. The results are given in table 9.

MEGILLE (1954) found the following coefficient of transport (CT) during ground skidding and high skidding of logs and full length stems with the butt end forward.

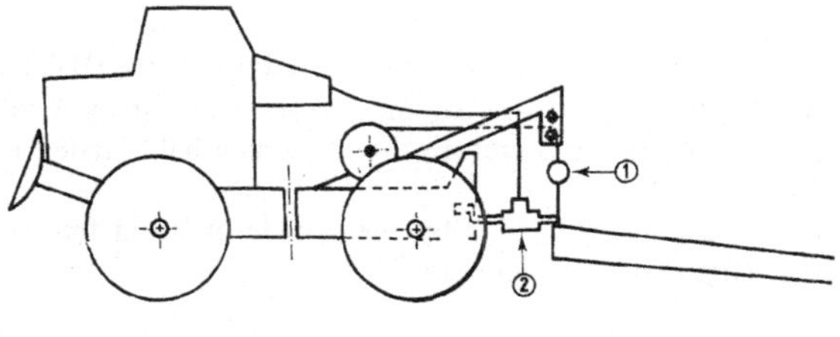

Figure 68. The arrangements of the measurements of friction coefficients during skidding.
1. The measurement of vertical forces.
2. The measurement of horizontal forces.

Table 9. Friction coefficients and transport coefficients when skidding stems and trees of Norway Spruce *(Picea abies)*.

Type of load	Butt end or top end forward	Friction coeff.		Transport coeff.		Skidding coeff.	
		Ground skidding f_{gu}	High skidding f_{go}	Ground skidding CT	High skidding CT	Ground skidding CL	High skidding CL
Trees	Butt end	0.66	0.88	0.66	0.43	1.38	1.00
	Top end	0.61	0.67	0.61	0.50	1.32	1.02
Stems	Butt end	1.00	1.25	1.00	0.74	–	–
	Top end	0.96	0.97	0.96	0.76	–	–

Table 10. Transport coefficient during skidding with front end forward (after MEGILLE 1954).

Surface	Ground skidding	High skidding
Dry clay	0.80	0.65
Wet clay	0.64	0.50
Dry, sandy soil	0.70	0.55
Wet, sandy soil	0.72	0.57
Dry humus	0.76	0.61
Wet humus	0.57	0.46

Table 11. Skidding of trees with lifted front end on ground without vegetation. Transport coefficient when the front end is lifted 76 cm above the ground (after BENNETT 1962).

Surface	Spruce, pine	Balsam fir
Dry, sandy clay	0.626	–
Dry sand	0.506	–
Wet clay. Some organic material	0.500	0.757
Wet humus. 20% clay	0.343	0.452
Compact snow-road	0.170	0.170

As shown in table 9 the coefficients of friction and the coefficients of transport are less during full tree skidding than full length skidding of stems. BENNETT (1962) found the coefficients of transport (CT) during high skidding of unbranched trees with the butt end forward (table 11).

On hard compacted snow-roads the coefficient of transport is low. During high skidding of stems the coefficient of transport often varies between 0.05 and 0.15.

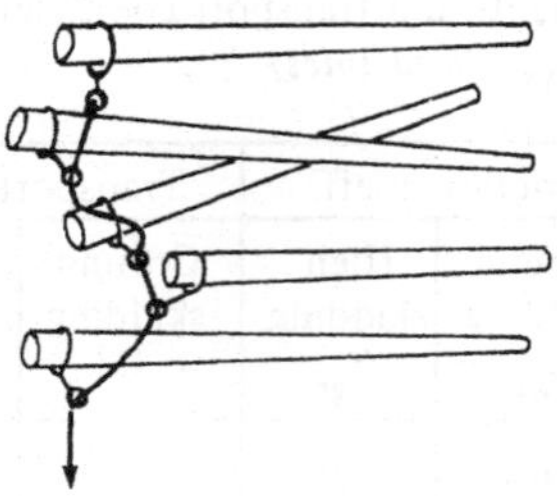

Figure 69. Winching in a load with the butt end forward.

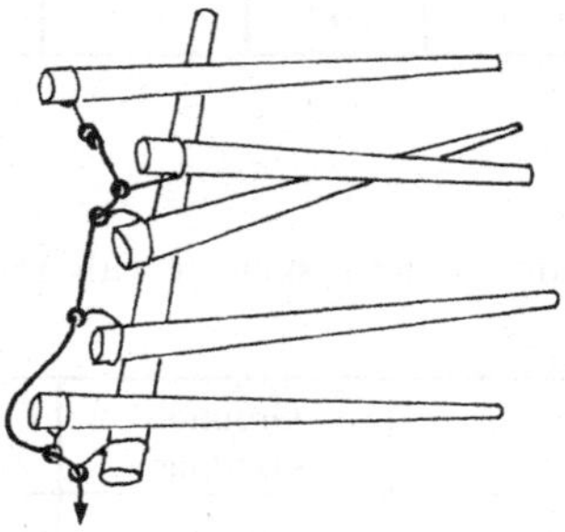

Figure 70. Winching in a load with the butt end forward on uneven and difficult terrain.

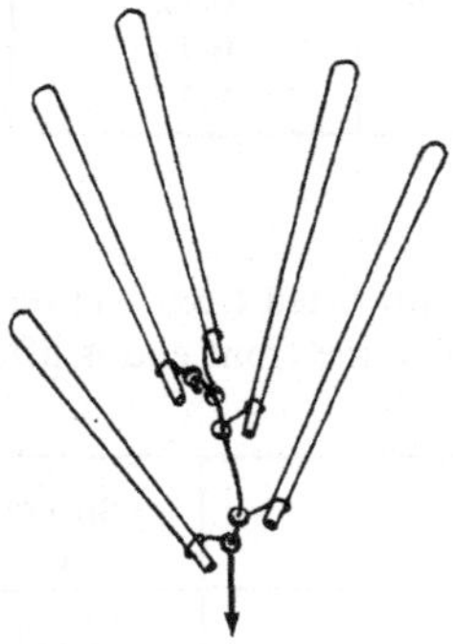

Figure 71. If there are few trees per hectare in difficult terrain, it may be easier to collect the trees by hauling in with the top end forward.

The pulling force requirement during skidding in soft snow varies according to the depth of the snow and the snow conditions. The force required, when skidding unbranched trees along a ground covered with wet, soft snow, may be very high. The same is true when the trees or stems have been snowed down. It may even be difficult for the choker setters to find the logs or the stems. In such cases the cable skidding should be carried out simultanrously with the felling.

It is an advantage to pull in the load perpendicularly to the direction of the stems (figure 69). This method makes it easier to loosen logs or stems which are frozen to the ground. The same applies when skidding logs over an uneven and stony ground. It may be an advantage to fell the first stem in the direction of skidding and use this as a rail for the rest of the load as illustrated in figure 70.

In very difficult terrain it may be an advantage to fell the trees in the skidding direction and to set the chokers at the top end as illustrated in figure 71. In this case less weight is transferred to the carriage which increases the pull during high skidding.

When skidding the loads with the top end forward, the heaviest part of the stem is dragged along the surface, and the most valuable part of the stem may be damaged by dirt, gravel and stones. High skidding of stems or trees with the butt end forward should always be preferred during the snow-free season.

5.0 MAIN, HAUL-BACK AND HOIST LINES

In a cable crane the main, haul-back and hoist lines carry out the dynamic part of the work. They transfer the forces from the winch drum or the pulley to the point where the force is required. The wire ropes are specially constructed in order to transfer pulling forces but they cannot transfer any pushing forces. The winch, therefore, must be constructed so that it can return the carriage or rig to the terrain, i.e. by means of a haul-back line. If not, the crew must carry out this heavy work.

During the operation the pulling force always induces torsional forces in the lines. These are being kept to a minimum by using preformed wire ropes and by different constructions of the rope itself. Torsional forces in the rope, can be eliminated by using swivels. At high tensions the torsional forces may cause parallel lines to twist themselves around each other. Light, empty carriages or rigs may even be thrown around the skyline due to such torsional forces, especially if there is transversal vibration in the lines. The carriage, therefore, should not be too light, and its center of gravity should be well below the skyline. We say that the carriage should have a deep keel.

The ropes may be subjected to bending forces beyond the elastic limit when passing sharp edges such as trees, stones etc.

These problems reduce the possibility of using automatic couplings or other attachments to the ropes in the terrain. With earlier cable crane constructions, automatic devices on complicated carriages were often used. With modern cable cranes all movable parts such as carriages, riggs etc. are constructed as simple as possible. Instead, the winch is made more complicated having extra lines etc., in order to solve the problems found in the terrain. It is important that all moveable parts for use in the terrain are as simple as possible in order to reduce breakdowns and repair work in the terrain where the working conditions are difficult.

5.1. Salvage winch on a vehicle

The purpose of a single drum winch on a terrain vehicle is often to pull up the vehicle when it has got stuck in the terrain. The required pulling force should be 25 % more than the weight of the vehicle. The main line should have a factor of safety of s = 3.0.

Example 17.

The weight of the tractor:	**30 kN**
The line pull:	**30 kN · 1.25 = 37.5 kN**
The braking load:	**37.5 kN · 4 = 150 kN**

With a wire rope construction of 6 x 37 + 1 and a breaking stress due to tension of 1 770 N/mm² the diameter of the wire rope becomes: D = 16 mm.

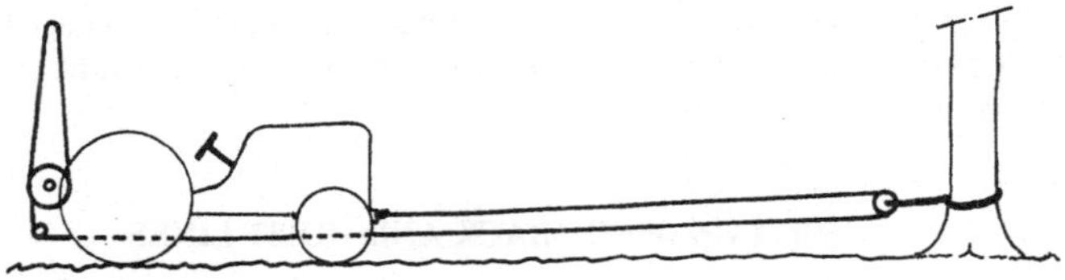

Figure 72. The skidding winch may be used as a salvage winch for the tractor.

The winch may have several functions. Besides being a salvage winch it may also be used for skidding purposes. A salvage winch would be too powerful and too slow for the skidding work. Instead, a winch is constructed for the skidding purpose. A tackle arrangement may be used in situations where the winch has to be used for salvage purposes (figure 72). It is an advantage if the salvage winch has approximately the same line speed as the tractor on the lowest gear. The line speed should be 10 – 15 % lower than the speed of the tractor due to the slip of the tractor wheels.

5.2. Ground skidding

During ground skidding the entire length of the load has contact with the terrain and the drawbar pull may be calculated as follows:

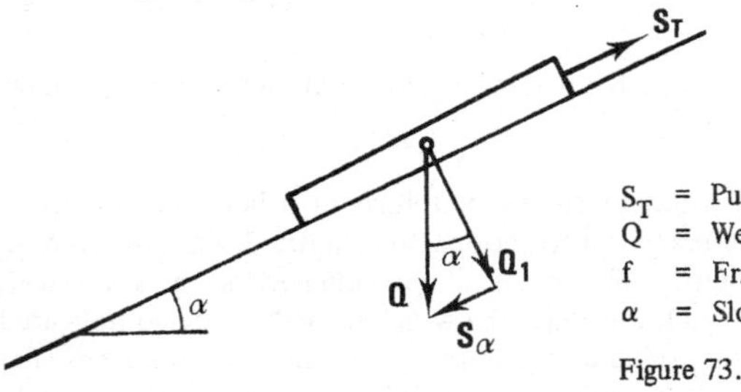

S_T = Pull in N
Q = Weight of the load in N
f = Friction coefficient between the load and the ground
α = Slope in degrees

Figure 73.

5.2.1. Uphill skidding

$$S_T = Q_1 \cdot f + S_\alpha$$
$$S_\alpha = Q \cdot \sin\alpha$$
$$Q_1 = Q \cdot \cos\alpha$$
$$S_T = Q \cdot \cos\alpha \cdot f + Q \cdot \sin\alpha$$

$$\boxed{S_T = Q \cdot (f \cdot \cos\alpha + \sin\alpha)} \qquad (31)$$

The slope which gives maximum pull (S_{max}) may be found by derivation:

$$\frac{dS_T}{d\alpha} = Q(-f \cdot \sin\alpha + \cos\alpha)$$

$$\frac{d^2S_T}{d\alpha^2} = -Q(f \cdot \cos\alpha + \sin\alpha)$$

The maximum pull takes place when $\frac{dS_T}{d\alpha} = 0$:

$$f \cdot \sin\alpha = \cos\alpha$$

$$\boxed{f = \frac{\cos\alpha}{\sin\alpha} = \operatorname{cotg}\alpha} \qquad (32)$$

The maximum pull takes place at the slope angle where cotgα has the same value as the coefficient of friction (figure 74). When the slope angle is 90° the load is lifted vertically and the maximum pull is the same as the weight of the load ($S_T = Q$).

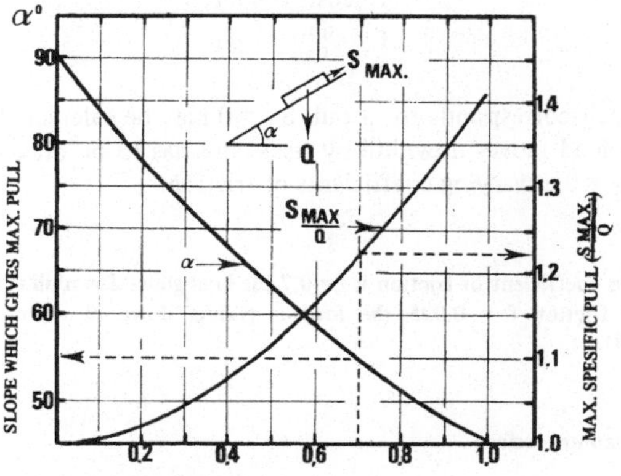

Figure 74. Slope corresponding to maximum pull during uphill ground skidding.

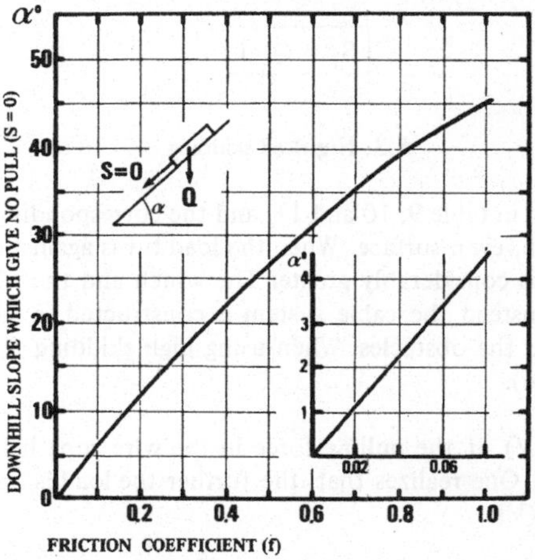

Figure 75. The relation between the coefficient of friction and the slope where the load starts to slide downhill due to gravity.

Examle 18:

The weight of the load: Q = 20 kN
Friction coefficient. f = 0.7

According to figure 74 the maximum pull takes place at a slope angle $\alpha = 55°$. At this moment the specific pull $S_T/Q = 1.23$ and the maximum pull:

$$S_T = 20 \cdot 1.23 = 24.6 \text{ kN}.$$

5.2.2. Downhill skidding

$$\boxed{S_T = Q \cdot (f \cdot \cos\alpha - \sin\alpha)} \qquad (33)$$

$$S_T = 0 \text{ when}$$
$$f \cdot \cos\alpha = \sin\alpha$$
$$f = \frac{\sin\alpha}{\cos\alpha} = \text{tg}\alpha \qquad (34)$$

The slope angle which corresponds to a pull $S_T = 0$ may be calculated by formula 34. If the slope is steeper the load moves downhill by itself and has to be braked. Figure 75 shows the slope angle when $S_T = 0$ at varying coefficients of friction.

Example 19. **If the coefficient of friction is f = 0.7 the load glides downhill if the slope angle is $\alpha \geqslant 35°$. If the coefficient of friction f = 0.025 the load is gliding downhill when the slope is steeper than $\alpha = 1.43°$ (see figure 75).**

5.2.3. Skidding on horizontal surface

If the slope is $\alpha = 0°$, $\cos\alpha = 1$ and $\sin\alpha = 0$. The pulling force may be found as follows:

$$\boxed{S_T = Q \cdot f} \qquad (35)$$

5.3. Highlead skidding

The coefficient of friction in table 9, 10 and 11, and the corresponding pulling force are valid for skidding over a smooth, even surface. When the load turns against obstacles like stumps or stones, the pulling force is considerably greater.The winch and the wire ropes are not dimensioned for these forces. Instead the cable system is constructed so as to lift the load or the front end of the load over the obstacles. When using high skidding the guide block is placed high up on a spar (figure 76).

The vertical component (V) of the pulling force in the wire rope helps to lift the front end of the load over obstacles. One realizes that, the further the load is from the spar, the smaller is the vertical lifting force (V).

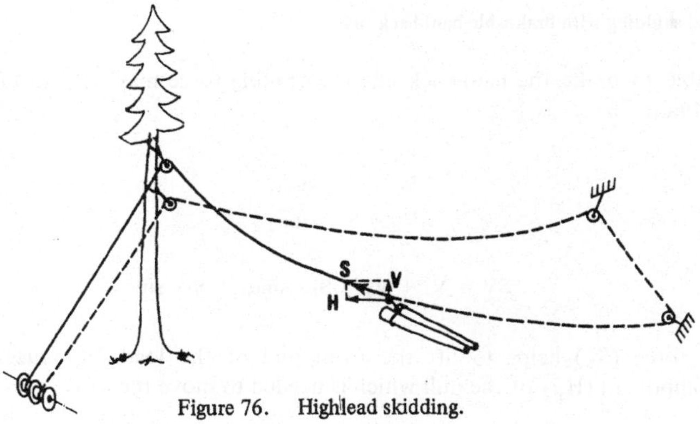

Figure 76. Highlead skidding.

5.3.1. High skidding without haul-back line

If the winch has a main line only and a high located guide block, the vertical and horizontal forces may be calculated according to the following formulae (figure 77):

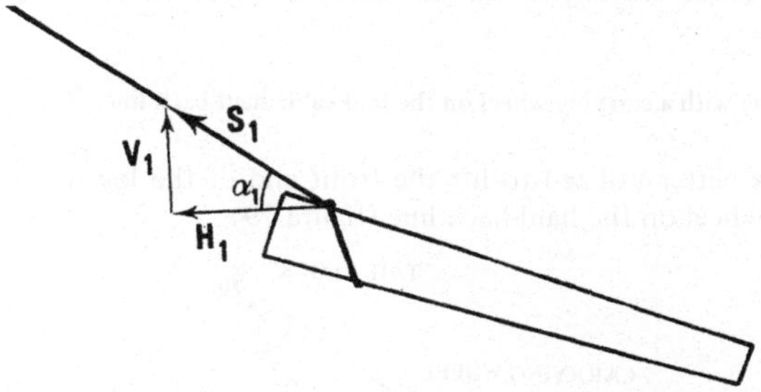

Figure 77. The vertical force when pulling the load with main line only.

$$\boxed{V_1 = S_1 \cdot \sin\alpha_1} \quad (36)$$

$$\boxed{H_1 = S_1 \cdot \cos\alpha_1} \quad (37)$$

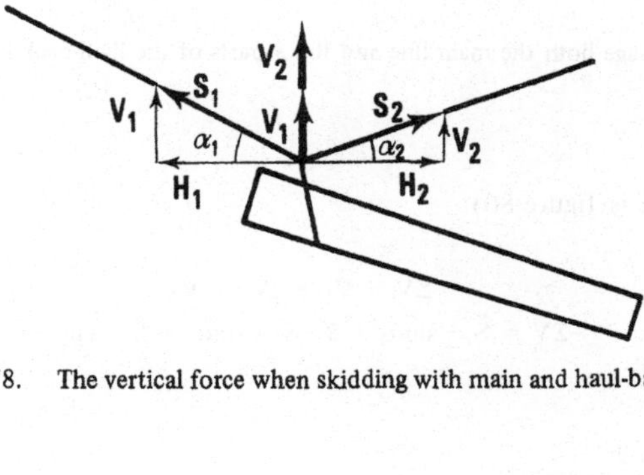

Figure 78. The vertical force when skidding with main and haul-back lines.

5.3.2. Highlead skidding with brakeable haul-back line

If it is possible to brake the haul-back line the braking force may help to lift the front end of the load (figure 78)

$$V_1 = S_1 \cdot \sin\alpha_1$$
$$V_2 = S_2 \cdot \sin\alpha_2$$

$$\Sigma V = V_1 + V_2 = S_1 \cdot \sin\alpha_1 + S_2 \cdot \sin\alpha_2 \qquad (38)$$

The braking force (S_2) helps to lift the front end of the load. It reduces, however, the horizontal component (H_T) of the pull which is needed to move the load:

$$H_T = H_1 - H_2$$

If $H_1 = H_2$ no pull remains to move the load. Braking force may be reduced by using as large angles (α_1 and α_2) as possible since this gives greatest possible lifting force. Under uneven and very difficult terrain conditons (boulders, cliffs etc.) the winch should be installed in such a way that both the guide block and the tail block are located at a height in relation to the logging area.

5.3.3. Highlead skidding with a carrying wheel on the brakeable haul-back line

The braking force is better utilized to lift the front end of the log if the hoist wheel is being lifted by a carrying wheel on the haul-back line (figure 79).

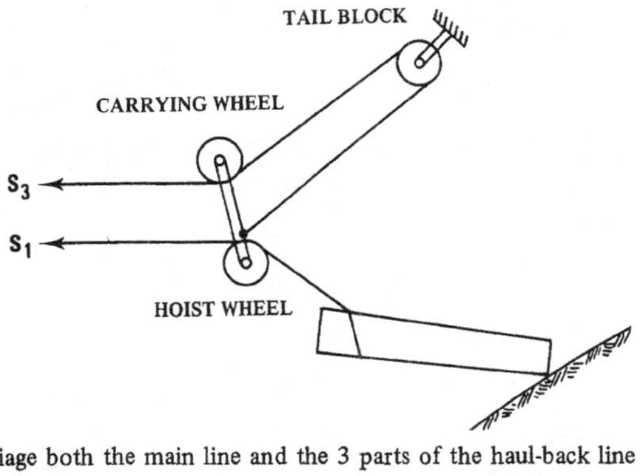

Figure 79. With a carriage both the main line and the 3 parts of the haul-back line carry the load (running skyline).

The total lifting force is (figure 80):

$$\Sigma V = V_1 + 2V_2 + V_3$$
$$\Sigma V = S_1 \cdot \sin\alpha_1 + 2 \cdot S_2 \cdot \sin\alpha_2 + S_3 \cdot \sin\alpha_1 \qquad (39)$$

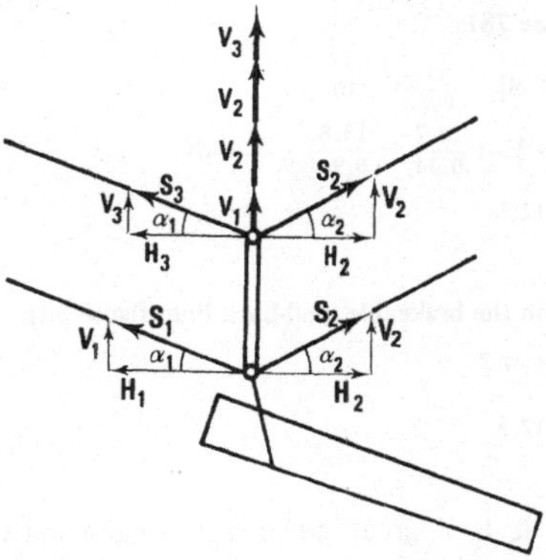

Figure 80. The vertical force in a running skyline system caused by the main line and the 3 parts of the haul-back line.

In the cases where $\alpha_1 = \alpha_2$ the formulae may be simplified as follows:

Main line only:

$$\boxed{V = S_1 \cdot \sin\alpha} \qquad (40)$$

Main line and brakeable haul-back line:

$$\boxed{V = (S_1 + S_2) \cdot \sin\alpha} \qquad (41)$$

Main line and carrying wheel on the brakeable haul-back line:

$$\boxed{V = (S_1 + 3 \cdot S_2) \cdot \sin\alpha} \qquad (42)$$

Example 20. **A 2 cu.m. load is being pulled in, with the butt end forward, by a winch. The load weighs Q = 16 kN. 2/3 of the load has to be lifted in order to pass obstacles: V = <u>10.7 kN.</u>**
According to table 9 the coefficient of transport on a vaccinium type forest floor is CT = 0.74.

The horizontal component of the pull which moves the load is:

$$H_T = Q \cdot CT = 16 \cdot 0.74 = 11.8 \text{ kN}$$

The main line pull is:

$$S_T = \frac{H_T}{\cos\alpha} = \frac{11.8}{0.94} = \underline{12{,}5 \text{ kN}}$$

With brakeable haul-back line and $\alpha_1 = \alpha_2 = 20^\circ$ the main line pull is:

$$S_1 = S_2 + \frac{H_T}{\cos\alpha}$$

Main line only (figure 77):

$$S_1 = \frac{V}{\sin\alpha} = \frac{10.7}{0.34}$$

$$S_1 = \underline{31.5 \text{ kN}}$$

This force is needed to lift the front end of the load when turning against an obstacle such as a stump. With no obstacles the force necessary to move the load is S_T = 12.5 kN.

Main and haul-back line (figure 78):

$$V = (S_1 + S_2) \cdot \sin\alpha = (2 \cdot S_2 + \frac{H_T}{\cos\alpha}) \cdot \sin\alpha$$

$$S_2 = ½ \cdot (\frac{V}{\sin\alpha} - \frac{H_T}{\cos\alpha}) = ½ \cdot (\frac{10.7}{0.34} - \frac{11.8}{0.94}) = 9.4\ kN$$

$$S_1 = S_2 + \frac{H_T}{\cos\alpha} = 9.4 + 12.5$$

S_1 = <u>21.9 kN</u>

Main line and carrying wheel on the brakeable haul-back line (figure 80):

$$S_2 = ¼ \cdot (\frac{V}{\sin\alpha} - \frac{H_T}{\cos\alpha}) = 4.7$$

$$S_1 = S_2 + \frac{H_T}{\cos\alpha} = 4.7 + 12.5$$

S_1= <u>17.2 kN</u>

Example 20 shows that it is a great advantage to use a carrying wheel on the haul-back line, since this reduces the braking force needed, to lift the load past obstacles, to a minimum. This principle is being systematized in some cable crane constructions where interlock winches and running skylines are being used (figure 81 B).

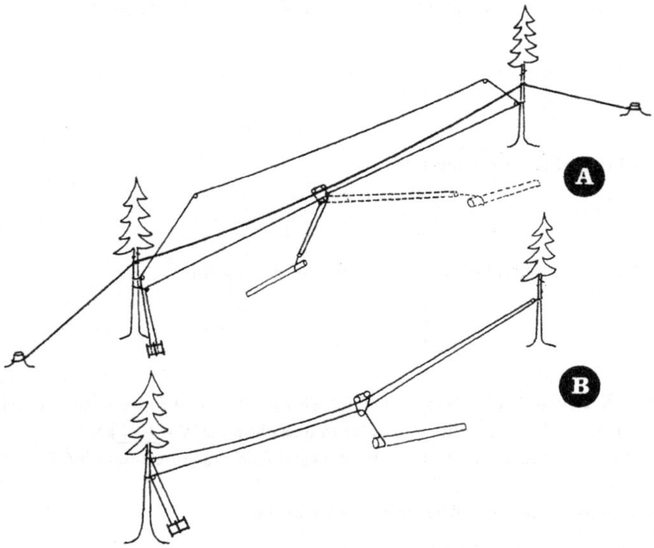

Figure 81. A. Cable crane with skyline.
B. Cable crane with the haul-back line as running skyline.

5.4. Braking forces in the haul-back line

During lateral skidding the load is being ground skidded or high skidded from each side of the skyline. In some constructions the carriage is fixed to the skyline with a stop device during lateral skidding. In other constructions the carriage is being kept in position by braking the haul-back line. The braking forces vary with the direction of the lateral skidding and some angles give very high forces.

In figure 82 A and 82 B the skidding force S_1 reduces the braking force S_2. If lateral skidding takes place in a direction perpendicular to the skyline, the forces S_1 and S_2 are equal. If lateral skidding is carried out in a direction opposite to the transport direction of the cable crane (figure 82 C) the braking force in the haul-back line (S_2) is greater than the main line pull. This is especially the case if a double hoist line and falling block are being used (figure 82 D).

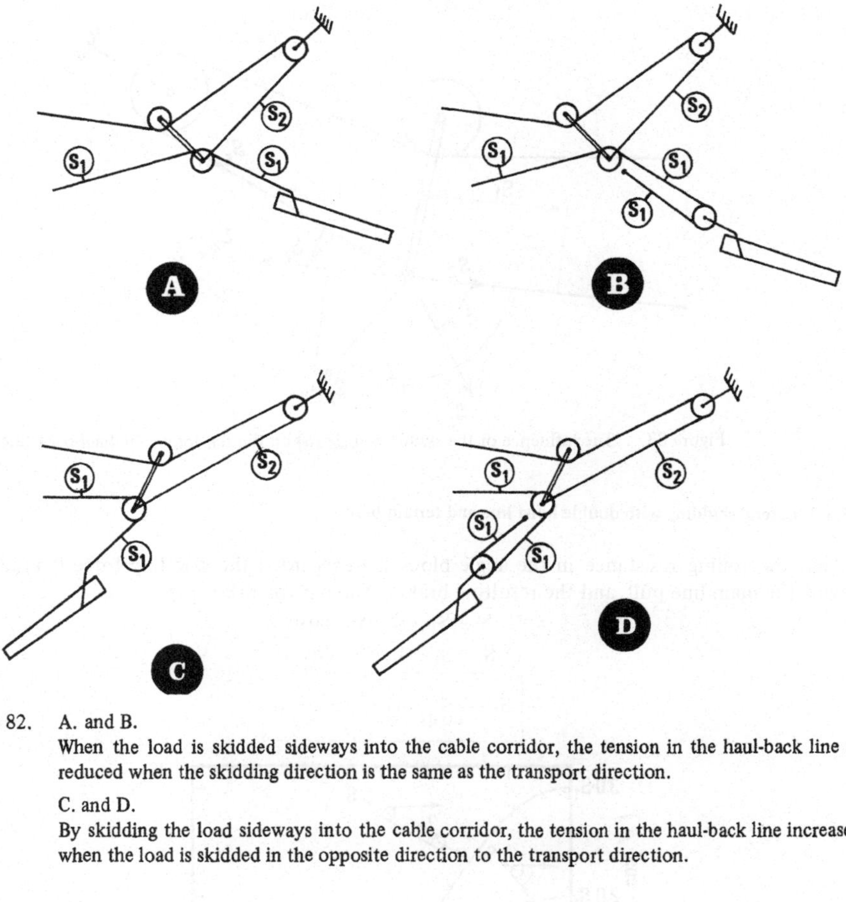

Figure 82. A. and B.
When the load is skidded sideways into the cable corridor, the tension in the haul-back line is reduced when the skidding direction is the same as the transport direction.
C. and D.
By skidding the load sideways into the cable corridor, the tension in the haul-back line increases when the load is skidded in the opposite direction to the transport direction.

5.4.1. Lateral skidding with single hoist line

The relationship between the braking force in the haul-back line (S_r) and the main line pull (S_1) is (figure 83):

$$S_r = \frac{S}{\cos\beta}$$

In many cases the angle β is small and $\cos\beta \rightarrow 1$. With a configuration as in figure 83, it is seen that:

$$S_2 = S_1 \cdot \cos\alpha$$

This force is positive or negative depending on whether ∠ α is bigger or smaller than 90°. The resulting braking force is:

$$S = S_l + S_l \cdot \cos\alpha$$

$$S = S_l \cdot (1 + \cos\alpha)$$

$$\boxed{S_r = \frac{S_l}{\cos\beta}(1 + \cos\alpha)} \qquad (43)$$

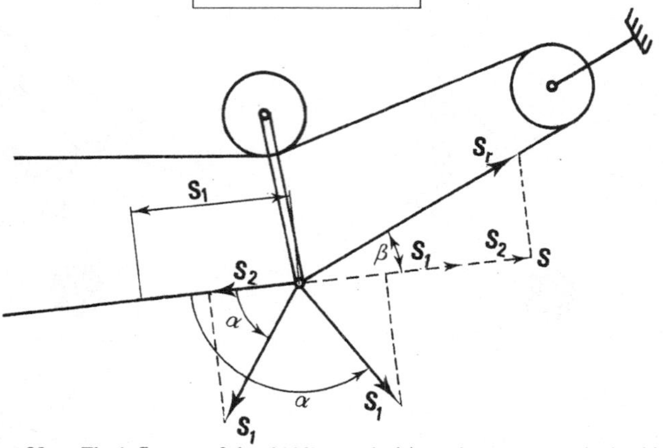

Figure 83. The influence of the skidding angle (α) on the tension in the haul-back line (S_r).

5.4.2. Lateral skidding with double hoist line and terrain block

When the rolling resistance in the angle block is disregarded the skidding force is practically twice the main line pull, and the resulting braking force is the following:

$$S = S_l + 2 \cdot S_l \cdot \cos\alpha$$

$$S = S_l \cdot (1 + 2 \cdot \cos\alpha)$$

$$\boxed{S_r = \frac{S_l}{\cos\beta} \cdot (1 + 2 \cdot \cos\alpha)} \qquad (44)$$

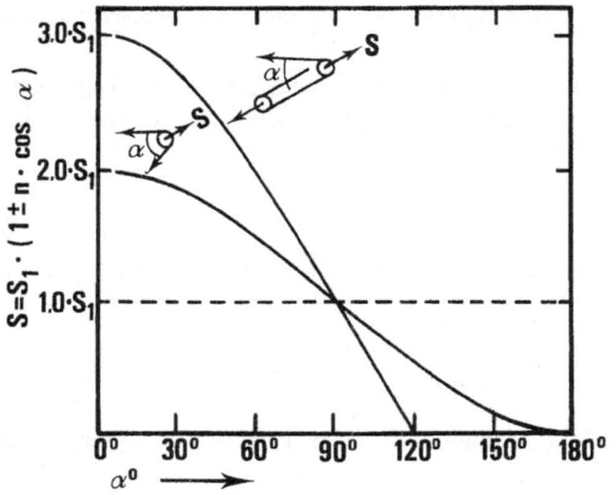

Figure 84. The influence of the skidding angle on the tension in the haul-back line with single and double hoist line. When the angle is 90°, the force in the main line equal the force in the haul-back line.

Figure 84 illustrates how the braking force in the haul-back line (S) varies with the angle of the lateral skidding.

Some cable crane constructions have carriages equipped with hoist drum. The gearing of the hoist drum may give larger skidding forces than the double hoist line. Lateral skidding may therefore cause very high forces on the stop device, on the skyline or on the haul-back line if the stop device is not functioning. When using such constructions it should always be observed that the direction of lateral skidding is being carried out in the same main direction as the transport direction of the cable crane, so that the skidding angle (α) is higher than 90° (figure 83).

Example 21: **A cable crane has a double hoist line.**

Main line: **D = 10 mm 6 x 19 + 1FC with a breaking force 59.8 kN (1 770 N/mm²)**

Haul-back line: **D = 8 mm 6 x 19 + 1FC with a breaking force 38.3 kN (1 770 N/mm²)**

The lateral skidding angle is 30° and the main line pull is S_1 = 15 kN (the corresponding skidding force is approximately 30 kN). The factor of safety in the main line:

$$s = \frac{S_{Br}}{S_1} = \frac{59.8 \text{ kN}}{15 \text{ kN}} = \underline{4.0}$$

The braking force in the haul-back line (figure 84):

$$S = S_1 \cdot (1 + 2 \cdot \cos\alpha) = 15 \cdot (1 + 2 \cdot \cos 30°)$$
$$S = 41 \text{ kN}$$

The factor of safety is:

$$s = \frac{S_{Br}}{S} = \frac{38.3}{41} = \underline{0.93}$$

In this case the haul-back line would break.

5.4.3. Skyline and carriage with double hoist line or hoist drum

The rolling resistance of carriage sheaves on skyline is somewhat higher on slack than on tight skylines. The coefficient of friction between the load and the forest floor during ground skidding is, however, much higher than the rolling resistance, often 20 – 50 times higher. Accordingly, forces and energy are saved if the load which is being transported is lifted free from the ground, for example on a fixed skyline (figure 85) or on a running skyline (figure 86). This is illustrated in example 22.

Example 22. **A light cable crane is used for high skidding. The relative height of the trees is h_r = 1.2 and the breast height diameter $D_{1.3}$ = 25 cm. According to table 7 the distance from the butt end of the trees to the center of gravity is 36% of the tree length.**

The weight of the load is Q = 16 kN. The weight of the butt end is $Q_r = Q \cdot 0.64$ = 10.2 kN, and the weight of the top end is $Q_t = 0.36 \cdot 16$ = 5.8 kN.

According to table 9 the coefficient of friction between the load and the forest floor is f_{go} = 1.0. The rolling resistance between the carriage sheave and the skyline is approximately f_r = 0.025. The corresponding main line pull is:

Top end forward:

$$S_1 = Q_t \cdot f_r + Q_r \cdot f_{go}$$
$$S_1 = 0.145 + 10.2 = \underline{10.3\ kN}$$

Butt end forward:

$$S_1 = Q_r \cdot f_r + Q_t \cdot f_{go}$$
$$S_1 = 0.255 + 5.8 = \underline{6.1\ kN}$$

The load is carried by the carriage and the skyline:

$$S_1 = Q \cdot f_r$$
$$S_1 = 16 \cdot 0.025 = \underline{0.4\ kN}$$

In addition to these forces are the possible braking forces necessary to lift the load and possible forces required to overcome resistance due to slope.

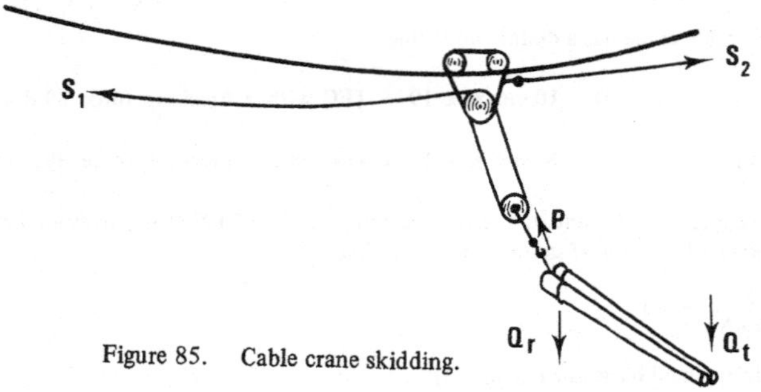

Figure 85. Cable crane skidding.

If the skyline carriage is equipped with double hoist line or hoist winch, it is heavy for the terrain crew to pull out the hook in the terrain. By simple cable crane constructions the haul-back line may be directly connected to the hook in order to overcome this problem. The arrangements are illustrated in figure 86.

On modern cable crane constructions, however, the hoist line is being pulled out by the winch by means of a slack-pulling line, or other automatic devices.

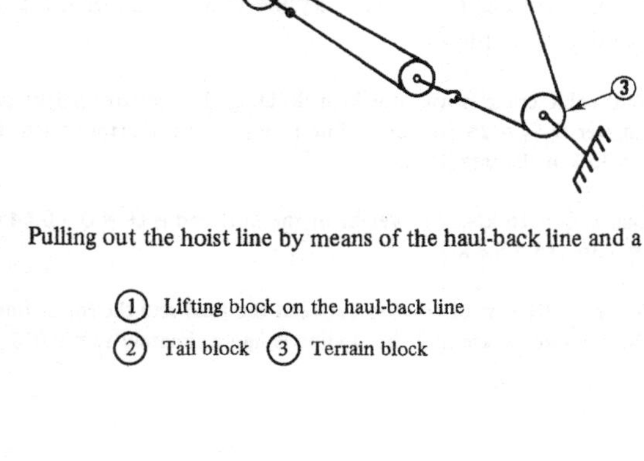

Figure 86. Pulling out the hoist line by means of the haul-back line and a terrain block.

(1) Lifting block on the haul-back line

(2) Tail block (3) Terrain block

6.0. ANCHORS, SPARS AND INTERMEDIATE SUPPORTS

Most of the cable cranes and cable ways in forestry are light equipment and should be easy to move from one place and rig up on a new logging site. It is an advantage to use green trees or stumps as spars, supports or anchors. Very often these should be preset by the foremen before the equipment and the cable crew arrives at the new logging area. It is a responsible job, to choose trees or stumps for this purpose, which should be undertaken by the responsible leader.

In some cases it may be difficult to find stumps or trees placed appropriately with relation to the location of the cable. It may be necessary to use earth anchors or rock bolt anchors. In some cases these anchors may be used again, for example in connection with cultivation work, or possible thinnings. They should be made strong and should be clearly marked to make it easy to find them later on. In connection with the construction of feeder roads it may be an advantage to place anchors on good installation points.

6.1. Anchors

6.1.1. Green trees or stumps as anchor

The wire rope should be fixed close to the root of the tree. It is an advantage if the tension line (guyline, skyline etc.) approaches the anchor tree or stump with approximately the same slope as the terrain. If this is not possible, the tree must be secured against the vertical component of the line pull.

If an anchor tree is exposed to heavy wind it is better to cut it, for example at a height of one meter.

The strength of green trees or stumps varies with the depth of the soil, the moisture content and the tree species. Usually hardwood trees and pine trees are stronger than spruce and fir trees. One should avoid trees growing on shallow ground (rock face with little soil). The same is the case with trees on marsh land with high ground water level. It is best to choose healty trees growing on deep and fresh soil. Dry or rotten stumps should as a rule be avoided. PESTAL (1961) has suggested the following formula for the maximum forces which may be taken up by green spruce and pine stumps on fresh and deep soil:

$$\boxed{S_{max} = \frac{D_{1.3}^2}{20}} \qquad (45)$$

Example 23: **A spruce tree on deep, fresh soil has a $D_{1.3}$ = 40 cm. The maximum horizontal pull is:**

$$S_{max} = \frac{40^2}{20} = \underline{80\ kN}$$

When anchoring a guyline or a skyline, torsional forces on the stump should be avoided. Together with the vibrating pulling forces, the torsional forces may cause gradual loosening of the stump. Often several stumps after each other are being used as a tail anchor for the skyline (figure 87). If the guyline approaches the stump at an angle in relation to the terrain it is necessary to watch that the vertical component of the pull together with the vibration are not loosening the stump (figure 88).

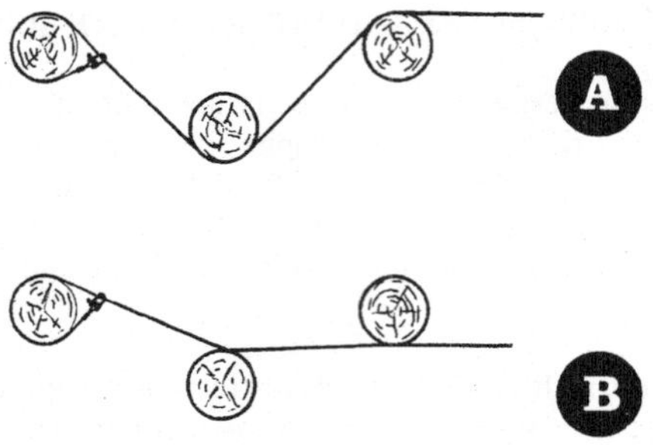

Figure 87. A. This anchoring is wrong due to the extra forces on the stumps or trees.
B. Correct anchoring.

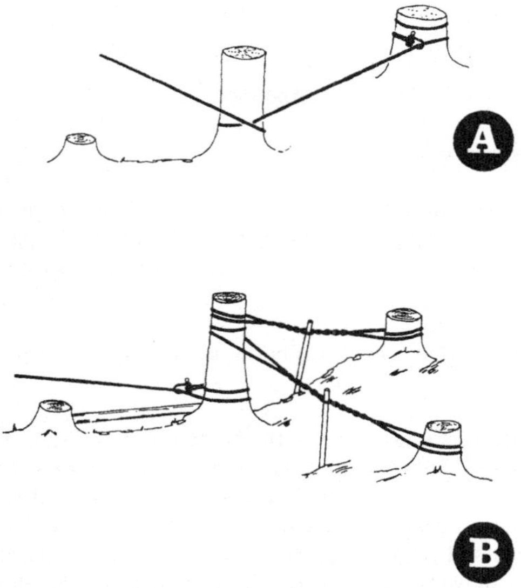

Figure 88. A. Wrong anchoring gives extra forces on the anchoring tree.
B. Correct anchoring. The anchoring tree is secured with guylines.

Table 12. Cable anchoring in green trees. Maximum horizontal pull at stump height in kN (after PESTAL 1961).

DBH cm	Max. pull under normal conditions in kN	Max. pull under optimal conditions in kN
10	3	5
20	13	20
30	30	45
40	53	80
50	83	125
60	120	180

Stationary cable ways may have spiral cables as skylines. Light movable cable cranes have wire ropes as skylines, because they are easier to handle. 6 x 7 + 1 or 6 x 19 + 1 have been used successfully. Sometimes the tail end of the skyline has been fixed directly to the stumps as shown in figure 87. This should be avoided in order to save the skyline from damage. Instead a tag guyline or an anchoring strap should be used. This saves the skyline, and the anchoring strap may be preset before the installation of the cable crane starts. The tail end of the skyline should be equipped with a reliable eye splice. The anchoring strap is connected to the eye splice of the skyline with a shackle (figure 89).

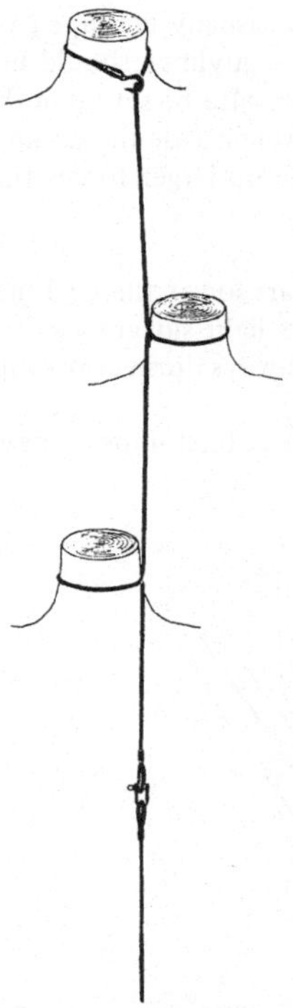

Figure 89. If the cable crane or the cable way is often moved from one corridor to another, an anchoring strap should be used because it can be preset before rigging up the crane in the new corridor.

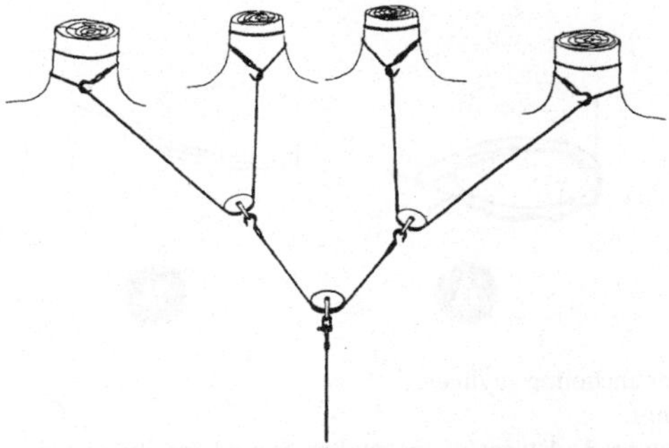

Figure 90. Distribution of the tension in the skyline to several stumps or trees.

When fixing guylines or skylines to several stumps after each other as shown in figure 89, the first stump (closest to the span) may receive the largest forces. During the operation it may loosen and stump number 2 takes gradually over a bigger part of the pull. It is therefore better to distribute the guyline or skyline pull between several stumps beside each other as shown in figure 90.

Sometimes the line pull may be distributed between two trees by means of a log anchor (figure 91). Table 17 gives informations concerning the dimensions of the log anchor.

Modern cable crane constructions are usually equipped with a steel spar on the winch. If the spar is higher than 10 meter, 2 or 3 guylines should be used in a direction opposite to the skyline. In most cases a guyline should also be set up in the direction of the skyline in order to prevent the steel spar from tipping over in case the skyline breaks. The guylines opposite to the skyline direction usually have to take up larger forces than the skyline tension. See formulae 47.

On lighter cable cranes with steel spars lower than 10 meter, one main guyline may be used, sometimes with the addition of two light sideguys or two hydraulic side feet (side anchors) in order to avoid the spar tipping side-ways during operation.

A knot may be used for fixing light guys to stumps or trees (figure 92).

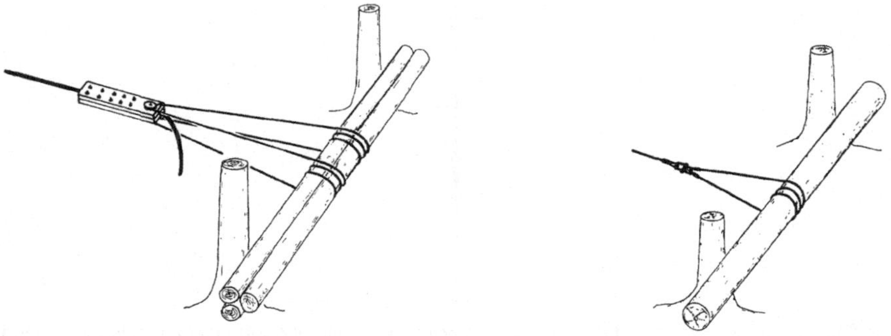

Figure 91. Log anchor.

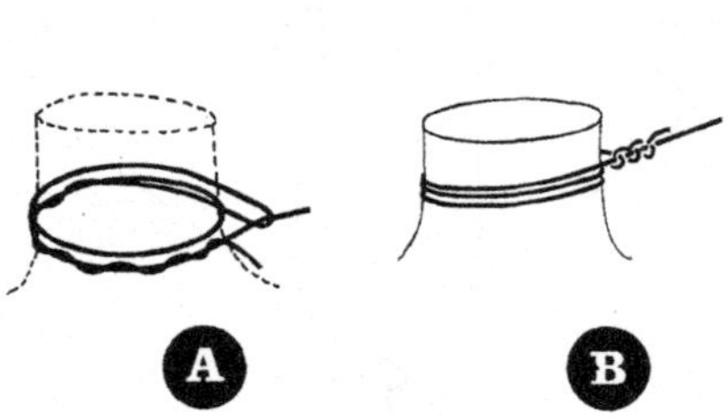

Figure 92. Knots used for anchoring guylines.
A. Forest knot.
B. Anchoring by 3–4 turns of the guyline around the stump or tree with 3 halfhitches around the guyline.

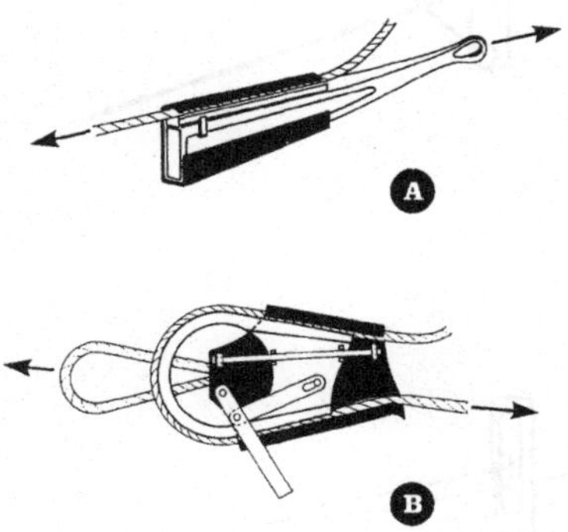

Figure 93. A. Single wedge grip.
B. Double wedge grip.

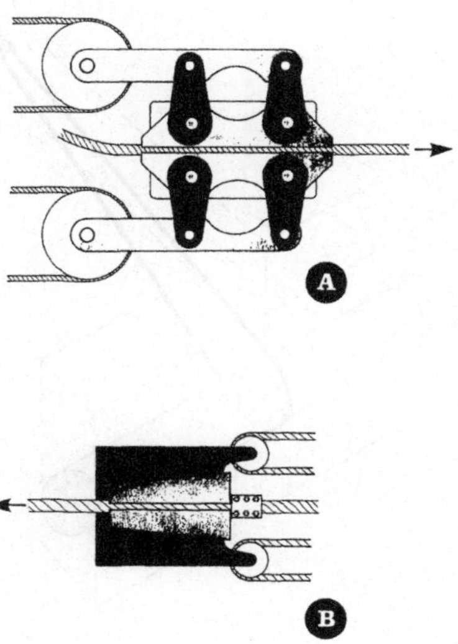

Figure 94. A. Eccentric pressure grip.
B. Wedge pressure grip.

There exist many practical devices for anchoring light guylines etc. Cable clamps, clips or plate clamps should be avoided partly because they are time consuming and partly because it is necessary to look after them and tighten them since the wire rope is gradually getting thinner during the operation. Single wedge grip and double wedge grip are useful equipment (figure 93).

The skyline tension is often very high. An eccentric pressure grip or a wedge pressure grip may be used in this case (figure 94).

Table 13. Earth anchor with log of green wood (»Dead man»). The mid.diameter of the log in dry, good soil (after PESTAL 1961).

Pull in kN	Mid. diameter in cm	Length of log in m
10	16	2
30	22	2
50	32	4
100	40	4
150	50	5
200	56	6

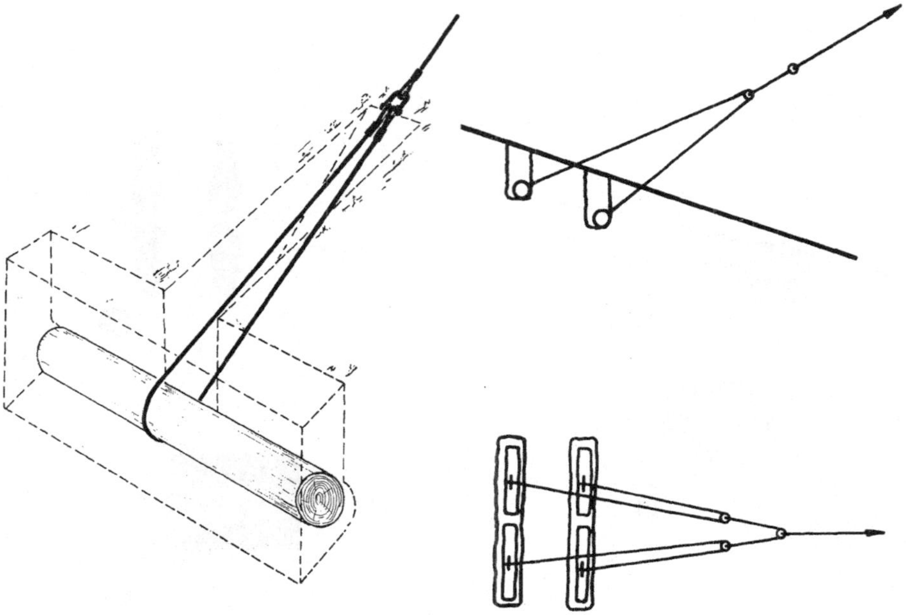

Figure 95. Left: Earth anchor with one log (»dead man anchor»).
Right: Distribution of high skyline tension on several earth anchors by means of equalizing blocks.

6.1.2. Earth anchor

If no stumps or trees are available the guyline or the skyline may be fixed to an earth anchor, the so-called »dead man» (figure 95).

If the soil is good it is sufficient to dig a 1.5 meter deep ditch. After the log has been placed in the ditch and the anchoring strap has been fixed to the log, the ditch should be filled with soil. It is an advantage if the angle between the guyline and the ground is as small as possible. If the guyline angle is very large it is necessary to dig the ditch deeper.

The anchoring strap should go around the log half a turn. This makes it easier to pull out the anchoring strap afterwards. Table 13 gives some information concerning the log sizes to be used.

6.1.3. Rock bolt anchor

When a rock bolt is used as an anchor it is important to see that the rock is of good quality and that there are no cracks in the rock which may give way because of the vibrating tension during the operation. Norwegian rock is usually primary rock of high quality. Accordingly we are using shorter and thinner bolts than in Central Europe where the rocks often are eroded on the surface. PESTAL (1961) suggest that the bolt should have a length of 750 mm and a thickness of 60 mm for a cable tension of 100 kN.

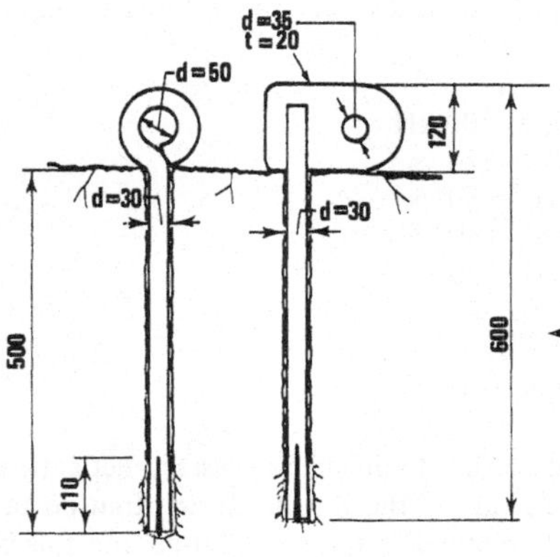

Figure 96. Rock bolt anchor.

Figure 97. Under high tension the bolt may be bent until the eye is in contact with the rock.

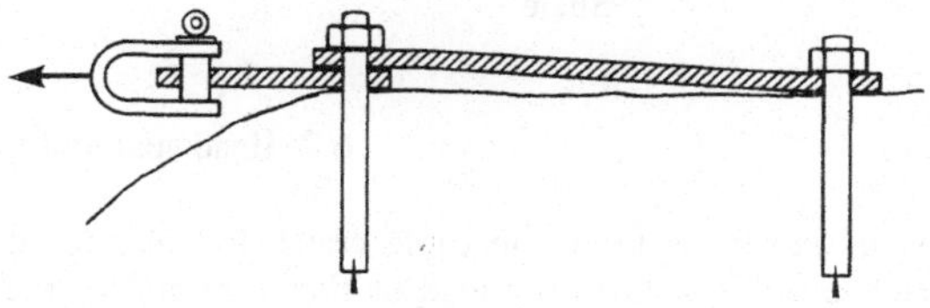

Figure 98. Distribution of the tension to several bolts.

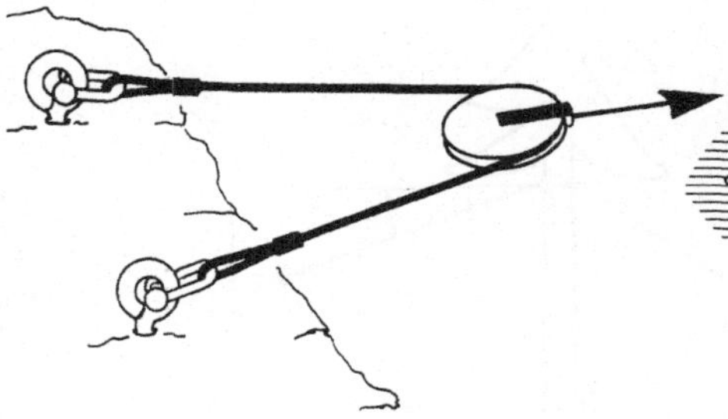

Figure 99. Due to the variation in the rock structure, it is often preferable to place the bolts beside each other in relation to the direction of the skyline.

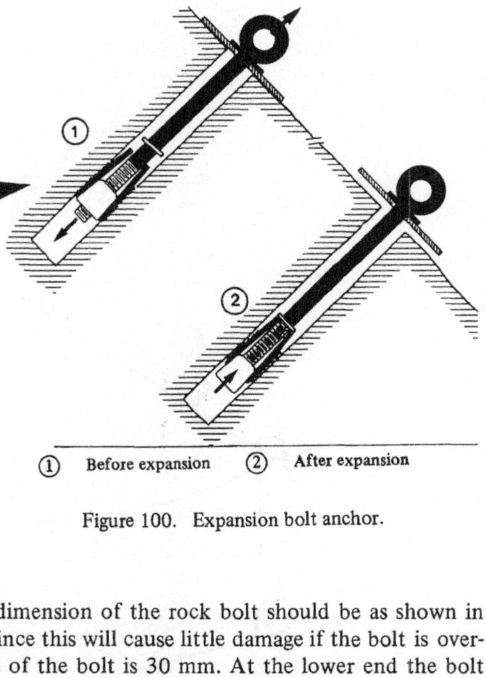

① Before expansion ② After expansion

Figure 100. Expansion bolt anchor.

According to Norwegian experiments the dimension of the rock bolt should be as shown in figure 96. Mild steel should be used (st 37) since this will cause little damage if the bolt is overloaded and bent (figure 97). The thickness of the bolt is 30 mm. At the lower end the bolt is split to a length of 110 mm. A 6 mm thick wedge is used. When the bolt reaches the bottom of the drilled hole, the wedge will widen out the bolt and press or break it fixed to the sides in the lower part of the hole (figure 96).

A 33 mm bore should be used to drill holes for a 30 mm steel bolt. This may be used on a portable bore hammer.

If the maximum pull in the skyline is S = 120 kN two 30 mm thick bolts may be used. The rock is always the minimum factor. The two bolts should be placed with a good distance between them (according to the local rock conditions) for example as illustrated in figure 99.

If the cable tension is longitudinal to the direction of a 25 mm thick bolt, the following anchor strengths are representative for various types of rocks:

Granite and bassalt	200–300 kN
Sandstone and quartzite	150–220 »
Limestone and marble	120–180 »
Firm sandy shale and gypsum	100–150 »
Shale	80–120 »

6.2. Head spar and tail spar

Spars are being used on cable cranes in order to allow the skyline the necessary deflection and in order to keep the load or the front end of the load off the ground during transportation. The head spar is located near or on the winch. Sometimes a tail spar is used and this is located at the outer end of the span. Sometimes spar trees are used, but modern cable cranes are mostly equipped with steel spars.

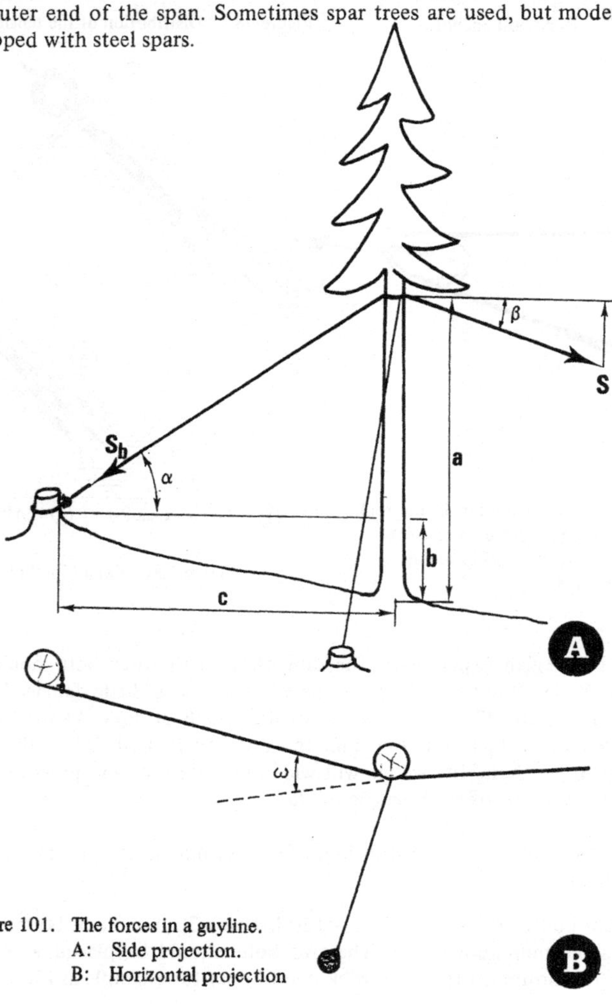

Figure 101. The forces in a guyline.
A: Side projection.
B: Horizontal projection

6.2.1. Guyline tension

Whether the spar is a spar tree or a steel spar it has to be kept in position by guylines. When a spar tree is used at the end of the span the skyline may be used as guyline. The skyline passes over a cable shoe or a block in the tree and continues down to the anchoring point (figure 101). If the skyline changes direction when it passes the spar tree (angle ω in figure 101 B) it may be sufficient with one guyline. The skyline tries to bend the tree and these forces should be taken up by the guyline. The change in the skyline direction should preferably be less than 5° ($\omega < 5°$).

Example 24: **The horizontal component of the skyline pull is H = 100 kN. With a change in the skyline direction of ω = 5° the bending force on the tree is:**

$$H \cdot \sin\omega = 100 \cdot 0.087 = 8.7 \text{ kN}$$

The force 8 700 N must be taken up by a guyline in the opposite direction.

The guyline tension may be calculated by the following formulae using the symbols in figure 101.

$$\text{tg}\alpha = \frac{a \pm b}{c} \qquad (46)$$

b is negative when the foot of the spar is lower than the anchoring point and positive if the foot of the spar has a higher location than the anchoring point.

$$S_b = \frac{S \cdot \cos\beta}{\cos\alpha} \qquad (47)$$

Since $\cos\alpha$ always is less than 1, S_b is bigger than $S \cdot \cos\beta$. Normally the skyline angle α should be between 30° and 45°. The smaller the guyline angle α, the lower the guyline tension. The same applies to the vertical forces on the spar.

The following limiting values for the rope deflection angle ($\gamma = \alpha + \beta$) are recommended. (See also figure 138):

16° . Maximum rope deflection angle with stationary cable ways.
22° : Maximum rope deflection angle with light (movable) cable cranes when the main line is approximately parallel to the skyline.
90° : Maximum rope deflection angle at head spars and tail spars.

Table 14. The vertical pressure of the skyline on the support (in kN) with various skyline tensions and rope deflection angle between the skylines on each side of the support (after PESTAL 1961). $V = 2 \cdot S \cdot \sin\frac{\gamma}{2}$

Rope deflections angle in degrees (°)	Skyline pull in kN						
	40	60	80	100	120	140	160
5	3	5	7	9	10	12	14
10	7	10	14	17	21	24	28
16	11	17	22	28	33	39	45
22	15	23	30	38	46	53	61
30	21	31	41	52	62	72	83
45	31	46	61	77	92	107	122
60	40	60	80	100	120	140	160
90	57	85	113	141	169	197	226

Table 15. The mid. diameter in cm of a mast tree, tail tree or strut support at various vertical forces (buckling load). (After PESTAL 1961).

Vertical force (buckling load) in kN	Height in metres to cable shoe/crossbeam					
	<5 m	6 m	8 m	10 m	12 m	15 m
10	14	15	17	19	21	24
20	16	17	20	22	25	28
30	18	20	22	25	27	30
40	19	21	24	27	29	33
50	20	22	25	28	31	35
70	–	24	27	31	34	38
100	–	29	30	34	37	41

The information given in table 14 and 15 may be used for dimensioning the spar tree.

Example 25: **Using the symbols in figure 101 a guyline has the following specification:**

$S = 100$ kN
$a = 8$ m
$b = +2$ m (the ground has a downhill slope from the spar to the anchor)
$c = 15$ m
$\beta = 15°$

$$tg\alpha = \frac{a + b}{c} = \frac{8 + 2}{15} = \underline{0.67}$$

$$\alpha = 34°$$

$$S_b = \frac{S \cdot \cos\beta}{\cos\alpha} = \frac{100 \cdot 0.97}{0.83} = \underline{117 \text{ kN}}$$

The vertical force on the spar is:

$$x = S \cdot \sin\beta + S_b \cdot \sin\alpha$$
$$x = 100 \cdot \sin 15 + 117 \cdot \sin 34$$
$$x = 91.3 \text{ kN}$$

The rope deflection angle is $\alpha + \beta = 49°$. The results we achieved corresponds well with the figures given in table 14. Table 15 shows that the spar tree should have at least 29 cm mid.diameter. There are good green stumps with 30 cm diameter which may be used as anchors. Table 12 shows that the guyline tension (S_b = 117 kN) must be distributed between three stumps. If a »dead man» has to be used as anchor, a 4 meter long log with 43 cm mid.diameter should be used.

6.2.2. Spar tree

The spar tree is a green tree, preferably spruce or pine tree. It should be delimbed up to the height where the cable shoe and the guylines shall be fixed. When there are green and thick branches they may be cut off approximately 25 cm away from the stem in order to help the crew when climbing the tree.

Earlier, climbing shoes where used to climb the tree. The steel spur on the climbing shoe must be long and sharp and formed as a chisel. The climber must use a safety belt during the climbing.

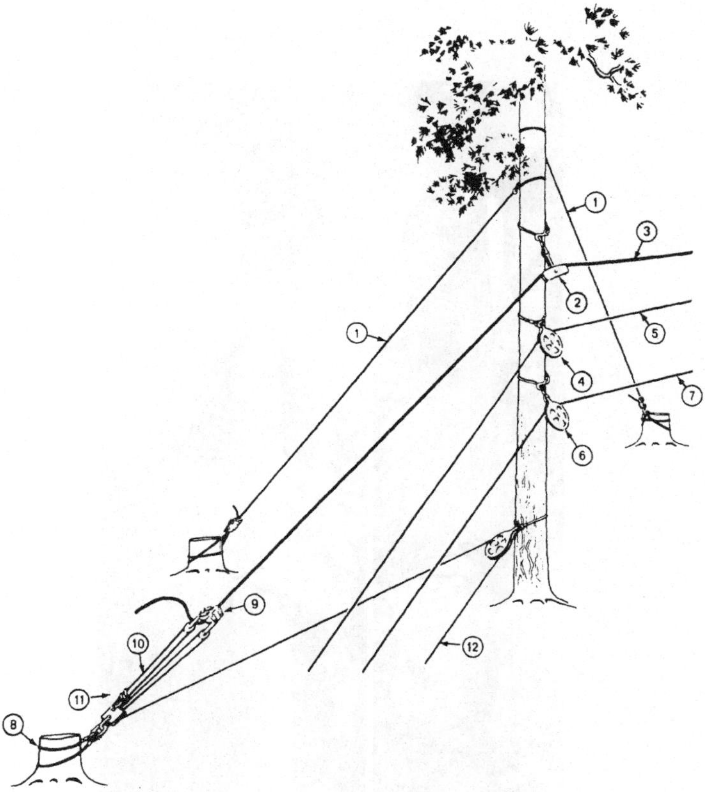

(1) Guyline (2) Cable shoe (3) Skyline (4) Guide block for the main line (5) Main line
(6) Guide block for the haul-back line (7) Haul-back line (8) Tailhold (9) Buffalo grip or wedge grip
(10) Tackle (11) Buffalo grip to lock the tacle (12) Pulling line from auxiliary winch

Figure 102. Spar tree.

It is, however, both safer and lighter for the crew to use a rig up ladder (figure 104). This is a light aluminum ladder built in 3 meter sections. Each section is fixed to the tree by a light chain around the tree at the top end of each ladder section.

The spar tree must be kept in position by the guylines (figure 102). For complete anchoring 3 guylines are needed. Very often, however, one or two guylines are placed in the opposite direction of the bi-secting line of the angle between the main and the haul-back lines.

The skyline may be tightened by means of a tackle. It is connected to the skyline by means of an eccentric pressure grip (figure 94). As soon as the skyline is tightened a buffalo grip locks the tackle. If a large volume has to be transported by the cable crane an extra safety strap between the skyline and the anchor should be used.

Figure 103. Climbing shoe used to rig up a spar tree. Håøya, 1955.

6.2.3. Steel spar

Modern cable crane constructions are usually equipped with a steel spar. This is built together with a winch on a vehicle. During the transport from one logging area to another the steel spar is taken down, and two different methods have been used, *tip spar* and *telescope spar*.

The tip spar may be laid down over the vehicle during transport. On the new logging site the steel spar is lifted into up-right position for example by means of hydraulic cylinders.

Telescope spar. The spar may be shortened telescopically before transporting the winch from one logging area to another. On the new logging area the spar is lengthened telescopically by means of hydraulic cylinders. Sometimes a combination of the two systems is used, and this may be called a *tip/telescope spar.*

Figure 104. Rig up ladder in a finger support. Kviteseid, 1978.

A complete tip/telescope spar is shown in figure 105. There are four guyline drums which may give the guyline 25 kN pre-tension. As soon as the guylines have been connected to the anchoring straps they are tightened by means of the guyline winches. Each winch is then locked by a safety device in order to take up large tensions during the logging operation.

Modern cable cranes are usually equipped with rig up drums with rig up lines and straw drums with straw lines. In order to simplify the installation of the crane on an area these are usually placed at the lower end of the steel spar. The guide blocks for the skyline, the main line, the haul-back line and the slack pulling line, are located at the top end of the spar. These guide blocks should be constructed in such a manner that the vehicle with the winch can be placed in any direction relative to the direction of the cables. This makes it easier to find a good position for the yarder under various terrain conditions.

Swing boom is a specially shaped steel spar used on mobile swing yarders (figure 109). By means of a swing assembly on the crawler carrier or wheel carrier the winch unit with the A-frame and the boom may be swung around powered by a hydraulic motor. The boom is anchored to the A-frame by the boom guylines. The A-frame is kept in position during the operation by means of two guylines from the top of the A-frame to the anchoring points. The guide blocks are placed at the top end of the boom.

Lighter cable cranes with a tip spar shorter than 10 meter use only one guyline in the opposite direction to the main and haul-back lines etc. Side-tipping may be avoided by means of hydraulically operated side anchors (figure 106). This construction simplifies the rigging.

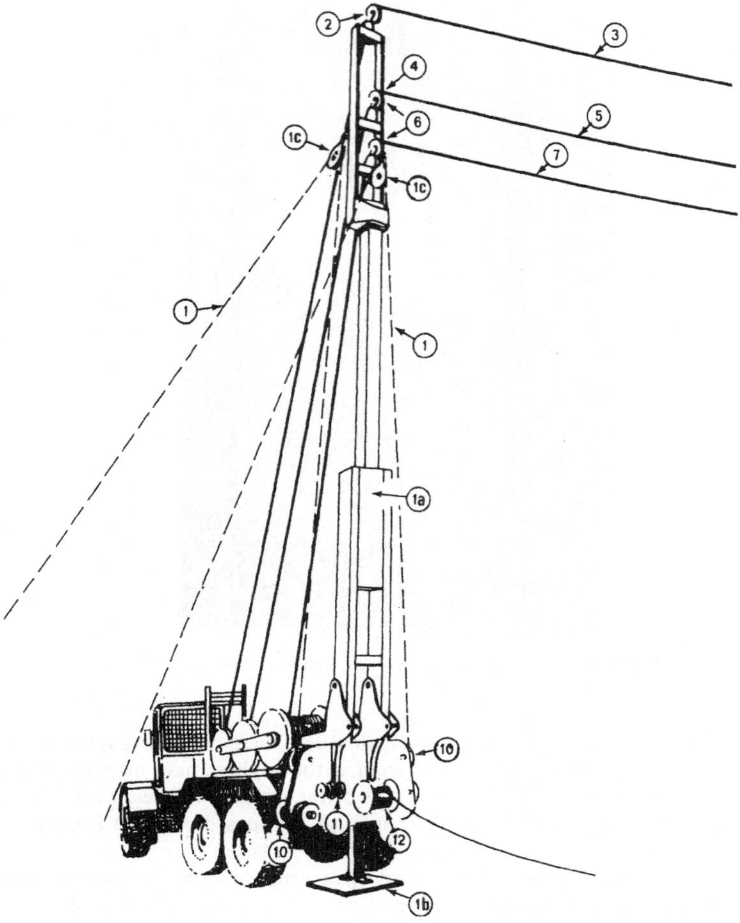

(1) Guyline (1a) Steel spar (telescope/tip spar) (1b) Foot for the steel spar (1c) Guide block for guyline

Alternative I:

(2) Guide block for skyline (3) Fix or live skyline (4) Guide block for haul-back line (5) Haul-back line
(6) Guide block for main line (7) Main line

Alternative II:

(2) Guide block for haul-back line (3) Haul-back line (running skyline) (4) Guide block for main line (5) Main line
(6) Guide block for slack-pulling line (7) Slack-pulling line (10) Guyline winches
(11) Erecting drums with erecting lines (12) Straw drum with straw line

Figure 105. Steel spar.

Figure 106. Owren Vario Winch with tip tower and hydraulic side anchors (hydraulic feet).

Figure 107. Front anchor instead of back guy on a skidder mounted winch and tower for short distant cable crane.

Figure 108. A. and B. Tip tower on the Owren Vario Winch.
C. and D. Tip tower on the Igland interlock winch. Hurdal, 1979.

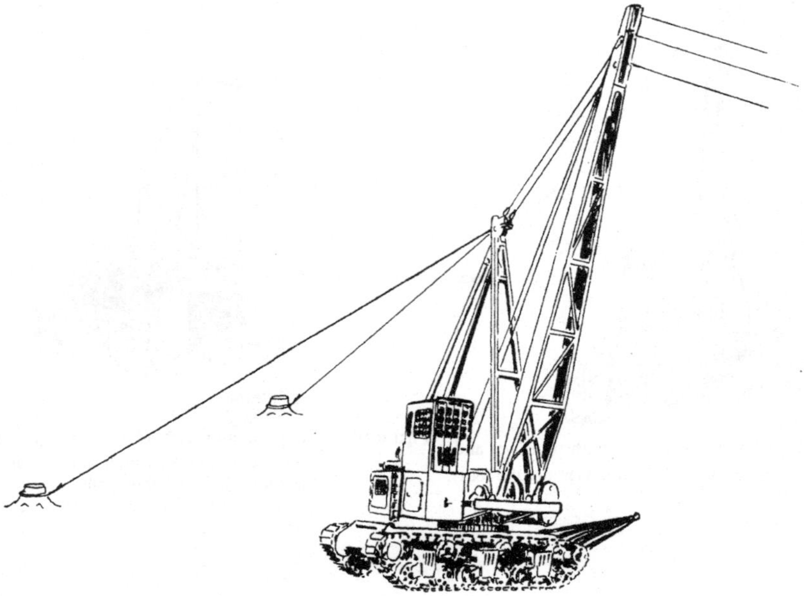

Figure 109. A swing boom yarder.

Some light cable cranes may use a shorter spar than 6 meter. If the winch with spar is placed on a heavy skidding tractor, the weight of the tractor may serve as anchoring. In this case an anchoring foot may be used instead of a guyline. The advantage is the very fast rig up time.

6.3. Intermediate supports

Intermediate supports should only be used when absolutely necessary. It is better to plan the location of the feeder roads in such a way that intermediate supports can be avoided. They are a hindrance to the operation, partly because they increase the rig up time, and partly because the transport speed along the skyline is reduced. In some terrain intermediate supports cannot be avoided and they may be installed on standing trees or as erected logs on the felling site.

6.3.1. Four strut supports

These are mostly used on stationary cable ways and are of little interest for the movable cable cranes. No guylines are usually needed (figure 198).

6.3.2. Two strut supports

The crossline support is the most commonly used two strut support. The crossline or the support line goes from a stump in the terrain over a cable shoe in the first tree and a block in the next tree. The jack with the cable shoe and skyline is attached to the crossline while on the ground. The skyline is lifted into position by tightening the crossline, often with a small hand-winch. The crossline should not be tightened too much. A loose crossline gives better flexibility and less tension.

The portal support is very seldom used (figure 111). The A-framed support, however, is often used where no support trees are available (figure 112).

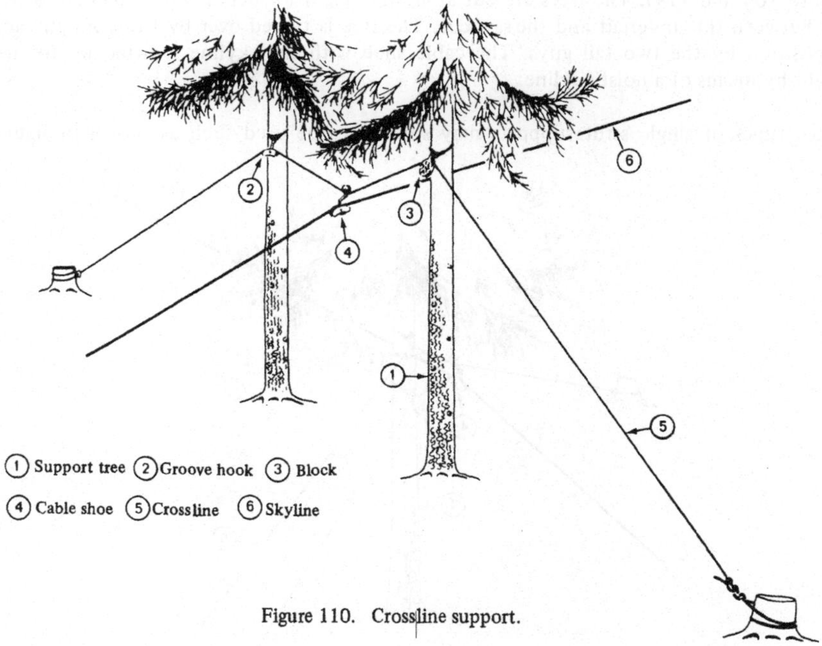

Figure 110. Crossline support.

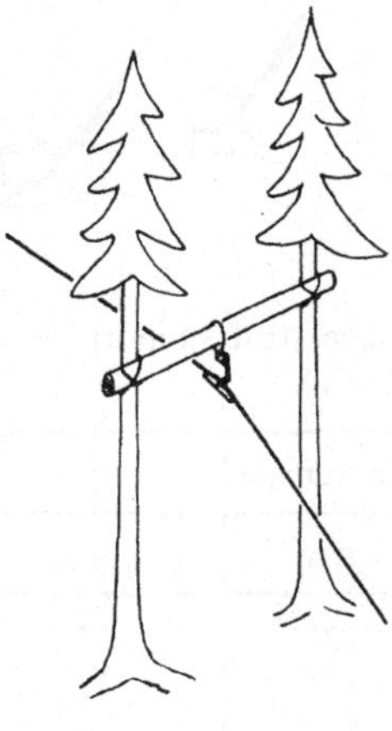

Figure 111. Portal support (double strut support).

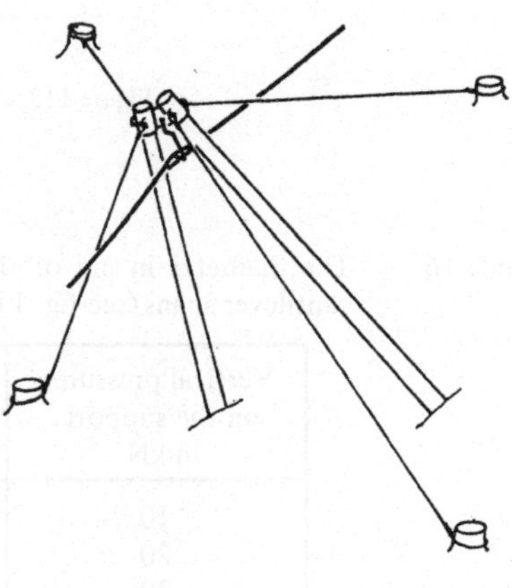

Figure 112. A-framed support (double strut support).

6.3.3. Single strut support

Single crossline support is a variation of the crossline support, and may be used in side slopes etc. (figure 113). The most interesting single strut support is, however, the *finger support* (figure 104 and 114). The trees are cut at stump height in such a way that 3 cm of wood is left between the undercut and the saw-cut. The tree is tipped over by the front guy and kept in position by the two tail guys. The cable shoe with the skyline is lifted up to the right height by means of a hoist guyline.

Other types of single strut supports may sometimes be used such as shown in figure 115.

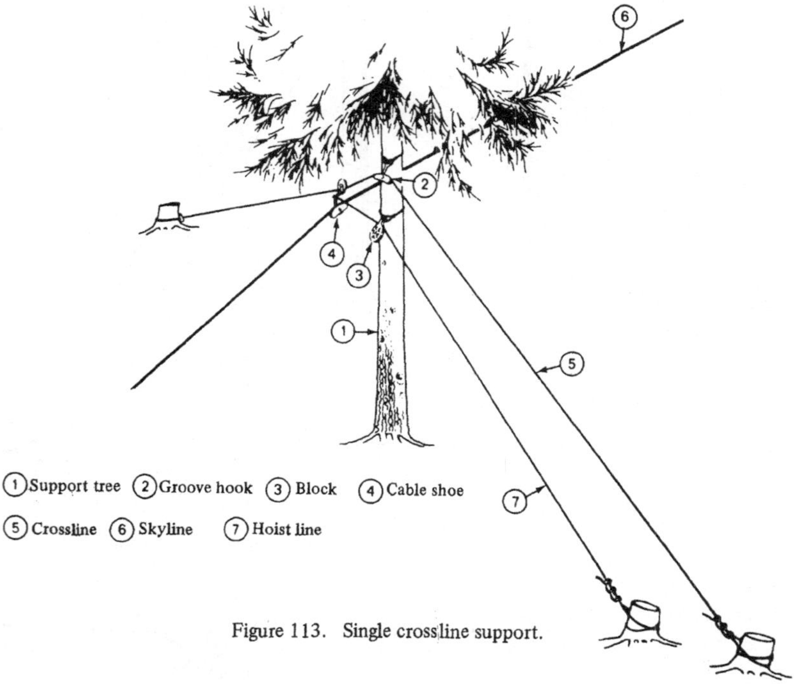

Figure 113. Single crossline support.

Table 16. The diameter in cm of the cantilever on a single strut support at various cantilever spans (see fig. 115). (After PESTAL 1961).

Vertical pressure on the support in kN	Cantilever span		
	0.5 m	1.0 m	1.5 m
10	17	22	25
20	22	27	32
30	25	32	36
40	28	35	40
50	30	37	42
100	37	47	53

Table 17. Diameter in cm of the crossbeam for a portal type support (double strut support). (After PESTAL 1961).

Vertical pressure on the support, kN	Distance between the struts under the crossbeam	
	1.5 m	2.5 m
10	16 cm	19 cm
20	20 cm	23 cm
30	23 cm	27 cm
40	25 cm	30 cm
50	27 cm	32 cm
100	34 cm	40 cm

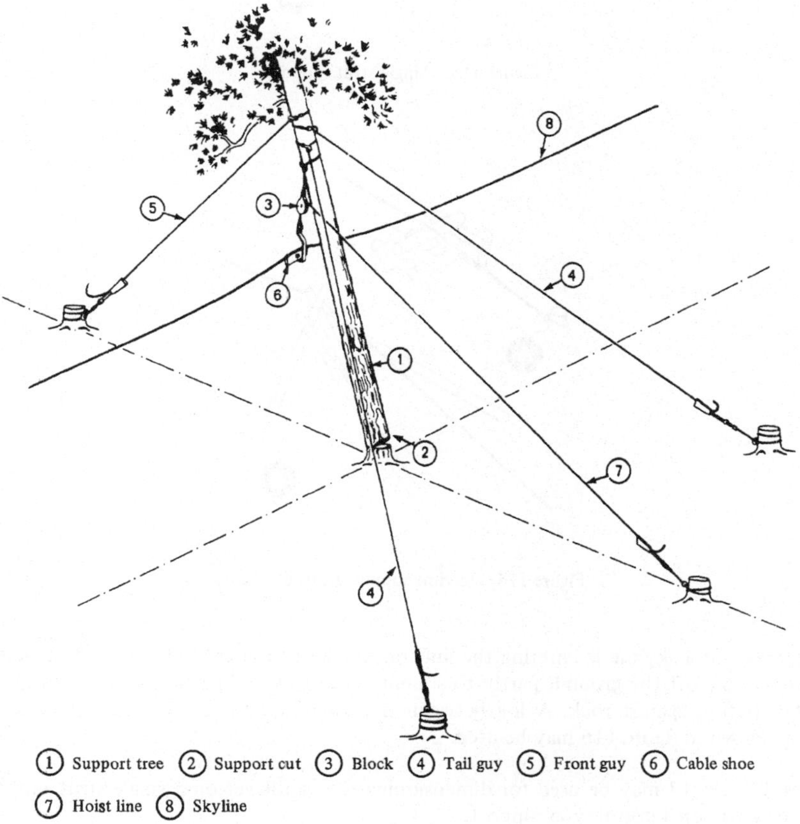

Figure 114. Finger support.

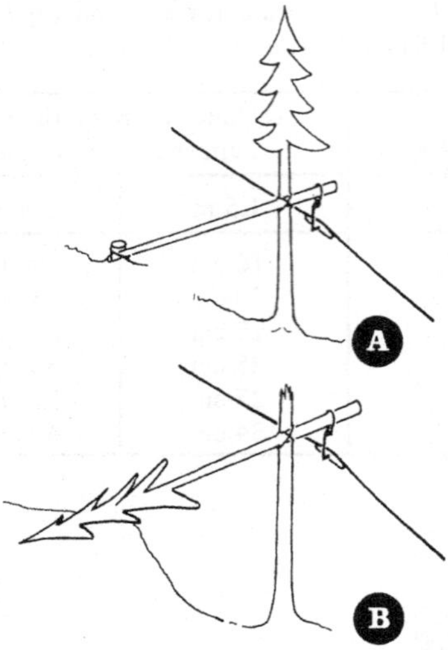

Figure 115. Single strut support.

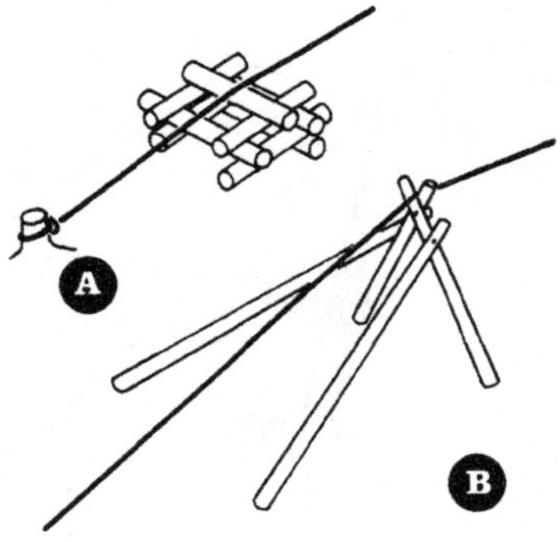

Figure 116. Skyline support near the anchor.

If a guyline or a skyline is entering the anchor parallel to the ground it may be an advantage to lift the line off the ground, partly to simplify the work and partly to avoid damage caused by the friction against rock. A log between the cable and the ground or a skyline support such as shown in figure 116 may be used.

Tables 16 and 17 may be used for dimensioning the cantilever on a single strut support and the crossbeam on a portal type support.

7.0. LINES IN LONG SPAN

7.1. Horizontal and vertical components of the tension

When a suspended line with uniformely distributed load is fixed at both ends it has the shape of a catenarian curve. It deviates little from the shape of a quadratic parabola when operating with the tensions which are common in cable cranes. Since the calculation is simpler when using the quadratic parabola and since this is sufficiently accurate for the purpose, .this formula will be used in the following.

The tension in a rope (S) may be decomposed into a horizontal force (H) and a vertical force (V). According to the law of the equilibrium of forces the horizontal forces will have the same magnitude along the entire cable. The vertical forces vary along the cable. The steeper the cable tangent, the larger the vertical force. Consequently, the vertical force is greatest near the upper end of the cable. The cable tension (S) is the resultant force of the horizontal (H) and the vertical (V) component forces:

$$\boxed{S = \sqrt{H^2 + V^2}} \tag{48}$$

Since the vertical force is greatest at the upper suspension point, the greatest tension in the cable occurs at the upper end of a free span. If the suspension points of a freely hanging cable have the same height the cable tension is the same at the two suspension points.

The angle between the tangent to the cable and the horizontal line is β. From figure 117 it is seen:

$$V = S \cdot \sin\beta$$

At the deepest point of the cable the tangent is horizontal: $\sin\beta = \sin 0 = 0$. The vertical force $V = 0$ and the cable tension is:

$$S = \sqrt{H^2 + 0} = H \tag{49}$$

At this point the cable tension is the same as the horizontal force. At any other point along the cable the tension is greater than the horizontal force.

The cable tension varies with the weight of the cable, the weight of the concentrated load (for example carriage and load), the length of the span and the rope sag.

For most cable cranes used in forestry the construction is given by the manufacturers. The purpose of our analysis is to find the size of the rope sag in order to operate with the necessary safety for a certain load and length of the span.With a concentrated load on the cable the deflection is greater than for the unloaded cable. The curve along which the load moves when moving along the skyline is called *the load curve* (see figure 131).

The deflection of the load curve is most important since it describes whether the load will pass freely over the terrain. In case of doubt the longitudinal profile of the terrain may be drawn. When the rope sag of the load curve is known, the load curve may be drawn in order to investigate whether there is sufficient clearance. If not, it is necessary to place an intermediate support and make a new calculation of the forces and deflections.

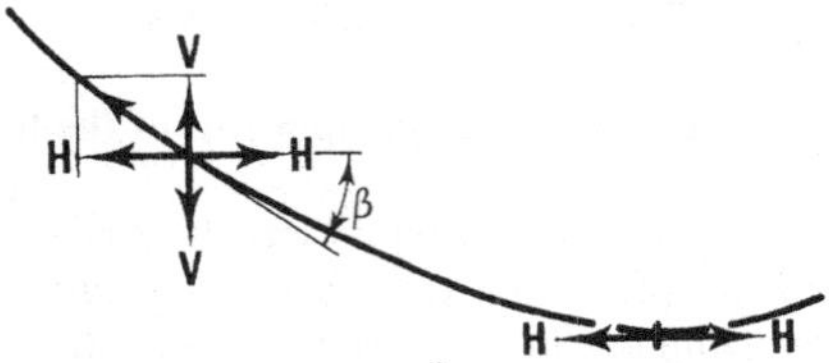

Figure 117. The horizontal component of the tension in the cable is the same along the skyline while the vertical component varies according to the angle of the cable tangent.

7.2. Cable tension weight

On stationary cable ways one end of the skyline, usually the lower end, is connected to a cable tension weight while the upper end is fixed to an anchoring arrangement. In this case the lower end of the cable passes a sheave on the top of a support and runs vertically down to a tension weight which can move up and down. The cable tension at the lower end of the skyline will be the same as the tension weight. When a load moves along the skyline, it will sink and the tension weight will be lifted. The cable is never subjected to a larger tension at the lower end than that caused by the tension weight, even if there are one or more concentrated loads on the cable.

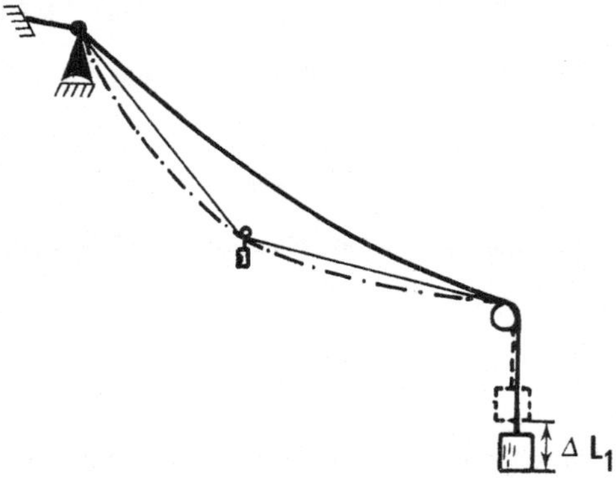

Figure 118. The tension in the skyline is kept constant by means of a tension weight.

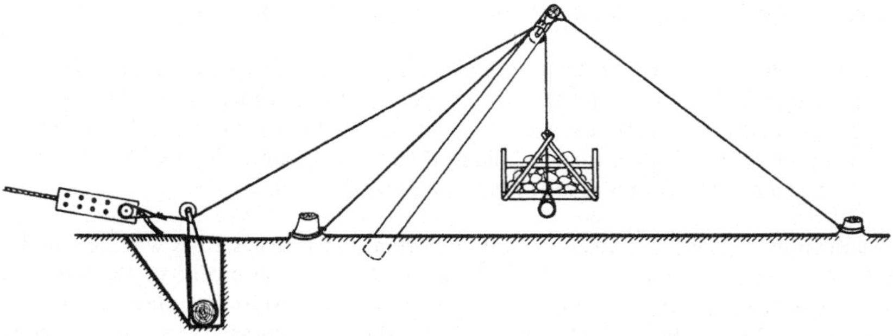

Figure 119. Arrangement of a skyline anchored to a tension weight.

Figure 120. Constant tension in the main and haul-back line by means of a tension weight.

The necessary lengthening of the cable (ΔL_1) in order to get sufficient deflection is created by lifting the tension weight (figure 118). ΔL_1 should be 3 ‰ for $L_1 \sim 400$ m and 4 ‰ for $L_1 \sim 700$ m. One should add 1 ‰ for temperature variations and 2 ‰ when using a new cable since it may stretch a little during the running in period. Accordingly ΔL_1 should be approximately 3 m for a 500 meter span.

7.3. Elastic elongation of the cable

Forestry cable cranes are usually firmly anchored at both ends (figure 121).

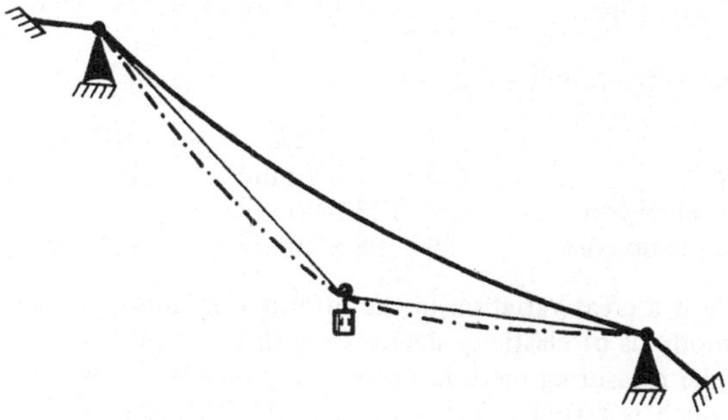

Figure 121. Fixed anchoring of a skyline. (Tight skyline).

Modern cable cranes are usually more mobile than the previous constructions. The winch is placed on a mobile chassis and equipped with steel spar and guyline winches. The skyline is fixed to a skyline drum which may give sufficient pull for pre-tensioning the skyline. The skyline drum may be locked in order to take up the greatest tensions during the load transport. In other cases a running skyline system is used. The line tensions are maintained by transferring a torque between the main and haul-back drums in the interlock winch. Overloading of the running skylines is avoided by an arrangement as shown in figure 44. It shows the tension continuously during the operation. The measuring device may also trigger a safety device which reduces the power or the tension as soon as a certain limit has been reached. It is, however, also of interest to calculate the forces and deflections in the cable system in this case. When the skyline is tightened by means of the skyline drum, the situation is the same as if it had been firmly anchored at both ends.

When the load moves along a skyline which is firmly anchored at both ends, the load curve has a larger deflection than the pre-tensioned cable without any payload. The concentrated load causes higher cable tension and the cable is straightened between the load and the anchoring points. The cable also is lengthened due to its elasticity:

$$\varepsilon = \frac{\Delta L_l}{L_l} = \frac{\sigma}{E_r}$$

The specific elongation or the elasticity is in inverse ratio to the modulus of elasticity as long as the tension is lower than the yield point of the steel. A high quality steel has a modulus of elasticity E = 220 kN/mm².

In a spiral laid cable the wires are shaped like spirals. Accordingly it is more elastic than conpact steel.

A wire rope with several parts is even more elastic, especially if the rope has fibre core.

The reason is that the wires are longer than the rope. When the rope is tensioned, the wires are being pressed together.

CZITARY (1962) and SCHNEIGERT (1966) give the following relation between the modulus of elasticity in steel and steel rope:

Spiral rope:	E_r = 0.60–0.85 · E = 129–183 kN/mm²
Stranded wire rope:	E_r = 0.35–0.66 · E = 75–129 kN/mm²

PESTAL (1961) refers to the following values:

	WYSS (1956)	HUDLER (1936)
Spiral rope:	123–142 kN/mm²	158–167 kN/mm²
Wire rope, steel core:	98–123 kN/mm²	
Wire rope, hemp core:	69– 98 kN/mm²	139–147 kN/mm²

This shows that there is a great variation in the information about the modulus of elasticity for wire ropes. The modulus of elasticity decreases with the number of wires in the rope. It is even probable that the measuring method plays a certain role in the result, especially if the length of the rope which is tested is too short. HUDLER's information refer to used ropes under high tension.

Table 18. Modulus of elasticity for some steel ropes (E_r).

Construction	E_r kN/mm²
6x 7+1 polypropylene core	93.2
6x 7+1 steel core	98.1
6x19+1 polypropylene core	86.3
6x19+1 steel core	93.2
6x37+1 polypropylene core	72.6
6x37+1 steel core	78.5
Spiral strand, 7 w.	144.2
Spiral strand, 19 w.	131.5
Spiral strand, 37 w.	124.6

The moduli of elasticity in table 18 are good practical average values and may be recommended for cable crane calculations in forestry.

The Norwegian Forest Research Institute carried out an analysis of the tension and deflection in a spiral cable. If the spiral cable had 19 wires and the diameter was d = 13 mm the average modulus of elasticity was:

$$E_r = 143.34 \pm 6.54 \text{ kN/mm}^2$$

This corresponds to:

$$E_r = 0.67 \cdot E$$

This investigation gave a very interesting result (figure 122):

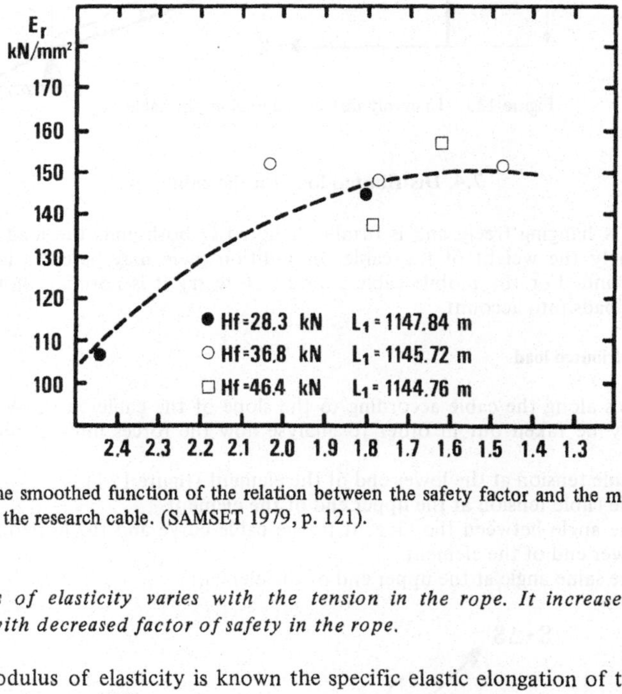

Figure 122. The smoothed function of the relation between the safety factor and the modulus of elasticity in the research cable. (SAMSET 1979, p. 121).

The modulus of elasticity varies with the tension in the rope. It increases with increased tension and with decreased factor of safety in the rope.

When the modulus of elasticity is known the specific elastic elongation of the cable may be found (P_E in meter per kN):

$$\boxed{\varepsilon = \frac{\Delta L_l}{L_l} = \frac{\sigma}{E_r}} \qquad (50)$$

$$\Delta L_l = \frac{S \cdot L_l}{A \cdot E_r}$$

$$\frac{\Delta L_l}{S} = \frac{L_l}{A \cdot E_r}$$

$$\boxed{P_E = \frac{L_l}{A \cdot E_r}} \qquad (51)$$

Example 26: **For a skyline the specific elastic elongation is (formula 51):**

$$P_E = \frac{1\ 145\ \text{m}}{100\ \text{mm}^2 \cdot 143.34\ \text{kN/mm}^2}$$

$$P_E = 0.080\ \text{m pr kN}$$

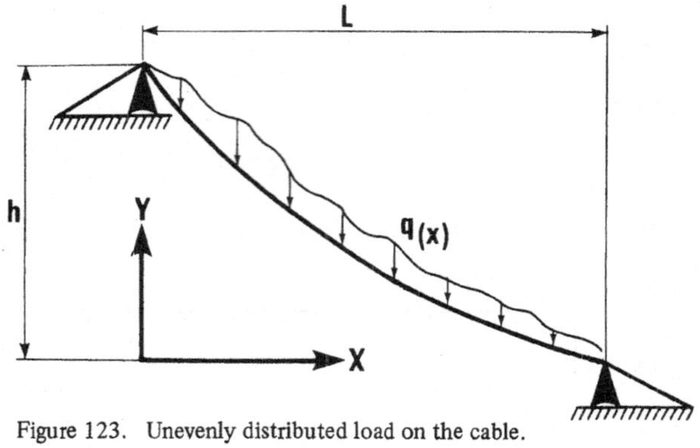

Figure 123. Unevenly distributed load on the cable.

7.4. Distributed load on the cable

When a skyline is hanging freely and is firmly anchored at both ends the load is uniformely distributed, namely the weight of the cable. In addition there may be extra load caused by icing, snow or wind. For the mobile cable cranes in forestry it is normally not necessary to take these extra loads into account.

7.4.1. Unevenly distributed load

The tension varies along the cable according to the slope of the cable curve. A small element of the cable may be taken out in order to analyse how the forces interact on the element:

S = Cable tension at the lower end of the element (figure 124).
$S + \Delta S$ = The cable tension at the upper end of the element.
β = The angle between the tangent to the cable curve and the horizontal line at the lower end of the element.
$\beta + \Delta\beta$ = The same angle at the upper end of the element.

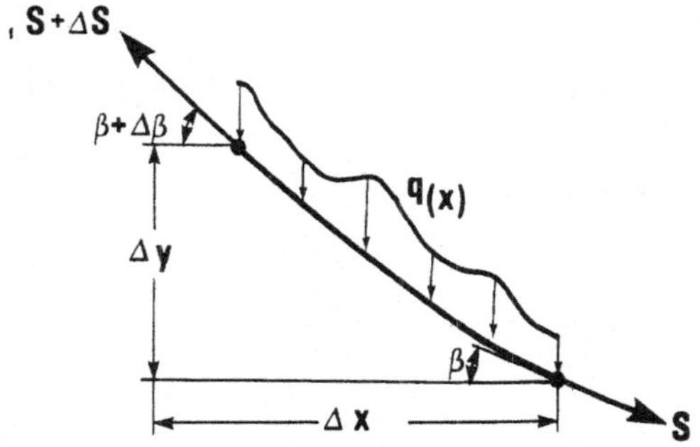

Figure 124. Unevenly distributed load on an element of the cable.

The horizontal force (H) is equal at both ends of the element (figure 124).

$$H = S \cdot \cos\beta = (S + \Delta S) \cdot \cos(\beta + \Delta\beta)$$
$$(S + \Delta S) \cdot \cos(\beta + \Delta\beta) - S \cdot \cos\beta = 0$$

By dividing with Δx and by taking the limit value when the length of the element is being reduced to $\Delta x \rightarrow 0$ we get:

$$\frac{d\,(S \cdot \cos\beta)}{dx} = 0 \qquad (52)$$

If formula 52 is correct, it is seen that $d(S \cdot \cos\beta) = 0$. This means that there is no difference in the horizontal component $H = S \cdot \cos\beta$ in the two ends of the element. *The horizontal component of the tension is the same along the entire cable.*

The vertical forces increases from $V = S \cdot \sin\beta$ at the lower end of the element to $V + \Delta V = (S + \Delta S) \cdot \sin(\beta + \Delta\beta)$ at the upper end of the element. The increase corresponds to the weight of the distributed load: $q_{(x)} \cdot \Delta x$ and the equilibrium of the vertical components of the forces is:

$$-S \cdot \sin\beta + (S + \Delta S) \cdot \sin(\beta + \Delta\beta) - q_{(x)} \cdot \Delta x = 0$$

By dividing with Δx and reducing the length of the element in figure 124 to $\Delta x \rightarrow 0$ we get

$$\frac{d\,(S \cdot \sin\beta)}{dx} = q_{(x)} \qquad (53)$$

This equation may be integrated:

$$\int d\,(S \cdot \sin\beta) = \int q_{(x)}\,dx + C_1$$
$$S \cdot \sin\beta = \int q_{(x)}\,dx + C_1 \qquad (54)$$

The slope at any point along the cable is:

$$\frac{dy}{dx} = tg\beta = \frac{\sin\beta}{\cos\beta}$$

Since $S = \dfrac{H}{\cos\beta}$ the equation (54) can be written:

$$\frac{dy}{dx} = \frac{1}{H} \int q_{(x)}\,dx + C_1$$

The function for the unevenly distributed load on the skyline is found by integration:

$$\boxed{y = \frac{1}{H} \int\int q_{(x)}\,dxdx + C_1x + C_2} \qquad (55)$$

C_1 and C_2 are constants which may be found by integration of a certain length of the cable and inserting the limit values.

7.4.2. Evenly distributed load from the weight of the cable

Concerning portable cable cranes in forestry unevenly distributed load from snow, icing or wind is of no importance, and is not taken into account in the calculations. The evenly distributed load consists of the weight of the cable per meter ($\bar{q}$ in N/m). The cable weight must be transformed to weight per meter of the horizontal distance. We introduce a small error when using the length of the span, L_2 instead of the length of the cable curve, L_1, but this error is of no importance. The cable weight per meter of the horizontal distance is:

$$\boxed{q = \frac{\bar{q}}{\cos\alpha}} \tag{56}$$

By substituting $q_{(x)}$ in equation 55 with q we get:

$$y = \frac{1}{H} \iint q\,dx\,dx + C_1 x + C_2$$

$$y = \frac{1}{H} \int q \cdot x\,dx + C_1 x + C_2$$

$$y = \frac{1}{2H} q x^2 + C_1 x + C_2 \tag{57}$$

The value of the constants C_1 and C_2 is found by inserting the limit values when integrating from $y = 0$ to $y = h$:

$$y = h \text{ when } x = 0$$

$$C_2 = h$$

$$y = 0 \text{ when } x = L$$

$$C_1 = -\frac{2Hh + qL^2}{2HL}$$

By substituting the value of C_1 and C_2 in the equation (57) we get:

$$y = \frac{q}{2H} \cdot x^2 - \frac{2Hh + qL^2}{2HL} \cdot x + h \tag{58}$$

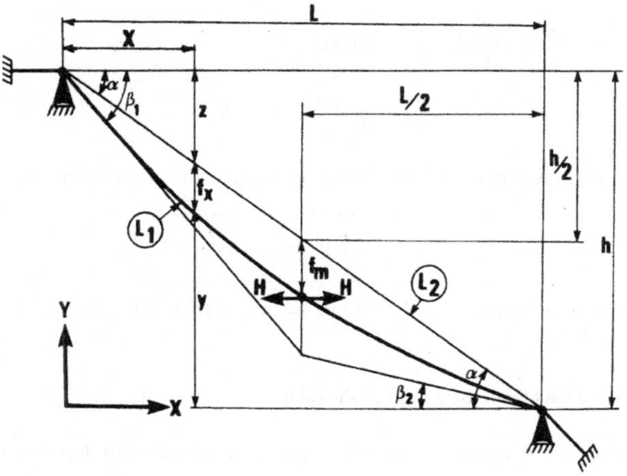

Figure 125. Evenly distributed load on the cable.

According to figure 125 the deflection of the cable is:

$$f_x = h - y - z$$

$$z = \frac{hx}{L}$$

By substituting the value of y and z in the equation for f_x we get:

$$f_x = h - \frac{q}{2H} \cdot x^2 + \frac{2Hh + qL^2}{2HL} \cdot x - h - \frac{hx}{L}$$

$$\boxed{f_x = \frac{q \cdot x}{2H}(L - x)} \qquad (59)$$

The equation (59) shows that we get the same result whether x is measured from the upper or the lower end of the cable.

The maximum value of f_x in formula (59) takes place when $x = \frac{L}{2}$. This means that the highest deflection is in the middle of the cable span. The largest deflection (f_m) is called the *rope sag*. By substituting $x = \frac{L}{2}$ in formula (59) the formula for the rope sag is simplified to:

$$\boxed{f_m = \frac{q \cdot L^2}{8H}} \qquad (60)$$

The horizontal component of the cable tension for a certain rope sag is:

$$\boxed{H = \frac{q \cdot L^2}{8 \cdot f_m}} \qquad (61)$$

The cable tension is higher than the horizontal force. The tension at the upper end of the cable is (figure 125):

$$\boxed{S' = \frac{H}{\cos\beta_1}} \qquad (62)$$

and at the lower end:

$$\boxed{S'' = \frac{H}{\cos\beta_2}} \qquad (63)$$

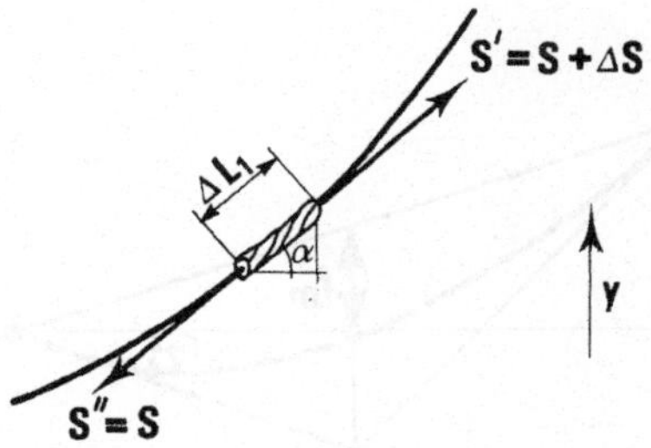

Figure 126. The tension in an element of the cable.

It is of interest to find out how much greater the tension is at the upper than at the lower end of a cable. From figure 126 it is seen:

$$\Delta S = S' - S'' = \bar{q} \cdot \Delta L_1 \cdot \sin\alpha$$

ΔL_1 is a small part of the cable which is acted on by the forces S′ and S″. At the limit, when $\Delta L_1 \rightarrow 0$, we get:

$$dS = \bar{q} \cdot dL_1 \cdot \sin\alpha = \bar{q} \cdot dy$$

$$\int_{S''}^{S'} dS = \int_{y=0}^{y=h} \bar{q} \cdot dy$$

$$S' - S'' = \bar{q} \cdot h$$

With a freely hanging cable the difference between the cable tensions at the lower and upper end is equal to the product of the cable weight in N/m, and the height difference. This is, however, only valid for an empty skyline and not when a consentrated load is applied.

The equation (62) shows that the larger the angle to the cable curve, the greater the cable tension, and the greatest tension is at the upper end where the cable is steepest.

The tangent angle at the upper end may be found as follows (figure 125):

$$\mathrm{tg}\beta_1 = \frac{\frac{h}{2} + 2f_m}{\frac{L}{2}} = \frac{h}{L} + 4\frac{f_m}{L} \qquad (64)$$

By insubstituting $\mathrm{tg}\alpha = \frac{h}{L}$ and the value for the rope sag f_m equation (64) may be simplified:

$$\mathrm{tg}\beta_1 = \mathrm{tg}\alpha + \frac{qL}{2H} \qquad (65)$$

$$\mathrm{tg}\beta_2 = \mathrm{tg}\alpha - \frac{qL}{2H} \qquad (66)$$

If $\frac{qL}{2H} > \mathrm{tg}\alpha$, $\mathrm{tg}\beta_2$ and also the angle β_2 is negative. This means that the cable curve goes uphill towards the lower end of the cable. The deepest point of the cable is lower than the lowest suspension point. The value of the angle is found as follows:

$$\mathrm{tg}\beta_2 = -\left(\frac{qL}{2H} - \mathrm{tg}\alpha\right) \qquad (67)$$

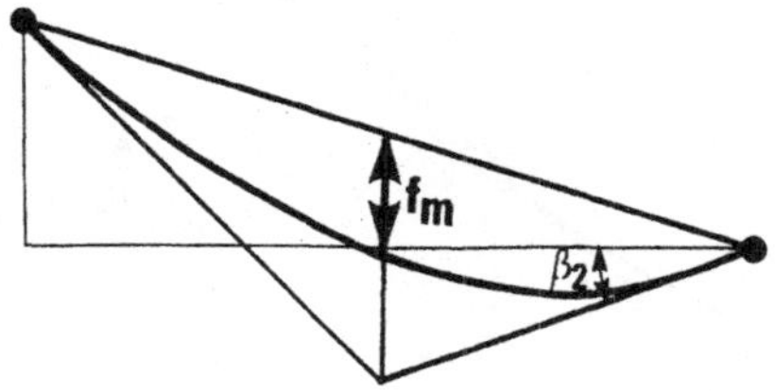

Figure 127. The situation when the cable is hanging lower than the lowest cable support.

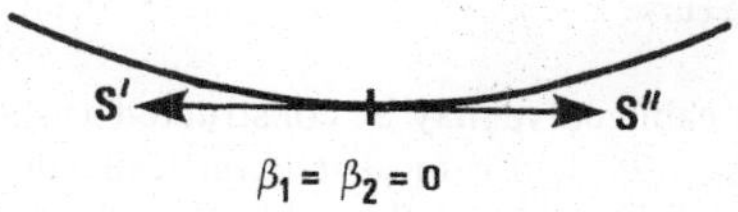

Figure 128. At the lowest point on the cable there are no vertical components and the cable tension is the same as the horizontal force.

When one of the suspension points is lower than the other the lowest point of the cable will be located somewhere between the middle of the span and the lower suspension point. Sometimes we may wish to find where this point is.

It may be found by means of equation (58). Usually it is, however, sufficient to make a drawing of the cable (figure 129) and find the lowest point of the cable graphically.

The cable is horizontal at the lowest point and the angle between the tangent to the cable curve and the horizontal line is $\beta_1 = \beta_2 = 0$ (figure 128).

$$S' = \frac{H}{\cos\beta_1} = \frac{H}{\cos\beta_2} = S''$$

$$S' = S'' = H$$

At this point the cable tension is the same as the horizontal force and there is no vertical force (figure 128):

$$V = S \cdot \sin 0 = S \cdot 0 = 0$$

The main line and haul-back line on a cable crane are freely hanging ropes. The same is the case with an unloaded skyline, that is, a skyline without carriage or payload. The tension in a freely hanging cable is called »the pre-tension», and the stress due to this tension is called »the pre-stress».

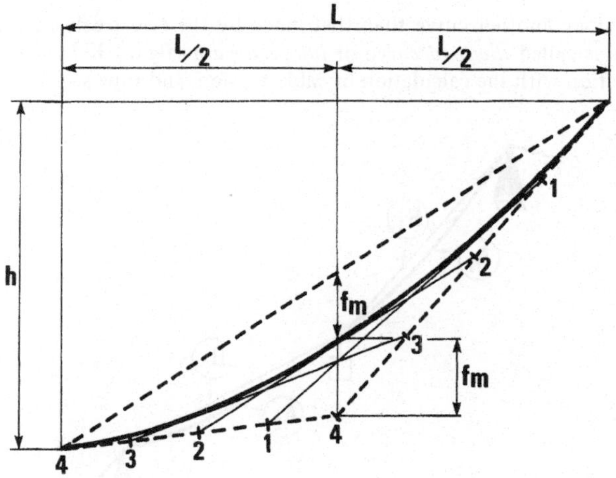

Figure 129. The construction of the cable curve when the maximum rope deflection (rope sag) is known. This construction also gives a good approach for the load path.

7.4.3. The construction of a cable curve

If the rope sag is known the cable curve may be constructed as shown in figure 129. First the double length of the rope sag ($2 \cdot f_m$) is measured vertically down from the middle of the span between the two suspension points of the cable. Two lines are drawn from each of the two suspension points to the end of the double rope sag. Each of these two lines are divided into the same number of equal parts. The tangent lines to the rope curve are the lines between the points 1 – 1, 2 – 2, 3 – 3 and 4 – 4. The cable curve may be drawn so that it touches the middle part of these tangent lines.

If the rope sag of the load curve is known (equation 72) the load curve may be constructed in a similar way.

A cable profile must be constructed with the same scale on the height and length co-ordinates.

7.5. Cable tension during load transport

7.5.1. Concentrated load on the cable

As the carriage moves along the skyline the cable has to carry two vertical forces, namely the weight of the cable and the weight of the carriage with its load. In addition are the vertical components of the main and haul-back line tensions.

When the load rests on the cable the skyline takes the shape of two freely hanging cables with the spans B_1 and B_2 (figure 130). Since the cable tension increases due to the concentrated load, the rope sag in each of the two spans is being reduced correspondingly. The shape of the skyline with the concentrated load at the middle of the span is especially interesting, for example when the length of the cable is to be found. Such a calculation may be undertaken in two parts.The length of each of the two freely hanging cable curves (A_1 and A_2 in figure 130) is calculated separately and added together afterwards.

The load moves along another curve than that made by the two freely hanging parts of the cable. This curve is called *the load curve* or *the load path* (figure 131). The load path is of interest in connection with the calculation of cable tension, and rope sag.

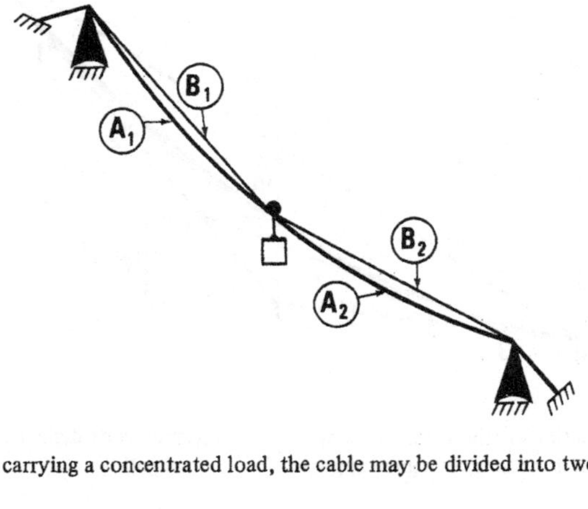

Figure 130. When carrying a concentrated load, the cable may be divided into two parts (A_1 and A_2).

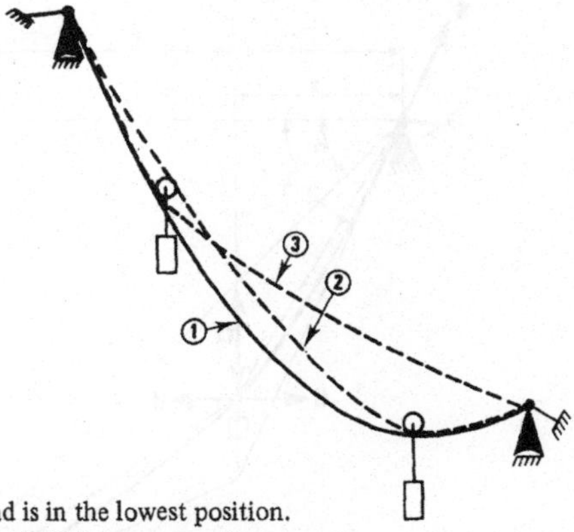

Figure 131. The load path.
1. The load path.
2. The skyline when the load is in the lowest position.
3. The skyline when the load is in the upper position.

The relationship between the weight of the concentrated load (Q), the horizontal distance between the two suspension points of the cable (L), the horizontal component of the cable tension (H) and the deflection of the cable at a distance x from the left suspension point may be deduced from figure 132.

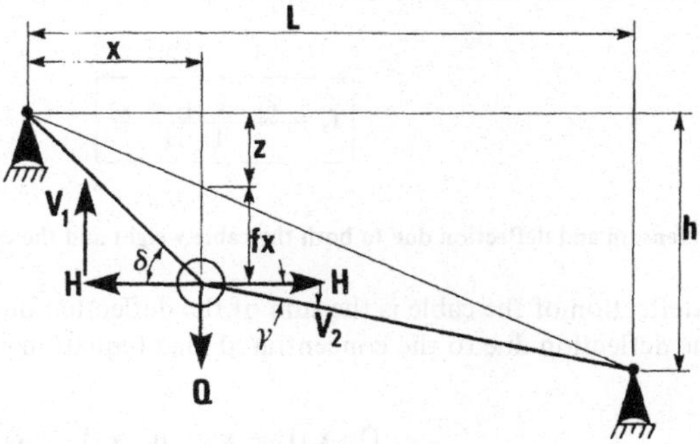

Figure 132. The influence of a concentrated load (Q) on the cable.

$$\mathrm{tg}\delta = \frac{z + f_x}{x}$$

$$\mathrm{tg}\gamma = \frac{h - z - f_x}{L - x}$$

$$V_1 = H \cdot \mathrm{tg}\delta = H \cdot \frac{z + f_x}{x}$$

$$V_2 = H \cdot \mathrm{tg}\gamma = H \cdot \frac{h - z - f_x}{L - x}$$

$$V_1 - V_2 - Q = 0$$

$$H \cdot \frac{z + f_x}{x} - H \cdot \frac{h - z - f_x}{L - x} = Q$$

$$(z + f_x) \cdot (L - x) - (h - z - f_x) \cdot x = \frac{Q}{H} \cdot (L - x) \cdot x$$

$$f_x + z - \frac{hx}{L} = \frac{Q}{LH} (L - x) \cdot x$$

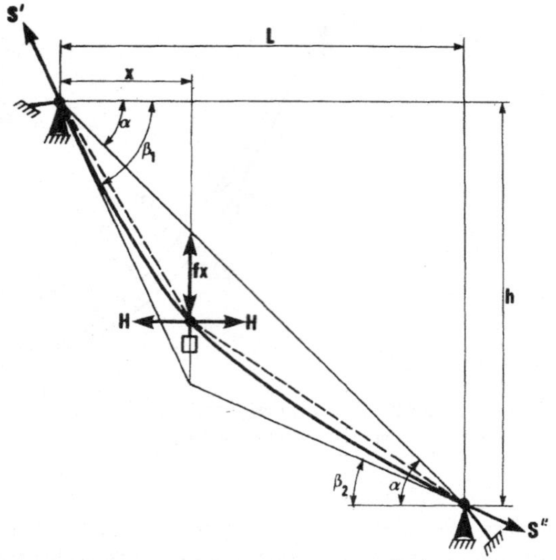

Figure 133. The influence of the evenly distributed load (q) and the concentrated load (Q) on the cable.

The equation for the deflection is obtained by substituting the value of $z = \frac{hx}{L}$ in the last formula:

$$f_x = \frac{Q \cdot x\,(L - x)}{L \cdot H} \tag{68}$$

7.5.2. Cable tension and deflection due to both the cable weight and the concentrated load

The total deflection of the cable is the sum of the deflection due to the cable weight (equation 59) and the deflection due to the concentrated load (equation 68).

$$f_x = \frac{Q \cdot x\,(L - x)}{H \cdot L} + \frac{q \cdot x\,(L - x)}{2H} \tag{69}$$

$$f_x = \frac{Q + \frac{1}{2}q \cdot L}{H} \cdot \frac{x\,(L - x)}{L} \tag{70}$$

In this formula q is the weight of the skyline in N per meter of the horizontal distance and Q is the weight of the concentrated load in N which consists of the weight of the carriage, the payload and the vertical components of the main and haul-back line tensions. Usually, however, these vertical components are estimated. This is especially the case when the main and haul-back lines touch the ground during transportation.

If the main and haul-back lines are hanging freely along the entire span the tensions may be taken into account as follows (CZITARY 1962) (figure 134):

$$f_x = \frac{Q + \frac{1}{2}q \cdot L + \frac{1}{2}q_2 \cdot L + \frac{1}{2}(q_1 - q_2)\,x}{H} \cdot \frac{x \cdot (L - x)}{L} \tag{71}$$

In the formula Q is the weight of the payload and the carriage in N, q is the weight of the skyline in N per meter of the horizontal distance and q_1 and q_2 are the similar weights for the main and haul-back lines.

Sometimes one of the lines may not be used in the cable system. If, for example, there is no main line in the span L – x this may be taken care of by using the weight $q_2 = 0$ in equation 71. Some cable systems have an extra line before or after the carriage. In the calculation this is counted as a single line the weight of which is the sum of the two lines. If, for example, the main line weighs 2.32 N/m and a hoist line weighs 2.60 N/m the total weight is:

$$\overline{q}_2 = 2.32 + 2.60 = 4.92 \text{ N/m}$$

$$q_2 = \frac{4.92}{\cos\alpha} \text{ N/m of the horizontal distance}$$

The cable has the largest deflection when the load is in the middle of the span. This deflection is the same as the rope sag of the load path. The rope sag is specially interesting because the maximum cable tension occurs when the load is in the middle of the span. The equation for the rope sag is found by substituting the value of $x = {}^1/_2\,L$ in the equation (71):

$$f_m = \frac{Q \cdot L}{4H} + \frac{q \cdot L^2}{8H} + \frac{q_2 \cdot L^2 + q_1 \cdot L^2}{16H} \qquad (72)$$

If there are no main or haul-back lines or if their vertical components at the carriage are estimated and added to the weight of the loaded carriage (Q) the formula is simplified as follows:

$$f_m = \frac{Q \cdot L}{4H} + \frac{q \cdot L^2}{8H} \qquad (73)$$

The tension is largest at the place where the cable is steepest, namely just below the upper cable shoe. If the angle between the cable curve tangent and the horizontal line is known, the cable tension may be calculated from the horizontal force (equation 62 and 63).

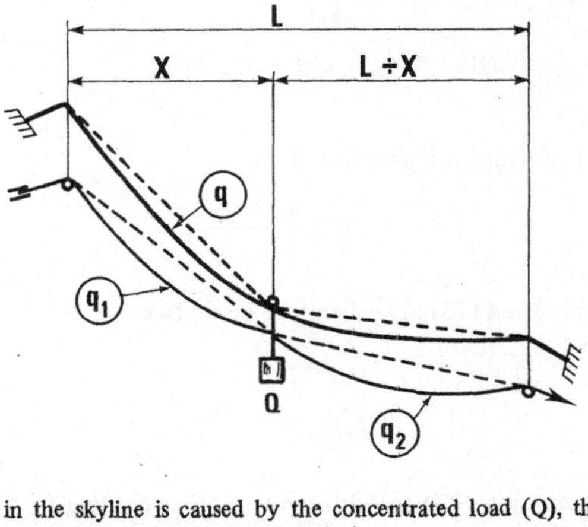

Figure 134. The tension in the skyline is caused by the concentrated load (Q), the weight of the skyline (q), the main line (q_2) and the haul-back line (q_1).

At the upper cable shoe in a span $S' = \dfrac{H}{\cos\beta_1}$

At the lower cable shoe in a span $S'' = \dfrac{H}{\cos\beta_2}$

The angle between the cable curve tangent and the horizontal line may be calculated from the following equations:

At the upper cable shoe
$$\boxed{tg\beta_1 = tg\alpha + \frac{Q + \frac{1}{2}qL}{H}} \qquad (74)$$

At the lower cable shoe, if the cable goes downhill towards the shoe:
$$\boxed{tg\beta_2 = tg\alpha - \frac{Q + \frac{1}{2}qL}{H}} \qquad (75)$$

At the lower cable shoe when $tg\beta_2$ is negative which means that the cable is hanging uphill towards the cable shoe (figure 127 and equation (67)):

$$\boxed{tg\beta_2 = -\left(\frac{Q + \frac{1}{2}qL}{H} - tg\alpha\right)} \qquad (76)$$

7.5.3. Simplified calculation of the cable tension

a) Simplification by the constants in table 19.

The calculation of f_m and S′ may be easier if the equations (60), (72) and (73) are rearranged as follows:

Equation (60): $f_m = \frac{1}{8H} \cdot (qL^2) = \frac{1}{8H} \cdot A$

Equation (73): $f_m = \frac{1}{8H} \cdot (2QL + qL^2) = \frac{1}{8H} \cdot A$

Equation (72): $f_m = \frac{1}{8H} \cdot [2QL + L^2 (q + q')] = \frac{1}{8H} \cdot A$

Here, q′ is the weight of main and haul-back lines:

$$q' = \frac{q_1 + q_2}{2}$$

The equations (60), (72) and (73) may therefore be written:

$$\boxed{f_m = \frac{1}{8H} \cdot A} \qquad (77)$$

The constant A in equation (77) has different values according to the various alternatives. This is shown in table 19.

Table 19. The calculation of the constant »A» in the various alternatives.

Alternative	A
Empty cable	$q \cdot L^2$
Cable with load	$q \cdot L^2 + 2QL$
Cable with load, haul-back line and main line	$q \cdot L^2 + 2QL + q'L^2$

The equation (64) may be rearranged as follows:

$$tg\beta_l = \frac{h}{L} + \frac{4}{L} \cdot f_m = \frac{h}{L} + \frac{4}{L} \cdot \frac{A}{8H}$$

$$tg\beta_l = \frac{h}{L} + \frac{A}{2HL} \qquad (78)$$

$tg\beta_l = \frac{V'}{H}$ and $V' = \sqrt{S'^2 + H^2}$ is subsituted in equation (78):

$$\frac{h}{L} + \frac{A}{2HL} = \frac{\sqrt{S'^2 - H^2}}{H} \qquad (79)$$

Sometimes it is of interest to find the horizontal force (H) when the cable tension at the upper suspension point (S′) is known. ($S_{Br} = s \cdot S'$). Equation (79) may be converted as follows:

$$\left(\frac{h}{L} \cdot H + \frac{A}{2L}\right)^2 = \left(\sqrt{S'^2 - H^2}\right)^2$$

$$H^2 \cdot \frac{h^2}{L^2} + \frac{Ah}{L^2} \cdot H + \frac{A^2}{4L^2} = S'^2 - H^2$$

$$H^2\left(\frac{h^2}{L^2} + 1\right) + H \cdot \frac{Ah}{L^2} + \frac{A^2}{4L^2} - S'^2 = 0$$

$$H^2 + H \cdot \frac{Ah}{L^2\left(\frac{h^2}{L^2} + 1\right)} + \frac{\frac{A^2}{4L^2} - S'^2}{\left(\frac{h^2}{L^2} + 1\right)} = 0 \qquad (80)$$

A new constant B may be found as follows:

$$\boxed{B = \frac{h^2}{L^2} + 1} \qquad (81)$$

This constant is substituted in equation (80) which is solved as an ordinary quadratic equation:

$$\boxed{H = -\frac{Ah}{2L^2B} + \sqrt{\frac{A^2h^2}{4L^4B^2} - \frac{A^2 - 4L^2S'^2}{4L^2B}}} \qquad (82)$$

The equation (82) can be used to find the horizontal force in the span when the maximum cable tension at the upper suspension point (S′) is known. Example 29 includes a calculation of the constants A and B as well as the equation (82).

b) Simplification by the constant $C = \frac{1}{\cos\beta_1}$

In many cases it is sufficient to analyse the cable tension at the upper end of the skyline in order to check the factor of safety.

Equation (73) may be written:

$$H = \frac{Q \cdot L}{4 \cdot f_m} + \frac{q \cdot L^2}{8 \cdot f_m} \tag{83}$$

The tension at the upper suspension point of the cable may be found from equation (62):

$$S' = H \cdot \frac{1}{\cos\beta_1}$$

By introducing $\bar{q}$ in equation (83), we receive:

$$H = \frac{Q \cdot L}{4f_m} + \frac{\bar{q} \cdot L^2}{8f_m} \cdot \frac{1}{\cos\alpha}$$

It is understood that $\frac{\bar{q}}{\cos\alpha}$ is giving too low a value of q since the cable is hanging in a curve. This error is very small. It would be a simplification to use $\frac{\bar{q}}{\cos\beta_1}$ in spite of the fact that this gives too high a value of the cable weight per meter of the horizontal distance. This error is small and of no practical importance:

$$\frac{\bar{q}}{\cos\beta_1} > q > \frac{\bar{q}}{\cos\alpha}$$

By substituting $C = \frac{1}{\cos\beta_1}$ in the equations (62) and (83) we get:

$$H = \frac{QL}{4f_m} + \frac{\bar{q} \cdot L^2}{8f_m} \cdot C \tag{84}$$

$$S' = H \cdot C \tag{85}$$

The value of $C = \frac{1}{\cos\beta_1}$ is calculated for various slope angles and rope sags in per cent of the horizontal distance. The results are given in figure 135. The factor C may be found from the diagram in figure 135. The cable weight in N/m of the cable may be obtained from a wire rope catalogue. The calculation of the cable tension may be carried out by means of the following formulae:

Horizontal force caused by the load:

$$\boxed{H_1 = \frac{Q \cdot L}{4 \cdot f_m}} \tag{86}$$

Horizontal force caused by the cable weight:

$$\boxed{H_2 = \frac{\bar{q} \cdot L^2}{8 \cdot f_m}} \tag{87}$$

Total horizontal force:

$$\boxed{H = H_1 + H_2 \cdot C} \tag{88}$$

Tension in the cable at the upper suspension point:

$$\boxed{S' = H \cdot C} \tag{89}$$

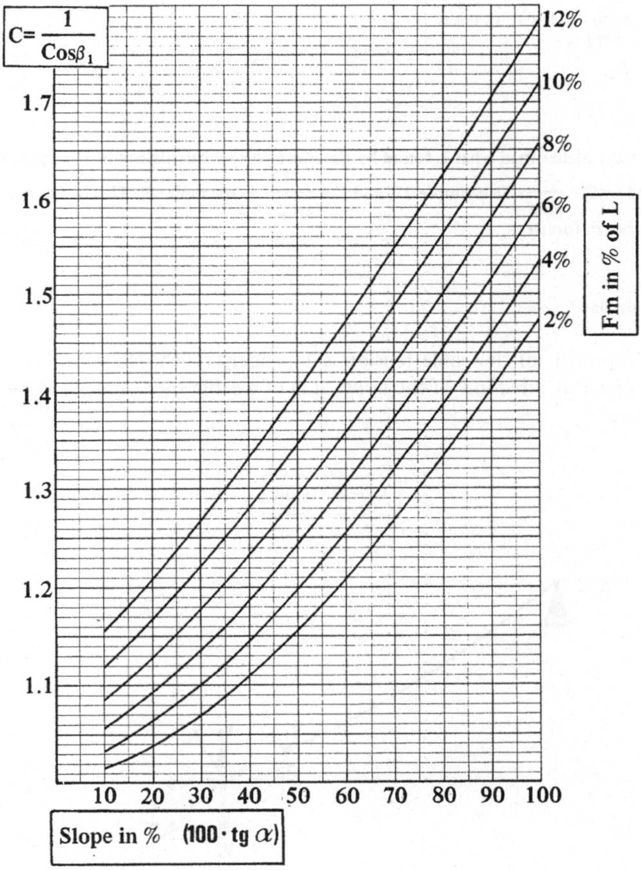

Figure 135. Diagram for the constant C in the formula (88) and (89).

Example 27: **(The same cable crane as in example 29). The cable crane had the following specifications:**

$L = 550$ m
$\bar{q} = 11.48$ N pr. m
$Q = 20\,000$ N
$f_m = 60.8$ m): 11.05 % av L
$S_{Br} = 194$ kN
$tg\alpha = 0.5636$

$$H_1 = \frac{Q \cdot L}{4 \cdot f_m} = \frac{20000 \cdot 550}{4 \cdot 60.8} = 45230 \text{ N}$$

$$H_2 = \frac{\bar{q} \cdot L^2}{8 \cdot f_m} = \frac{11.48 \cdot 550^2}{8 \cdot 60.8} = 7140 \text{ N}$$

Since the slope is 56.4%, C can be found in figure 135:

$$C = \frac{1}{\cos\beta_1} = 1.418$$

$$H = 45230 + 7140 \cdot 1.418 = 55354 \text{ N}$$
$$S' = H \cdot C = 78492 \text{ N}$$

$$s = \frac{194}{78.492} = 2.5$$

H and S′ is somewhat higher than what is found by the more correct calculation in example 29. The reason is that H_2 is adjusted by $C = \frac{1}{\cos\beta_1}$ in example 27, and $C = \frac{1}{\cos\alpha}$ in example 29. This difference is unimportant for light cable cranes in forestry.

7.5.4. Multispan skyline

The vertical component of the cable tension is dependent on the slope of the cable. In figure 136, it has an upwards direction of the upper side of a cable shoe, and a downward direction on the lower side.

$$\boxed{V = S \cdot \sin\beta} \tag{90}$$

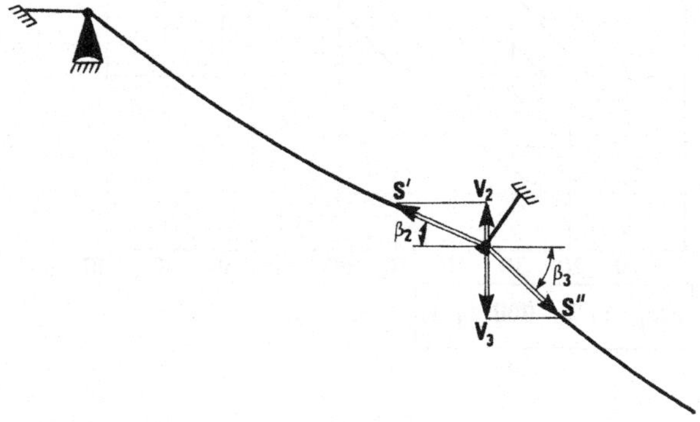

Figure 136. The vertical forces on a support.

$$V_2 = S' \cdot \sin\beta_2$$
$$V_3 = S'' \cdot \sin\beta_3$$
$$S' \sim S''$$
$$\boxed{\Sigma V = V_3 - V_2 \sim S''\,(\sin\beta_3 - \sin\beta_2)} \qquad (91)$$

If ΣV is positive the vertical force on the lower side of the cable shoe is higher than on the upper side. The cable rests on the cable shoe.

If ΣV is negative the vertical force is higher on the upper side than on the lower side of the cable shoe, it is necessary to use *a pull down cable shoe* in order to keep the cable down. It may also be necessary to anchor the support in order to avoid it being lifted.

When the load path is known, the curve may be drawn on the longitudinal profile of the terrain in order to check that there is sufficient distance between the cable and the terrain. If not, it is necessary to rig up a support in the span. The sum of the vertical components of the cable tension at the support should have a downward direction or an upward direction, both for the loaded and the unloaded skyline. To install the support in such a manner that the vertical components have an upward direction with the unloaded cable and a downward direction with the loaded skyline must be avoided.

Experienced crew will usually foresee this situation and move the support to a higher point in the terrain.

In case of doubt it is possible to calculate the rope sag in the longest of the the two spans as well as a span covering the total length of the skyline. Both cable curves may then be drawn in order to see if the situation is similar to the illustration in figure 137.

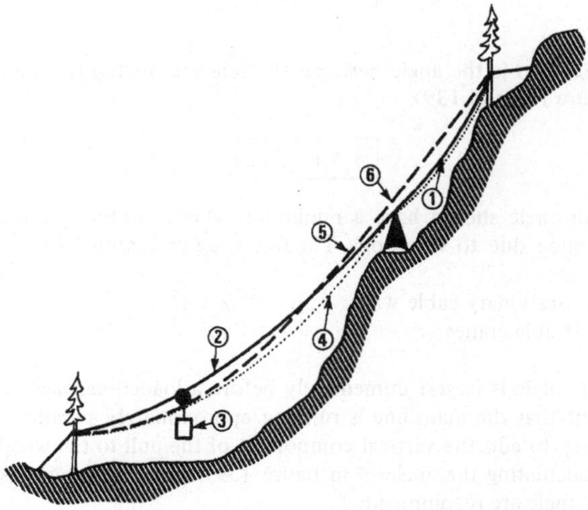

Figure 137. If the support is too low, the skyline may be lifted from the support when the load is away from the support.
1. Empty cable in upper span.
2. Empty cable in lower span.
3. The load.
4. The load path.
5. The loaded cable is lifted from the support (6).

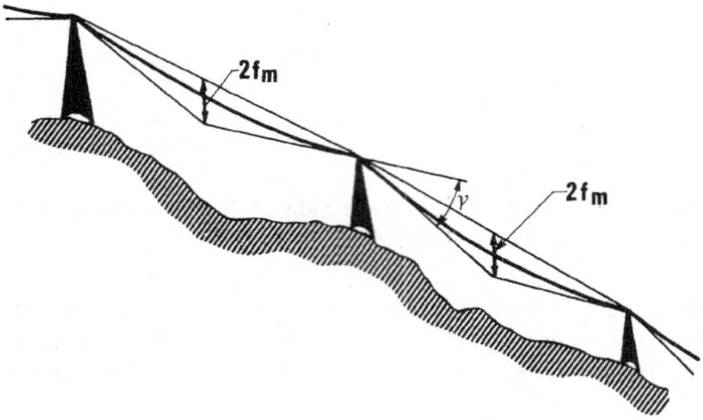

Figure 138. The angle between the cable tangent on two sides of a support γ.

It is also possible to try out the situation with the aid of a computer using equation (70) for the load path over the total length and equation (59) for the span between the load and the spar tree.

When the skyline has several spans it is usually sufficient to calculate the rope sag and the cable tensions with the load in the longest span. If it is necessary to find the elastic elongation of the cable it is necessary to take the entire length of the cable into account $\Sigma L_1 = L_1$ (in the first span) + L_1 (in the second span) etc. The elastic elongation is then:

$$\boxed{P_E = \frac{\Sigma L_1}{A \cdot E_r}} \qquad (92)$$

Rope deflection angle (γ) is the angle between the tangents to the cable curves on each side of a cable shoe (figure 138 and 139).

$$\boxed{\gamma = \beta_2 - \beta_1} \qquad (93)$$

The rope deflection angle should have a minimum value in order to avoid the cable being thrown out of the shoe due to vibrations. The following minimum values are recommended:

Stationary cable ways:	$\gamma > 1°$
Cable cranes:	$\gamma > 2°$

The rope deflection angle is largest immediately before a loaded carriage arrives at the cable shoe. It is important that the main line is running approximately parallel with the skyline. If not, it is necessary to add the vertical component of the pull to the weight of the concentrated load when calculating the angle (β in figure 139). The following maximum values for the rope deflection angle are recommended:

Stationary cable ways:	$\gamma < 16°$
Movable cable ways:	$\gamma < 23°$
Cable cranes:	$\gamma < 29°$
Maximum angle between the skyline and the guyline at the tail tree:	$\gamma < 90°$

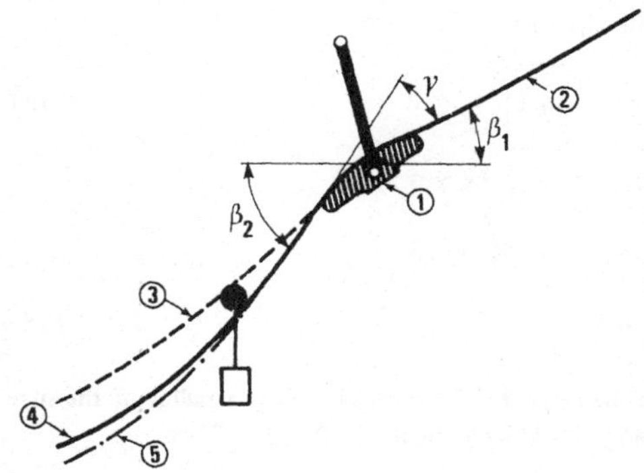

Figure 139. 1. Cable shoe.
2. Empty skyline in the upper span.
3. Empty skyline in the lower span.
4. Loaded skyline in the lower span.
5. The load path.

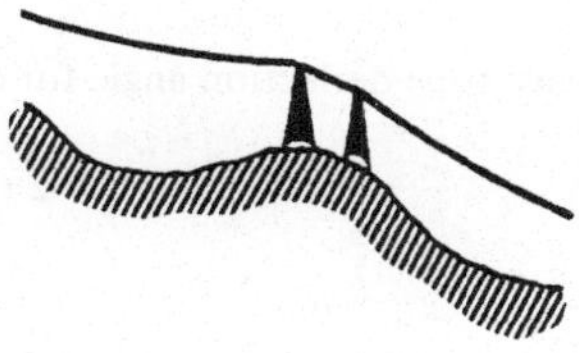

Figure 140. A double support reduces the angle between the cable tangents.

The rope deflection angle should always be as small as possible especially when using cable cranes on long spans where the carriage is running at high speeds.

If the rope deflection angle is larger than the recommended maximum values a double support should be installed (figure 140).

If the cable shoe is too short the carriage sheave may bend the cable the moment it mounts the shoe. The ends of the cable shoe should be rounded off in order to reduce the problem.

The longitudinal profile of the cable shoe should be shaped as part of a circle with a minimum diameter 80 times the diameter of the cable:

$$R \geq 80 \cdot d$$

The length of the cable shoe should be sufficient to take up the whole rope deflection angle.

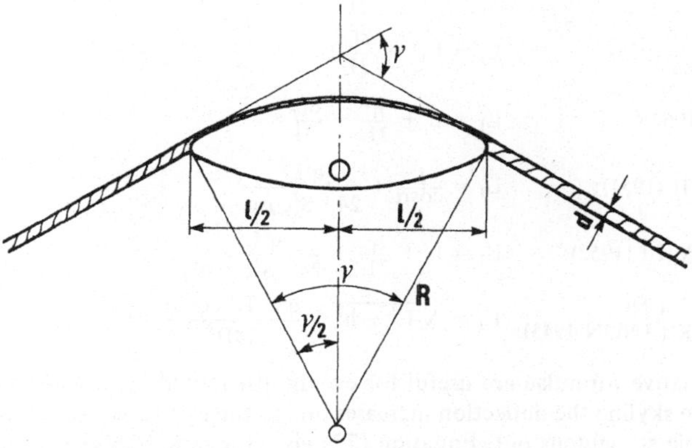

Figure 141. The dimensions of the cable shoe.

According to figure 141 the length of the cable shoe is:

$$\boxed{l = 2 \cdot R \cdot \sin \frac{\gamma}{2}} \tag{94}$$

The largest rope deflection angle for cable cranes should be $\gamma = 29°$:

$$l = 2 \cdot R \cdot \sin 14.5 = 2 \cdot R \cdot 0.25$$

$$\boxed{l = 0.5 \cdot R} \tag{95}$$

Example 28: **A cable crane has a skyline with diameter d = 18 mm. This gives a radius of the shoe profile R = 1.8 · 80 = 144 cm and the length of the shoe should be minimum l = 0.5 · R = 72 cm.**

7.6. The length of the skyline, elastic elongation and pre-tension

7.6.1. Equations for calculating the cable length

An equation for calculating the length of a cable may be found by observing the length of a small element of the cable and intergrating it over the entire span of the cable:

$$dL_l = \sqrt{dx^2 + dy^2} = dx \sqrt{1 + \frac{dy^2}{dx^2}} \tag{96}$$

$$\boxed{L_l = \sqrt{h^2 + (2\frac{H}{\bar{q}} \cdot \sinh \frac{L \cdot \bar{q}}{2H})^2}} \tag{97}$$

The equation 97 is accurate. A computer for calculating the hyperbolic functions is, however, needed [$\sinh x = {}^1/_2(e^x - e^{-x})$]. The equation (97) may also be solved by approximation. Many authors have published approximative formulae which they have found by means of Taylor's expansion (SCHNEIGERT 1966).

BIELAYEV et.al. (1964): $$L_l = L(1 + \frac{8f_m^2}{3L^2}) \tag{98}$$

CAPPELONI (1955): $$L_l = L(1 + \frac{8f_m}{3L})^2 \tag{99}$$

CZITARY (1962): $$L_l = L + \frac{h^2}{2L} + \frac{8f_m^2}{3L} \tag{100}$$

DUKIELSKII (1951): $$L_l = \frac{L}{\cos\alpha} + \frac{\bar{q}^2L^3}{24 \cdot S^2\cos\alpha} \tag{101}$$

GULISASHVILI (1952): $$L_l = L + \frac{h^2}{2L} + \frac{\bar{q}^2L^3}{24 \cdot S^2\cos^2\alpha} \tag{102}$$

WIRE ROPE, ENG. HANDBOOK (ANON 1943): $$L_l = \sqrt{L^2 + h^2} + \frac{q^2 \cdot L^3 \cos^3\alpha}{24H^2} \tag{103}$$

These approximative formulae are useful for evenly distributed load. When a loaded carriage moves along the skyline the deflection increases due to the elastic elongation of the cable and because the cable straightens out. Equation (72) gives the ratio between the horizontal force and the rope sag.

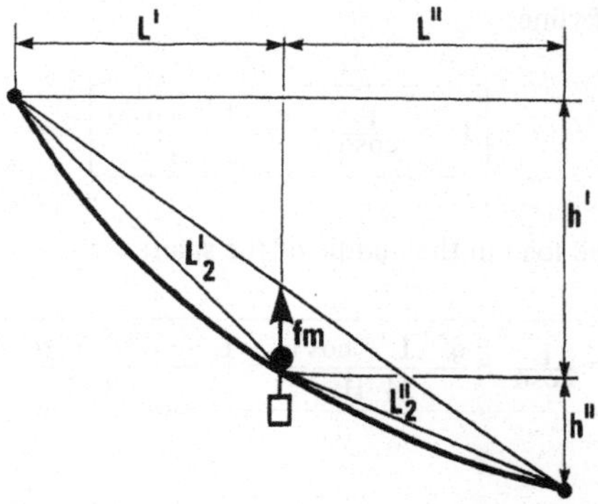

Figure 142. When a concentrated load is applied to the cable, the calculation of the cable length may be carried out separately for each of the two parts, L'_2 and L''_2.

The maximal tension and deflection occurs when the concentrated load is in the middle of the span. Any of the formulae (98) – (103) may be used to calculate the length of the cable when the concentrated load is in the middle of the span (figure 142):

$$\boxed{L_2' = \sqrt{L'^2 + h'^2}} \qquad (104)$$

$$\boxed{L_2'' = \sqrt{L''^2 + h''^2}} \qquad (105)$$

The rope sag and the length of the cable curve may be computed within each of the two spans and the length of the whole cable is:

$$L_1 = L_1' + L_1''$$

This is a laborious method. SCHNEIGERT (1966) has suggested the following equation for the calculation of the length of a cable with evenly distributed load as well as a concentrated load:

$$L_1 = L + \frac{h^2}{2L} + \frac{\bar{q}^2 L^3}{24S^2 \cdot \cos^2\alpha} + \frac{x\,(L - x)}{2L \cdot S^2} \cdot Q\,(Q + \frac{\bar{q}L}{\cos\alpha}) \qquad (106)$$

The last part of this formula is the addition to the length caused by the concentrated load. When the carriage is in the middle of the span the formula for this addition may be simplified by substituting $\bar{q} = q \cdot \cos\alpha$, and $S \sim \frac{H}{\cos\alpha}$:

$$\Delta L_1(\text{according to Q}) = \frac{x\,(L - x)}{2L \cdot S^2} \cdot Q \cdot (Q + \frac{\bar{q}L}{\cos\alpha})$$

$$\Delta L_1\,(\text{according to Q}) = \frac{L \cdot \cos^2\alpha}{8H^2} \cdot Q \cdot (Q + qL) \qquad (107)$$

The Norwegian Forest Research Institute has investigated the length of a 1,145 meter long skyline (height difference 382.4 meter) measuring accurately the cable length, rope sag and tensions (SAMSET 1979). The accurate length was compared with the length calculated by means of the various approximative formulae and it was found that the following two equations for the length of the cable gave the best result. The error varied with the size of the load and the pre-tension of the cable from 0.5 – 1.2 ‰ of the cable length. Formulae (108) and (109) are therefore recommended for calculating the length of an unloaded and loaded skyline:

Empty (pre-tensioned) skyline:

$$\boxed{L_l = \frac{L}{\cos\alpha} + \frac{q^2 \cdot L^3 \cdot \cos^3\alpha}{24 \cdot H^2}} \tag{108}$$

Skyline with concentrated load in the middle of the span:

$$\boxed{L_l = \frac{L}{\cos\alpha} + \frac{q^2 \cdot L^3 \cdot \cos^3\alpha}{24 \cdot H^2} + \frac{L \cdot \cos^2\alpha \cdot Q\,(Q + qL)}{8 \cdot H^2}} \tag{109}$$

7.6.2. The pre-tension of the cable

When the elastic elongation of the cable is known the pre-tension of the empty skyline can be calculated. The elastic elongation can be found by means of equation (51). If the skyline has several spans *the calculation of the pre-tension should be carried out for the longest of the spans.* In this case equation (92) must be used to find the elastic elongation.

The length of the pre-tensioned empty skyline $L_1(q)$ can be found by means of equation (108). In order to find the total length of the loaded skyline, $L_1(qQ)$ one must add the elastic elongation caused by the increased tension due to the concentrated load in the middle of the span:

$$\Delta L_1 = P_E(H - H_f)$$

This gives the following equation for the total length of the cable:

$$L_l\,(qQ) = P_E\,(H - H_f) + \frac{L}{\cos\alpha} + \frac{q^2 \cdot L^3 \cdot \cos^3\alpha}{24 H_f^2}$$

The equation may be written:

$$\boxed{\frac{q^2 \cdot L^3 \cdot \cos^3\alpha}{24} \cdot \frac{1}{H_f^2} - P_E \cdot H_f = L_l\,(qQ) - (P_E \cdot H + \frac{L}{\cos\alpha})} \tag{110}$$

The first step is to find the tension and the rope sag when the load is hanging in the middle of the span. The next is to see that the tension has the necessary factor of safety. The pre-tension in the empty skyline may be calculated using formula (110) as follows:

(1) $$P_E = \frac{L_l\,(qQ)}{A \cdot E_r}$$

(2) $$k_1 = L_l\,(qQ) - (P_E \cdot H + \frac{L}{\cos\alpha})$$

(3) $$k_2 = \frac{q^2 \cdot L^3 \cdot \cos^3\alpha}{24}$$

Afterwards the following equation is used:

$$\boxed{\frac{k_2}{H_f^2} - P_E \cdot H_f = k_1} \tag{111}$$

The left side of the equation is computed for various values of H_f. The equation is drawn in a co-ordinate system with H_f on the abscissa. The value of H_f which gives the computed value of k_1 can be found graphically. The method is shown in example 29.

The calculation of the pre-tension is elaborate. In addition the calculation cannot be exact since the value of the modulus of elasticity is uncertain. It is better to install an extensometer in the skyline, or to measure the tension by means of a stop watch as shown in chapter 7.8.

Example 29. **A cable crane in Telemark with single span skyline:**

The horizontal distance:	L	=	550 m
Height difference:	h	=	310 m

A wire rope skyline 6x19+1 Seale, with the following specifications was used:

Diameter:	d	=	18 mm
Cross-section:	A	=	124 mm^2
Weight:	$\bar{q}$	=	11.48 N/m
Breaking load:	S_{Br}	=	194 kN

The maximum load was Q = 20 kN including carriage and vertical components of main and haul-back lines. The safety factor of the skyline should be s = 2.55. The slope between the upper and lower suspension points of the skyline:

$$tg\alpha = \frac{h}{L} = \frac{310}{550} = 0.5636$$

$\alpha = 29.4^\circ$ and $\cos\alpha = 0.8712$.

The cable weight per meter of the horizontal distance:

$$q = \frac{\bar{q}}{\cos\alpha} = \frac{11.48}{0.8712} = 13.18 \text{ N pr. m}$$

The maximum tension at the upper suspension point should be:

$$S' = \frac{194\ 000}{2.55} = 76.0 \text{ kN}$$

The horizontal force can be computed by means of equation (82).

$$A = 2 \cdot Q \cdot L + q \cdot L^2$$

$$A = 2 \cdot 20000 \cdot 550 + 13.18 \cdot 550^2$$

$$A = 23986950$$

$$B = \frac{h^2}{L^2} + 1$$

$$B = \frac{310^2}{550^2} + 1$$

$$B = 1.317686$$

$$H = -\frac{A \cdot h}{2L^2B} + \sqrt{\frac{A^2h^2}{4L^4B^2} - \frac{A^2 - 4L^2S'^2}{4L^2 \cdot B}}$$

$$H = 53629 \text{ N}$$

The rope sag for the load path is found by means of equation (73):

$$f_m = \frac{Q \cdot L}{4H} + \frac{q \cdot L^2}{8H} = 60.5 \text{ m}$$

The angle between the horizontal line and the cable tangent at the lower suspension point can be found by equation (75):

$$tg\beta_2 = tg\alpha - \frac{Q + \frac{1}{2}qL}{H} = 0.5636 - \frac{23625}{53629}$$

$$tg\beta_2 = 0.5636 - 0.4405 = 0.1231$$

$$\beta_2 = 7.02°$$

The angle between the horizontal line and the cable tangent at the upper suspension point (equation (74)):

$$tg\beta_1 = 0.5636 + 0.4405 = 1.0041$$

$$\beta_1 = 45.12°$$

The cable tension at the lower suspension point (equation (63)):

$$S'' = \frac{H}{\cos\beta_2} = \frac{53\,629}{0.9925} = 54\,034 \text{ N}$$

The length of the span:

$$L_2 = \frac{L}{\cos\alpha} = \frac{550}{0.87115} = 631.35$$

Usually the calculation stops here. The cable tension is sometimes measured for a few loads in order to find the correct pre-tension of the skyline.

The pre-tension may also be calculated. The length of the cable with the concentrated load in the middle of the span is first computed by the equation (109):

$$L_1 = \frac{L}{\cos\alpha} + \frac{q^2 \cdot L^3 \cdot \cos^3\alpha}{24 \cdot H^2} + \frac{L \cdot \cos^2\alpha}{8H^2} \cdot Q\,(Q + q \cdot L)$$

$$L_1 = \frac{550}{0.87115} + \frac{13.18^2 \cdot 550^3 \cdot 0.8712^3}{24 \cdot 53629^2} + \frac{550 \cdot 0.8712^2}{8 \cdot 53629^2} \cdot 20000\,(20000 + 13.18 \cdot 550)$$

$$L_1 = 631.34 + 0.28 + 9.89$$

$$L_1 = 641.51 \text{ m}$$

The elastic elongation of the skyline in meter per N of the tension can now be computed. Since the skyline is a wire rope 6x19+1 Seal with 18 mm diameter the cross-section is A = 124 mm^2 and the modulus of elasticity E_r = 86.3 kN/mm^2.

$$P_E = \frac{L_1}{A \cdot E_r} = \frac{641.51}{124 \cdot 86300} = 0.000060 \text{ m pr. N}$$

The pre-tension of the unloaded skyline may now be found by solving the following equation with respect to H_f (equation (110)).

$$\frac{q^2 \cdot L^3 \cdot \cos^3\alpha}{24} \cdot \frac{1}{H_f^2} - P_E \cdot H_f = L_1 - (P_E H + \frac{L}{\cos\alpha})$$

This is a cubic equation. A simple way is to solve the equation graphically.

$$k_1 = L_1 - (P_E \cdot H + L_2) = 641.51 - 0.000060 \cdot 53629 - 631.35$$

$$k_1 = 6.95 \text{ m}$$

$$k_2 = \frac{q^2 \cdot L^3 \cdot \cos^3\alpha}{24 \cdot H_f^2} = \frac{13.18^2 \cdot 550^3 \cdot 0.8712^3}{24 \cdot H_f^2} = \frac{796\,271\,640}{H_f^2} \text{ m}$$

The value of k_1 and k_2 is inserted in the equation 111:

$$\frac{k_2}{H_f^2} - P_E \cdot H_f = k_1$$

$$\frac{796\,271\,640}{H_f^2} - 0.000060 \cdot H_f = 6.95 \text{ m}$$

The left hand side of the equation is solved for different values of H_f (Table 20).

Table 20.

H_f	$\frac{796{,}271{,}640}{H_f^2}$	$-$	$0.000060 \cdot H_f$	Answer
12,000	5.52	–	0.72	4.80
11,000	6.58	–	0.60	5.92
10,000	7.96	–	0.60	7.36
9,000	9.83	–	0.54	9.29
8,000	12.44	–	0.48	11.96

The answers in table 20 are now drawn graphically with H_f on the vertical axis.

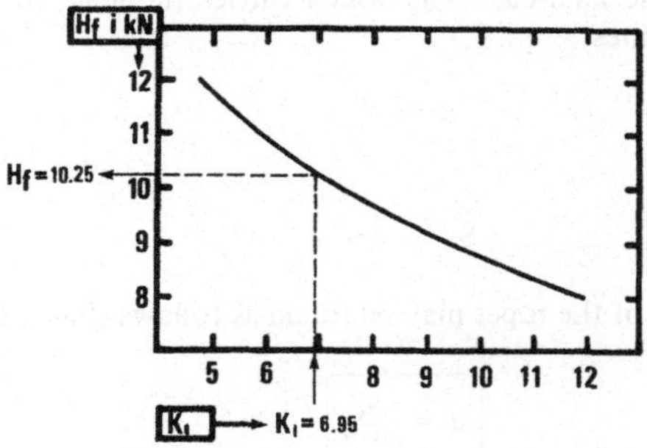

Figure 143. The relationship between the values of K_1 and H_f according to table 20.

The figure 143 shows that the value of H_f which corresponds to K_1 = 6.95 meter is

H_f = 10,250 N = 10.25 kN

This is the pre-tension of the skyline. The corresponding rope sag is (equation 60):

$$f_m = \frac{q \cdot L^2}{8 \cdot H} = \frac{13.18 \cdot 550^2}{8 \cdot 10250} = 48.6 \text{ m,): } 8.8\ \% \text{ of the horizontal distance.}$$

The freely hanging empty skyline may be drawn graphically as illustrated in figure 129.

The tension at the upper suspension point of the empty skyline can be calculated:

$$\mathbf{tg\beta_1 = tg\alpha + \frac{q \cdot L}{2H}}$$

$$\mathbf{tg\beta_1 = 0.5636 + \frac{13.18 \cdot 550}{2 \cdot 10\,250} = 0.9172}$$

$$\mathbf{\beta_1 = 42.53^\circ}$$

$$\mathbf{S' = \frac{H}{\cos\beta_1} = 13.909\ kN\ (s = 13.9)}$$

The tension at the lower suspension point of the skyline:

$$\mathbf{tg\beta_2 = 0.5636 - 0.3536 = 0.2100}$$

$$\mathbf{\beta_2 = 11.86^\circ}$$

$$\mathbf{S'' = \frac{H}{\cos\beta_2} = 10.474\ kN}$$

7.7. Running skyline

Some cable cranes do not have any fixed or live skylines. The load is carried by the main and haul-back lines. This is called *a running skyline system*. The main and haul-back lines have the same dimensions and in some cases the haul-back line is thicker than the main line. The slack-pulling line is usually hanging rather loose during the transport and is not helping to lift the load.

7.7.1. Line tensions due to the rope weight (q) and the load (Q)

The line tensions can be computed by means of the equations (62) or (63) and (73). Since it is both the main line and the haul-back line which carries the load, the tension has to be distributed between the two lines:

At the upper end of the span: $S_2' = \frac{1}{2}S'$

At the lower end of the span: $S_2'' = \frac{1}{2}S''$

The average weight per meter of the ropes may be found as follows (figure 144):

2-drum winch:

$$q = \frac{2\bar{q}}{\cos\alpha} \tag{112}$$

3-drum winch:

$$q = \frac{2\frac{1}{2}\bar{q}}{\cos\alpha} \tag{113}$$

The equation for the maximum tension in one of the ropes at the upper end of the span may be found from equation (73):

$$S_2' = \left(\frac{Q \cdot L}{4f_m} + \frac{q \cdot L^2}{8f_m}\right) \cdot \frac{1}{2\cos\beta_1} \tag{114}$$

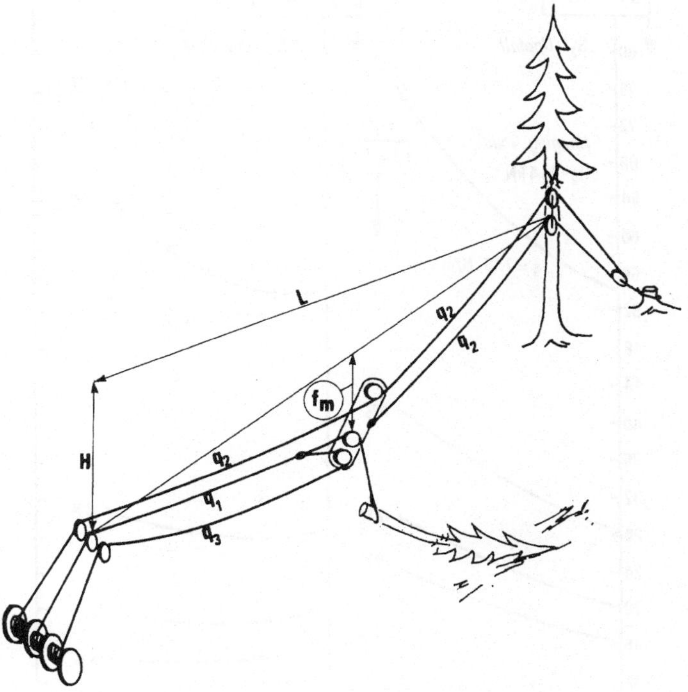

Figure 144. Running skyline with interlock winch.

The rope tension in the lower part of the span (S_2'') may be found using formula (114) if the angle β_2 is used instead of β_1.

The rope sag may be given as a percentage of the horizontal distance:

$$x = 100 \cdot \frac{f_m}{L}$$

The angle β in the equation (114) may then be found as follows:

Upper end of the span: $$tg\beta_1 = tg\alpha + 4 \cdot \frac{x}{100} \tag{115}$$

Lower end of the span: $$tg\beta_2 = tg\alpha - 4 \cdot \frac{x}{100} \tag{116}$$

Horizontal span: $$tg\beta_0 = 4 \cdot \frac{x}{100} \tag{117}$$

The line tension at the upper end of the span (S_2') is calculated in order to control the factor of safety (s).

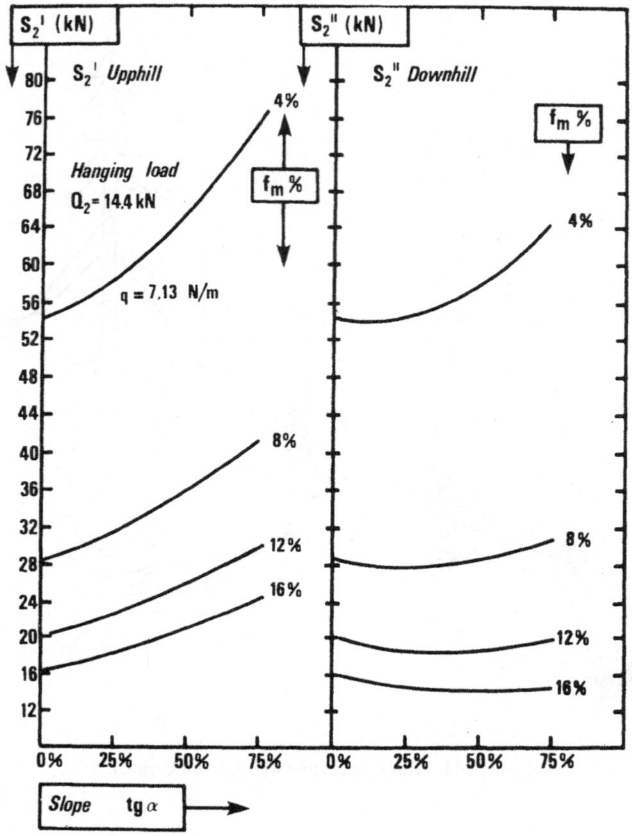

Figure 145. The cable tension at the upper support (S_2') and the lower support (S_2'') when the load is hanging in the middle of the skyline.

Example 30. **A cable crane with running skyline has the following specifications:**

Horizontal distance	:	**L**	=	**400 m**
Carriage and load	:	**Q**	=	**14.4 kN**
Main, haul-back, slack-pulling lines	:	**d**	=	**14 mm, 6x31+1FC**
		σ_{Br}	=	**1960 N/mm²**
		S_{Br}	=	**126 kN**
		$\bar{q}$	=	**7.13 N/m**

The maximum tension in the main line and in the haul-back line with safety factor $s = 3$, is $S_2 = 42$ N/m, and with safety factor $s = 4$, is $S_2 = 31.5$ N/m.

The tension in each rope with the load hanging in the middle of the span may be found using the formulae (115), (116) and (114). The result is shown in figure 145, which illustrates how the tension due to the weight Q and the weight of the rope q varies with the slope of the cable (tgα%) and the rope sag (f_m%).

7.7.2. Line tension due to load resistance

The line tensions due to load resistance is discussed in the chapters 4,0 and 5,0. If the load is hanging clear off the ground, then the tension is high in the running skylines. The factor of safety may be low which reduces the life of the wire ropes (figure 43). Accordingly, it is an advantage to transport the load with as great a rope sag as possible in order to reduce the line tensions. Usually the load should touch the ground and the transportation is carried out as a high skidding (see figure 65 and 66).

When the load is hanging clear of the ground the pull may be calculated using the equations (31) and (33):

$$S_T = Q \cdot (f_r \cdot \cos\beta \pm \sin\beta)$$

In the middle of the span $\beta \sim \alpha$. During downhill transport $f_r \cdot \cos\beta$ may be smaller than $\sin\beta$. This happens at slope angle $\beta = 3°$ when the rolling resistance between the carriage wheels and the rope is $f_r = 0.05$ (see figure 75). If the cable is steeper gravity may move the load in a downhill direction and might have to be braked.

If the span is horizontal gravity might move the load at the outer end of the span while the motor has to pull it when it gets close to the winch.

Uphill transportation creates greater tension and it is greatest when the carriage is close to the spar.

The pull due to load resistance (S_T) represents an addition to the tension in the running skylines due to load and cable weight. The winch operator should therefore, drive with as low a tension as possible (large rope sag) especially immediately before the load approaches the spar at the landing during uphill transport (figure 146).

It is an advantage to use high skidding which means that the running skylines carry only part of the load while the rest of the load touches the ground (figure 147). The pull due

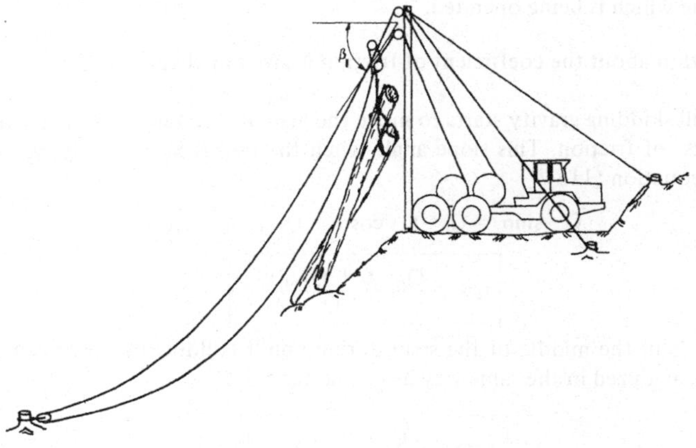

Figure 146. The greatest cable tension usually appears when the load is near the spar, where the rope is steepest.

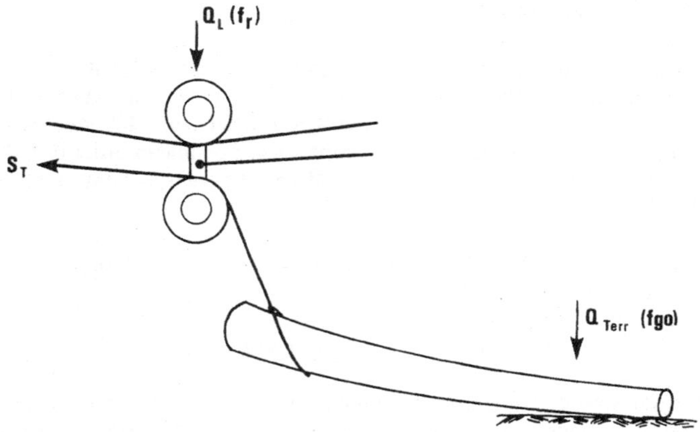

Figure 147. High skidding with cable crane.

to load resistance may be computed by means of the equations (31), (33) and (35). These may be converted as follows:

Sloping terrain:

$$\boxed{S_T = Q_L \cdot f_r \cdot \cos\beta + Q_{Terr} \cdot f_{go} \cdot \cos\alpha \pm (Q_L \cdot \sin\beta + Q_{Terr} \cdot \sin\alpha)} \quad (118)$$

Horizontal terrain (in the middle of the skyline):

$$\boxed{S_T = Q_L \cdot f_r + Q_{Terr} \cdot f_{go}} \quad (119)$$

The equation (118) is valid on condition that the load is practically in the middle of the span and the terrain has approximately the same slope as the span. Usually this is not the case. The terrain profile varies according to the topographic conditions and the variation in the terrain surface. Some places the load has to be lifted over cliffs, blocks etc. and the reduction in the rope sag increases tension. The equation (118) is, however, interesting because it makes it possible to study the variation in the tension due to the type of load, loadsize, slope, rope sag and the way the winch is being operated.

Some information about the coefficient of friction is given in chapter 4.2.

During downhill skidding gravity starts to move the load at a certain slope angle depending on the coefficients of friction. This slope angle when the pull is $S_T = 0$ may be found by the conversion of equation (118):

$$Q \cdot \sin\alpha = Q_L \cdot f_r \cdot \cos\alpha + Q_{Terr} \cdot f_{go} \cdot \cos\alpha$$

$$\boxed{\mathrm{tg}\alpha = \frac{Q_L \cdot f_r + Q_{Terr} \cdot f_{go}}{Q}} \quad (120)$$

When the load is in the middle of the span during uphill skidding the slope giving maximum tension may be analyzed in the same way as in chapter 5.2.1.

$$\boxed{\mathrm{cotg}\alpha = \frac{Q_L \cdot f_r + Q_{Terr} \cdot f_{go}}{Q}} \quad (121)$$

The corresponding maximum pull may be calculated using equation (118).

Example 31. **The cable crane with running skyline which is described in example 30, transports a 1 cu.m. load of unbranched spruce trees (full tree transport). The average breast height diameter is $D_{1.3}$ = 25 cm, and the relative height h_r = 1.2. High skidding is used and the chokers are set at the butt end of the trees (butt end forward). The weight of the carriage is 2.5 kN.**

$$\text{Payload} = \frac{4.86\ \text{kN/tree}}{0.409\ \text{m}^3/\text{tree}} = 11.9\ \text{kN (Table 7 and 8)}$$

$$Q_{Terr} = 0.38 \cdot 11.9 = 4.5\ \text{kN (Table 7)}$$

$$Q_L = 7.4 + 2.5 = 9.9\ \text{kN}$$

$$Q = 4.5 + 9.9 = 14.4\ \text{kN}$$

The tension when high skidding on horizontal ground (table 9):

$$S_T = Q_L \cdot f_r + Q_{Terr} \cdot f_{go}$$

$$S_T = 9.9 \cdot 0.05 + 4.5 \cdot 0.88$$

$$S_T = 0.50 + 3.96 = 4.46\ \text{kN}$$

If the load was hanging clear of the ground and was carried by the running skylines, the pull at the middle of the span during horizontal transport would be:

$$S_T = 14.4 \cdot 0.05 = 0.72\ \text{kN.}$$

The cable crane may be used to transport the wood down from a hillside. If high skidding is used, gravity will move the load if the slope angle is above a certain limit. One has to brake the load, for example by the motor. This angle may be calculated using equation (120):

$$\text{tg}\alpha = \frac{Q_L \cdot f_r + Q_{Terr} \cdot f_{go}}{Q}$$

$$\text{tg}\alpha = \frac{9.9 \cdot 0.05 + 4.5 \cdot 0.88}{14.4} = 0.3094$$

$$\underline{\alpha = 17.2^\circ}$$

During uphill skidding the slope angle which gives maximum pull can be calculated using formula (121):

$$\text{cotg}\alpha = 0.3094$$

$$\underline{\alpha = 72.8^\circ}$$

The corresponding maximum pull can be found from equation (118):

$$S_T = 0.15 + 1.17 + 13.76 = 15.08\ \text{kN}$$

When the load approaches the steel spar (figure 146) the rope deflection has to be increased in order to maintain the high skidding.

The most critical situation during uphill skidding take place immediately before pulling the load in onto the landing, especially in the cases where the landing terrain is steep (figure 148). It may be difficult to lift the front end of the load: (equation (31)):

$$S_T = Q \cdot f_{gu} \cdot \cos\alpha + Q \cdot \sin\alpha$$

The information in table 9 indicates that usually $f_{gu} < f_{go}$. During skidding according to the situation in figure 148, the humus layer will soon be damaged and scraped away. The load therefore, will be ground skidded directly against the mineral soil where the coefficient of friction is higher (see table 10 and 11). The slope angle when the maximum pull takes place may be found from equation (32):

$$\text{cotg}\alpha = f_{go}$$

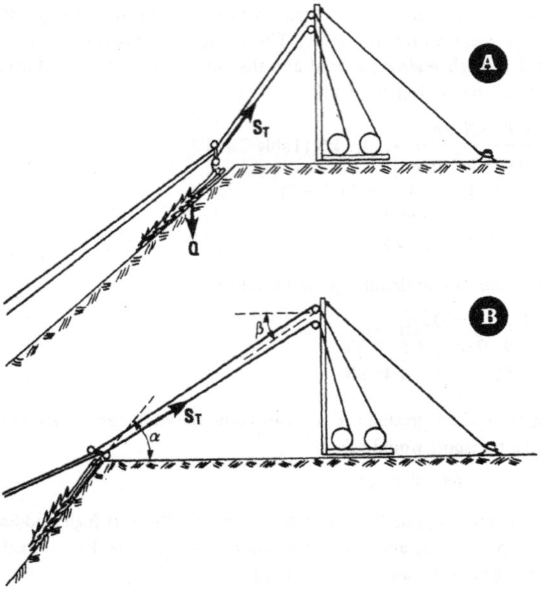

Figure 148. With top mounted winch it is necessary to place the spar in a position where the load is lifted over the upper edge of the terrain.

Example 32. **The cable crane used in example 30 and 31 is being used for uphill skidding. The slope angle (α) for the maximum pull (S_T) when the load is ground skidded into the landing as shown in figure 148 A is the following:**

$$\cot g\alpha = f_{go} = 0.88$$

$$\alpha = 48.7°$$

$$S_T = 14.4 \cdot 0.88 \cdot 0.66 + 14.4 \cdot 0.75$$

$$S_T = 8.36 + 10.80 = 19.16 \text{ kN}$$

The result in example 32 shows that when pulling the load up over a hill crest as shown in figure 148, the pulling force may be very high. With a 2 cu. m. load the pull is 38.3 kN in this case. This force adds to the main and haul-back line tensions caused by the weight of the load. In a situation like this it is an advantage to tighten the lines in order to lift the load or the front end of the load when pulling it over the hill crest. It is important to see that there is sufficient room for the deflection when rigging up the yarder. The situation in figure 148 B must be avoided where $\alpha > \beta$. In this case the load is being pressed against the ground during the pulling which increases the friction forces.

7.7.3. Total line tension due to Q, q in S_T and μ (μ = efficiency)

The total tensions in the main and haul-back lines is the sum of the tension caused by the weight of the load (Q), the weight of the steel rope (q) and the pull (S_T) caused by the load resistance. The situation is simplified in figure 149 where an endless line is used and a pulley

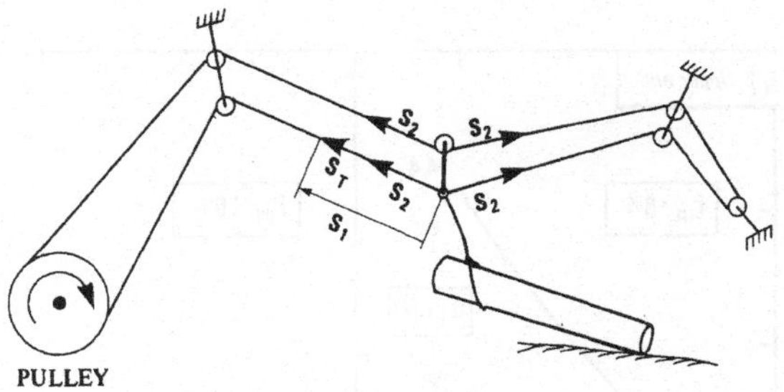

Figure 149. The total tension in the main line (S_1) is the sum of the pulling force (S_T) and the tension (S_2) needed to lift the load.

moves the carriage with load forward. An addition to the pull is needed in order to move the lines around in the system. This is caused by the rolling resistance in blocks and sheaves. Usually this resistance is accounted for in the coefficient of efficiency:

$$\bar{S}_T = \frac{S_T}{\mu}$$

ANDERSSON (1977) suggest that the loss of efficiency in a running skyline is 12.5 % if the lines are hanging clear of the ground. This corresponds to an efficiency $\mu = 0.875$. The total tension in the main line may be calculated from equation (122):

$$\boxed{S_1 = S_2 + \frac{S_T}{\mu}} \qquad (122)$$

Example 33. **The running skyline described in example 30, with additional information in example 31, is being used for transport or high skidding of one cubic meter green timber of spruce. The part of the load which is being lifted by the running skylines (Q_L) varies with the transport method as follows:**

Hanging load full tree transport:	Q_L	=	**14.4 kN**
High skidded load, full tree skidding	Q_L	=	**9.9 kN**
High skidding of stems, butt end forward	Q_L	=	**7.9 kN**
High skidding of stems, top end forward	Q_L	=	**5.1 kN**

The tension caused by load resistance, the tension caused by the load and cable weights and the total tension has been calculated using the formulae (114), (118) and (122). The result of the calculations are shown graphically in the figures 150, 151 and 152.

The line tension at the upper end of the haul-back line (S_2') is illustrated in figure 150. The steeper the terrain, the higher the tension in the line. The tension, however, is being considerably reduced by increasing the deflection.

If the cable crane was equipped with a skyline this would carry the load caused by the weight of the load and the carriage. This would reduce the wear on the main and haul-back lines. The running skyline system is, however, easier to rig up on a logging site and easier to move from one site to another. Usually a running skyline system is better on shorter cable cranes while the long cable cranes should preferably be equipped with a fixed or a live skyline.

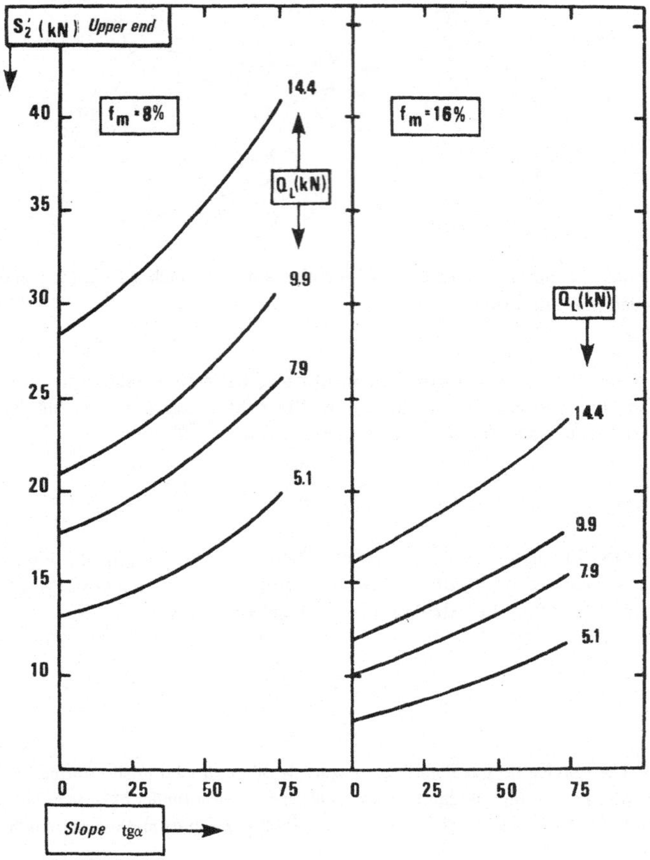

Figure 150. Tension in the haul-back line at the upper end (S'_2) with the load in the middle of the span (Q_L is described under example 33).

Figure 151 and 152 illustrate the total tension in the main line calculated using formula (122). Uphill transport gives the highest tensions and it is advantageous to use high skidding with rather slack lines in this case. When skidding full length stems, both calculations and practical experience shows that it is preferable to set the chokers on the butt end of the stem (butt end forward).

It is rather easy for the winch operator to over-load the running skylines. This must be avoided in order to reduce the wear and to increase the life of the wire ropes. The factor of safety should be at least s = 4 (figure 43) and the winch should be equipped with a safety device (figure 44) whenever possible, which makes it impossible to overload the lines.

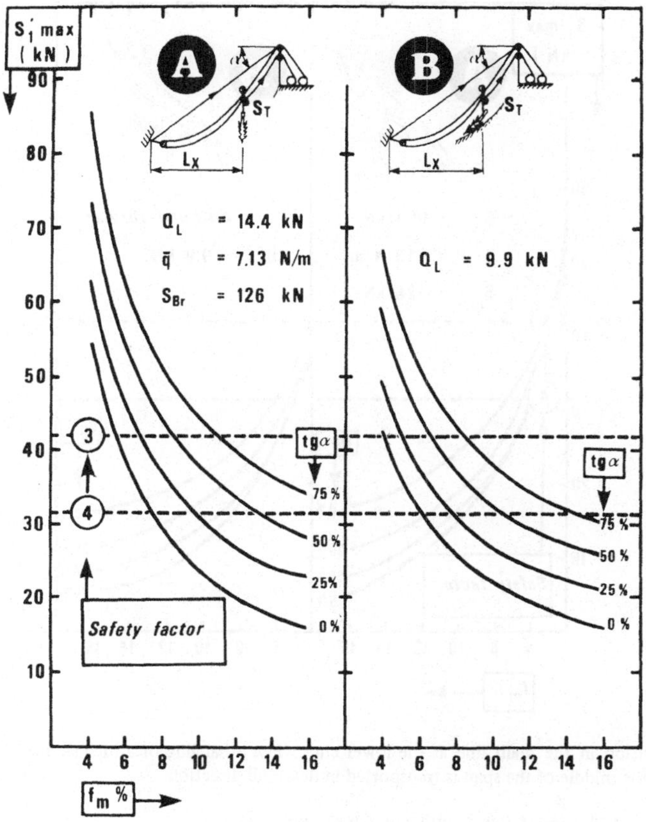

Figure 151. Tension in the main line at the upper end, when a hanging or high skidded load of full-trees in the middle of the span is transported in uphill direction.

$$S_1' = S_2' + \frac{S_T}{\mu}$$ (Q_L is described under example 33).

From figure 144 it can be seen that the outer end of the main line and the haul-back lines is being used more than the inner end which is wound up on the drums. The outer end of the haul-back line is passing the tail block arrangement when transporting each load and the outer end of the main line is passing the sheaves or drums in the carriage. Experiments carried out by the Norwegian Forest Research Institute show that it is possible to double the life of the wire ropes by turning their direction approximately at the middle of the lifetime. The lines are pulled out of the drums, turned and pulled in on the drums again with the outer end first (SØRHAGEN 1983).

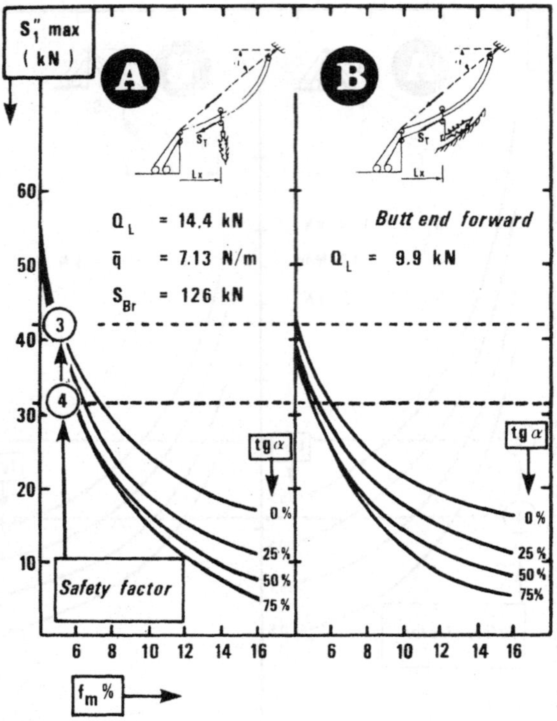

Figure 152. Tension in the main line at the lower end, when a hanging or a high skidded load of full-trees in the middle of the span is transported in downhill direction.

$S_1'' = S_2'' + \frac{S_T}{\mu}$ (Q_L is described under example 33).

7.7.4. Power in the lines and transmissions

The power (P_1, P_2 and P_T) is expressed in kilowatt (kW). Kilowatt is the product when multiplying the pull or the rope tension (S_1, S_2 and S_T) with the speed (v) (kW = kN · m/sec).

$$P_1 = S_1 \cdot v$$

$$P_1 = (S_2 + \frac{S_T}{\mu}) \cdot v$$

$$\boxed{P_1 = S_2 \cdot v + \frac{S_T \cdot v}{\mu}} \qquad (123)$$

$$P_1 = P_2 + P_T$$

$$\boxed{P_T = P_1 - P_2} \qquad (124)$$

The equation (124) illustrates that the power which moves the load forward is the difference between the powers in the main line and the haul-back line.

Table 21. An example of power consumption in kilowatt (kW) when transporting unbranched Norway spruce with interlock winch on Myrtillus type of forest floor. Speed 6 meters per second.

Rope sag f_m %	Slope tgα %	Uphill transport: Hanging load P_1	Hanging load P_2	Hanging load P_T	High skidding P_1	High skidding P_2	High skidding P_T
		kW					
8	0	175	171	4	156	129	27
	25	216	191	25	192	145	47
	50	262	219	43	230	168	62
	75	310	255	55	268	195	73
16	0	101	97	4	101	74	27
	25	137	112	25	134	87	47
	50	173	130	43	164	102	62
	75	207	152	55	191	118	73
Rope sag f_m %	Slope tgα %	Downhill transport: Hanging load P_1	Hanging load P_2	Hanging load P_T	High skidding P_1	High skidding P_2	High skidding P_T
		kW					
8	0	175	171	4	156	129	27
	25	144	161	– 17	126	121	5
	50	128	163	– 35	107	122	– 15
	75	128	177	– 49	103	134	– 31
16	0	101	97	4	101	74	27
	25	68	85	– 17	70	65	5
	50	44	79	– 35	44	59	– 15
	75	29	78	– 49	29	60	– 31

In order to illustrate the problem the power in the running skylines are analysed for the cable crane described in the examples 30 - 33. The speed is v = 6 m/sec. The results are given in table 21, which shows that large amounts of power is transmitted through the wire ropes when the transport is carried out with a running skyline cable crane.

The running skylines illustrated in figure 149 used an endless line and the load was moved forward by a pulley. The part of the cable tension caused by the weight of the load and the skyline is added to the pre-tension in the lines.

Some of the Janpanese cable cranes use this system. It simplifies the yarder construction. A special arrangement is needed on the yarder in order to adjust the length of the endless line to the variation in yarding distances.

When the yarder must be moved frequently from one place to another (small logging volume on each spot) it is unpractical to work with such a system since the yarding distance varies

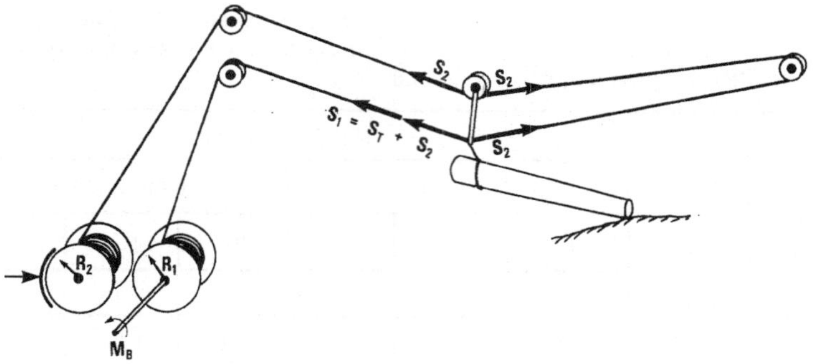

Figure 153. Running skyline system when braking the haul-back drum of a double-drum winch.

from place to place. Instead two drums are used, a main drum and a haul-back drum (figure 153). If one intend to lift the log it is necessary to brake the haul-back drum simultaneously as the load is being pulled in by the main drum. The winch motor has in other words to overcome both the power necessary to move the load and the power caused by the braking of the haul-back line, namely that part which is needed to lift the load ($P_2 = S_2 \cdot v$). The necessary driving power is:

$$\boxed{P_1 = P_T + P_2} \tag{125}$$

The power needed to lift the load (P_2) is considerably greater than the power needed to move the load (P_T). This is illustrated in table 21. If the rope sag is f_m = 16 % of the horizontal distance and the slope of the cable crane is 25 %, the power needed to move the load uphill is 25 kW, while the power needed to keep the load clear of the ground is 112 kW. Braking the haul-back drum resulted in a power of 137 kW in spite of the fact that it takes only 25 kW to move the load uphill.

Instead of wasting part of the power by braking, it is possible to connect the two drums to each other, for example as illustrated in figure 154. The two drums rotate in different directions which means that the haul-back drum is letting out the same amount of wire rope as the main drum is pulling in. The part of the tension in the haul-back line which lifts the

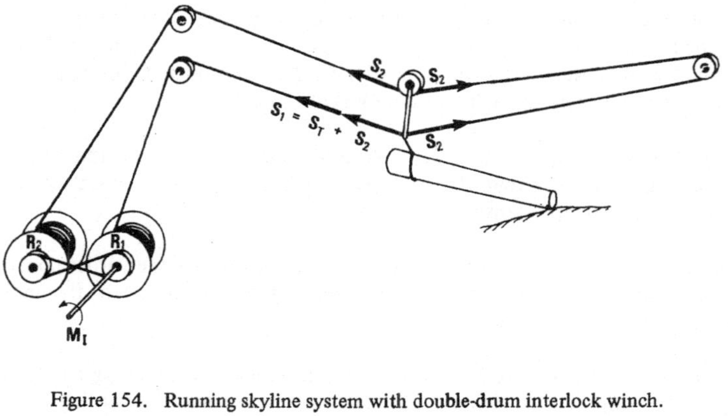

Figure 154. Running skyline system with double-drum interlock winch.

load results in a torque at the haul-back drum ($S_2 \cdot R_2$). This is transferred as a driving torque to the main drum (see figure 154). The same is the case with the braked power (P_2). In reality the double-drum winch in figure 154 works in a similar way as the pulley in figure 149. The difference is the following: In figure 149 the rope tension needed to carry the load is being transferred from the haul-back line to the main line by means of the tension in the rope itself, which is laid around the pulley. In figure 154 the tension is transferred through the transmissions between the two drums. The torque ($S_2 \cdot R_2$) work as a braking torque on the haul-back drum and an equal driving torque on the main drum, which means that the two torques eliminate each other. The same is the case with the power (P_2). The winch illustrated in figure 153 may be called a haul-back braked double-drum winch while figure 154 illustrates an interlocked double-drum winch. *With an interlocked winch the torque as well as the power is regenerated.*

7.7.5. The torque and the variation of the drum speed with an interlocked winch

If the running skyline is an endless line only driven by a pulley (groove wheel pulley or capstan) as illustrated in figure 149, there is no possibility of lifting or lowering the load. The carriage and the load follows the load path where the rope sag is the deflection when the load is in the middle of the span. This type of running skyline works in a similar way as a fixed skyline. The payload may be lifted to the carriage by a hoist line, for example if there is a hoist drum in the carriage.

One advantage of using a pulley is that the radius of the pulley is constant. The speed of the haul-back line is the same as the main line. The disadvantage is that the endless line is fixed and it is difficult to change the length of the cable way. One possibility is to arrange a side-bend on the endless pulling line, which can be tightened by means of a tightening line. A carriage which runs on the endless pulling line may be lifted or lowered by pulling in or giving out the tightening line. An other possibility is to use two sets of grooved wheels with storage drums as illustrated in figure 49. The deflection can be varied by using clutch and brake on the grooved wheels. The advantage of the system is that the line speed is constant as long as the rotational speed of the grooved wheel is constant, since it has the same radius all the time.

If an ordinary double-drum is used instead of pulleys the line speed varies with the coil radius on the drum. When pulling in on the main drum the coil radius increases simultaneously as the coil radius on the haul-back drum decreases. This must be taken care of by adjusting the rotational speed of the two drums relative to each other during the transport. The adjustment to the drum speeds must be carried out continuously when driving.

One of the following methods may be used to power a running skyline system:

1. Groove wheel pulley (endless running skyline)
2. Capstan (endless running skyline)
3. 2 groove wheel pulleys with storage drums
4. 2 capstans with storage drums
5. Main drum and haul-back drum where the drum speeds can be varied in relation to each other:
   a) Slipping clutch interlock
   b) Planetary gear in each drum
   c) Belt variator between the drums
   d) Hydrostatic drive
   e) Variable hydraulic pumps and fixed motor to each of the drums

The torque at haul-back braked or interlocked winches are illustrated in figure 153 and 154.

The haul-back braked double-drum winch needs the following torque from the motor if we do not take the loss of efficiency into account:

$$M_B = R_1S_1 = R_1(S_2 + S_T)$$

$$\boxed{M_B = R_1S_2 + R_1S_T} \qquad (126)$$

If an interlock double-drum winch (figure 154) has a planetary gear in each drum it is possible to adjust the drum speed according to the changes in the coil radius of the two drums. The torque required to pull the lines is the following when loss of power due to friction etc. is eliminated:

$$M_I = R_1S_2 + R_1S_T - R_2S_2$$

$$M_I = (R_1 - R_2)S_2 + R_1S_T$$

The haul-back drum lets out the haul-back line and the main drum pulls in the main line simultaneously. The power needed to maintain the tension is regenerated. This means that the power is being transferred through the transmission from the haul-back drum to the main drum. The difference between the torque which the motor needs to pull a haul-back braked winch or an interlocked winch is the following:

$$M_B - M_I = R_1S_2 + R_1S_T - R_1S_2 - R_1S_T + R_2S_2$$

$$\boxed{M_B - M_I = R_2S_2} \qquad (127)$$

$$\text{If } R_1 = R_2: \quad M_I = R_1S_T$$

$$\text{If } R_1 > R_2: \quad M_I > R_1S_T$$

$$\text{If } R_1 < R_2: \quad M_I < R_1S_T$$

This indicates that if the main and haul-back drums have the same amount of wire rope and the coil radius (R_1 and R_2) are equal, the torque from the motor is used to move the load along if the loss of power due to friction in blocks and transmissions are disregarded. If the main drum radius is bigger than the haul-back drum radius the motor torque needed is somewhat larger. If the radius of the haul-back drum is bigger than the main drum the torque needed is somewhat smaller than that required to move the load. These variations are small.

The variation in the drums rotational speed which is necessary in order to adjust for the variation in the drums coil radius must be done by means of planetary gear or other adjustment devices. If not, the lines will tighten if the main drum radius is bigger than the haul-back drum radius. Such adjustment is also necessary in order to avoid that the lines become slacker and the load lowered if the the haul back drum radius (R_2) is bigger than the main drum radius. When returning the carriage back to the logging site, the contrary situation exists.

7.8. Measurement of cable tension

7.8.1. Measurement by means of extensometer

It is of importance to work with correct tension in the skyline. This is difficult in forestry where the cable cranes often are moved from one logging site to another with varying lengths and slopes. The skyline is often tightend by experience and this may lead to a wrong tension in the cable during operation. Too low tension gives extra wear when passing cable shoes, and too high tension reduces the life of the cable.

The inclusion of an extensometer as a part of the cable equipment is recommended. By measuring the tension now and then, the crew's judgment is continuously verified. The extensometer should be installed on the skyline (fix or live) by means of two wedgegrips (figure 93) as illustrated in figure 155.

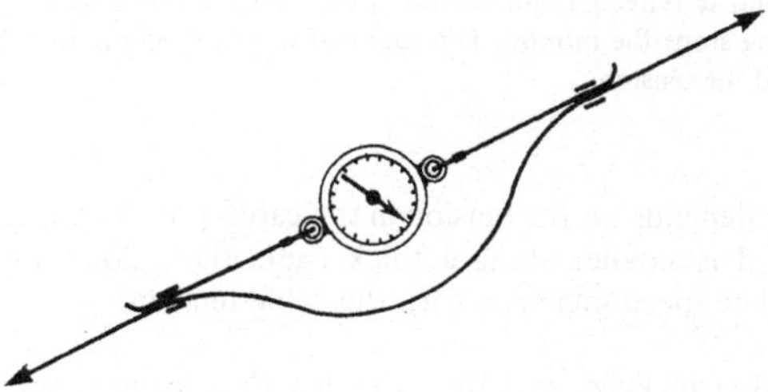

Figure 155. The installation of a dynamometer in a skyline.

An extensometer is often based on a spring element, for example a spring dynamometer. With time the spring element may gradually get fatigue which leads to errors. The extensometer should therefore be calibrated, for example every half year.

Moveable cable cranes may have the skyline wound up on a skyline drum, and the tension may be adjusted during operation (live skyline). This is an advantage for the work with rigging up or down the cable crane as well as during the operation itself. In this case it is absolutely necessary to include some kind of measuring device in the cable arrangement. The same is the case for cable cranes with running skylines (figure 44). The display of the measuring instrument should be placed where the yarder operator can see it when operating the winch enabling him to continuously adjust the tension of the skyline. It is also a great help for the operator if he has an instrument measuring the distance between the steel spar and the carriage. The two gauges for the skyline tension and the distance to the carriage helps the operator towards efficient driving and increased productivity.

7.8.2. Tension measurement by means of watch and cable oscilliations

If the skyline is hit by a small log near one of the suspension points a disturbance of the cable equilibrium occurs. A wave springs up which moves along the cable to the suspension point at the other end of the cable (figure 156). At this point the wave is reflected and returns back to the starting point where it is reflected again. The wave continues to move forward and back between the two suspension points of the cable. The wave is gradually decreased by the damping of the oscillation until it stops. If one touches the cable with a hand, one can feel when the wave returns, usually several times.

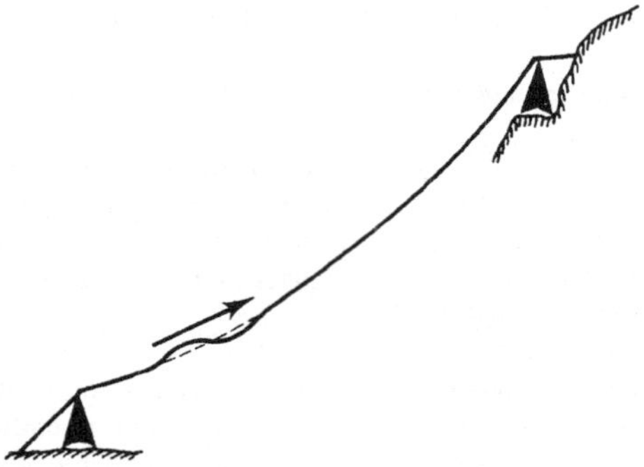

Figure 156. When hitting the cable (for example with a small log), a wave starts to move along the cable to the other end, is reflected and returns again. This forward and backward movement continues until damping stops the motion. The speed of the wave along the cable depends on the mass of the cable and the tension.

The speed of the wave depends on the tension in the cable and the oscillating mass (CZITARY 1961). The wave speed is smaller along a thick cable than along a thin cable of the same material and tension. The speed increases with the cable tension.

If the mass of the cable in kg/m and the cable length is known, the cable tension may be analysed by measuring the time for a reciprocating wave. It is more accurate to measure the time for several waves, for example with a stop watch, and then figure out the average time for the reciprocating movement of the wave.

The horizontal component of the cable tension can be calculated by means of equation (129) when the frequency of the wave (v in Hz), the horizontal distance (L) and mass of the cable is known.

$$v = \frac{1}{2L} \cdot \sqrt{\frac{H}{m}} \tag{128}$$

$$H = 4 \cdot m \cdot L^2 \cdot v^2 \tag{129}$$

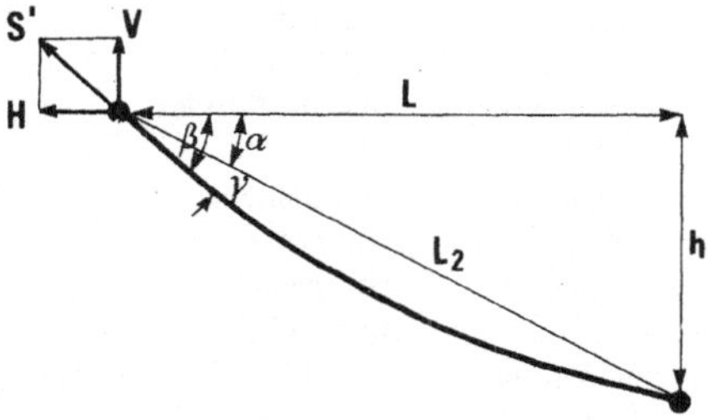

Figure 157. The skyline.

Example 34. **The horizontal distance of a cable with D = 13 mm was 1,075.8 m. The time of 5 reciprocating waves was measured by stop watch to 48 seconds:**

$$L = 1{,}075.8 \text{ m}$$
$$h = 382.4 \text{ m}$$
$$\bar{q} = 8.44 \text{ N/m}$$
$$q = \frac{\bar{q}}{\cos\alpha} = \frac{8.44}{0.9422} = 8.96 \text{ N/m}$$
$$m = \frac{q}{9.81} = \frac{8.96}{9.81} = 0.91 \text{ kg/m}$$
$$t = \frac{48}{5} = 9.6 \text{ sec.}$$
$$v = \frac{1}{9.6} = 0.104 \text{ Hz}$$
$$H = 4 \cdot 0.91 \cdot 1{,}075.8^2 \cdot 0.104^2$$
$$H = 45.6 \text{ kN}$$

Similar measurements were made on a research cable where the cable tension also was measured by means of a calibrated spring dynamometer.

Table 22 shows that we receive a very good result by this method.

It is of importance to hit the cable close to the suspension point. When hitting the cable at a certain distance from the suspension point a wave starts to move in both directions. They will interfer with each other and the result is no good.

The skyline should hang freely in the span which is being measured. If there is a carriage in the span the wave will be reflected against it and one does not succeed in measuring the time for the reciprocating wave between the suspension points. It is necessary to pull the carriage towards one of the ends of the skyline.

Since the wave reflects against the carriage this may be used to measure the cable tension when a load is hanging in the middle of the span. This can only be done if it is possible to find the horizontal distance or the length of the span between the load and the suspension point of the cable.

Table 22. Comparison between the forces in a research cable measured by dynamometer and by means of the time measured by stop watch of a forward- and back-running wave after a stroke on the cable.

Center rope sag	t (sec.)	$v = \frac{1}{t}$ (Hz)	H (kN)	
			Formula (129)	Dynamometer
I (2.5%)	9.6	0.104	45.56	45.36
II (3.1%)	10.6	0.094	37.22	36.82
III (4.0%)	12.2	0.082	28.33	28.28

If one continues to hit the cable there will be many waves after each other. If one succeeds in striking in time with the natural oscillation of the cable, there may be a continuous oscillation along the entire cable. Such oscillation may lead to extra load on the fixing points at the end of the cable.

Heavy wind may cause similar types of oscillations in the cable. It is possible to see such oscillations when looking along the cable, and sometimes one may hear them as a singing sound.

If the cable is fixed at the end by means of a casting in a socket the alternate load caused by the continuous oscillation may in time lead to fatigue failure. This problem is serious for stationary cable ways, for example those used for passenger transport. This problem is, however, of little interest for forestry cable cranes since they are often moved from one place to another with differences in cable lengths, slopes, rope sag etc., and also differences in the oscillation pattern.

Usually, one is more interested in the cable tension (S) than in the horizontal component of it, and an approximate value may be worked out as follows:

$$H \sim S \cdot \cos\alpha$$
$$L \sim L_2 \cdot \cos\alpha$$
$$m \sim \frac{\bar{m}}{\cos\alpha}$$

$$\boxed{S = 4 \cdot \bar{m} \cdot L_2^2 \cdot v^2} \qquad (130)$$

Equation (130) is correct if $\alpha = \beta$ (figure 157). Usually $\beta > \alpha$ and the cable tension at the upper suspension point is always somewhat bigger than the tension according to equation (130). With high tensions and small rope sags the difference is small. It is, however, necessary to be aware of this, especially with high deflection of the cable where there are big differences between the angles β and α.

If the time for one oscillating wave is measured in seconds (t) the equation (130) may be converted to:

$$\boxed{S = \frac{4 \cdot \bar{m} \cdot L_2^2}{t^2}} \qquad (131)$$

If the time is measured in cmin., i.e. min./100 (μ):

$$\boxed{S = \frac{11{,}11 \cdot \bar{m} \cdot L_2^2}{\mu^2}} \qquad (132)$$

A practical rule:

The equation (131) may be converted to:

$$\boxed{t = \frac{2L_2}{\sqrt{\frac{S}{\bar{m}}}}} \qquad (133)$$

As long as the braking stress due to tension is the same, the ratio between the braking load (S_{Br}) and the mass of the wire rope per meter ($\bar{m}$) is approximately constant for the various wire rope constructions and dimensions (PESTAL 1961). If the factor of safety is taken into account the ratio between the allowable tension and the mass per meter is:

$$\frac{S_{Br}}{s \cdot \bar{m}} = \frac{S}{\bar{m}} \sim \text{Constant}$$

The ratio has the units N/kg per meter of the wire rope (or kN/kg per meter). If the braking stress due to tension is 1,770 N/mm², and d = 20 mm, 6 x 19 + 1FC:

$$\frac{S_{Br}}{s \cdot \bar{m}} = \frac{216 \text{ kN}}{4 \cdot 1.34 \text{ kg/m}} = 40 \text{ kN/kg/m}$$

In this case the time of a reciprocating wave according to equation 133 is:

$$t = \frac{2L_2}{\sqrt{\frac{S_{Br}}{s \cdot \bar{m}}}} = \frac{2L_2}{\sqrt{\frac{216\,000}{4 \cdot 1.34}}} = \frac{L_2}{100}$$

Table 23 illustrates that equation 134 gives a good approximation for the time in seconds of a reciprocating wave for all the analysed wire rope dimensions when σ_{Br} is 1,770 N/mm² and the factor of safety is s = 4:

$$\boxed{t = \frac{L_2}{100} \text{ (seconds)}} \qquad (134)$$

If one hit the skyline near the cable shoe in a span and the time for a reciprocating wave in seconds is equal to the length of the span in 100 meters, the factor of safety is 4.

Table 23. Examples on the relation $\frac{S}{\bar{m}}$ and the time in seconds of a resiprocating wave on the cable for some rope constructions and dimensions, calculated on the basis of information in Norse wire rope catalogue (ANON 1975). (σ_{Br} = 1,770 N/mm² and s = 4).

D mm	Construction	$\frac{S_{Br}}{s \cdot \bar{m}}$ = (kN/kg/meter)	t (sec.) = (Equation 133)
18	6x 7+1FC	42	$\frac{L_2}{100}$
20	6x19+1FC	40	$\frac{L_2}{100}$
20	6x19+IWRC	39	$\frac{L_2}{98}$
14	6x31+1FC	39	$\frac{L_2}{99}$
14	6x31+IWRC	38	$\frac{L_2}{97}$
20	6x31+1FC	40	$\frac{L_2}{99}$

8.0. THE YARDING CREW

The productivity is usually higher the smaller the crew number. One man operation gives a higher production per man than if the same job is carried out by two men. The same is the case with mechanized operations. It is good organization of the work when one machine operator carries out a specific job with his machine, independent of other machine operators and their machines.

Cable operations can only be performed by means of several men co-operating as a crew. Some of them are the field workers and the others work at the landing. It is an advantage to keep the number of men in the crew as low as possible since this always gives the highest production per man. This is especially the case with light equipment. With heavy and high production equipment it may be necessary to increase the crew in order to utilize the production potential of expensive equipment.

In some cases felling and conversion of trees to stems or logs is carried out by a separate felling crew. This is especially the case for tree dimensions larger than one cubic meter per tree. In other cases it is an advantage to combine felling, conversion and cable yarding in one sequence, and the yarding crew also takes care of the felling and conversion. This usually applies for tree dimensions less than 1 cubic meter per tree.

When comparing production and cost figures both felling, conversion and yarding should be taken into account.

The number of workers in a yarding crew depends on the size of the equipment. Light cable cranes suitable for 200 – 300 meter and with tree dimensions less than 1 cu.m./tree are often operated by a two men crew, one winch operator and one field worker. They are rigging up the cable crane, and during the yarding the field worker sets the chokers while the winch operator also undertakes the chasing. If cutting and yarding is carried out in one transport sequence, the field worker will fell the trees, limb them on the upper side and set the chokers while the winch operator chase the chokers, limbs the remaining branches and bucks stems into assortments.

With heavier equipment there are usually more men in the yarding crew. Even in this case it has become more common to combine felling and yarding in one transport sequence. The limbing, bucking and sorting can be carried out by a harvesting machine at the landing. Accordingly, the yarding crew partly undertake the underway functions (load transport and returning the empty carriage) and partly the terminal functions (loading , unloading, conversion etc.).

It is of importance to distinguish between the planning procedure, the rig up operation and the yarding itself. The planning work as well as part of the rig up work should be carried out by the foreman (the hooktender) before the yarder with the yarding crew arrives at the logging area. This minimizes the installation time for the yarding crew. Such preparatory work may consist of pre-setting the guyline anchors and the tail anchor, pulling out the rig up line etc. and clearing the cable corridors. The foreman may be assisted by one of the field workers during the preparation before the crew starts to rig up the crane.

8.1. Organization of the yarding crew

A complete yarding crew or cable crane crew consists of the following members:

The foreman (the hooktender) is the leader of the yarding crew. After planning, based on aerial photographs, longitudinal profiles etc., he leads the installation of the cable crane as well as the operation itself. He also should have the necessary skill to lead, as well as repair and maintain the equipment.

The working foreman carries out the preparatory work as mentioned above before the yarding crew and the equipment arrives at the logging site. In difficult terrain some of this preparatory work may be too hard for one man and the foreman requires help from one of the crew members.

As soon as the cable crane has been rigged up on one yarding strip, the crew starts the yarding. In the meantime the foreman prepares the installation on the next yarding strip by laying out the straw lines, preparing the anchors etc. This simplifies the work involved in moving the cable crane from one yarding strip to the next.

The winch operator operates the yarder. He should have the necessary knowledge to carry out repair and maintenance work.

The field crew set the chokers and connect them to the rig or the hoist line. Very often the cable crane has a higher production than the field crew. This may be taken care of by having more than one loading point along the skyline corridor, for example one low down on the hillside, and another higher up on the hillside. The winch operator alternates between them, taking one turn from the first choker setter and the next from the second choker setter etc.

The field crew may have the following functions:

The choker setter sets the chokers and fixes them to the rig or the hoist line. In some cases he pre-sets the chokers which may be attached to a tagline. When the carriage arrives he unhooks the empty tagline and connects the pre-set tagline to the hoist line.

The signaller gives instructions to the winch operator. *The cutter* carries out felling and maybe some of the conversion work, for example limbing the upper side of the stems. Some times each of the field crew carries out all the three functions, felling, setting the chokers and signalling. If the cable crane is radio controlled he may even operate the cable crane by radio signal during the choker setting and lateral skidding.

The chaser man unhooks the chokers at the landing. He may do some of the cutting work, for example limbing remaining branches and bucking. When yarding timber less than 1 cu.m./ tree with short distance cable cranes the winch operator and the chaser man is the same person.

When logging big trees in virgin forests, such as on the West Coast of America and Canada, or in the tropics, the cutting and conversion of trees to logs is done prior to the yarding by a felling crew.

Heavy equipment is needed for the yarding operation. The machine costs are high, and terminal times like choker setting and chasing should be reduced to a minimum. The chokers are heavy, and usually there is one choker setter for each choker.

The rigging slinger is the head man on the rigging crew. He is responsible for choker setters and chaser, and he selects logs to make up a turn.

8.2. Ergonomic work load

The field crew work under difficult terrain conditions. When the manual work takes place in steep terrain the work load increases. The body weight has to be lifted when walking from the valley to the work place high up on the hillside. Cliffs, rocks and other difficulties also increases the work load for the men.

The work load for Norwegian forest workers during cable crane operations has been investigated (VIK 1980). The work capacity is a measure of the amount of energy a human body can exert over a shorter period. It is not possible to utilize this amount of energy during continuous work. It is recommended that the maximum work load should be less than 50% of the work capacity during continuous work. The percentage of the work capacity which was used by the research workers during the test is a good measure of the total work load on their body (figure 158).

Cutting was investigated in ordinary forest terrain and in steep cable terrain (slopes of 50%–70%). Deep snow (70 cm) gave a higher work load than the steep terrain.

Setting the chokers is a hard work but not as hard as cutting. The choker setters had on average 132 heart beats per minute during 18% of the work day. When it is 70 cm snow or more on the ground setting the choker is very hard work. Choker setting is, however, interrupted by waiting during which time the choker setter may recover after a heavy work load.

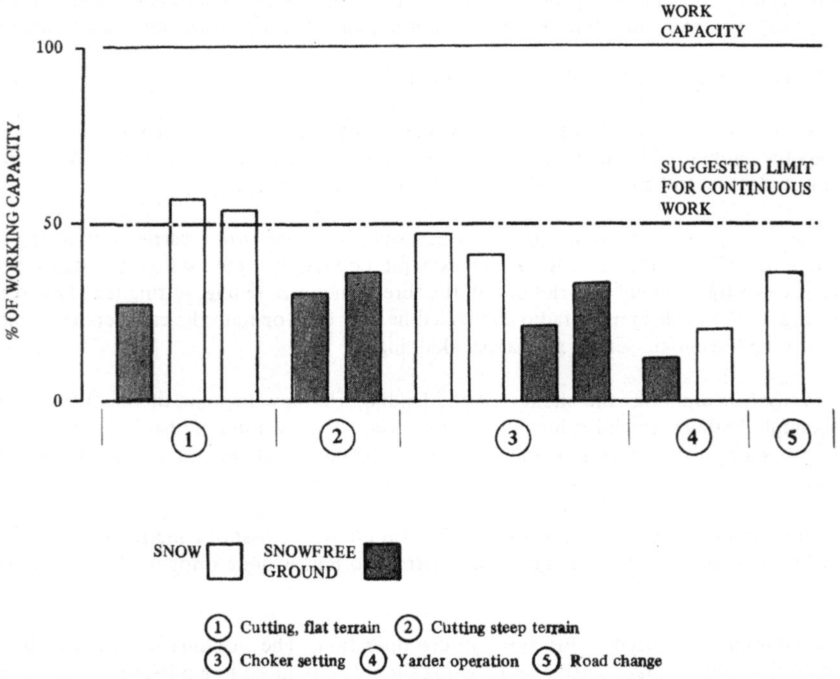

Figure 158. Relative work capacity for some work operations in steep terrain.

Table 24. Motor-manual work in flat and in steep terrain. Cutting and delimbing Norway Spruce to stems by chainsaw (VIK 1980).

Terrain	Work load in % of an aerobic work capacity	The workers energy consumption in kJ/m^3
Even and flat terrain	27%	236
Steep cable crane terrain	29.5%	815

Winch operation is an easy job from an ergonomic viewpoint. For the operator of the Moxy winch the work load was only 12% of the work capacity.

Rig up work is heavy work for the yarding crew, especially the rig up work of earlier cable crane constructions. Modern cable cranes are equipped with auxiliary equipment which makes the rig up work lighter for the workers.

The work load is dependent on the working speed. Figure 158 illustrates that even the heaviest operations had a work load of less than 50% of the anaerobic work capacity. The reason is that the workers adjusted themselves to the heavy conditions by increasing the working time per tree. This is illustrated in table 24. A group of research workers felled and delimbed Norway spruce to stems partly in flat terrain and partly in steep cable crane terrain with rough surface (70% steepness). The work load was only 27%–29.5% of the anaerobic work capacity. The workers' energy consumption, however, measured in kilojoule per cubic meter produced timber was much higher in steep than in flat terrain. *With the same energy consumption the forest worker could cut 3.5 times as much per day in flat than in steep terrain.* Investigations of the forest workers energy consumption and production under steep forest conditions in India and in Central Europe have given similar results (STREHLKE 1971).

It is important to reduce the work load for the field crew under steep and difficult working conditions. The cutting of trees less than 1 cu.m./tree should be reduced to felling only, while operations like limbing, bucking, sorting etc. should be carried out by harvesting machine at the landing. By limiting the cutting to felling only in the terrain it is also easier to co-ordinate felling and choker setting in one work sequence.

8.3. Signal systems

There are several workers in the yarding crew who must depend on each other during the operation. It is of importance that they understand each other and a good signal system simplifies the work. It increases both safety and efficiency of work.

The crew operating the former primitive forestry cable cranes, used simple signalling systems such as hitting the skyline with a small log.

8.3.1. Field telephone

Field telephone has been in use on former cable cranes. The telephone cable was a disadvantage. The telephone had to be placed outside the logging area which led to extra walking for the crew. It also happened that the telephone cable was damaged during the operation.

New development introduced loudspeakers on the field telephones.This led to a practical and good comunication system. This Austrian development is shown in figure 159. There are 3 loud-speakers, one at the landing, one at the winch and one centrally located at the logging area in the field. They are connected with single telephone cables. The winch operator or the chaser man may contact the field crew through the loud-speaker in the terrain when they are within a hundred meters distance. The field crew can talk to the winch operator or the chaser man using a small pocket radio transmitting through the loud-speaker in the terrain,provided he is less than a hundred meters away from the loud-speaker.

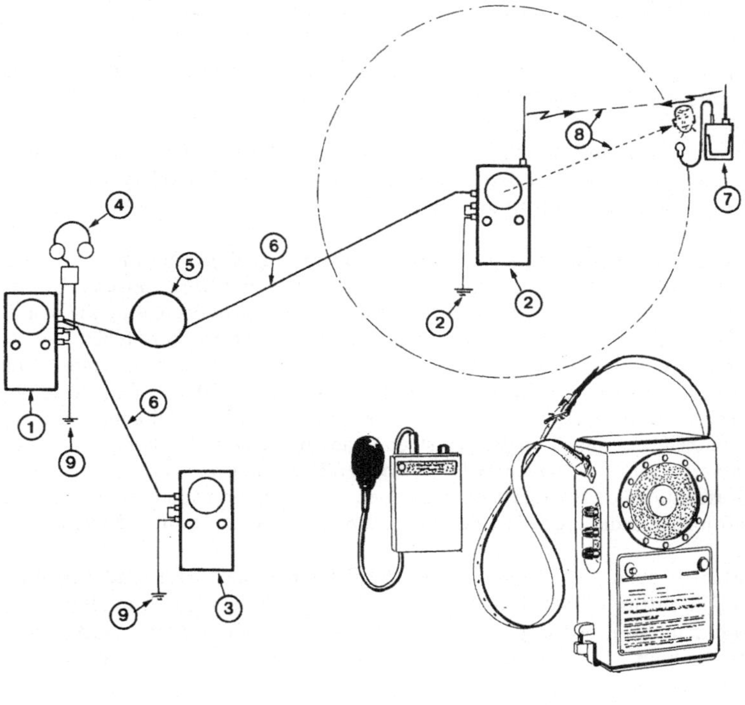

Figure 159. An Austrian field telephone with loud-speakers.

1. Loud-speaker at the yarder (type FT-22)
2. Radio transmitter and receiver with loud-speaker centrally located on the logging area (type FT-23/FB)
3. Loud-speaker at the landing (type FT-22)
4. Head-phone
5. Drum with cable
6. Cable with one channel from yarder to the centrally located radio/loud-speaker at the logging area and from yarder to landing
7. Radio transmitter with microphone
8. Radio distance and hearing distance from loud-speaker approximately 100 m
9. Earth connection

8.3.2. Short distance communication with radio

The short distance communication radio represents a further development since the disadvantages of the telephone cable is avoided (figure 160). The short distance radio (walkie-talkie) is of great help in connection with repair and maintenance work. The same is the case when rigging up or down cable cranes longer than 200 meters. A short distance radio should always be a part of the cable equipment, even when other signalling systems are used during the operation itself.

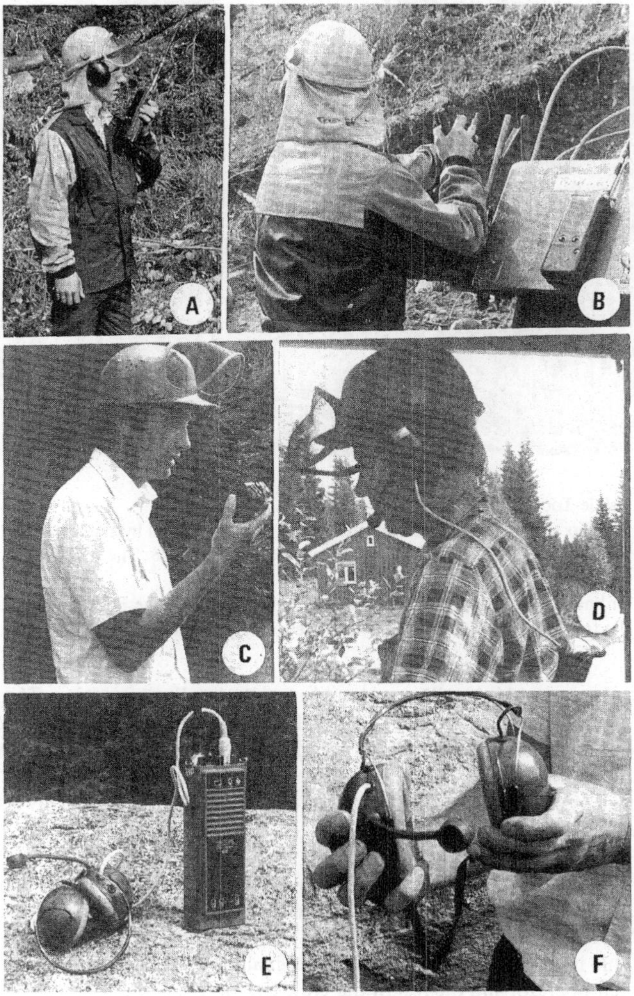

Figure 160. Short distance radio for communication between the winch crew.
A. and B. Ordinary walkie-talkie.
C. – F. Portable VHF FM communication radio (type BDR 1606) with microphone and earphones.

Figure 161. The Talkie-Tooter.

A. A radio signal from the transmitter in the field is received by the receiver at the winch. This actuates a horn on the winch which gives a sharp whistle. The whistle can be heard by the winch operator as well as on the felling site.

B. Ordinarily the transmitter gives a signal which actuates the horn on the winch. By turning the transmitter upside down the field crew may give voice information to the winch operator.

Walkie-talkie is a less suitable signalling system during the yarding operation. The conversation method is not precise and may often be disturbed by noice from the operation. Many research operations have shown that acoustic signals are more safe and efficient than conversation in a two-way radio. The talkie-tooter is an American system (figure 161). Each of the field crew have a transmitter. A radio signal can be sent to a receiver at the winch. This actuates a horn on the winch which gives a sharp whistle.

The cable crane may be remote controlled. Each of the yarding crew has a transmitter, and can send a radio signal via a receiver on the winch to the various winch controls. It is then the choker setter in the terrain or the chaser man at the landing who can control the winch to suit his work. This is the most efficient and safe signalling system since any misunderstanding between the winch operator and the rest of the crew is avoided.

8.3.3. Acoustic signals

Acoustic signalling is a good system where the visibility between the winch station and the field is poor or when the terrain conditions prevent any visibility at all. Even when the winch operator can see the choker setter the acoustic signal is efficient since the choker setter can give the signal while carrying on with other work. Various types of signals have been tried out. Such as »dot and dash» signals. It is, however, often difficult to distinguish between the dots and the dashes. The signal illustrated in figure 162 B is recommended.

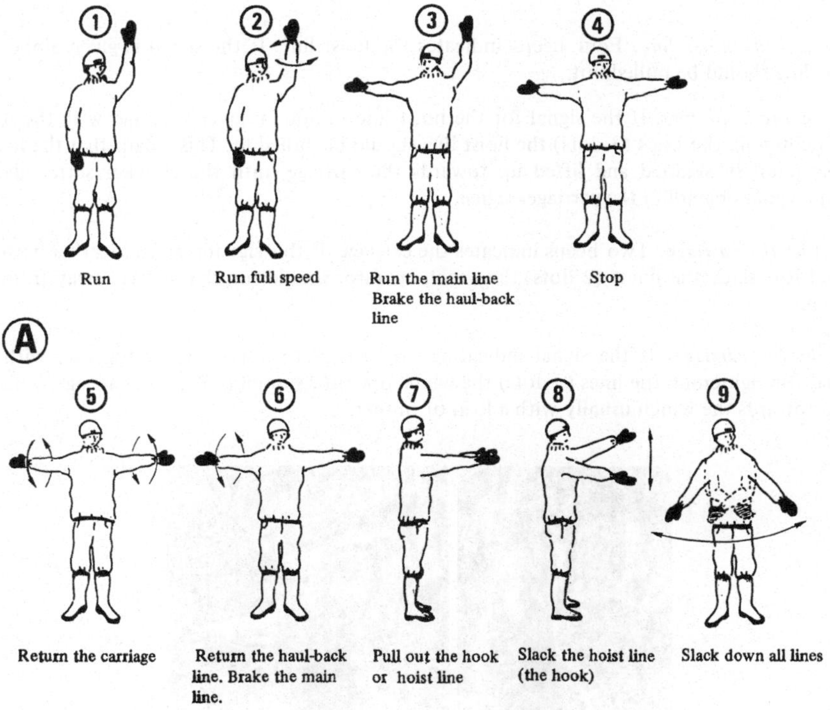

B

STOP	●
SLACK THE LINES	●●
TIGHTEN THE LINES	●●●
PULL OUT THE HOIST LINE	●●●●
PULL IN THE HOIST LINE	●●●● ●●●
PULL OUT THE CARRIAGE (AWAY FROM WINCH)	●● ●●
PULL IN THE CARRIAGE (TOWARDS THE WINCH)	●● ●●●

Figure 162. A. Arm signals.
B. Sound signals.

The stop signal consists of one beep (one dot). The operator stops the winch but maintains the line tension.

Slack the lines is indicated by two beeps and the winch operator slackens all lines.

Tighten the lines is indicated by three beeps. The winch operator tightens the lines until the signal »stop» is given.

Pull out the hoist line. Four beeps indicates the hoist line. If the signal is given alone, the hoist line should be pulled out.

Pull in the hoist line. If the signal for the hoist line (4 dots) is given together with the signal for tightening the lines (3 dots) the hoist line should be pulled in. This means that the load is being laterally skidded and lifted up towards the carriage until the chocker setter gives a »stop» signal or »pull in the carriage»-signal.

Pull out the carriage. Two beeps indicates the carriage. If this signal is given together with the signal for »slack the lines» (2 dots) the winch operator should run the carriage away from the winch.

Pull in the carriage. If the signal indicating the carriage (2 dots) is given together with the signal for tightening the lines (3 dots) the winch operator should pull in the carriage from the field towards the winch,usually with a load of timber.

Figure 163. Arm signals.

If the winch operator does not understand a signal, he should wait for a new signal. The acoustic signal is recommended for short distance cable cranes as well as for big cable cranes.

8.3.4. Arm signals

With light cable cranes working up to 200 – 300 meter, arm signals is a good system when the visibility is good. The signals illustrated in figure 162 A has proven to be safe and efficient and is recommended, especially for cable yarding shorter than 300 meter.

9.0. THE FUNCTION OF CABLE SYSTEMS

The »complete» or theoretical cable way consist of components as illustrated in figure 164. Some cable ways may have more than one component. The pendulum cable way has 2 skylines and 2 carriages. In other cable systems the function of one of the components may be taken over by another component. A highlead yarder has no skyline. Instead of the skyline carrying the load, it lays on the ground and is carried by the forest floor during the transport.

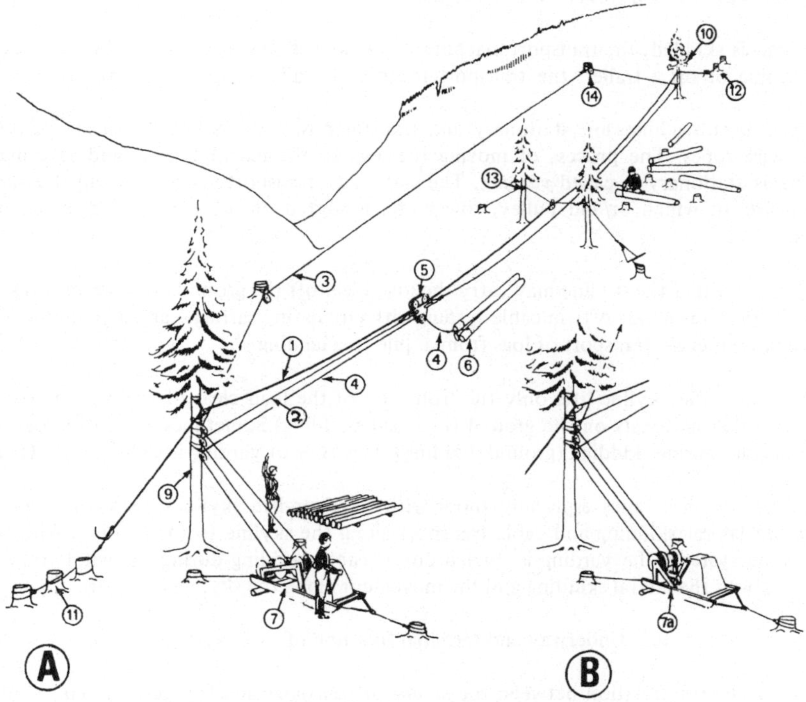

(1) Skyline (2) Main line (3) Haul-back line (4) Hoist line (5) Carriage
(6) Chokers (If desired: Tagline with chokers) (7) Tree-drum winch
(7a) Winch with a capstan or a groove wheel and a winch drum (9) Head spar
(10) Tail spar (11) Heelhold (lower anchorage) (12) Tailhold (upper anchorage)
(13) Support (crossline support) (14) Terrain block

Figure 164. The »complete» cable way.

The main and haul-back drum on the winch may be replaced by a pulley. Instead of a separate hoist line this may be a part of the main line. There may be an automatic locking device in the carriage, and sometimes this is combined with a carriage stop on the skyline.

The skyline is the general name of the line which carries the load. In some of the cable systems, the skyline is a *fixed skyline*. After the skyline has been tightened to a pre-tension the skyline is anchored at both ends. Other cable systems have a *live skyline*. Normally the tail end is anchored while the inner end is attached to a winch or another tightening device. The skyline may be lowered or tightened during the transport. In some cable systems it is the running lines, usually the main and haul-back lines, which carries the load. This is called a *running skyline*. The line tensions are maintained through inter-locking between the main and haul-back drums. It is the torque between the two drums which carries the load.

9.1. Cable skidding or cable transport

Some transport machines travel on the ground to the logs, and then transport the logs to a concentration point. Instead of logs, other assortments may be transported, such as stems (full length logging) or felled trees (full tree logging).

If the load is skidded, the transport machine is a *skidder*. If the load is carried on the back of the machine or on a trailer, the transport machine is called a *forwarder* or a *transporter*.

Some transport machines are stationary and transfer power to the logs by means of one or several wire ropes. The process of moving the logs to the machine or a landing while the machine is stationary is called *yarding*. The stationary transport machine is called a *yarder*, and consists of winch or/and pulley, tower etc. mounted on a self propelled or moveable chassis.

During the yarding the skyline may carry the load clear off the ground (airial transport). This is called *cable transport*. A twin cable system with circulating carriages and a pendulum cable way carries out cable transport of logs from a pile to a landing.

In other cases the skyline lifts only the front end of the load clear of the terrain obstacles while the other end rests on the ground (highlead skidding).Sometimes the entire load rests on the ground during skidding (ground skidding). This type of yarding is called *cable skidding*.

Some of the cable cranes especially those with fixed or live skylines, uses cable skidding during the lateral skidding and cable transport along the skyline to the landing. With other cable crane systems the yarding is carried out as cable skidding during the whole transport sequence, both the lateral skidding and the movement along the skyline corridor.

9.2. Underway and terminal function of cable systems

There is a close correlation between the *means of transport* and the *lines of transport* in a transport system. Cable cranes and cable ways are intergrated parts of the transport system in forestry.

The means of transport as well as the lines of transport within a cable transport system may be devided into underway or terminal functions:

Underway functions. All the components in a cable crane which moves the load or return the empty carriage are underway functions.

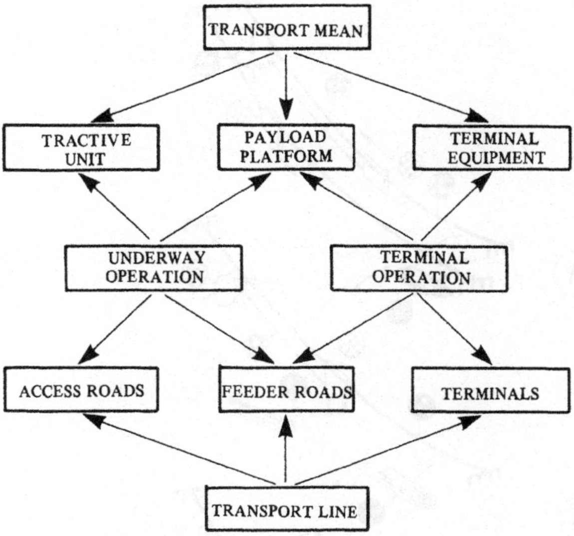

Figure 165. The relationship between the underway and terminal functions for the transport mean and the transport line.

Line of transport: The skyline carries the load and the empty carriage, and is therefore a line of transport.

Mean of transport: The haul-back line, the haul-back drum, the pulley etc. pull the carriage along the skyline, and are examples on means of transport.

Terminal functions. This is the loading, unloading etc. Cable yarding with modern equipment often includes conversion of trees to stems or assortments. In this case the conversion is a terminal function of the cable system.

The lines of transport: The hoist line carries out a terminal function on a cable crane. It pulls the logs together and move the load to the carriage.

The means of transport: Unloading equipment and harvesting machine at the landing are examples of transport means which carry out terminal functions.

Combined terminal and underway functions. Some components in the cable crane carries out both underway and terminal functions.

The lines of transport: The main line and some times the haul-back line have terminal functions during the lateral skidding. They have underway functions when the carriage is being pulled along the skyline.

The means of transport: The carriage, as well as the winch drums have terminal functions during lateral skidding, unloading etc. When the winch pulls the carriage along the skyline it has an underway function.

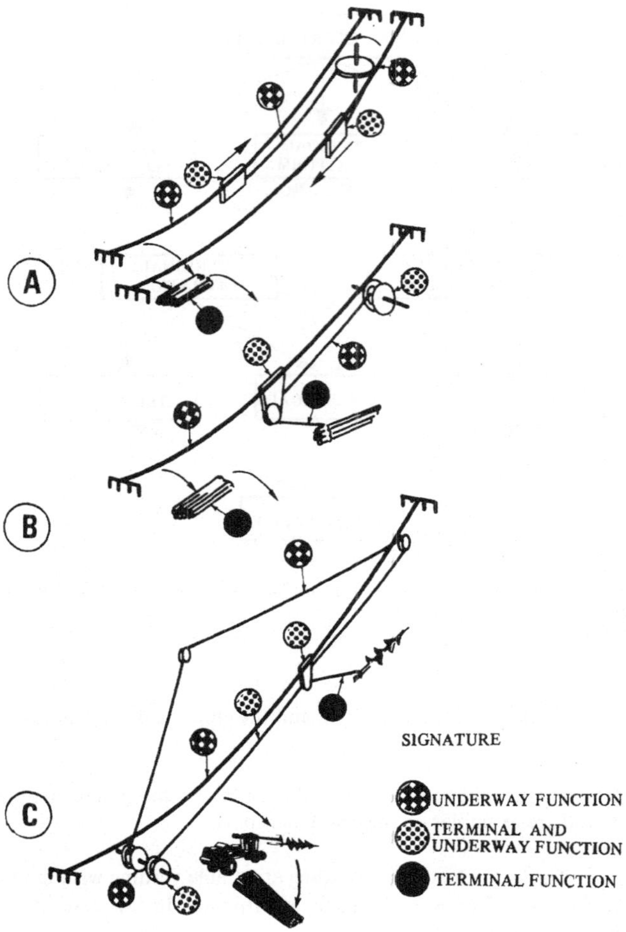

Figure 166. Underway and terminal functions for some types of cable ways.
A. Pendulum cable way.
B. Cable crane with top mounted winch.
C. Cable crane with valley mounted winch and limbing machine at lower landing.

An example of underway and terminal functions on various types of cable ways, is illustrated in figure 166.

9.3. Access roads and feeder roads

The transport network in forestry consist of *access roads, feeder roads* and *terminal places. The access roads* link the forest areas through a network for long distance transport. Public truck roads, railways and float ways are typical access roads. In steep terrain the access roads have to overcome height differences. They are good quality roads but may be steep (up to 20%).

The feeder roads branches out from the access roads and run mainly parallel to the contour lines in the terrain. The feeder roads are generally truck roads, but may be tractor roads. In steep terrain the cable yarding takes place on the area between the feeder roads. Consequently the terrain conditions and the yarding possibilities is the deciding factor for the layout of the feeder roads.

Terminal places. These are places along the feeder roads suitable for a winch station, a landing, a loading macine, a limbing/bucking machine etc. If the soil and the terrain conditions are suitable a wide feeder road may serve as a continuous terminal place. The cable yarding is being carried out on narrow cutting strips, perpendicular to the feeder road. The cable crane is moved sideways on the logging area 20 meters at a time. The cable yarding results in a continuous pile of for example unbranched trees along the feder road. A harvesting machine may follow the yarder, convert the trees to logs and pile them on the other side of the feeder road.

When yarding big timber from virgin forests with high production equipment the landing must be large enough both for the yarder, for the loader and for the truck. The truck loading must go on simultaneously with the yarding. If not, the landing may be filled up of logs and block up the operation.

In some cases, a double-road is laid on parallel to each other and with 50 meters spacing in the steep hillside as landing. The yarder operates on the upper road, while logs are landed on the lower road, where the loader operates and the truck transportation takes place.

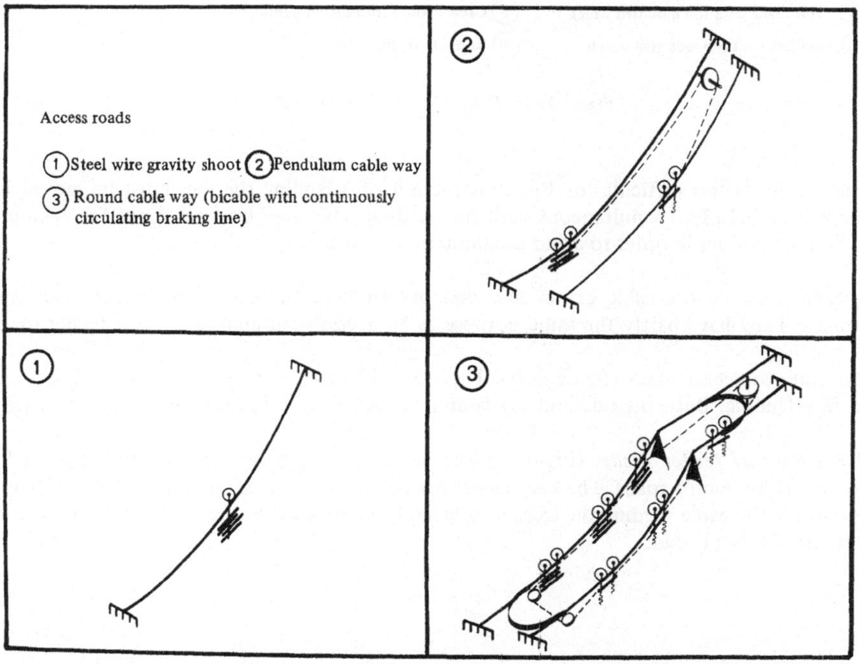

Figure 167. The cable ways' access roads.

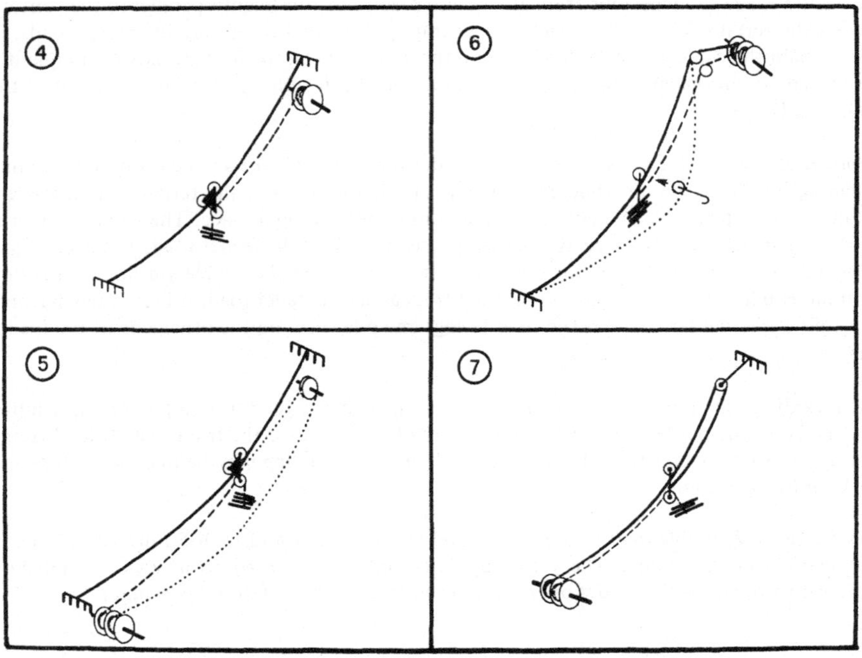

(4) Cable crane with top mounted winch
(5) Cable crane with valley mounted winch
(6) Slackline with top mounted winch
(7) Running skyline system

Figure 168. The cable ways' feeder roads.

If the terrain is too difficult for the construction of a landing the logs must be moved to another terminal place simultaneous with the yarding. This may be carried out for example by a grapple-skidder in order to avoid accumulation of logs near the yarder.

The cable ways or the cable cranes branches out from the access roads, feeder roads and terminals. They have partly the same purpose as an access road and partly as a feeder road.

Cable ways as access roads (figure 167). These are the cable ways which move logs from a pile to a landing. Both the pile and the landing represents the terminal places in this case.

Cable cranes as feeder roads. (Figure 168). The cable cranes branch out from the feeder truck roads or tractor roads. They are feeder roads of secondary order, as illustrated in figure 168. During the cable yarding the trees may be collected from the entire area on both sides of the truck or tractor road.

10.0. CABLE WAYS AS ACCESS ROADS

The purpose of these cable ways is to transport timber from one point on a high mountain plateau to another point usually located in the valley. The skyline represents a simple transport line in order to overcome very difficult terrain (figure 191). Most of the access cable ways are stationary and the investment is written off over a long period. These cable ways represent the access to plateaus with large amount of mature timber. The logs may be yarded to the mountain station of the cable way by tractors, cable cranes etc.

Some of the light cable ways are movable and give access to rather small plateaus with mature forest which are distributed over a steep hillside. The cable way is moved from one plateau to another. Today a cable crane would replace the movable cable ways of this type.

In most of the cases where the former access cable ways were used one would today build an access truck road, since this would not only cover the transport of mature timber, but also enable other operational activities such as cultivation, thinnings etc.

Under very difficult terrain conditions rocks, boulders, cliffs etc. may hinder road construction. Today helicopters are sometimes used for the timber transportation. Some of the new hybrid vehicles such as the Helistat or the Cyclo-Crane may possibly be used in such cases in future. The same is the case when road building is prohibited due to environmental concern and political solutions.

In the end access roads are always better solutions. Roads facilitate future silvicultural work as well as the work needed to maintain and protect the forest environment and the nature.

10.1. Wire or wire rope for gravity transport

This is the simplest form of cable way used as access road (figure 4). It is a skyline only, without any braking line or haul-back line to control the speed. Single steel wires with 5 mm diameter were used first and later on wire rope, usually 6x7+1 was used. Instead of carriages, wooden hooks were used for transport of fuel wood or small sized wood from thinnings downhill to a landing in the valley. The heat which is generated from friction between the wooden hook and the wire rope carbonizes the wooden surface. This reduces the friction due to the lubricating effect of the carbon. The wooden hook is reliable, and may be re-used several turns. A steel hook would be cut and destroied after a short running distance along the skyline. The unloading may be carried out by a throw-off arrangement attached to the skyline or manually after the load has stopped where the skyline flattens out.

This type of cable way is seldom of interest any longer.

10.2. Stationary pendulum cable way

The stationary pendulum cable way was used when large volumes of timber had to be transported from a mountain forest to the valley. The cable way used to stay at the same location for several years.

The pendulum cable way transports the timber down the skylines by gravity. It consists of two skylines running parallel to each other (figure 169). At the mountain station the distance

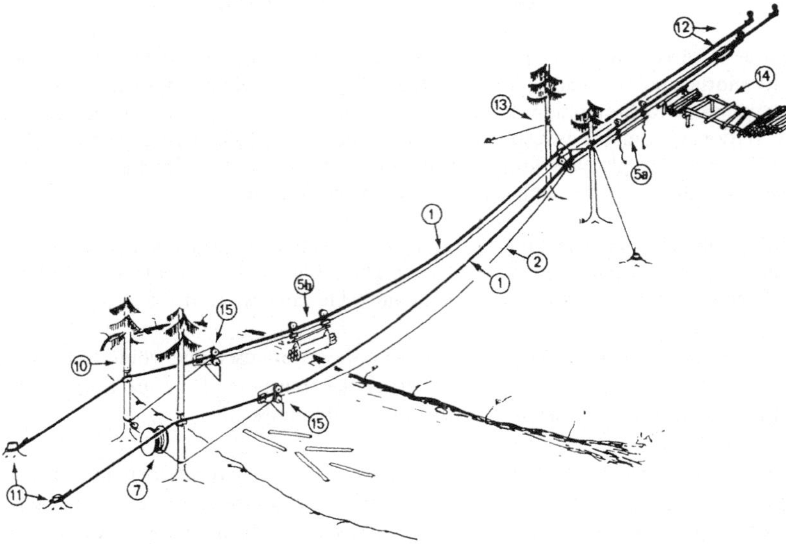

(1) Skyline (2) Braking line (5a) Empty carriage (5b) Loaded carriage
(7) Braking capstan (10) Tail spar (11) Lower anchorage (12) Upper anchorage with tail block
(13) Support (14) Loading platform (15) Carriage stop for automatic unloading.

Figure 169. Stationary pendulum cable way.

between the two skylines is 0.5 – 1.0 meter in order to load the two carriages from the same pile. There is one carriage on each skyline. They are connected to an endless line and the operator controls the speed by a brake (figure 170 and 171). While the loaded carriage runs downhill by gravity it pulls the empty carriage uphill by means of the endless line. As soon as the loaded carriage reaches the unloading point (the landing) the empty carriage arrives at the loading platform. The operator keeps the two carriages in position until the loaded carriage is unloaded and the empty carriage is filled with a new load, whereafter this moves downhill by gravity to the landing and brings the other carriage uphill to the loading platform.

10.2.1. The equipment

It consists of two skylines of the same dimension. 19 wires, galvanized spiral cable has been used. When the normal size of the load is 1 cu.m. (solid) the diameter of the skyline is d = 13 mm and the cross-section = 100 mm^2 · σ_{Br}= 1 420 N/mm^2 and the breaking load S_{Br} = 134.4 kN.

The endless braking line passes a tail block at the top end of the cable way and a brake at the valley station. The dimension is d = 10 mm 6x19+ 1.With a breaking stress σ_{Br} = 1.570 N/mm^2, the breaking load is S_{Br}= 56.9 kN.

These dimensions are suitable for a normal load of 10 kN and a maximum load 20 kN.

Figure 170. The pendulum cable way at Spjotsodd, 1951.
A. View from the valley station towards the top of the cable way.
B. The brake (capstan brake) at the valley station of the cable way.

Some constructions uses a capstan brake (figure 170) and others a grooved wheel brake (figure 171). The brake has to be air-cooled and should always have a safety brake beside the brake normally used during the operation. The best results are obtained when there are no supports along the skyline as illustrated in figure 178. Many places supports are needed (figure 179). In this case a cable shoe arrangement as illustrated in figure 172 has given good results. Since the endless line is lifted clear of the ground, it is possible to pass the support at high speeds, normally between 5 and 6 m/sec.

Figure 171. The groove wheel brake at the pendulum cable way in Voss, 1952–53.

Figure 172. The pendulum cable way in Voss, 1952–53.
A. and B. Portal support.
C. Crossline support.

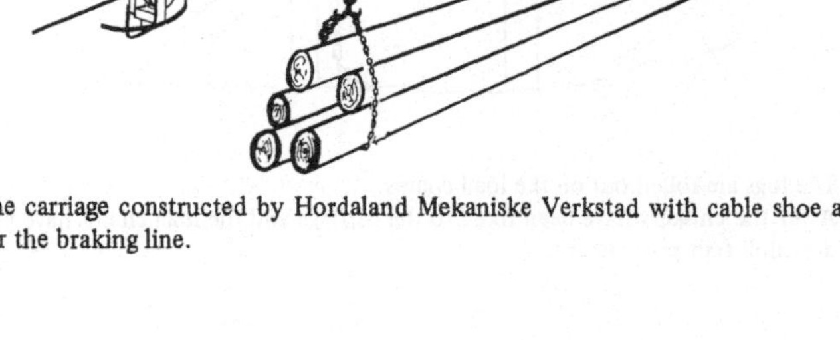

Figure 173. The carriage constructed by Hordaland Mekaniske Verkstad with cable shoe and carrying wheel for the braking line.

10.2.2. The operational method

The normal crew is four men:

1 brake operator
2 choker setters at the loading platform
1 chaser man at the landing.

The organization of the work on the loading platform at the mountain station is the deciding factor for the production. It is of importance to plan the installation and the work in such a way that it is a short distance between the pile and the skylines. The timber may be transported to the pile for example by tractor or cable crane.

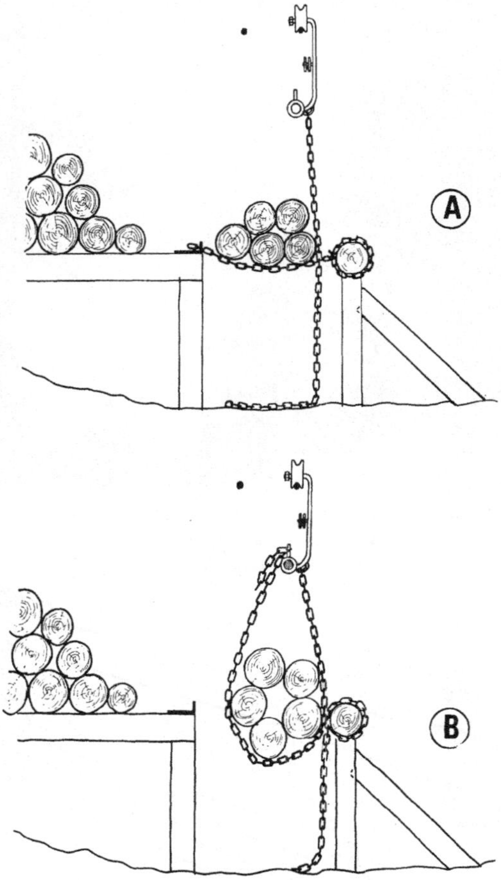

Figure 174. A. The logs are rolled out on the load chains.
B. After the chokers have been fixed to the carriage and the load chains have been released, the downhill transport starts.

Figure 175. The top station of the pendulum cable way in Spjotsodd, 1951.
A. and B. While the empty carriages return from the valley station, the logs are rolled out on the load chain.
C. and D. After the load has been fixed to the carriage by the chokers, the load chain is released and the downhill transport can start.

An arrangement as shown in figure 174 has given good result. While the loaded carriage moves downhill and at the same time pulls the empty carriage uphill, the two choker setters are rolling logs from the pile to the load chains. When the empty carriage arrives at the loading platform the two chocker setters have prepared the new load. While the carriage stands still the two workers put the chokers around the load and release the load chain. This takes approximately the same amount of time as the chaser man takes to unhook the chokers at the landing, and the downhill transport of the new load can start.

In many instances it is possible to use a carriage which automatically unhooks the load when it hits the carriage stop at the landing (figure 176 and 177). The arrangement saves unloading time and increases the efficiency. Usually no man is needed at the landing,so the total crew is 3 men: 1 brake operator at the landing and 2 loaders at the loading platform.

The speed is controlled by the brake operator. Figure 178 shows the average speeds for a 1 210 meter long cable way with a height difference of 403 meter, and no supports in the span. Figure 179 shows another example. This pendulum cable way is 1.250 meters long with a height difference of 435 meters. There are 12 supports along the skyline. In both cases the speed was normally between 5 and 6 meters/sec. and was seldom higher than 8 meters/sec. The driving time with load and the time to return the empty carriage from landing to the pile had no influence on the production. The chaser man usually used shorter time for unhooking than the choker setters to prepare the load at the mountain station. The organization of the work at the loading platform, therefore, should be done in such a way that the two choker setters are ready with a new load as soon as the empty carriage return to the platform. Table 25 gives an example of the distribution of the loading time on elements time. In addition to these effective times comes the delay times such as waiting for carriage, maintenance etc. (This amounted to 38% of the effective time in the Spjotsodd operation).

Figure 176. The carriage constructed by Sørkås for automatic unloading of the logs into the floatway approaches the carriage stop.

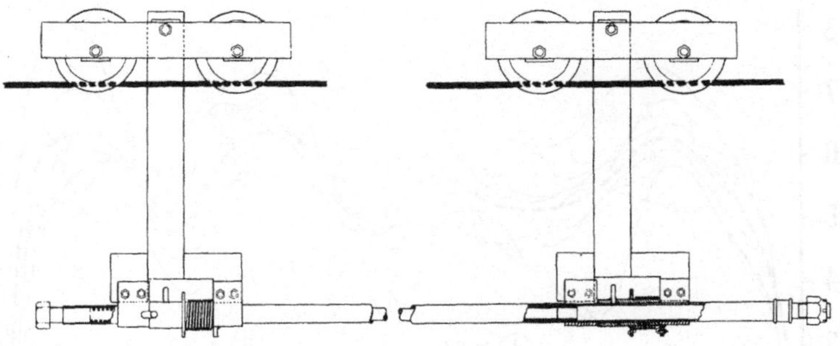

Figure 177. The carriage for automatic unhook on pendulum cable way constructed by Sørkås.

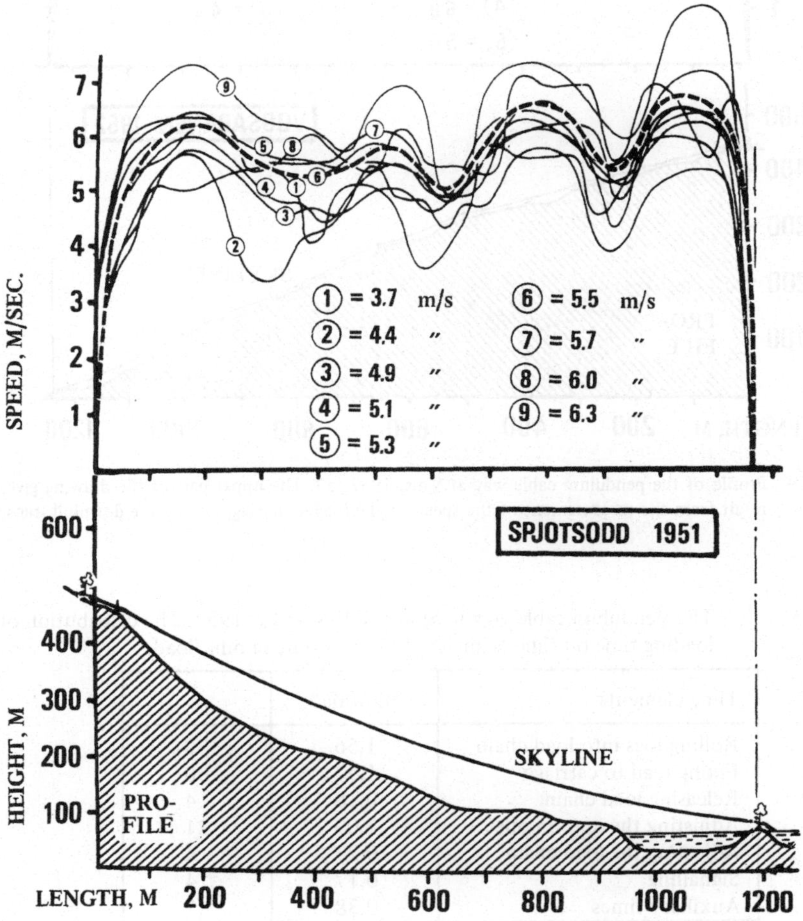

Figure 178. Profile of the stationary pendulum cable way in Spjotsodd, 1951. The upper part of the drawing shows the results from the measurements of the carriage speed during downhill transport.

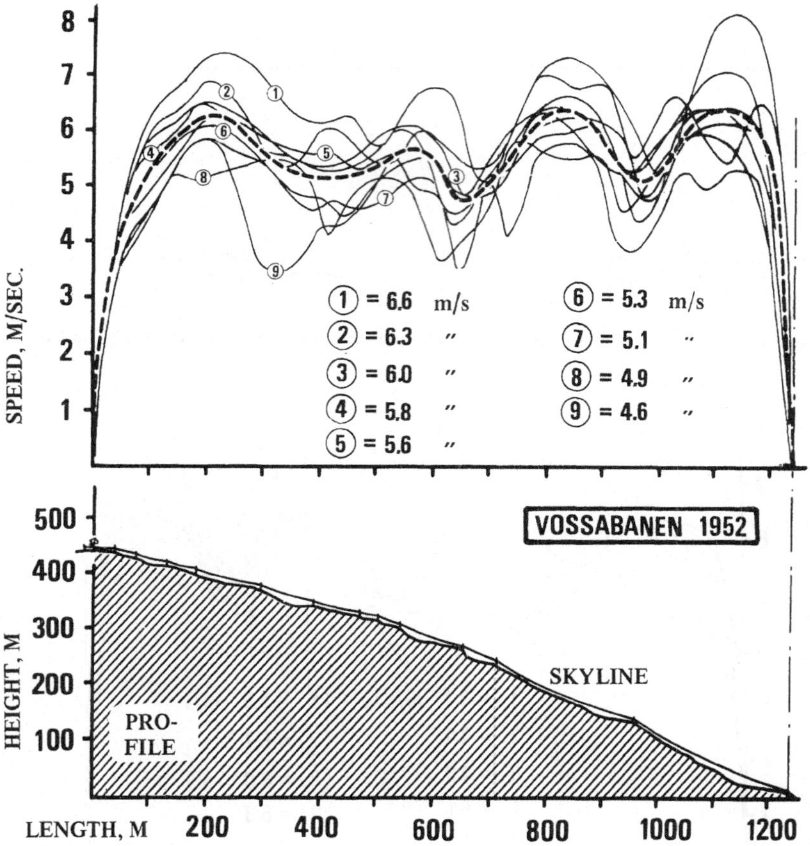

Figure 179. Profile of the pendulum cable way at Voss, 1952–53. The upper part of the drawing gives the result from the measurement of the speed of the loaded carriage during the downhill transport.

Table 25. The pendulum cable way in Spjotsodd, Norway 1951. The distribution of loading time on time elements. Effective time in min./load.

Time elements	Min./load	%
Rolling logs into load chain	1.56	30
Fixing load to carriage	1.57	30
Releasing load chain	0.18	4
Adjusting the logs on pile	1.09	21
Adjusting the load chain	0.23	4
Signalling	0.17	4
Auxiliary times	0.38	7
Total	5.18	100

Table 26. Performance in m^3/hr. of the work place time.(Effective time + delay time). 1 m^3/load. Spjotsodd, Norway 1951.

Rolling distance	m^3/log			
m	0.1	0.2	0.3	0.4
2	9.5	9.8	10.2	10.7
4	8.0	8.2	8.5	8.8
6	6.2	6.5	6.7	6.8

): 3 m^3/man/hr. 0.33 hr./m^3.

10.2.3. The production

Table 26 gives an example of an expected production based on time studies in Spjotsodd. The rolling distance is the distance from the pile to the carriage. The performances is given in cubic meter per hour of the work place time. With 6.5 work place hour per day - the daily production would be 45 – 50 cu.m. per day, and with 8 work place hours per day - the daily production would be around 65 cu.m. per day.

These production figures are typical for 1 cu.m. loads, when the tree dimensions varies between 0.1 and 2.0 cu.m. per tree. A pendulum cable way has high productivity. When transporting timber from bigger tree dimensions the production may increase linearily with the load size, if the cable, the brake, the anchors etc. are dimensioned according to the weight of the loaded carriage.

The production for a stationary pendulum cable way varies little with the transport distances between 300 and 1.200 meters. The productivity for the transportation of logs from the pile to the landing is 3 cu.m./man/hour of the work place time, which corresponds to 0.33 man-hours per cu.m.

The pendulum cable way is a simple way to overcome extreme terrain difficulties. If it is too difficult to build an access truck road to a mountain plateau a pendulum cable way may be used, for example in combination with an articulated frame steered skidder working on a plateau. In this case the brake should be placed at the top end of the cable way close to the landing platform. The carriages should have an automatic unloading arrangement in order to avoid a chaser man at the landing.

The tractor operator may haul the load from the forest to the top end of the cable way. At the loading platform of the cable way he transfers the tractor load to the carriage on the cable way. He then transports the load from the top end to the valley by controlling the brake on the cable way. He stops the carriage at the landing where the load is automaticly unhooked. Due to the pendulum system, the empty carriage has arrived at the loading platform on the other skyline. The operator fixes the brake and returns with his skidder back to the forest to fetch a new load. There is no extra worker on the pendulum cable way since this is controlled by the tractor operator.

10.3. Movable pendulum cable way

The brake is placed at the top end of the cable way and close to the loading platform as illustrated in figure 180. The two skylines have different dimensions, one heavy for down-hill load transport and one light for returning the empty carriages uphill. The crew has to lift the carriages from the heavy to the light skyline after unloading at the landing and from the

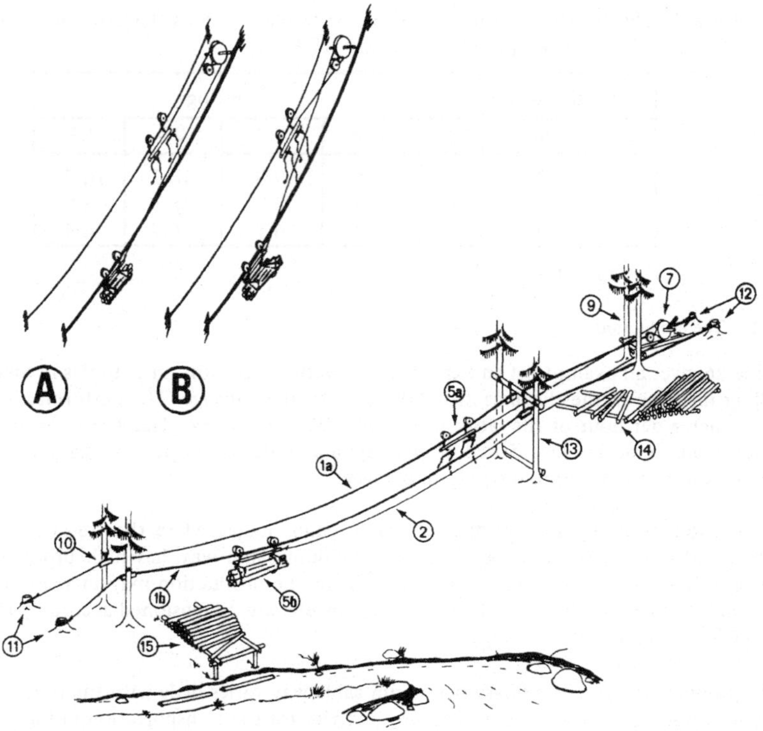

(A) First load: The braking lines are parallel to each other

(B) Second load: The braking lines are crossing each other.

(1a) Light skyline (return cable) (1b) Heavy skyline (loadcarrying cable) (2) Braking line

(5a) Empty carriage (5b) Loaded carriage (7) Groove wheel with brake (9) Head spars

(10) Tail spars (11) Lower anchorage (12) Upper anchorage (13) Support

(14) Loading platform (15) Log deck

Figure 180. Movable pendulum cable way.

light to the heavy skyline before loading at the loading platform. The braking line is not endless, it runs from one carriage on one skyline to the top mounted brake and back again to the carriage on the other skyline. On one turn the braking lines runs parallel to each other. On the next turn, the braking lines cross each other. This prohibits the use of supports along the cable way and the movable pendulum cable ways can only be used where there is a free span from the landing in the valley to the mountain station, with the exception of a portal support at the edge of the cliff (figure 180).

The movable pendulum cable ways have been used to transport wood from small plateaus on a mountain side to the valley. It is useful for distances from 100 to 500 meters, and for tree dimensions up to 2 m^3/tree.

10.3.1. The equipment

This type of cable way can be used for loads up to 10 kN. Since the cable way must be easy to move from one place to another, wire ropes instead of spiral cables are used as skylines.

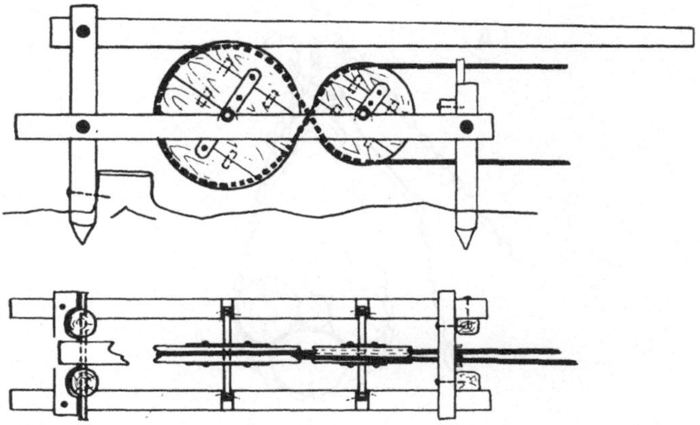

Figure 181. In the grooved wooden wheel, the fibres were placed perpendicular to the braking line.

Figure 182. Portable brake for movable pendulum cable way, Luksefjell 1953.

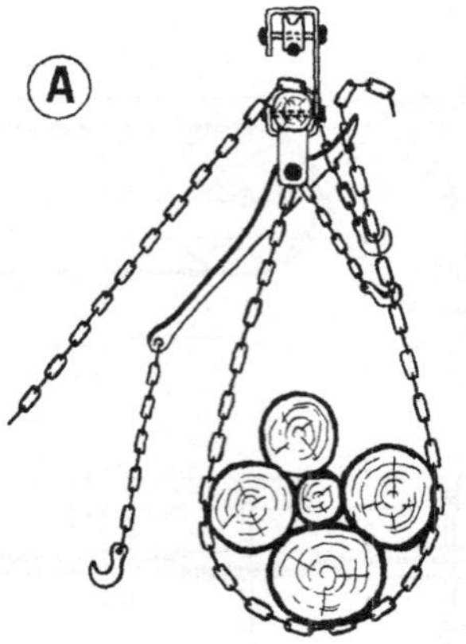

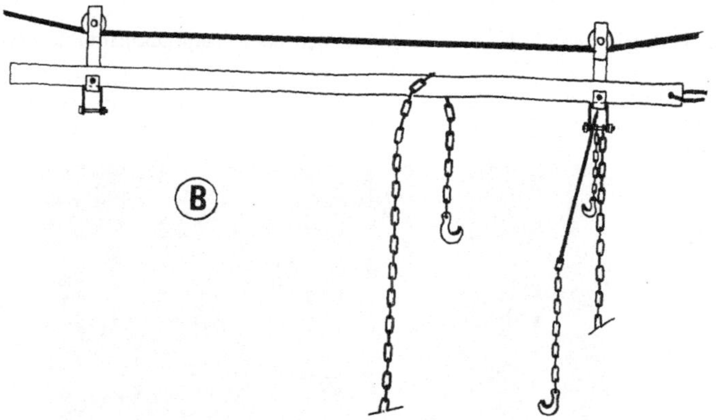

Figure 183. A. Beartrap binder for lifting the load from the ground to the carriage.
B. Light carriage for movable pendulum cable way.

Both the skylines and the braking line may have a steel quality equivalent to a braking stress of $\sigma_{Br} = 1{,}770\ N/mm^2$

The skyline:	d = 12 mm	6x 7+1	$S_{Br} = 84.7$ kN
The haul-back skyline:	d = 7 mm	6x 7+1	$S_{Br} = 28.8$ kN
The braking line:	d = 7 mm	6x19+1	$S_{Br} = 28.7$ kN

These dimensions can only be used where there is good deflection of the skyline. If there is problems with the clearance and the skyline has to be tightend, somewhat bigger dimensions should be used.

The top mounted brake may be constructed of wood as illustrated in figure 181. In the grooved wooden wheel the fibres must be placed perpendicular to the braking line in order to ensure even wear over the surface. A light portable brake is more commonly used (figure 182). The brake must be placed at a certain distance from the edge of the cliff (figure 184). This will give room for the pile of logs to be transported. At the loading platform the two skylines

Figure 184. Arrangement at the top of a movable pendulum cable way (Valebø, 1951).

must be rigged up parallel to each other and with a ground clearance of approximately 1.5 – 2.0 meters in order to facilitate the loading. The carriages are light and easy to carry in the terrain (figure 183). The front and rear part of the carriage is kept apart by a small log.

10.3.2. The operational method

The crew consists of three men:

At the loading platform:
1 brake operator (the foreman)
1 loader.

At the landing:
1 chaser man.

It is not possible to use automatic unloading with this type of cable way since the carriages have to be lifted between the two skylines after each turn.

When the empty carriage has reached the loading platform at the top end the brake operator and the loader lift the carriage from the return skyline to the heavy skyline. They roll out logs for a new load and fix them to the carriage by the chokers.

Figure 185. Movable pendulum cable way in Luksefjell, 1953.
A. Unloading at floating river.
B. Unloading at forest road.

In the meantime the chaser man at the landing unhooks the logs from the carriage. Afterwards he lifts the empty carriage from the heavy skyline to the return skyline.

As soon as the new load is ready at the top station, the brake operator controls the brake while the loader helps signalling during the downhill transport of the load.

10.3.3. The production

As an example of a production which may be expected we may refer to a production study in Luksefjell, Telemark 1953. The length of the cable way was 352 m and the the height difference was 82 m (figure 186). The average log size was 0.2 m^3/log and the average load, approximately 0.5 m^3/load. The average driving time (table 27) was 1.02 minutes corresponding to an average of 5.8 m/sec. The highest speed in the middle of the span was 9 m/sec.

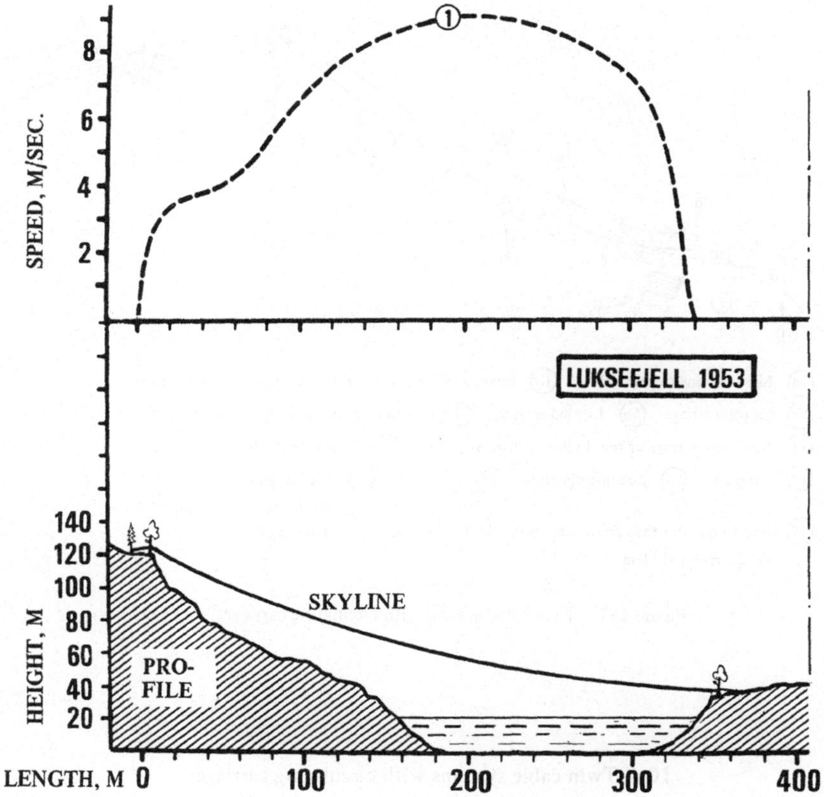

Figure 186. Profile of the pendulum cable way in Luksefjell, 1953. The upper part of the drawing shows the result from the speed measurements.

Table 27. Movable pendulum cable way (length 352 m). The distribution of the loading time on the time elements. Effective time (E_0) in min./load.

Time elements	Min./load	%
Rolling logs and moving carriage	0.58	21
Fixing the load	0.72	26
Starting (pushing) the load	0.35	13
Transport time	1.02	36
Receiving empty carriage	0.12	4
Total	2.79	100

The effective time per load was 2.8 minutes. When adding the delay times the total work place time per turn was 3.6 minutes. Since the average unloading time was 1.9 minutes per load it was the arrangement at the loading platform which limited the production. The average production was 7.3 m^3/hour of the work place time. This corresponds to 2.4 m^3 /man/hour and a productivity of 0.41 manhours/m^3 during the work place time.

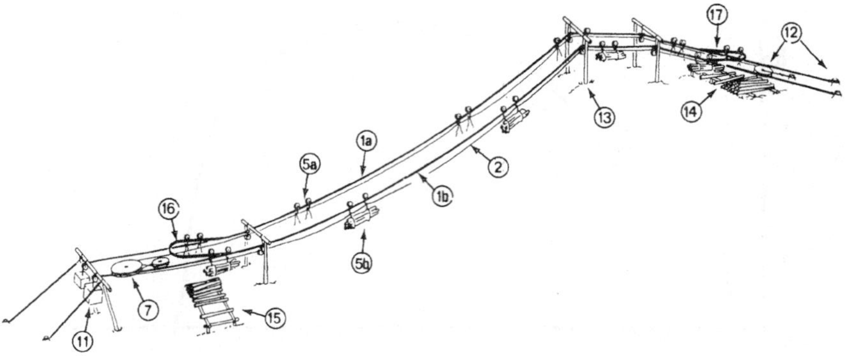

(1a) Light skyline (return cable) (1b) Heavy skyline (loadcarrying cable) (2) Braking line
(5a) Empty carriage (5b) Loaded carriage (7) Powered groove wheel with brake
(11) Lower anchorage of the skyline with gravity load (12) Upper anchorage with tail block
(13) Supports (14) Loading platform (15) Log deck (16) Rail to transfer the carriage from the loadcarrying to the light skyline
(17) Rail to transfer the carriages from the light to the loadcarrying skyline

Figure 187. Twin cable system with circulating carriages.

10.4. Twin cable systems with circulating carriages

Twin cable way with circulating carriages is the most stationary among the cable ways used as access roads and it has been installed where large volumes of timber had to be transported from a plateau or a high forest area during a period of years. In the cases where the twin cable system were rigged up, one would usually construct a truck road to-day.

A stationary pendulum cable way transports one load at the time, and high speeds, up to 10 m/sec., have to be used in order to achieve a sufficient production. On a twin cable system the driving speed is low, usually 0.5 – 1.0 m/sec. The low speed is compensated for by the many loads after each other on the skyline. The distance from one carriage to another is short (figure 187).

Usually the loads are transported downhill from a high plateau and the cable way is powered by gravity. The many carriages along the skyline makes it possible to overcome a height difference, for example over a ridge if the loading platform is located higher than the landing. The many loads on the downhill stretch are pulling the smaller number of loads running uphill and over the ridge (figure 187)

The twin cable system has normally two skylines, one heavy skyline for the transport of the loaded carriages and a haul-back skyline on which the empty carriages return from the landing and back to the loading platform. Usually the skyline tension is kept constant by means of a cable tension weight (figure 189).

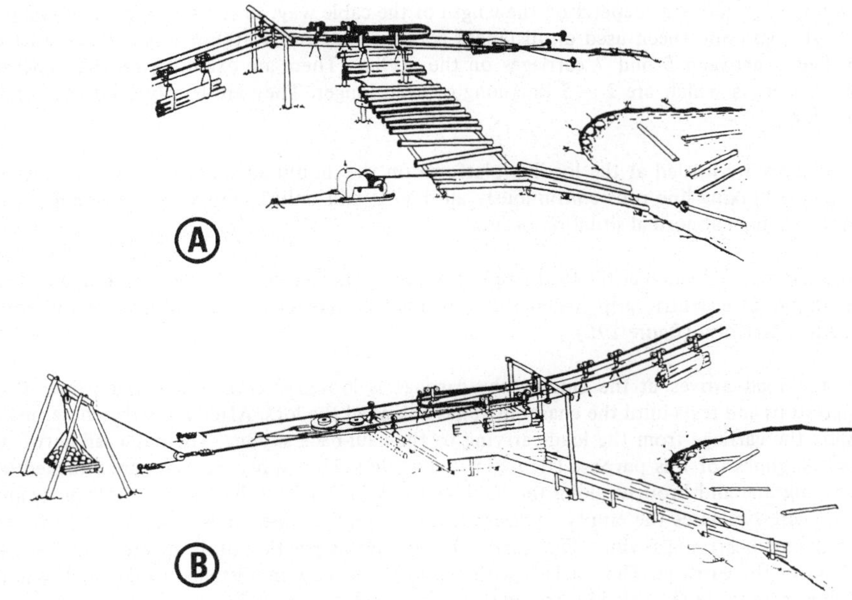

Figure 188. A. Arrangements at the top of a twin cable system with circulating carriages (the loading platform).
B. Arrangements at the valley station of a twin cable system with circulating carriages. After unloading the logs pass a dry chute down to the floatway.

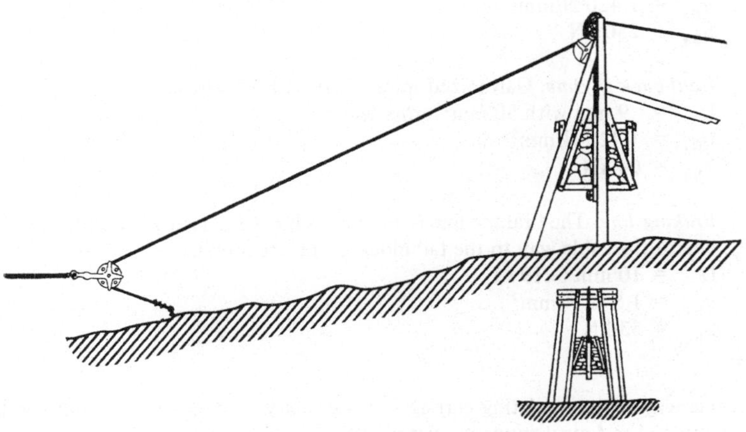

Figure 189. Tightening the skyline with cable tension weight. The twin cable system with circulating carriages from Stengestad, 1953.

The number of carriages depend on the length of the cable way. In forestry this cable system has most commonly been used on distances between 800 and 1,500 meters and there have often been between 5 and 7 carriages on the skyline. There are examples on this type of twin cable ways which are 2 – 5 kms long or even longer. They are hardly used any more in forestry.

The carriages are loaded at the loading platform on the mountain station of the cable way. The endless braking line runs continuously all day. It will only be stopped occasionally, for example when an abnormal situation occurs.

When a carriage is loaded at the loading platform it is attached to the endless braking line, by for example an eccentric grip, when the previous load has reached a certain distance from the loading platform (figure 195).

When the load arrives at the landing the carriage is loosened from the braking line. The carriage stops and rests until the chaser man has unloaded the logs. Afterwards the chaser man transfers the carriage from the load-carrying to the haul-back skyline. There is a rail between the two skylines for this purpose (figure 187 and 196). The empty carriages are attached to the braking line and brought along the haul-back skyline back to the loading platform again. Here the crew releases the empty carriages from the endless line, pulls them around on the rail to the load-carrying skyline. The carriage is kept in this position until the crew has fixed a new load to the carriage. The carriage with the load can now be fastened to the endless line and the transport along the load-carrying skyline to the landing can begin.

10.4.1. The equipment

For cable ways between 800 and 1,500 meters, carrying 5 – 7 loads of 10 kN the following lines have been used:

Load-carrying skyline. Galvanized spiral cable with 19 wires.
D = 18 mm and 190 mm² cross-section.
σ_{Br} = 1,420 N/mm²
S_{Br} = 250 kN

Haul-back skyline. Galvanized spiral cable with 19 wires.
D = 9 mm with 50 mm² cross-section
σ_{Br} = 1,420 N/mm²
S_{Br} = 64 kN

Braking line. The braking line is an endless line which runs from the brake at one end of the cable way to the tail block at the other end.
D = 10 mm 6x19+1
σ_{Br} = 1,570 N/mm²
S_{Br} = 53 kN

The twin cable way with circulating carriages is stationary, and so are the supports. Usually they are constructed as 4-strut supports (figure 198).

On the Norwegian twin cable ways the motor-driven brake of traditional type had a grooved wooden wheel pulley as illustrated in figure 192. If the cable way passes a ridge such as in Stengestad, Telemark, the pulley on the brake has to be powered by a 10 kW motor, in order

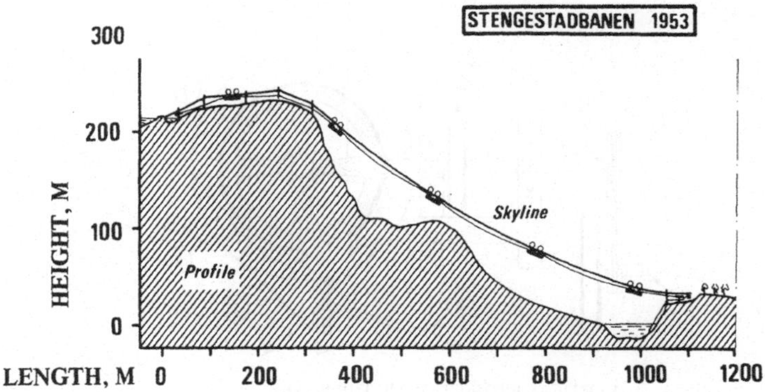

Figure 190. The profile of the twin cable with circulating carriages at Stengestad, 1953. (The scale of the height is twice the scale of the length).

Figure 191. The twin cable system with circulating carriages in Brakandalen, Telemark 1954–62.

to start the transportation. As soon as the loaded carriages were distributed along the skyline, gravitational forces took over and the motor could be stopped.

Figure 193 illustrates a grooved wheel brake used on a twin cable system with circulating carriages in Telemark. Figure 194 shows another interesting construction. Instead of a set of two grooved wheels (see figure 54) a grooved chain pulley around a barrel is used to pull or to brake the endless line. The barrel is shaped as a wide toothed wheel. Each link in the chain is adapted to the side grooves and may slide sideways when the barrel rotates. The grooved chain pulley runs 6 times around the barrel, the tail end of the chain passes a guide wheel at a distance from the barrel and is steered back again to the other end of the barrel. The endless line which shall be pulled or braked by the pulley lies on the grooved chain in as many turns as there are turns of the chain around the barrel (figure 194).

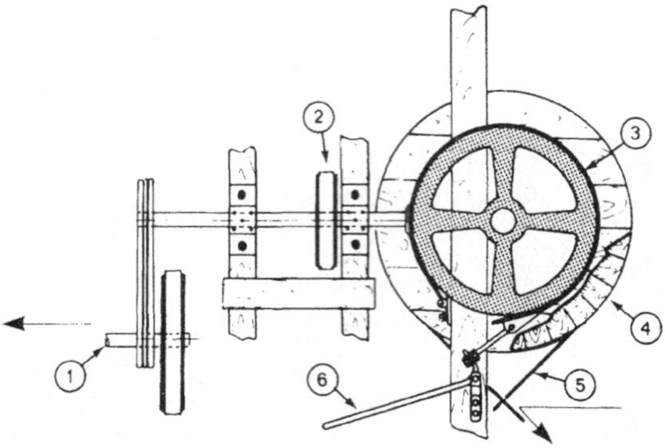

VERTICAL PROJECTION OF THE BRAKE

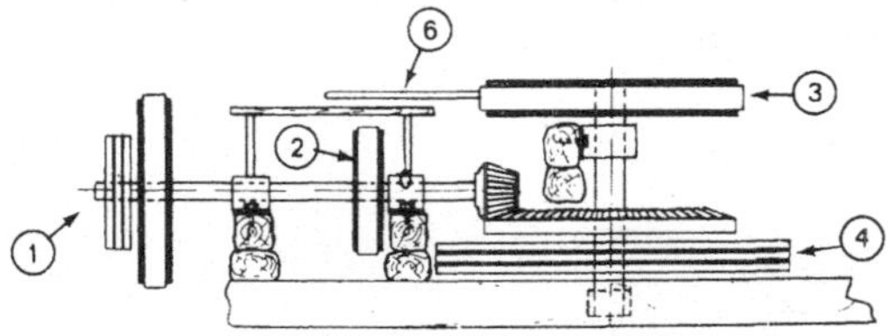

HORIZONTAL PROJECTION OF THE BRAKE

(1) Power take-off from motor (2) Auxiliary brake (3) Main brake (band brake)
(4) Wooden groove wheel (end wood outwards against the rope) (5) Endless braking line
(6) Brake lever

Figure 192. Motor-driven brake of traditional type for twin cable system with circulating carriages. In the grooved wooden wheel the fibres were placed perpendicular to the braking line in order to ensure even wear over the surface.

Figure 193. The twin cable system with circulating carriages in Brakandalen, Telemark 1954–62. The brake at the top of the cable way.

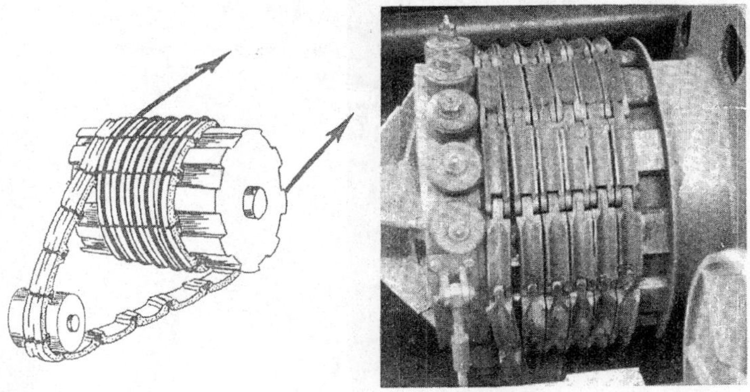

Figure 194. Grooved chain pulley, used at a twin cable system, Caucasus, USSR 1957. The chain at the right end of the barrel passes a guide-wheel and runs back again to the left end of the barrel. The endless line runs on the grooved chain.

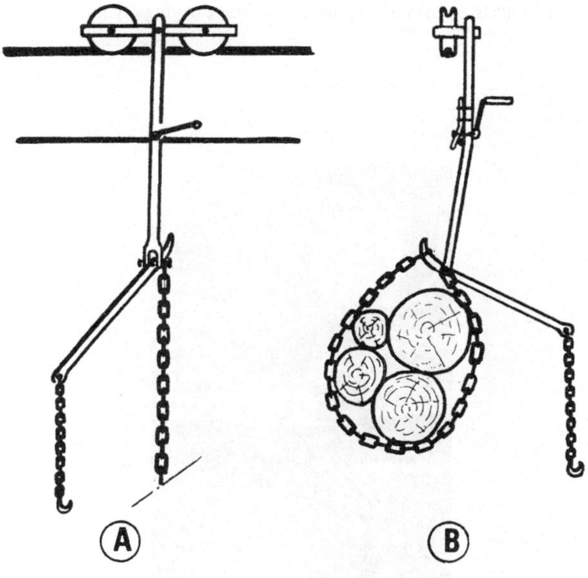

Figure 195. Carriage for a twin cable with circulating carriages.

10.4.2. The operational method

The cable crew needed to transport logs from a high location, sometimes over a ridge, to a landing in the valley consists of four men:

1 foreman who operates the brake
2 loaders on the loading platform (at the mountain station)
1 unloader (chaser man at the landing)

Figure 196. The top station at the cable way in Brakandalen, Telemark 1954–62. The carriages are transferred on the rails from the return cable to the load-carrying skyline.

Figure 197. The tail block at the top of a cable way.

On the loading platform at the mountain station the two loaders are using the time between two loads to roll the logs on the loading bunks (figure 199). The work is organized in such a way that there are always a supply of empty carriages on the rail at the loading platform (figure 196). One carriage consists of a front carrier and a rear carrier (figure 195). They are kept apart by the logs. The two loaders set the chokers from each of the two carriers around the load. The loading bunks are released and the load is hanging freely under the carriage.

Figure 198. The twin cable way with circulating carriages. Stengestad, Telemark, 1953.
A. Loading platform.
B. When the load bunks are released, the load hangs in the carriage.
C. The load on the load bunks before it is released.
D. and E. The load moves uphill on the skyline from the Stengestad lake.

During the loading the braking line moves continuously. As soon as the previous carriage with its load has reached a certain distance from the loading platform an automatic signal is given and the loader fastens the front carrier to the braking line by means of the eccentric grip or a screw grip. The load starts to move, pulled by the braking line, and the loader fastens the rear carrier of the carriage to the braking line. The load is then ready for the transport to the landing while the loaders can continue to prepare a new load.

At the unloading place (the landing) the chaser man receives the load (figure 200). He loosens first the front carrier (by releasing the eccentric grip) and afterwards the rear carrier of the carriage. The load is now resting at the landing and the chaser man can unhook the chokers in order to unload the logs. At the Stengestad cabel way the logs were dropped directly into a dry chute down to the Blestad lake in the floatway.

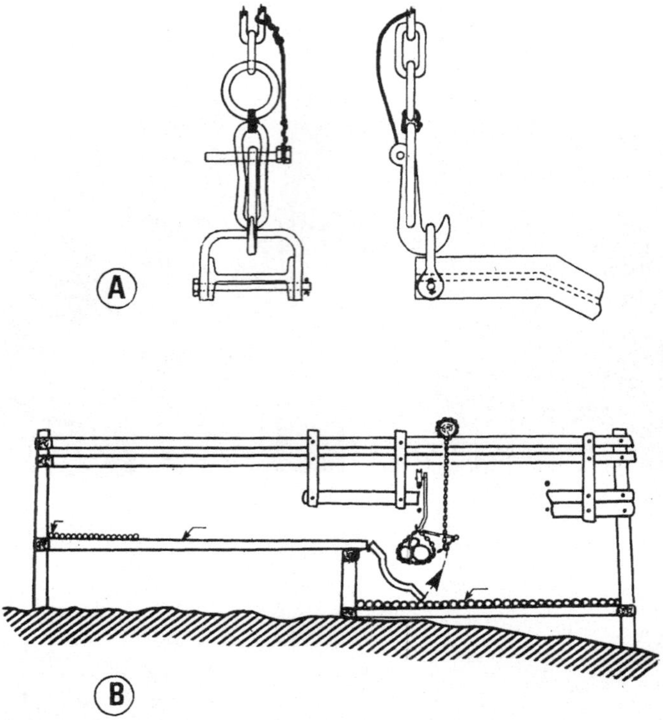

Figure 199. The loading platform at Stengestad. A. Arrangement for releasing the load bunks.
B. End view of the loading platform.

When the unloading was finished the chaser man transferred the empty carriages around the rail to the return skyline (figure 200). He fastened the front carrier and the rear carrier of the carriage to the braking line by means of the eccentric grip. The braking line pulls the empty carriages back to the mountain station along the return skyline.

10.4.3. The production

As an example of the productivity of a well organized operation may be mentioned the twin cable way with circulating carriages at Stengestad, Telemark. Detailed statistics for the operation were collected in the period 1953 – 1962.

The total length of the cable way was 1,093 m.There were 5 loads on the skyline, (totalling 10 carriages) and the average distance between the loads was 218 m. According to the time studies, the average time between two loads was 3.16 minutes (effective time and delay times) which corresponds to an average speed of 1.15 m per second. Table 28 shows the distribution of the loading time on the time elements. The total effective time was 2.44 minutes per load, the delay time has to be added (29.5% of the effective time).

Figure 200. The cable way from Stengestad lake to Blesa lake, Telemark 1953.

A. Three loads moving on the skyline from the highest point of the cable way to the unloading station.

B. and C. The load is received into a chute and goes by gravity into the Blesa lake.

D. After pulling the carriages around the rail, the worker is fixing them to the endless line. The carriages return to the top station.

Table 29 shows the performances in m^3 per hour of the work place time. With an average load of 0.8 m^3 per load the production of transporting logs from a high location downhill to the landing is 15.2 cu.m.per work place hour.With 6.5 work place hours per day the daily production is 99 cu.m. and with 8 hours per day, 122 cu.m. per day.

The production is 3.75 cu.m. per man per hour, corresponding to a productivity of 0.27 man-hours per cu.m. during work place time.

Table 28. The twin cable way at Stengestad, Norway. The distribution of the loading time on time elements. Effective time in min./load.

Time elements	Min./load	%
Rolling logs into load bunks	0.94	39
Regulating the logs	0.17	7
Adjusting the load	0.27	11
Pushing empty carriages to load bunk	0.20	8
Receiving empty carriages	0.20	8
Fixing load to carriage No. 1	0.37	15
Fixing load to carriage No. 2	0.17	7
Setting up the load bunk	0.12	5
Total	2.44	100

Table 29. Performances in m^3/hr. (under bark, solid volume) at the cable way in Stengestad, Norway 1953 (work place time).

m^3/load	m^3/hr. (work place time)
0.6	13.5
0.8	15.2
1.0	16.3

10.5. Single cable system with circulating main line and carriages

An Ukrainian cable way construction has been used in the Carpatian mountains and in Caucasus in the Soviet Union. This cable way is an access road and the operational method is similar to the cable way described in chapter 10.4.

The construction is different. There is only one skyline which carries the load. The endless main line (braking line) is above the skyline (figure 203). The carriage is fixed to the main line by means of a torpedo-shaped clamp (figure 202). The skyline is supported by an ordinary cable shoe. The main line passes two horizontal sheaves which impact against each other by means of springs. There are several loaded carriages after each other on the skyline in the transport direction. The carriages are carried on the main line during the return trip from the landing and back to the loading platform (figure 201). There must be a certain horizontal angle of the main line at the support in order to keep it in place. The horizontal angle should be between 3 and 15°. In the vertical plane the angle should be less than 6°.

A cable way of this type was in operation in Tioploje near Psebay in Caucasus.The total length was 1,900 meters and the average distance between the supports was 80 – 125 meters. A grooved chain pulley was used as shown in figure 194, and the rope dimensions were as follows:

Skyline: D = 25 mm (6x7+1)
Main line: D = 16 mm (6x19+1)

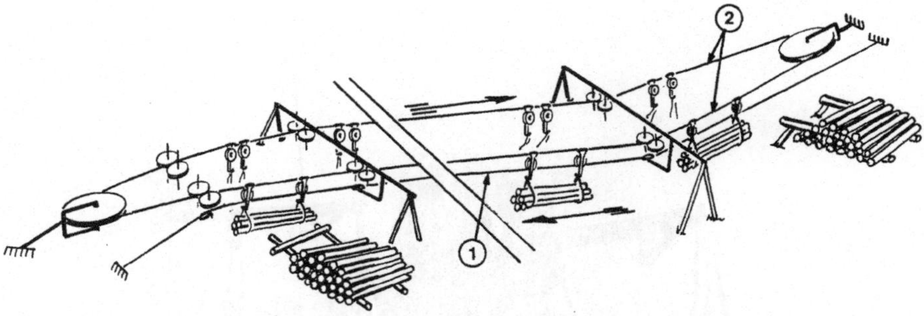

① Skyline ② Endless main/haul-back line

Figure 201. Single cable system with circulating carriages. The endless pulling line is rigged up above the skyline.

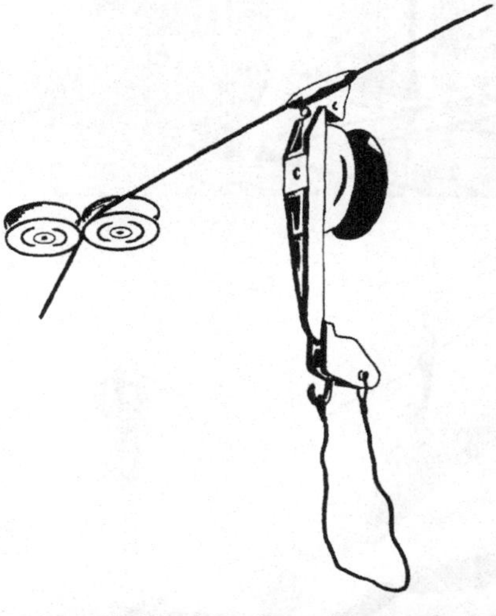

Figure 202. The carriage is attached to the endless pulling line (main/haul-back line) by means of a torpedo-shaped grip which can pass the horizontal sheaves in the support.

The distance between the 1–2 cu.m. loads was 50 meters. The speed was 0.5 meter per second. The production was approximately twice the production of the cable way described in chapter 10.4, mainly because of the bigger timber and loads.

11.0. CABLE CRANES AS FEEDER ROADS

A cable crane collects the timber from the logging area and transport the load to the landing, usually near a truck road or a tractor road.

The transport machine is stationary. It may be a yarder with one or several winch drums mounted on runners. Mobile yarders have winch drums, towers and other auxiliary equipment mounted on wheeled or tracklaying chassis. The mobile yarder may be a trailer or self-propelled. The yarder equipment is often mounted on a tractor or a truck.

Figure 203. The Ukranian cable way in Caucasus, USSR 1957.

The yarder moves the load of logs or stems along the cable corridor on a carriage or a butt rigging. The transport direction from the logging area towards the landing is the *main direction* or *the inhaul direction*. The opposite direction when returning the rigging or the carriage from the landing to the logging area is *the haul-back direction* or *the outhaul direction*. Sometimes the landing is located near the yarder and the inhaul of the load goes in the direction of the yarder. In other cases the winch or the yarder is located at the other end of the cable crane and the direction of the inhaul goes away from the yarder and the logging areas to the landing.

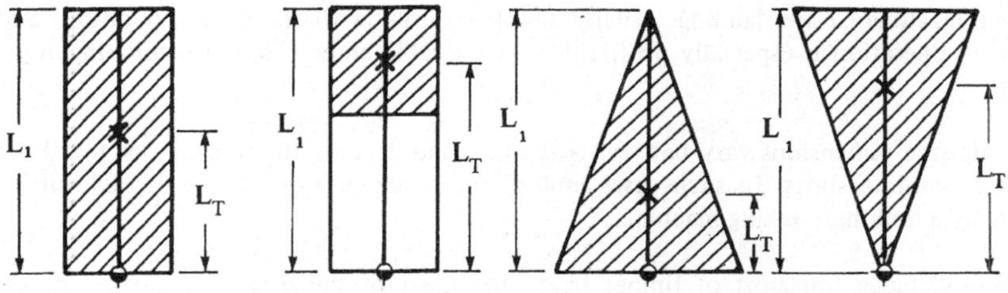

Figure 204. The relationship between the yarding distance and the length of the cable crane.

If the cable slope is steep enough, usually steeper than 50% (27°) the empty carriage or the load may be moved along the skyline by gravitational force.

For downhill transport of timber a *gravity inhaul system* may be used. In this case no main line is needed to move the load from the logging area to the landing.

For uphill transport of the timber, the gravitational force may be used to move the empty carriage from the landing and downhill to the logging area. A *gravity outhaul system* may be used, and no haul-back line is needed.

The length of a cable crane is the total length from the tail spar to the head spar (L_1). The cycle distance, sometimes called the »cydist», is the distance from the logging area where the chokers are set on the logs to the landing. The cycle distance consists of the skidding distance (sk) which is the distance from the place where the chokers are set on the logs to the skyline road (the latteral skidding). The transport distance is the distance along the skyline from the logging area to the landing (L_T). The cycle distance and the transport distance are not always half of the length of the cable crane. This is only the case in a rectangular layout when there is the same amount of timber along the entire length of the cable crane. For a rectangular layout the transport distance is more than 50% of the length of the cable crane if the timber on the area closest to the landing already has been logged for example by tractor (figure 204). If there is more timber near the landing than at the other end of the corridor, the opposite is the case. If the layout of the logging area is fan-shaped or triangular this also influences the relation between the transport distance and the length of the cable crane as illustrated in figure 204.

Most of the earlier cable cranes used *tight skyline*. On modern cable cranes *live skylines* have become more common. In this case the tension of the skyline may vary during the transport in order to keep the carriage closer to the ground. On *running skyline systems* it is the running lines which carry the load. This gives the most flexible type of skidding and transport during the yarding cycle. The latteral skidding is mostly ground or high skidding while the transport along the skyline is partly high skidding and partly cable transport, depending on the construction of the cable crane and the terrain conditions.

In most cases the timber has been felled and converted into logs or stems before the cable operation starts. This is also the case when logging big timber. If the tree dimensions are bigger than 3 cu.m. per tree the felling and conversion should always take place before the cable operation begins.

In small timber, less than 1 cu.m. per tree and when the tree dimensions are less than 0.5 cu.m. per tree especially, it is usually an advantage to combine felling and cable yarding in one operation. The limbing and bucking into assortments is carried out by a limbing/

bucking machine at the landing, usually simultaneously with the cable operation. Such a combined operation is especially profitable when a cable crane is used for thinning in young stands.

When the tree dimensions vary between 1.0 cu.m. and 3 cu.m. the tree may be felled before the cable yarding starts. In such cases limbing and bucking also may be carried out at the landing by a limbing/bucking machine.

The long distance transport of timber from the forest to the forest industry starts at the feder truck roads which branch out over the mountainous forest area. The cable cranes are feeder roads of secondary order and they collect the timber from the area between the feeder truck roads.

The earlier cable cranes were heavy and consisted of cable equipment which was put together in a more or less stationary cable crane system, suitable for the local terrain conditions. Since the rig up time was long, the cable crane should cover as large an area as possible for each set up, with lateral skidding of up to 80 meters or more. Todays cable cranes are more mobile and the rig up time has become more acceptable. A limbing and bucking machine as well as a loader, is part of the harvesting unit, and they are sometimes mounted on the same chassis.

A modern cable crane operation is a harvesting system for the timber on forest areas with difficult terrain between the truck feeder roads. Felling, yarding, conversion and loading are often elements of one and the same continuous transport sequence.

11.1. Yarding production cycle with cable cranes

Since the goal of the cable crane operation is to collect the timber from the logging area and transport it to a terminal point, the landing, each of the loads is a result of several work elements. These follow after each other in a continuous transport sequence which can be called the turn circuit.

11.1.1. The turn cycle

Some of the work elements are *underway operations,* namely the transport of the load along the skyline and the return of the empty carriage. These are marked with U in the following list. Other work elements are *terminal operations* at the loading or the unloading places. They are marked with T in the following list:

U_1 Return of empty carriage
T_1 Lowering the hoist line
T_2 Pulling out the hoist line to the logs
T_3 Setting the chokers
T_4 Moving into safety
T_5 Lateral skidding
T_6 Hoisting the load to the carriage
U_2 Load transport
T_7 Lowering the load to the landing
T_8 Unhooking the chokers
T_9 Hoisting the empty hook to the carriage
U_1 Return of empty carriage.

The underway operations vary with the construction of the cable crane. These work elements will be discussed for each of the systems. The height between the skyline and the terrain is greater for tight skyline systems than for live skyline or running skyline systems. This reduces the time for lowering and pulling in the hoist line. Other work elements such as part of setting the chokers or part of unhooking the chokers may take place while underway operations are performed and this will be discussed in connection with the description of the individual cable crane systems.

The continuous flow of work elements in a transport sequence is often interrupted by delays caused by mechanical breakdowns or caused by bad organization of the work. Terrain difficulties may cause other types of delays, for example when the load gets stuck. Through experience such delays have been reduced considerably and the same is the case with the terminal work elements. The underway elements have become faster with heavier loads.

11.1.2. The chokers and the rigging

It has been most common to use wire ropes as chokers. When yarding small sized wood they may be light, with a diameter 7 to 8 mm in thinnings and the choker setter can easily carry 10 chokers. In big timber the diameter of each choker may be 32 mm or more and the choker setter can only handle one choker. The rigging crew must consist of one man per choker. The simple cable systems such as the highlead system uses butt rigging which is a set of swivels and chain which is used to connect the main line and the haul-back line together (figure 205). The swivels are needed in order to eliminate the torsional forces in the lines. If a carriage is used instead of butt rigging the connection between the haul-back line and the carriage should also be made through a swivel for the same reason.

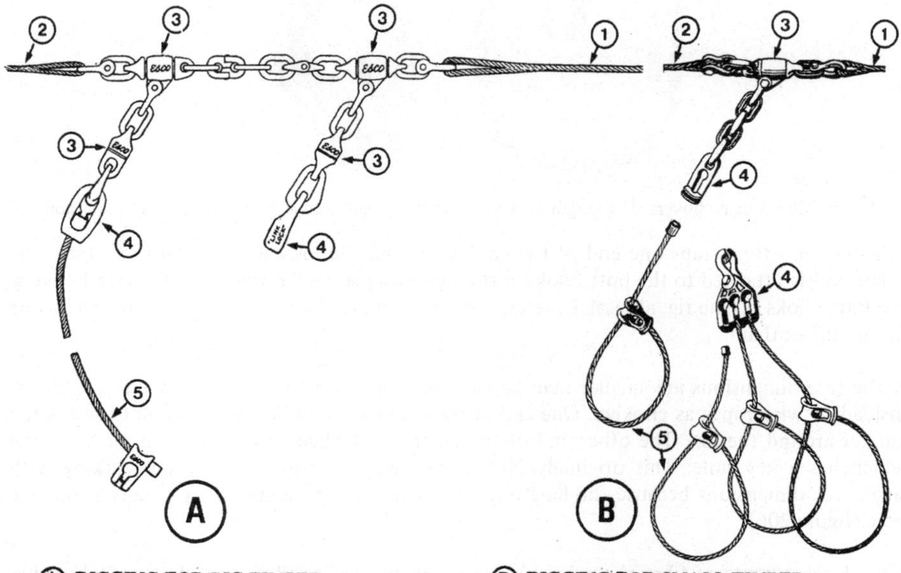

Ⓐ RIGGING FOR BIG TIMBER Ⓑ RIGGING FOR SMALL TIMBER

① Main line ② Haul-back line ③ Swivel ④ Butt hook ⑤ Choker

Figure 205. Butt rigging for highlead yarding.

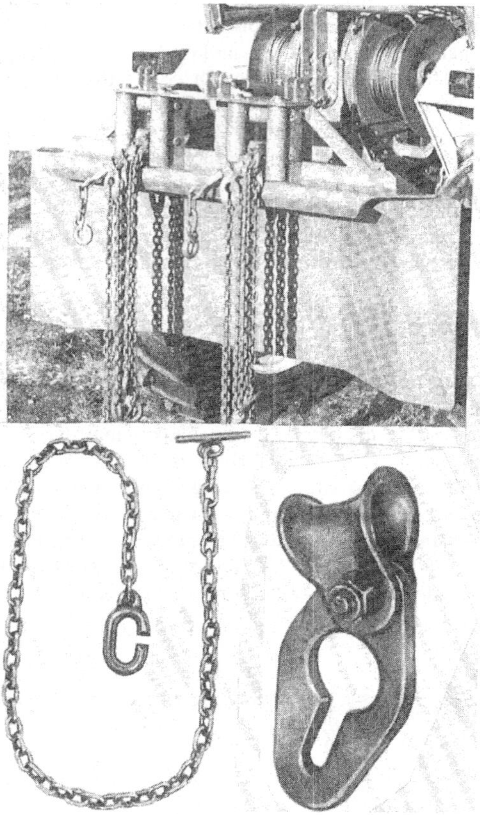

Figure 206. Chain chokers. The length of the choker can be adjusted according to the log dimension.

The choker setter wraps one end of the choker around the log and fastens it to a slider. The other end is attached to the butt hook on the rig. There should always be a distance between the butt hooks on the rig, at least 1 meter, in order to avoid the twisting of the empty chokers during the outhaul.

If the tree dimensions are smaller than 1 cu.m. per tree, it is often an advantage to use chains instead of wire ropes as chokers. One end of the chain has a hook which is used to fasten the choker around the log. The other end of the chain is attached to the slider on the main line which has a key hole. This originally Norwegian design is convenient when working with small tree dimensions because the length of the chain may be adjusted by means of the key hole (figure 206).

New developments of fibre ropes have made it possible to use fibre rope chokers. Igland has developed a special slider for this purpose (figure 207). The fibres in the rope consists of a mixture of polyester and polypropylene. The ropes are constructed by North Rope. If the tree dimensions are less than 1 cu.m. per tree a fibre rope with diameter D = 14 mm is suitable. The breaking strength is S_{Br} = 43 kN and weighs 1 N per meter.

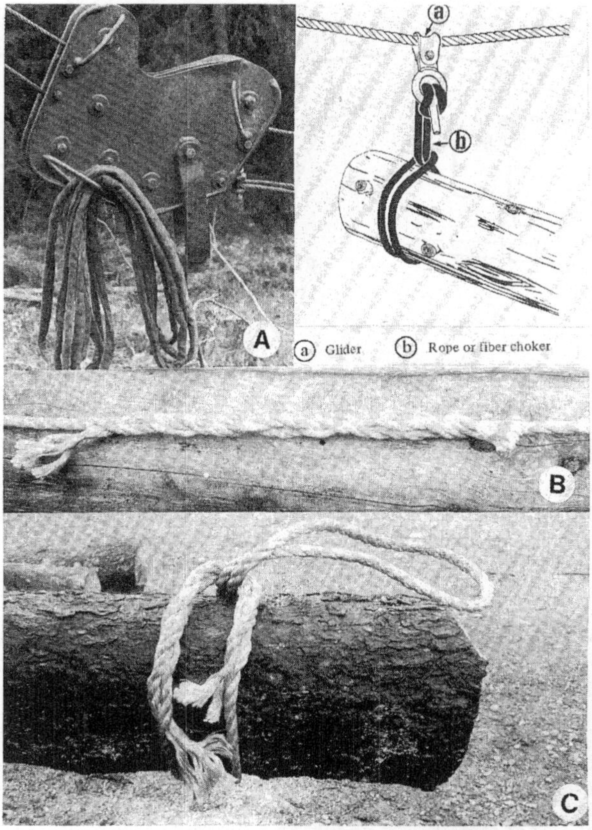

Figure 207. Fibre rope chokers. A. The fibres are laid parallel inside a cover in order to be more capable to resist against perpendicular strokes on the rope. B. and C. The fibres are laid as an ordinary rope.

The fibre rope is laid in a special way which makes it easy to splice the rope. The choker is an endless sling of the fibre rope which is wrapped around the log as shown in figure 207 and attached to the Igland slider. These chokers can easily be made by the choker setter from a length of rope. Two slings may be attached to each other to form a long choker for the oversized logs.

The fibre chokers are easy to work with and very handy especially when yarding in thinnings.

The disadvantage is that the fibre rope is more susceptible to damage from crushing. This may occur when the rope gets between a rock and the log.

11.1.3. Chokers for automatic unhooking

Unhooking the chokers on the landing can be a dangerous job, especially during cold decking, when the pile of logs, stems or unbranched trees is growing high. This work may be avoided by using a type of self releasing choker. The chaser man is not needed and the time of unhooking is reduced to a minimum.

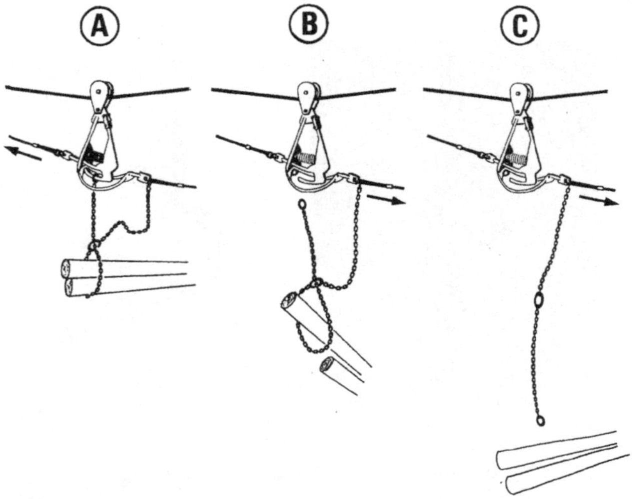

Figure 208. Automatic unhooking (Austrian system).
A. The hook on the carriage is closed by the tension in the main and haul-back lines.
B. With slack main line the hook in the carriage is opened by the pull in the haul-back line.
C. The choker chain goes with the empty carriage back to the felling site.

Figure 209. Carriage for automatic unhooking. Research operation in Kviteseid, 1978.
A. The empty carriage. The spring opens the hook when there is no tension in the lines.
B. The empty carriage with two chokers returns to the felling site.

Carriage with automatic unhooking. This is an Austrian idea (figure 208 and 209). The choker is made of a steel chain with a ring in the centre and at the end of the choker. The other end is attached to the carriage. Usually two chokers are used, one attached to the front end, and the other to the rear end of the carriage. The chokers are brought with the carriage when it returns to the logging area. When a choker is wrapped around the end of the log, the choker end is led through the centre ring of the choker and attached to the releasable hook on the carriage. When there is tension both in the main and in the haul-back lines the carriage hook will keep closed.

When the load has reached the landing, the winch operator reduces the main line pull to zero. The haul-back line returns the carriage towards the logging area, the carriage hook opens by the haul-back line pull and the chokers loosen from the carriage hook. The choker is pulled away from the timber as the carriage returns outward to the terrain.

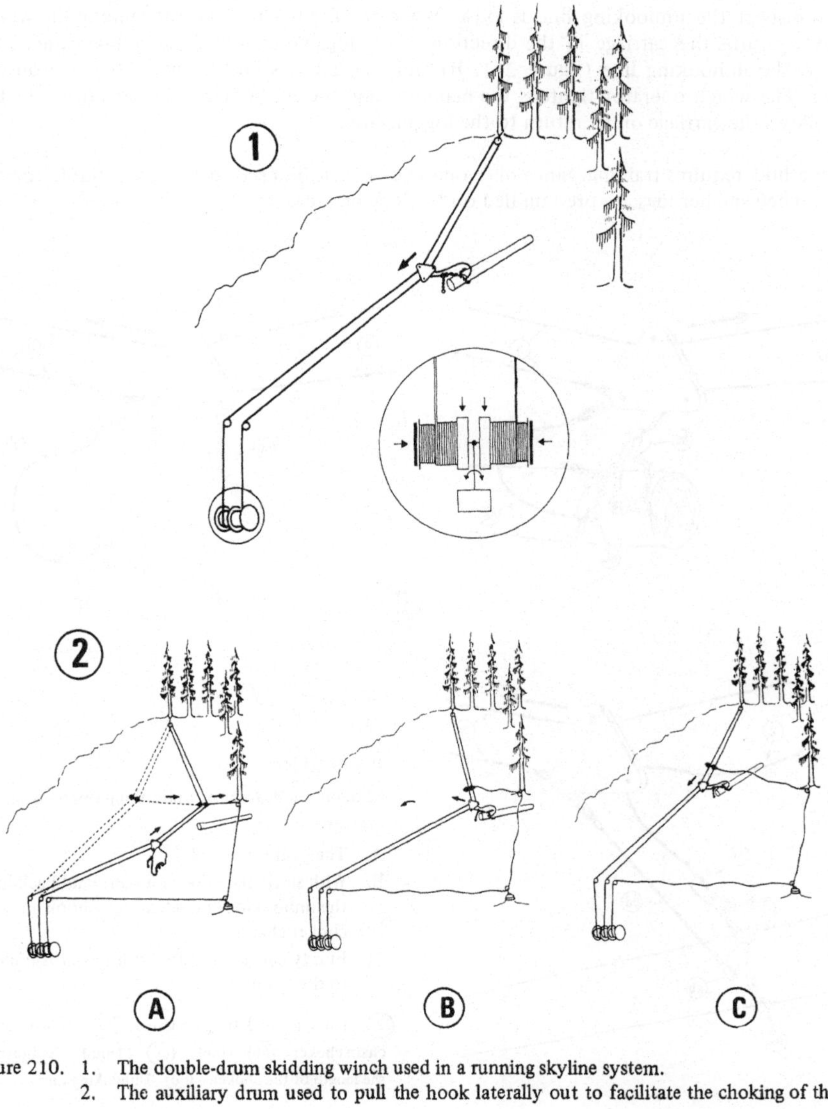

Figure 210. 1. The double-drum skidding winch used in a running skyline system.
2. The auxiliary drum used to pull the hook laterally out to facilitate the choking of the logs.

The method is especially designed for big timber. It is of no use for small tree dimensions because of small loads.

It is also heavy to work with. To overcome the problem an auxilliary line with a double block on the main and haul-back line may be used to pull the carriage sideways towards the logs, as illustrated in figure 210.

Chokerhook with unhooking line (figure 211). When setting the choker the choker man is pulling out the main line to the log and wrap the choker with the hook around the log. The log is then pulled to the carriage and the choker setter attaches the unhooking line to the carriage. As long as there is tension in both the haul-back line and the main line during the load transport the unhooking line is loose. When the log has reached the landing the winch operator returns the carriage in the direction of the logging area with loose main line. This tightens the unhooking line (figure 211, B) and the hook is pulled out of the ring on the choker. The winch operator tightens the main line and the choker chain loosens from the log and follows the carriage on its return to the logging area.

This method requires training. Since only one choker can be used, it is only suitable for big logs or when smaller logs are pre-bundled on the logging area.

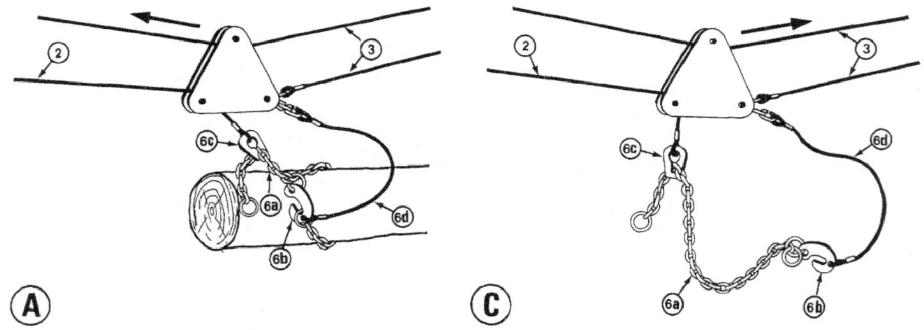

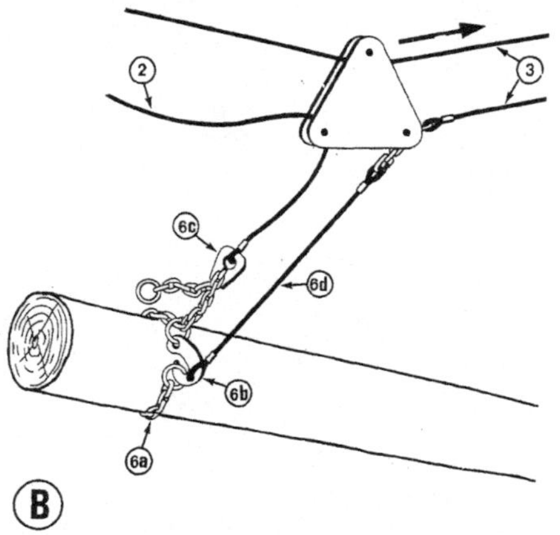

Figure 211.

Chokers for automatic unhooking (Norwegian system)

A. The load is hooked.
B. With slack main line and tight haul-back line the unhooking line releases the hook and the choker chain.
C. Empty carriage returns with the choker chain to the felling site.

(2) Main line (3) Haul-back line (6a) Adjustable chain choker (6b) Hook (6c) Thimble for adjusting the length of the choker (6d) Unhooking line

Figure 212. Automatic unhooking choker during a research operation in Hurdal, Norway 1978.
A. The load is choked and ready.
B. Transport.
C. Automatic unhooking.

The Igland choker for automatic unhooking (figure 213). The angle between the hoist line and the log is used to control the choker. When the hoist line is vertical or inclining forward in the load direction the ring is fastened in the choker hook. If, however, the hoist line inclines backwards, opposite to the direction of the load transport the ring at the end of the choker is released from the choker hook.

When the load has reached the landing, the yarder operator lowers the carriage and returns it towards the terrain. This change of the hoist line angle in relation to the timber loosens the choker. They are brought with the carriage back to the logging area. This method works well for one and two chokers. If there are three chokers they may twist and damage the choker hook. The method is therefore useful for big timber or for pre-bundled logs of smaller dimensions.

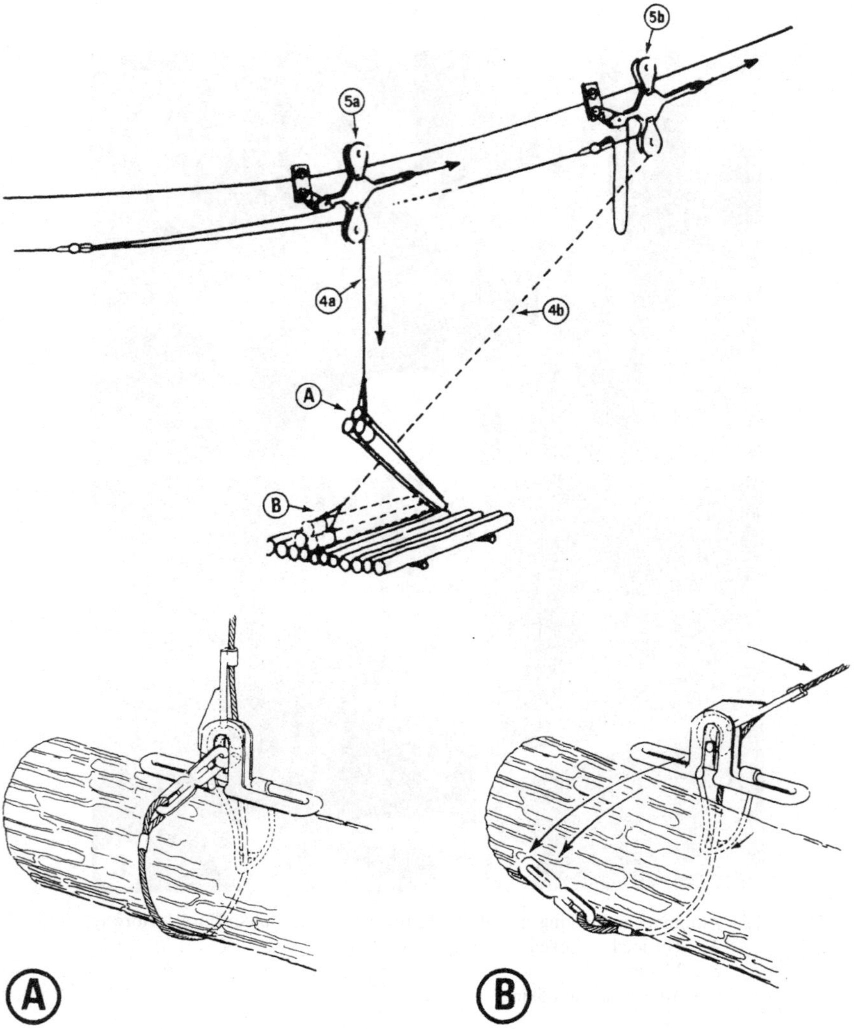

(A) The automatic choker before unhooking (B) The automatic choker during unhooking
(4a) The angle of the hoist line before unhooking (4b) The angle of the hoist line during unhooking
(5a) The position of the carriage before unhooking (5b) The position of the carriage during unhooking

Figure 213. The Igland choker for automatic unhooking of the logs on the landing.

Figure 214. Igland choker for automatic unhooking (see figure 213)

The Steyr self releasing choker hooks (fig. 215). This is an Austrian invention which is produced by Steyr. Inside the choker hook there is a spring which releases the hook and loosens the choker as soon as there is no tension in the hoist line. Inside the chooker hook there is also a small oil tank and a piston which works against a spring force. When the tension in the hoist line and the choker is being released the spring force will open the hook. This, however, takes 15 to 20 seconds since the oil working against the spring force has to pass a constriction in the tube.

If the load touches the ground during transport, the chokers will usually not be unhooked since the duration of the small tension is too short. At the landing the winch operator wait with loose hoist line 15 to 20 seconds. When he pulls in the hoist line after this waiting time the chokers are automatically unhooked.

Grapple yarding will be discussed in chapter 11.5.4.8. Both loading and unloading are automated and the terminal times in the turn are reduced to a minimum. The number of men in the crew can also be reduced. The method can only be used successfully in connection with 3-drum interlocked winches and the running skyline needs good deflection. Only one log can be yarded in each turn, so the method can only be used in big timber. Under such circumstances, however, it is an excellent method.

Figure 215. Steyr self releasing choker constructed by Aubeck, Österreichische Bundesforste, Austria 1980.

11.1.4. Collecting loads from small timber

When the trees are between 0.1 – 0.3 cu.m. per tree it is difficult for the crew to collect sufficient number of logs when setting the chokers. This is especially the case under very difficult and steep terrain conditions with uneven surface and many obstacles. Low volume per hectare tends also to reduce the size of the loads. As a result the load per turn is too low to keep sufficient production during the yarding. The production is proportional to the size of the load. Fish-hook yarding is a good method as illustrated in figure 216. The chokers may be set on the logs while the carriage transports the timber to the landing. After return of the carriage, the choker setter pulls out the main line to the most distant log and attaches the choker to the hook. The other chokers are attached to sliders along the main line.

A 10 to 15 meter long tagline may also be used (figure 217). When the carriage returns with tagline and empty chokers this is disconnected from the main line whereafter the choker setter attaches the end of the tagline to the main line by means of a quick coupling, for example a snap link.

Pre-bundling is another way of solving the problem. Pre-bundling can mainly be used in connection with clearfellings. The idea is to bring together sufficient number of logs to build a bundle which is suitable for the yarder. One choker may be used, and in this case it is an advantage to use a choker with automatic unhooking at the landing.

Much can be done by organizing the felling and the cutting in an appropriate manner. If the cutters are limbing the stems of the logs, gravity rolling may be used (figure 218). The cutters fell some trees in uphill direction. They are leaving 0.5 – 1.0 inch of wood so that the stems are fixed to the stump after the felling. 70 cm high stumps are used. When the cutters have limbed these trees they serve as runners for the rest of the wood. The cutters fell the trees over the runners, limb them and roll them downhill towards two or three buffer trees.

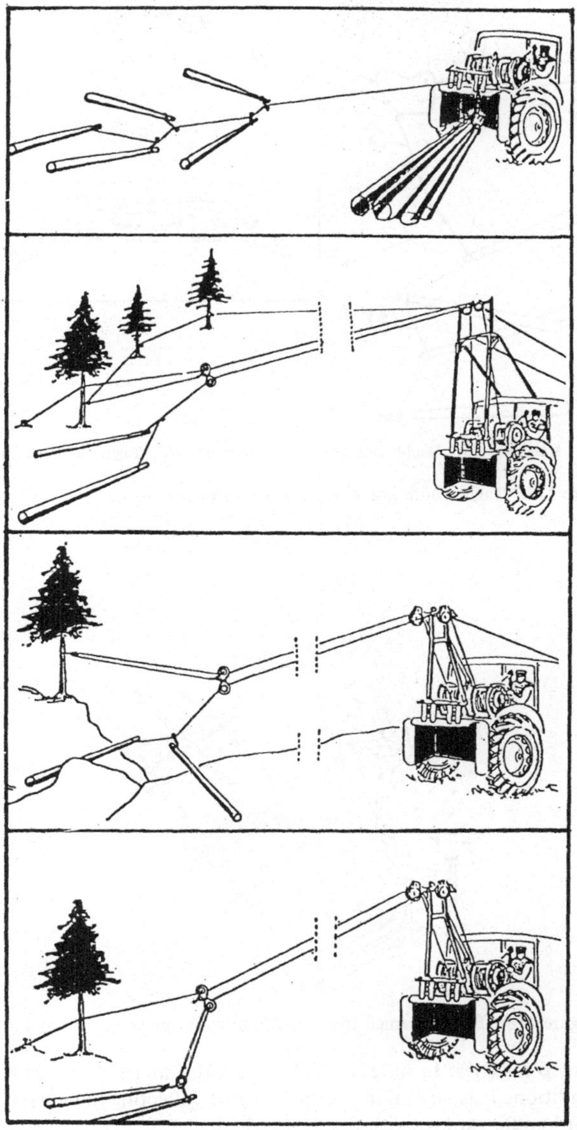

Figure 216. Fish-hook yarding.

In other cases the cutters may use a steep slope or a cliff to collect a bundle of trees, stems or logs by gravity.

Menzi Muck is an excavator which can move and operate on slopes with smooth surface up to 85 % (fig. 219). Experiments have been made to replace the shovel with a felling head. AROLA, R.A. (1981). The Menzi Muck then becomes a felling/bunching machine which can

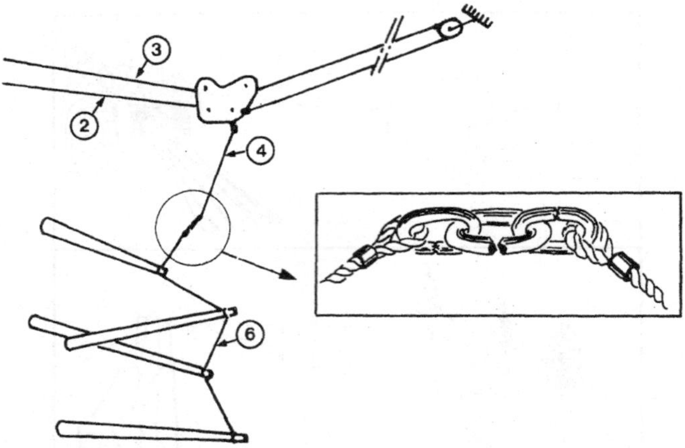

② Main line ③ Haul-back line ④ Hoist line ⑥ Tagline with chokers

Figure 217. Tagline with quick coupling for fish-hook yarding (Igland 1980).

Figure 218. Bundling small tree dimensions by means of gravity rolling.

move on rather steep hillsides. In spite of the fact that this method cannot be used under the most difficult conditions it is still an interesting way of collecting bundles of small wood and preparing them for bundle yarding with the cable crane.

A light portable yarder with radio controlled single-drum winch of Swedish design (the Norfor winch and the Kolpe winch) has been introduced to prebundle small sized timber in steep terrain. So far these experiments had limited success.

Thinnings are difficult to yard with cable systems, especially when the trees are between 0.02 and 0.2 cu.m. per tree. A running skyline system or a highlead system with lifting block on the haul-back line may be used. The main and the haul-back lines are attached to the winch by quick couplings (figure 220). Full tree logging should be used in connection with a chipping machine or a limbing machine working at the landing.

Figure 219. Bundling small sized wood with a feller/buncher. The Menzi Muck with felling head can move and operate on slopes up to 85%. (AROLA 1981).

The feller prepares bundles of unbranched trees. He fells them and sets the chokers which are attached to sliders on a tagline, as illustrated in figure 220. This work is being done while the tractor operator ground skids the previous load along the tractor road to the landing. When the empty tractor returns to the logging area the main and haul-back lines are attached to the main and haul-back drums by quick couplings. The yarding can start with the cutter as choker setter. He connects the loaded tagline to the main line. The load is then pulled into the tractor. The winch operator loosens the tagline from the main line and a new turn can start. When four or five bundles have been yarded to the tractor, the main and haul-back lines including the carriage, are disconnected and remain loose on the ground. The four or

Figure 220. Highlead yarding of small sized wood from thinnings, combined with tractor skidding of bundles. The cutter uses taglines with quick couplings and chokers to bundle the wood.

five bundles are then attached to the winch and the tractor operator ground skids the load along the tractor road to the landing. In the meantime the cutter continues his felling operation and by means of four or five taglines with empty chokers he prepares bundles for the next tractor load.

11.1.5. Collecting loads from big timber

When the tree dimensions are 6 to 9 cu.m. per tree or more setting the chokers is heavy work. After the tree has been felled the log may be pressed into the soil so it is difficult to wrap the chokers around the log. The chokers as well as the rigging is very heavy and cannot be easily handled by the workers.

The chokers should always be wrapped over the tree instead of under as illustrated in figure 221. In highlead yarding there is no hoist line and the yarder has to move the rigging with the

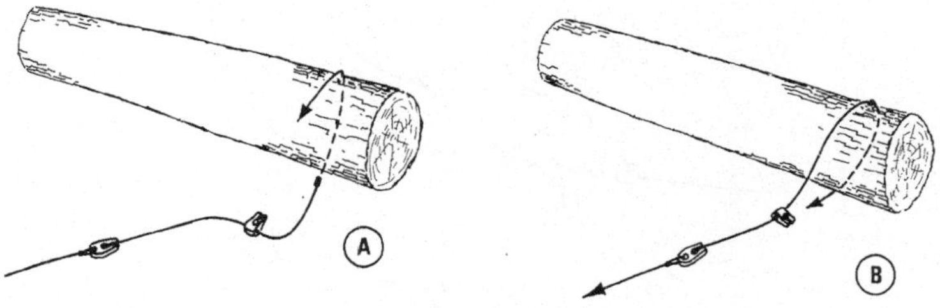

(A) Wrong (B) Correct

Figure 221. The chokers should be wrapped over instead of under big logs.

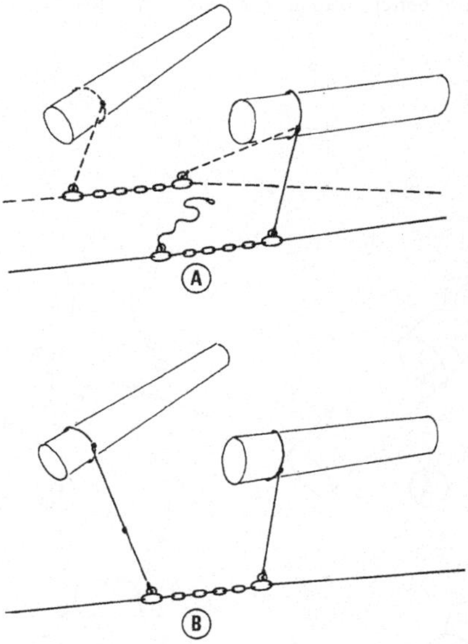

Figure 222. A. Moving the rigging closer to a log by means of a choker to a big log or a stump.
B. To reach a log by means of two chokers. (CONWAY 1978).

chokers close to the log by motor power. The rigging is too heavy to be moved by the workers. The positioning of the rigging must be accurate. If a log is too far away from the rigging, the rear choker may be fixed to a log or a stump (figure 222, A). When moving the main line ahead the rigging will be pulled sideways and closer to the second log. It is also possible to use a second choker (figure 222, B) but in this case it is better to take away the second choker as soon as the log has been pulled into the cable road.

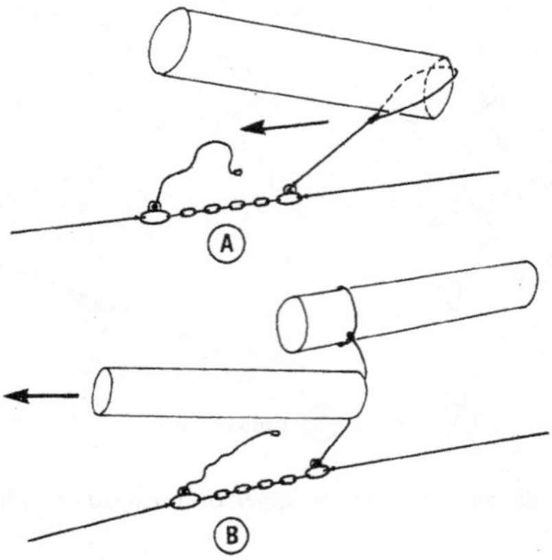

Figure 223. A. Moving a log by a choker around a log end to get better place under it for choker setting.
B. Moving a log into better position by means of the choker to the next log. (CONVEY 1978).

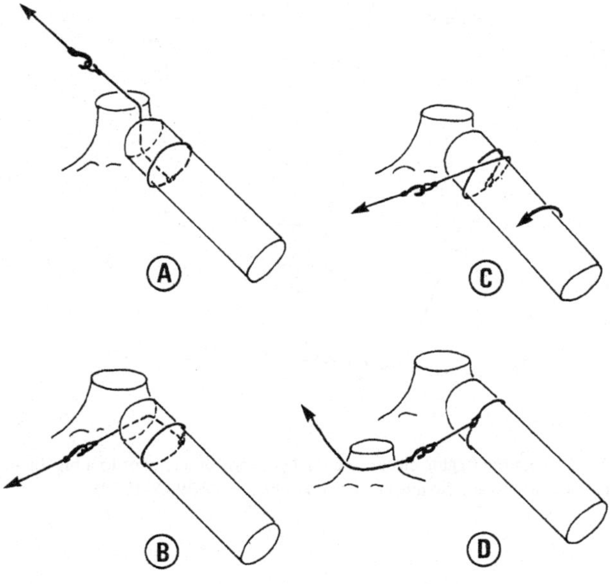

Figure 224. A. Lifting a log over a stump.
B. and D. Skidding a log sideways past a stump.
C. Rolling a log past a stump. (CONVEY 1978).

Sometimes the log is pressed into the soil and there is no space under the log for the choker. The choker setter has to dig a hole under the log. Sometimes dynamite has been used to do this work. An experienced rigging slinger or choker setter will move the log by wrapping one or two of the chokers behind the log and moving the rigging slowly forward. With the new position of the log it may be possible to find space enough for the choker underneath the end of the log (figure 223, A). Another way of doing it is illustrated in figure 223, B.

If a log gets stuck against a stump an experienced choker setter may place the choker in such a way that the log jumps off the stump when pulling in the main line. The log may be lifted over the stump (figure 224, A) or rolled sideways past the stump (figure 224, B and D). A very good method is to roll the log sideways past the stump as illustrated in figure 222, C.

Many practical methods and applications based on experience helps to overcome the heavy work load when setting chokers and ground skidding very big logs.It pays to let an experienced rigging slinger lead the rigging crew. This is quite different from the yarding of smallsized tree dimensions where a small number of choker setters are doing the job without any guidance of a rigging slinger.

The choker setting and the ground skidding or the lateral skidding is the bottleneck for all cable yarding systems. A good training for the choker setters is of importance in order to achieve the best possible utilization of the cable crane.

11.1.6. The chasing

The chaser unhooks the chokers at the landing. He may also saw knots off the logs or some remaining branches.During uphill yarding of big timber the rigging or the hoist line is tight and it can be difficult to loosen the chokers. In such cases he may use the haywire to hold the butt rigging in order to create enough slack for unhooking the chokers. When cold decking the landing may grow high which makes the chasing difficult. Instead of loosening the chokers from the logs, he may detach them from the hook and send fresh chokers back to the logging area. The chaser can use the time between two turns to release the chokers from the logs.

When yarding trees less than 2 cu.m. per tree, the winch operator may do the chasing. This is easier when the winch on the yarder has a remote control (radio or cable). When the tree dimensions are less than 1 cu.m. per tree, a two man crew operating a small yarder with remote control may carry out the felling, yarding and limbing in one operation. The choker man is felling and branching the upper side of the stems and he sets the chokers. The winch operator unhooks the trees, takes off the rest of the branches with his chainsaw and does the bucking. This all depends on the equipment and the method and will be described in the following chapters.

11.1.7. The landing

When the tree dimensions are less than 1 cu.m. and the trees are limbed and bucked before the yarding takes place, small landings may be placed close to the roadside. As the direction of the yarding is more or less perpendicular to the road the logs or stems should be unloaded parallel to the road. Each load has to be turned 90° at the landing (figure 225). The load is pulled ahead until it rests on the landing close to its centre of gravity. An experienced yarder operator may turn the load 90° by returning the carriage and pulling in the hoist line.

The size of the landing depends on how much wood is to be piled. Cold decking may lead to very large landings. Today such organization of the work is rare. Instead the yarding is

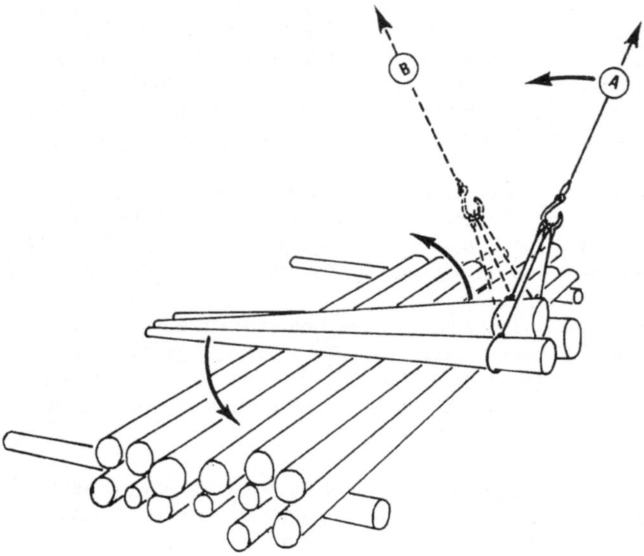

Figure 225. Working technique for moving the cable crane load 90° (see figure 298).
A. The yarder operator pulls the load until it rests near its centre of gravity.
B. The yarder operator returns the carriage. An experienced operator will then turn the load into its correct position on the landing.

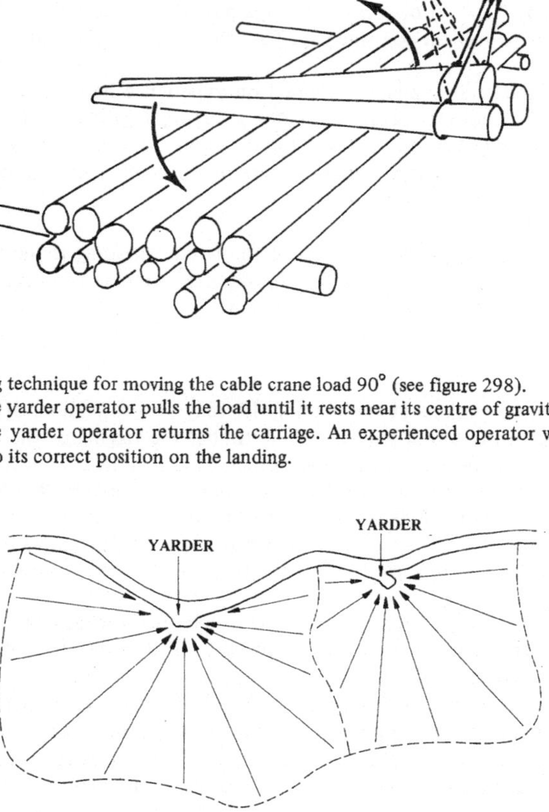

Figure 226. Terminals for cable yarding to a feeder road.

combined with loading on trucks (hot logging). Sometimes even the conversion of unbranched trees to logs is carried out at the landing or near the landing. The landing must have space both for the yarding machines and the loading machines. It may be necessary to plan and build landings beside or close to the truck road (the feeder road), where one landing can cover a large part of the logging area (figure 226). The landing should at least be 15 meters wide if the yarding of logs is combined with loading. In full tree logging with a limbing machine at the landing the size of the landing depends on the tree length (figure 227). In some cases it is an advantage to build a second road above the landing (figure 227, B).

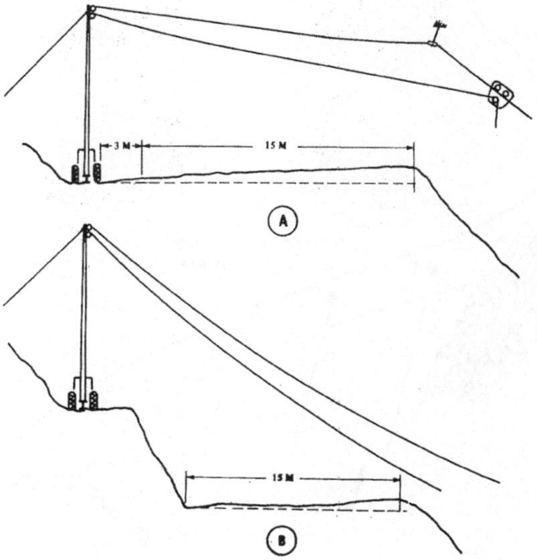

Figure 227. A. The yarder is rigged up on the landing. A support is often needed.
B. The yarder is rigged up on a side road above the landing.

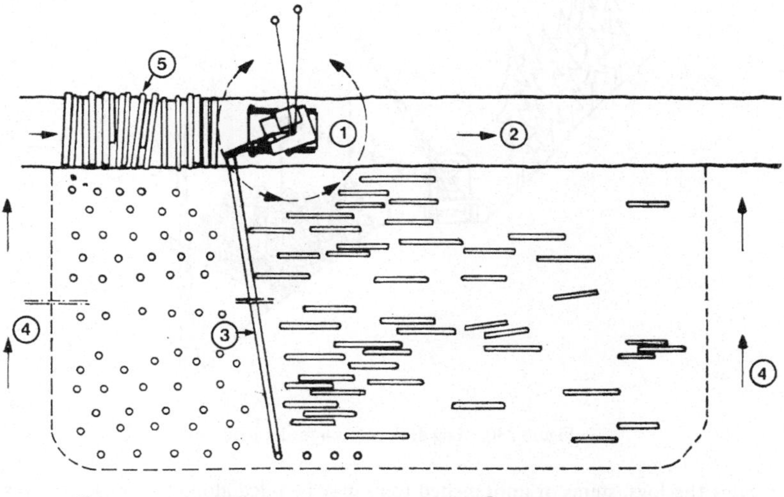

Figure 228. The feeder road used as landing during cable yarding with swing boom yarder.

The feeder roads are usually laid out parallel to the terrain contours. If the terrain conditions allow it, the feeder road may serve as an excellent landing. If the yarder has a swing boom the logs may be unloaded on the raod. The loading and trucking may start at the other end of the load deck, further down along the feeder road (figure 228).

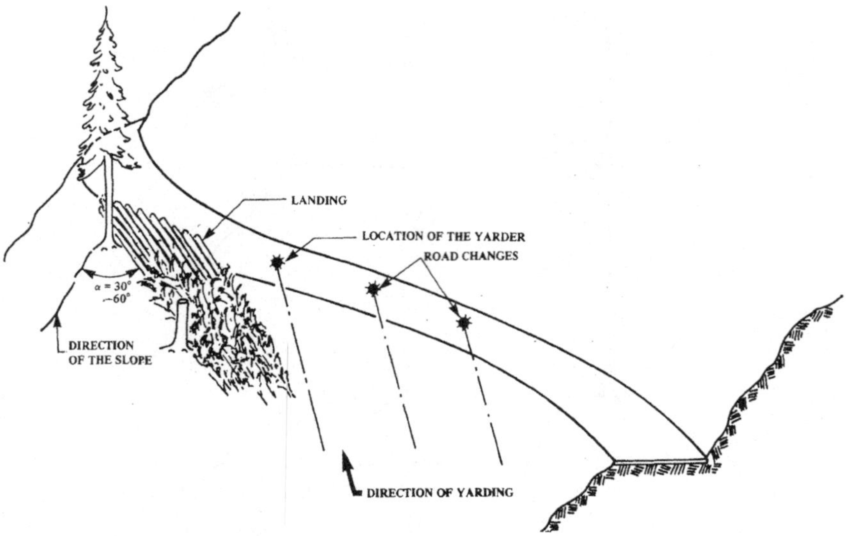

Figure 229. A feeder road used as yarder station during uphill yarding. The stems or logs rests sideways against high stumps or trees in order to avoid sliding downhill on steep slopes.

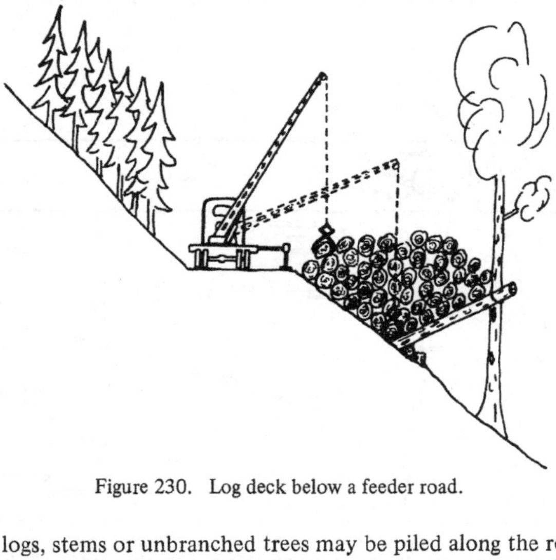

Figure 230. Log deck below a feeder road.

In other cases the logs, stems or unbranched trees may be piled along the roadside. This leads to a continuous pile (a windrow) along the roadside as the yarder and yarding moves forward along the feeder road. If full tree yarding is carried out, the limbing/bucking machines take the trees with its knuckleboom loader and places the converted logs on the other side or at the same side along the road, depending on the terrain conditions. During uphill yarding it may be difficult to keep logs in place on the landing due to steepness. The logs or stems may slide downhill past the reach of the loading machine. It may help to place the logs at an angle of 30° – 60° in relation to the direction of the slope. The operator places the first log at the correct angle for example against two stumps or trees. He continues to place the logs beside each other as the yarder moves along the feeder road (figure 229).

Figure 231. Log deck at feeder road. A truck mounted loading machine with hydraulic boom and hoist winch, hydraulically anchored in order to transfer heavy logs from the load deck to the truck. Hokaido, Japan 1968.

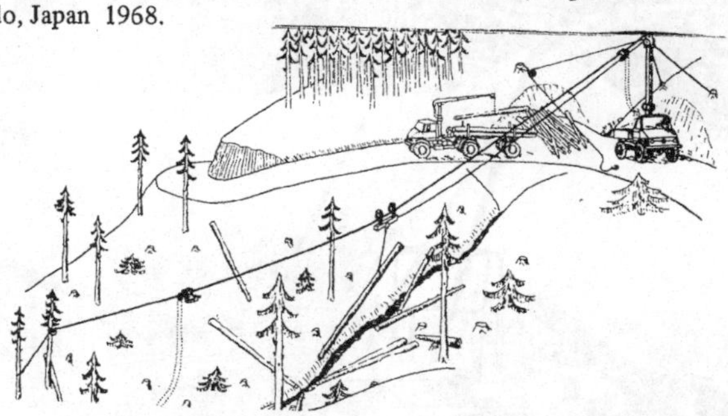

Figure 232. The logs are continuously moved away from the yarder station on a cable crane to a landing by means of a double-drum winch on the yarder. The truck loading may be carried out simultaneously with the cable crane yarding (HINTEREGGER 1980).

In Japan where the forest terrain is in many places extremely steep it is often difficult to find places for landings. Artificial load decks are often placed below the road as illustrated in figure 230. Special truck mounted loaders have been built for loading the logging trucks from such landings. It can be a hydraulic telescope boom with a single-drum winch. The logs are hoisted from the landing to the truck by the winch and by adjusting the slope angle and the length of the boom. The truck has to be securly anchored since the loader can exert a loading torque of 180 kN.m. on the truck.

If the space is narrow at the yarder it may be necessary to move the logs or the unbranched trees away from the yarder. HINTEREGGER in Austria has suggested to move the logs away from the yarder by winch as illustrated in figure 232. During uphill yarding with the Urus

Figure 233. Highlead yarding combined with loading on railway by means of a yarder with 5 winch drums. Caucasus, Soviet Union 1957.

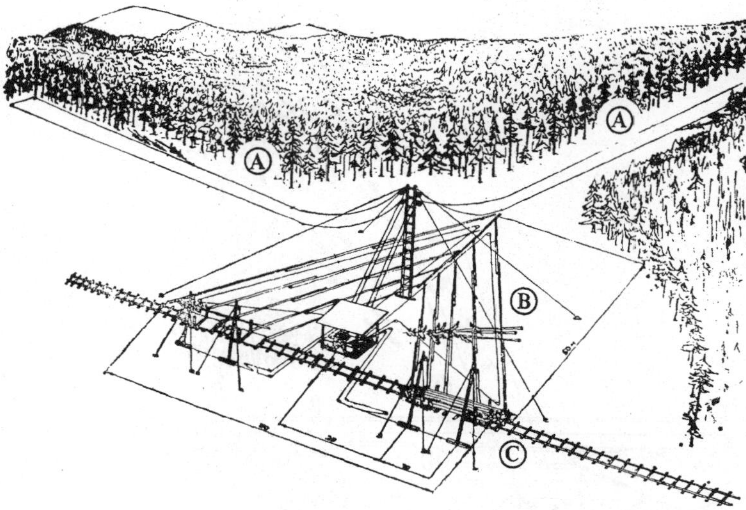

(A) Highlead yarding of full trees (top end first)
(B) Side movement of trees to loading station
(C) Loading on railroad cars

Figure 234. Highlead yarding combined with loading on railway cars. Soviet Union, 1957.

cable crane a double-drum winch may be used to move the logs away from the yarder, along the feeder road to a place where it is more convenient to load the trucks. Mayr Melnhof in Austria use a grapple skidder for the same purpose (figure 240).

11.1.8. Yarding and loading in one operation

SNIIME (The technical research institute in the Soviet Union) has developed a winch technique to do more of the terminal operations at the landing and the yarding simultaneously. A 5-drum winch was built for this purpose (figure 233). Two of the winch drums are used for yarding, two for moving the logs in position at the landing, and one drum for loading. One yarder engineer operates all the winches. Figure 235 shows how a counterweight can be used to pull out the hoist line when loading is carried out with a single-drum winch. Figure 236 shows the principle of simultaneous truck loading and yarding with the 5-drum winch and figure 234 shows highlead yarding and loading of railway carriages.

The virgin forests with mature and overmature stands often have tree dimensions averaging 6 – 9 cu.m per tree. Individual trees can be up to 20 cu.m per tree. Under such conditions the cutting has to be separated from the yarding operation. It must be carried out separately for safety reasons. The yarding production is very high, sometimes 200 – 300 cu.m. per day or more and it would be impossible to pile the logs on a landing. The logs have to be loaded simultaneously with the yarding operation. This requires good organization of the work and productive loaders. Figure 237 shows a goose-neck boom loader with a winch operated grapple. There must be sufficient space on the landing to allow for the loading and trucking at the same time as the yarding.

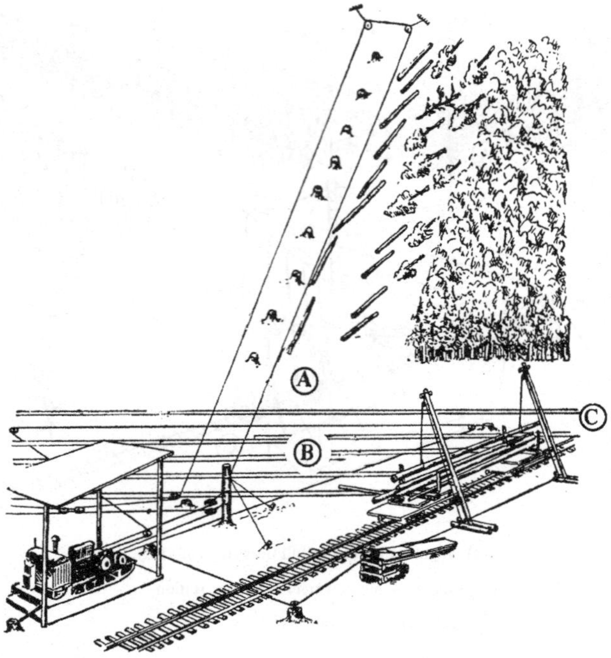

Ⓐ Highlead yarding of logs Ⓑ Side movement of logs Ⓒ Loading on railroad cars

Figure 235. Highlead yarding combined with loading on railway cars. Soviet Union, 1957.

Figure 236. Cable crane yarding combined with cable loading on trucks. Soviet Union, 1957.

Figure 237. Goose-neck boom loader transfers the logs from a cable crane to the truck in Crown Zellerback, Oregon, USA 1975.

11.1.9. Yarding and conversion in one operation

It may be an advantage to yard the stems in full length and buck them into assortments at the landing. Figure 238 shows a Urus cable crane yarding full length stem uphill at Hespa Domene in Austria 1977. The yarder is placed on the feeder road where there is little space for decking the logs. Bucking is carried out by chainsaw and loading by a hydraulic loader. It is an advantage to have a loader with a telescope boom which can pull the stem uphill from the roadside. The bucking of the stems into logs and the loading of logs on the truck take place during the time when the carriage returns to the logging area and brings a new load up to the landing.

Figure 238. Uphill yarding of full length stems by Urus cable cranes. The stems are bucked and loaded on trucks simultaneous with the yarding operation. Hespa Domene, Austria 1977.

When the tree dimensions are less than 3 cu.m. per tree and especially if they are less than 1 cu.m. per tree, full tree yarding is commonly used in modern cable crane operations. The coniferous trees in steep terrain have often many branches and the use of a limbing machine increases production and reduces costs. If the landing is large enough, the delimbing machine may operate near the yarder at the landing (figure 239). Sometimes the terrain does not allow building of large terminal places at the point where it is appropriate to place the yarder. A grapple skidder may be used to pull the trees along the road to a point where the limbing/bucking machine can operate (figure 240). Another method is to mount a limbing/bucking machine at the end of a knuckleboom. The skidder grapples the trees from the yarder, pulls it along the road to the landing and limbs and bucks the trees to logs which then are placed in the pile on the landing (figure 241). Such operations have to be synchronized so that the production of the yarder corresponds to the production of the skidder and the limbing/bucking machine.

Logging in mountainous forests means logging in difficult terrain. There is an obvious need to reduce the amount of work on the steep hillsides to a minimum, i.e. felling the trees and choker setting. The main part of the work load should be transferred to the road where the yarder, the skidder and the limbing/bucking machines can work.

Since the roads are the only places in very difficult terrain with easy working conditions one should have combined use of the feeder roads in mind when planning the road network constructing the roads and choosing the machines and the operational methods for harvesting as well as for the establishment of new forests.

Figure 239. Cable crane yarding of full trees combined with limbing machine at the landing.
A. Lateral skidding of the unbranched stems.
B. Full tree transport with Nestestog radio controlled cable crane. Logma T-300 telescope delimber at the landing. Kviteseid, Telemark, Norway 1980.

Figure 240. The Gösser cable crane during full tree yarding. A grapple skidder moves the unbranched trees away from the cable crane. A specially constructed harvesting machine converts the unbranched stems to logs on the feeder road. Meyr Melnoff, Austria 1975.

11.2. Continuous main line system. The mono cable

These cable systems consist of an endless main line which moves slowly, often 0.5 – 1.5 meters per second. The loads are attached manually to the main line by quick couplings. The distance between each load is short. The loads are unhooked at the landing and the empty carriages return with the main line back to the logging area (figure 242). A tractor driven or motor driven pulley powers the main line.

Basically, the mono cable system is similar to the twin cable system with circulating carriages (chapter 10.4) and the single cable system with circulating main line and carriages (chapter 10.5). The mono cable, however, has no skyline, and it is the main line which carries the load. The systems described in chapter 10.4 and 10.5 are access roads. The mono cable systems are feeder roads which collect wood from temporary piles along the main line road. The length of the endless main line has to be adjusted according to the logging area and the length of the cable way. This is done by splicing in or taking out lengths of wire rope.

11.2.1. Lasso cable conveyors

The Lasso cable conveyor is constructed and built in Switzerland. This cable way is built for various purposes for distances up to 10 kms and for the transport of small as well as heavy loads. In forestry the Lasso cable conveyor is mainly used in connection with selective cutting or thinnings since it is a smooth system which makes very little damage to the remaining trees in the stand during the operation.

Figure 241. Full tree yarding by the Steinkugel cable crane. A knuckleboom mounted limbing/bucking machine on a skidder moves the trees away from the yarding station, convert them to logs, and piles the logs on landings at the feeder road. Austria, 1980.

11.2.1.1. The equipment

The pulley consists of a large diameter friction wheel (figure 243) which is powered by a

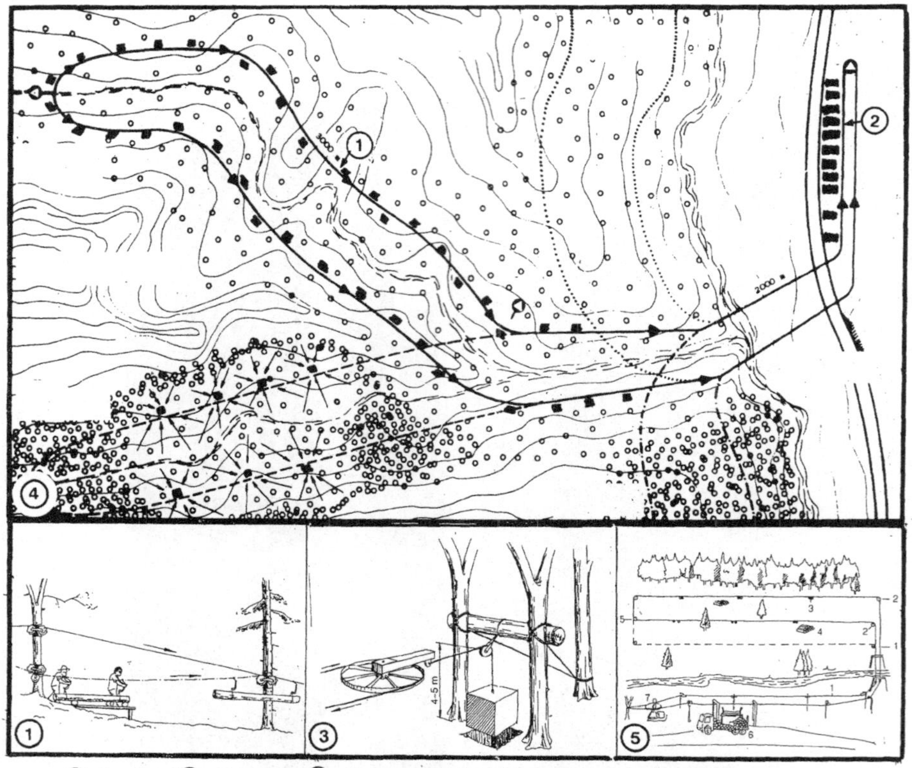

① Loading ② Unloading ③ Gravity weight for tightening the main/haul-back line

④ Alternative route ⑤ The system

Figure 242. The cable Lasso mono cable.

4 – 30 kW engine. The pulley may also be powered by the »power take off» on a farm-tractor. The line speed is 1.1 meters per second.

The load size. The size of the load is limited by the dimension of the main line and the supports on the trees. The equipment comes in two sizes:

Thinning in young stands: 1 kN loads suspended by two load hooks and with a spacing of 10 meters between the loads on the main line. The dimension of the main line is D = 12 mm.

Transport of heavier timber: Loads of 10 kN suspended by two load hooks on the main line and with 60 meters spacing between the loads. The dimension of the main line is D = 16 mm.

The support. The main line is kept a few meters above the ground by means of support sheaves. These are pivoting spiked sheaves (like toothed wheels) attached to trees by brackets. They are spaced at 40 – 60 meter intervals in the stand. When the main line has to pass ravines longer spans may be accepted.

The loads are hanging on the main line by means of U-shaped hooks and chains. The load hooks are designed to pass over the spiked support sheaves. The space between the spikes on the sheaves allows the hooks to pass while the spikes prevent the main line from jumping off.

Figure 243. The Lasso mono cable system. (Switzerland, 1975).

11.2.1.2. The operational method

The Lasso cable is operated by a 6 man crew:

1 foreman
1 drive unit operator
3–4 men for loading
1 man for unloading.

The drive unit may be remote controlled. In this case the drive unit operator may also carry out the unloading.

The Lasso cable conveyor is mostly used downhill where the required power is low. The steepest downhill yarding is 60° (173 %) and the steepest uphill yarding is 20° (36%). Before the cutting starts the line and support trees should be marked. The cutters must pile the wood along the main line corridor. When the support sheaves have been rigged up on trees, the main line is pulled around the loop by a small winch. When the endless main line has been spliced together it is laid around the pulley on the drive unit. The pulley is then tightened and securly anchored. Rig up is a slow process and requires approximately 30 mandays (5 days for a 6 man crew) for a 1,000 meters cable way.

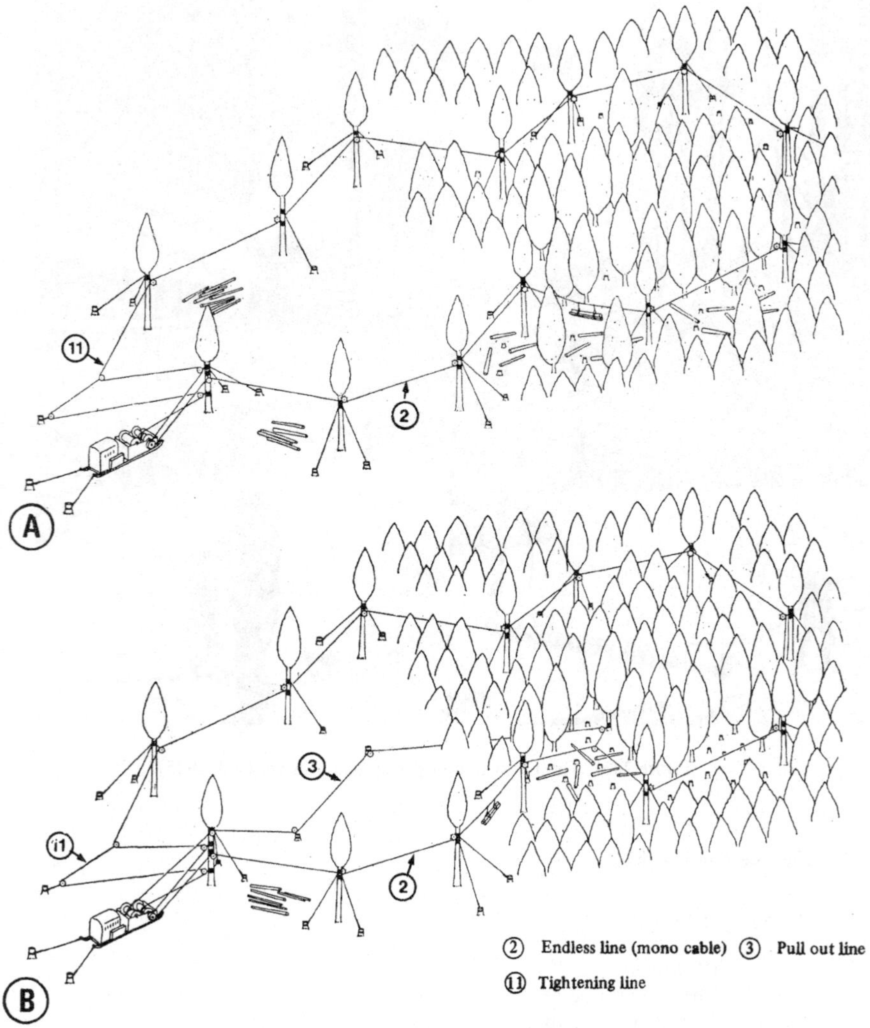

Figure 244. The Japanese mono cable.

During the operation the main line has to be stopped during loading and unloading. The loaders (choker setters) keep in contact with the unloader (the chaser) by means of radio. When the main line stops for unloading the loaders hook new loads on the main line. The productivity of such a system varies with the conditions and may be in the order of 0.5–1.0 cu.m. per man per hour with a labour consumption of 1–2 manhours per cu.m. When using the light cable in first thinnings the productivity is somewhat lower.

11.2.2. Nansei mono cable

The Nansei mono cable is constructed and built in Japan and the system is similar to the Lasso system. It is mainly used for thinnings in young stands. (Figure 244).

11.2.2.1. The equipment

The winch which is used to pull the main line is a standard Nansei KK-1, 3-drum winch which is used for several different cable systems. The winch has two ordinary drums for main and haul-back lines and one pulley. (Figure 245).

Power: 9 kW
Line speed: 0.7 m/s

The winch has a transmission system with four ratios and the speed of the the pulley can vary between 0.7 and 4.7 meters per second. With the Nansei mono cable the lowest speed is used.

The main line has a diameter of D = 12 mm, and the length of the main line is up to 3,000 meters. With a spacing of approximately 50 meters the usual load is 1 – 4 kN.

Figure 245. The yarder is equipped with a double-drum winch and a capstan. (Moroto Forest, Japan, 1981).

The support. The support sheaves are not spiked at one side. They have a special shape which allows passage of the load rope. Fibre ropes are used as chokers and the rope is attached to the main line by means of a special knot, which is easy to loosen during the unloading at the landing. The support sheaves are attached to standing trees by slings.

11.2.2.2. The operational method

The Nansei mono cable is operated by a 3 man crew:

1 yarder operator
1 hooker
1 unhooker

The Nansei mono cable is easy to rig up. Since the support sheaves are attached to the trees by slings they need not to be carefully aligned. The support sheaves also allow the use of sharp turns preferably between 16.5° and 49°. The main line has, however, to be well tightened and a special tightening line is used for this purpose. One of the winch drums on the yarder is often used for this purpose. The other drum may be used for a pull-out line in order to pull the endless main line sideways. This makes it easier to load the logs on the endless main line (the pull-out line is sometimes called the dutchman).

The loads normally vary between 0.4 and 1.0 cu.m. According to KONUMA and SHIBATA (1976) the productivity for transport only is approximately 1 cu.m./man/hour, corresponding to a labour consumption of 1 manhour per cu.m.

11.2.3. Steyr thinning-paternoster

The Steyr thinning-paternoster is an Austrian yarder specially constructed for thinnings. The continuous main line systems described under chapter 11.2.1. and 11.2.2. transported the logs clear of the ground and the main line changes direction at each support tree. In the thinning-paternoster a rectangular layout is used as illustrated in figure 246. The endless main line runs straight through the stand along the thinning corridor and returns along the next corridor with a 10 meters spacing between the two. The main line is divided into 20 meter lengths.

11.2.3.1. The equipment

The yarder is mounted on a truck chassis. A pulley moves the main line with a speed of 1.4 meters per second. The inhaul main line enters the yarder at the rear end of the truck and passes a block at the front end of the truck (figure 247). The distance between the two blocks on the truck is 10 meters corresponding to the distance between the thinning corridors in the stand. The endless main line consists of 20 meter lengths which are connected to each other by means of quick couplings and the maximum yarding distance is 400 meters.

The yarder has the following specification (fig. 246):

Total weight of yarder:	140 kN
Engine power:	88 kW
Maximum main line speed:	1.4 m/s
Maximum main line pull:	60 kN
Maximum yarding distance:	400 meters
Hight of guide blocks above ground:	4 meters

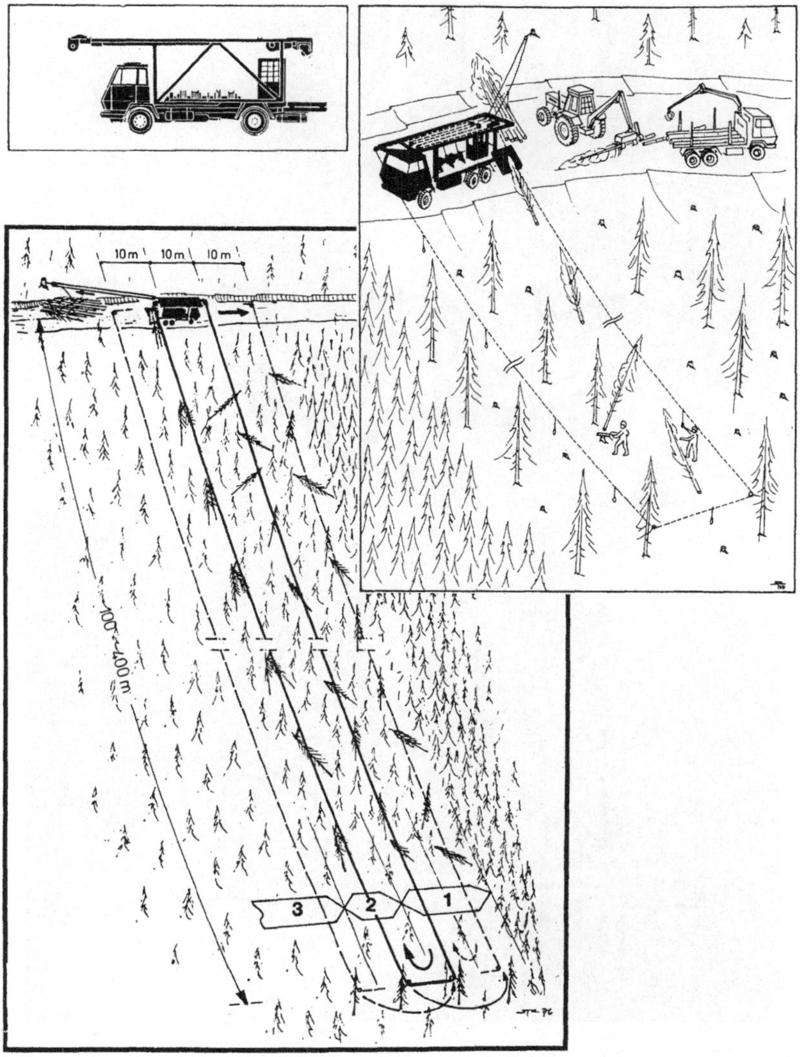

Figure 246. The Steyr thinning-paternoster, Österreichische Bundes Forste, Austria 1977.

11.2.3.2. The operational method

During the yarding the thinning-paternoster pulls the unbranched trees with the top end forward. It is ground lead skidding. This type of yarding can only work successfully when the ground is smooth with few obstacles. Rough terrain with cliffs, boulders etc. will reduce the production considerably.

The yarding crew consists of a yarding operator and 1–2 choker setters. In addition there is the felling crew and the operator on the limbing/bucking machine.

Figure 247. The Steyr thinning-paternoster, Österreichische Bundes Forste, Austria 1977.

The felling crew operates on a corridor ahead of the yarder. They fell the trees in the direction of main line corridor for the next set up (figure 246). One choker is attached to each 20 meters length. They are fastened around the tops of the trees so that the trees are yarded top first. When a turn of trees reach the yarder, the operator stops the main line, releases the turn and attaches it to an auxiliary line which yards the turn out of the way into a pile. Chokers with an automatic unhooking device are used. They unhook the turn automatically by returning the auxiliary line.

While the main line is stopped, the choker setter attaches a turn of trees to the next choker. The choker setter and the yarder operator coordinate their activities through the use of radio communication.

The chokers are free to move along each 20 meters section of the main line. The limbing/ bucking machine moves from one pile to the next and converts the trees into assortments which may be loaded onto the truck and transported along the feeder truck roads.

The thinning-paternoster produces good result for tree dimensions between 0.1 and 0.2 cu.m. per tree. With smaller dimensions the production is too low. According to information from Östereichische Bundesforste the productivity for yarding only is in the order of 2 cu.m. per man per hour with a labour consumption for the yarding of 0.5 manhours per cu.m. This production does not include the rig up time. The production is independent of the yarding distance. It is, therefore, an advantage to have as long yarding distances as possible, up to a maximum of 400 meters since this reduces the set up costs per produced unit of timber.

11.3. Cable cranes with top mounted yarders

Cable crane is the name of a cable system where a hoist line pulls the logs from the terrain laterally into the skyline corridor and lifts the load up to a carriage which can move along the skyline. The skyline is anchored at both ends to stumps in the terrain or sometimes to bolts in a rock or to a »dead man». The skyline may pass supports in order to keep the load off the ground during the transport. A carriage moves along the skyline between two carriage stops, one at the landing and the other at the place where the logs have to be collected.

The carriage stop in the terrain has to be moved from one place to another (approximately 15 meters each time) as the yarding goes on. With top mounted yarders the yarding starts at the top end of the hillside and continuous downward until the entire logging area around the skyline has been yarded.

The carriage stops are a disadvantage, since the collision between the carriage and the carriage stops often leads to extra wear which causes breakdowns and lost time during the operation. It also causes extra work when moving the carriage stop in the terrain from one logging point to another.

In modern cable cranes, therefore, the carriage stops have been replaced by an automatic clamp which keeps the carriage in position at the loading or unloading places.

In some cases the landing is placed in the valley and the transport direction is downhill. This is dowhill yarding and gravity powers the transport along the skyline if the slope is steeper than approximately 20%.

In other cases the landing is placed close to the yarder at the top end of the hillside, usually near, or at, the feeder road. This is uphill yarding and gravity may be used to power the return of the carriage along the skyline from the landing to the logging area (gravity outhaul system). These types of cable cranes may be used in very difficult terrain where expensive road construction limits the possibility to build feeder roads. Yarding distances up to 1,500 meters may be used. Due to the heavy work load for the workers (chapter 8.2) one tries to avoid such long distance cable cranes. With a denser network of feeder roads the yarding production increases. The possibility to take local forest and soil conditions into account in the silvicultural programme is also better.

These cable cranes may be divided into the following classes:

Short distance cable cranes:	100 – 300 meters
Medium distance cable cranes:	300 – 700 meters
Long distance cable cranes:	More than 700 meters

Figure 248. The layout of the road network and cable cranes for downhill cable crane yarding with top mounted winches. (WYSSEN 1983).

11.3.1. Downhill yarding. Gravity inhaul systems

This method was first developed in Central Europe where access roads to the Alps often made it possible to transport the light yarder to a high location. It is, however, often necessary to transport the yarder in the terrain to the top end of the cable crane. The yarder should have a low centre of gravity and be mounted on skids. The engine which powers the winch drum must be abel to run at steep angles. The main line of the winch is pulled out and fixed to a stump or other points in the terrain so that the yarder can pull itself uphill.

Such yarders are usually equipped with a single-drum winch. Originally such winches had two speeds only. The slow speed was used during lateral skidding of the load and hoisting it up to the carriage. The high speed was used to return the empty carriage from the landing in the valley and up to the logging area again. The load transportation is powered by gravity, and no main line is needed (figure 249). The main problem is to control the load with the brake on the yarder during downhill transport. Much heat is generated and a good cooling system on the brake is necessary. Modern winches have usually a fan brake in order to take the main braking power during high speeds (see chapter 3.4).

The carriage moves between the two carriage stops. The lower is placed over the landing and the upper at the loading place in the terrain. The upper carriage stop has to be moved from one loading place to another as the yarding goes on.

The carriage is pulled to the upper carriage stop and fastens automatically to it. By loosening the clamp on the carriage stop by a rope from the ground the carriage stop can be moved to the next loading point by moving the carriage along the skyline. At this point the carriage stop's clamps are fastened to the skyline again and the latteral skidding and yarding can continue.

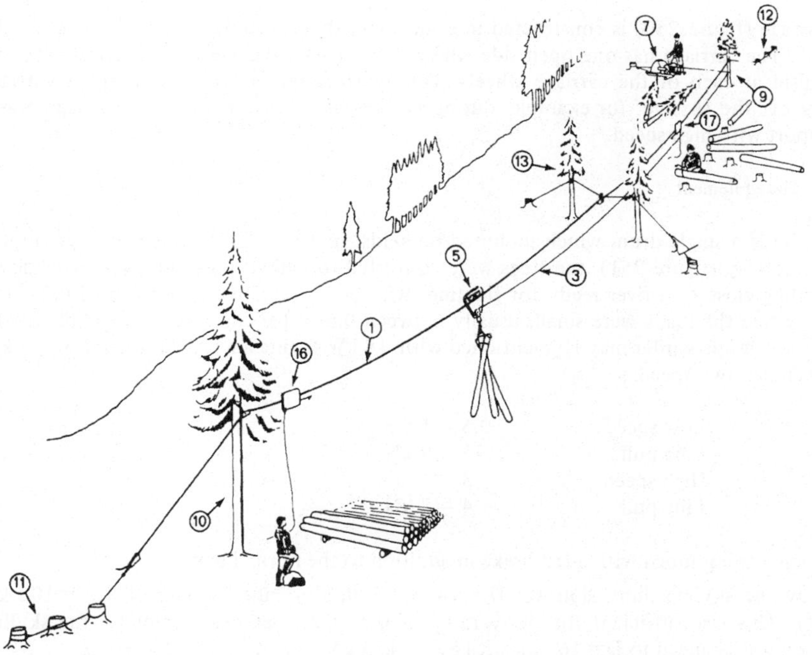

(1) Skyline (3) Haul-back line (5) Carriage (7) Single-drum winch (9) Head spar
(10) Tail spar (11) Tailhold (lower anchorage) (12) Heelhold (upper anchorage)
(13) Support (16) Lower carriage stop (17) Upper carriage stop

Figure 249. Cable crane with top mounted winch.

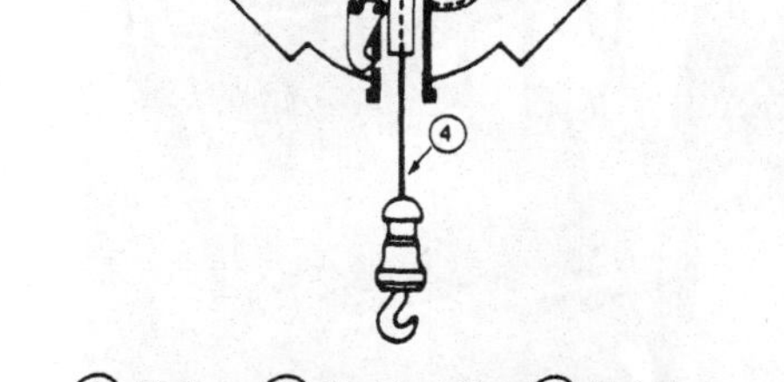

(1) Skyline (3) Haul-back line (4) Hoist line

Figure 250. Carriage for a cable crane with top mounted winch and two carriage stops, one at the felling site and one at the landing.

The carriage (figure 250) is constructed in such a way that it can pass the cable shoes on the supports. The carriage has one open side where it can pass the cable shoe. There is a locking mechanism at each of the carriage wheels. This is necessary in order to avoid the carriage jumping off the skyline, for example, during skyline vibrations or when the carriage passes the support with high speed.

11.3.1.1. The equipment

The yarder is a single-drum winch mounted on skids. In 1950–1960, when this system was used in Norway (figure 251), the trees were converted to barked logs and transported down to a landing close to a river ready for floating. With an average tree dimension of 0.2 – 0.5 cu.m. per tree the loads were small, usually between 0.5 – 1.0 cu.m. As an example of the yarders the Vinje yarder may be mentioned with 18 kW engine and a total weight of 4.5 kN. The winch had two speeds:

Low speed:	0.5 – 1.5 m/s
Line pull:	25 – 9 kN
High speed:	3 – 6 m/s
Line pull:	4 – 2 kN

The winch was equipped with a fan brake in addition to the normal brake.
Originally the skyline dimension was D = 14 mm with a braking strength of S_{Br} = 102 kN (6x7+1). This was sufficient for the yarding of 0.5 cu.m. per load. Later on the skyline dimension was changed to D = 16 mm with S_{Br} = 153 kN.

The dimension of the haul-back line was D = 8 mm with a braking strength of S_{Br} = 30.9 kN (6x19+1) but was later changed to D = 10 mm with a braking strength S_{Br} = 59.8 kN for the yarding of 1 cu.m. per load.

The development of operational methods changed the load weight. The logs were unbarked and yarding of unbranched trees with a limbing/bucking machine at the landing started. The loads which had to be yarded were heavier (up to 20 kN in Norway) and the yarder needed more engine power and more speed variations.

Figure 251. Vinje single-drum yarder for top mounted cable crane.

Figure 252. Above: Wyssen single-drum yarder.
Below: Nessler single-drum yarder with a fan brake.

In Central Europe, in Japan and on the American West Coast the tree dimensions are bigger than in Northern Europe. Two or three cu.m. per tree is common and in virgin timber 9 cu.m. per tree or more can be found. Accordingly the yarders had to be more powerful. As an example may be mentioned the **Wyssen** cable cranes where the yarders vary from 11 kW to 147 kW (figure 252).

Table 30. Examples on yarders from Wyssen for downhill yarding.

Model	W-30	W-200
Engine power	26 kW	147 kW
Load capacity	15 kN	50 kN
Transmission	6 forward	8 forward
	6 reverse	4 reverse
Speeds	0.5–6.7 m/s	0.6–9.4 m/s
Line pull	3.5-37 kN	12-149 kN
Yarder weight	12 kN	68 kN
Skyline	D=24 mm	D=31 mm
Haul-back line	D=10 mm	D=16 mm
Drum capacity	L=1,900 m	L=4,200 m

Both yarders can be equipped with other cable dimensions for other load capacities. Model W-200 can be used for load capacities up to 100 kN which is of interest for very heavy timber.

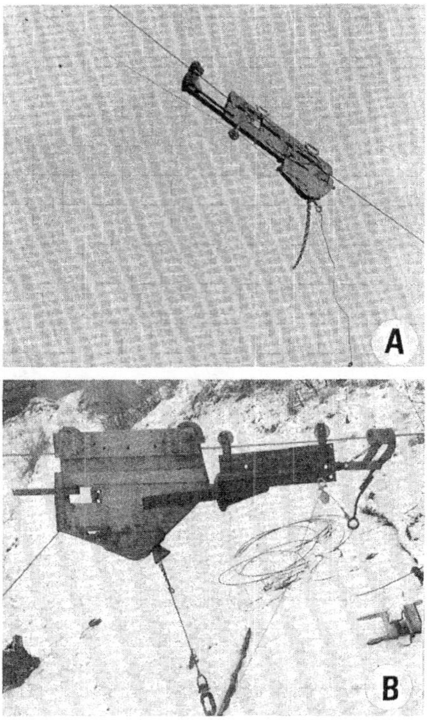

Figure 253. A. The carriage attached to the carriage stop in the terrain.
B. The carriage attached to the carriage stop at the landing.

The carriages. If two carriage stops are used a carriage of similar type as shown in figure 250 and 253 may be used.

During the first part of the yarding cycle, the carriage is being pulled uphill at maximum speed. When it arrives at the carriage stop, the carriage and the carriage stop are connected while the hook is loosened from the carriage (figure 254). The choker setter pulls the hoist line, which is the same as the haul-back line, laterally towards the timber. This work is not too heavy with top mounted yarder since the choker setter pulls the line downhill and gravity helps to pull out the line. When the chokers have been connected to the hook, the winch operator ground skids the load towards the skyline road and hoists the load up to the carriage. When the hook reaches the carriage, it is automatically fixed and the connection to the carriage stop is released. The downhill transport can start.

As soon as the load reaches the landing (figure 255), the carriage impact and fastens to the carriage stop.The load can be lowered under full control to the landing.

The chaser at the landing unhooks the chokers, and gives a signal to the winch operator who returns the haul-back line. When the empty hook reaches the carriage, it is automatically connected to it and the carriage is automatically disconnected from the carriage stop. The return trip uphill to the logging area can start.

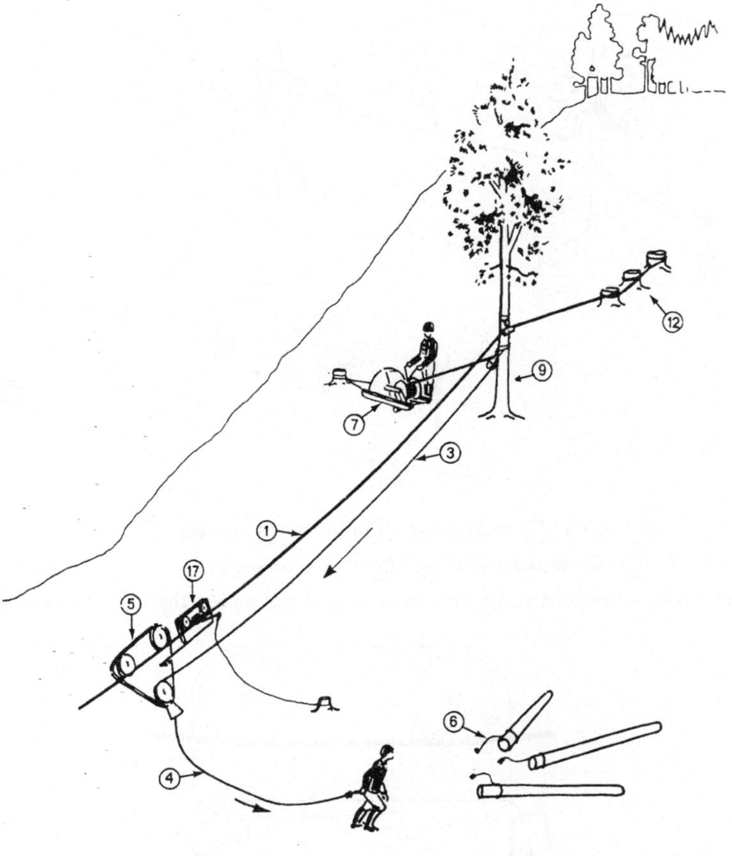

(1) Skyline (3) Haul-back line (4) Hoist line (5) Carriage (6) Chokers
(7) Single-drum top located winch (9) Head spar (12) Heelhold (upper anchorage)
(17) Upper carriage stop.

Figure 254. Pulling out the hoist line.

Sometimes on short cable cranes a carriage as illustrated in figure 256 may be used. Instead of a carriage stop a short log is fastened to the skyline over the landing. The loaded hook is loosened from the carriage by impact against the log. This works well when there are small height differences between the skyline and the landing (figure 257). When the height differences is big, the release by impact leads to uncontrolled unloading since the carriage moves uphill during the lowering sequence (figure 258).

Figure 253 shows a carriage of Norwegian construction for a 10 kN load capacity, with the two carriage stops (Bjønn, Norway). Figure 259 shows a Swiss construction of carriage and carriage stops for 20 kN load capacity (Wyssen, Switzerland).

A cable crane built in USSR was in operation in Chadichensky, Caucasus in 1960. The top mounted winch was powered by a 30 kW gasoline engine and weighed 20 kN. The maximum

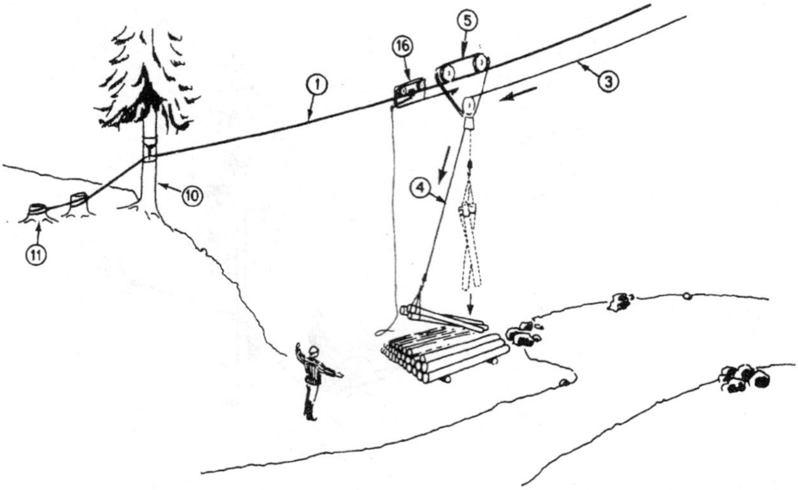

(1) Skyline (3) Haul-back line (4) Hoist line (5) Carriage (10) Tail spar
(11) Tailhold (lower anchorage) (16) Lower carriage stop

Figure 255. The carriage connected to the carriage stop at the landing. The load is lowered to the log pile.

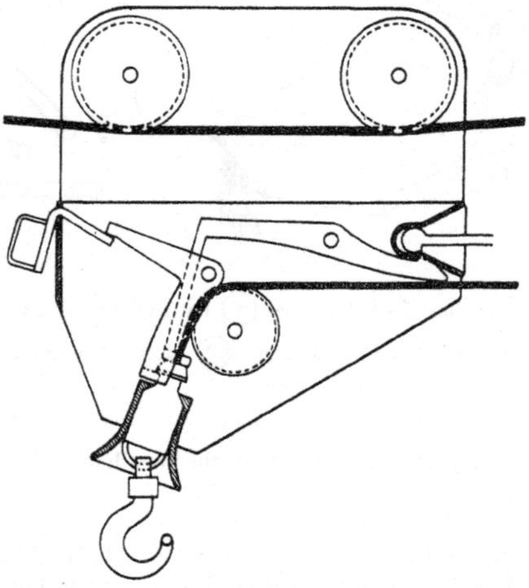

Figure 256. Carriage for a top mounted cable crane. (A log attached to the skyline instead of a carriage stop at the landing).

main line pull was 50 kN and the winch had two speeds, 1 and 6 meters per second. The main line diameter was D = 12 mm (6x7+1) and the skyline D = 25 mm (6x19+1) (figure 260).

An example of a carriage with automatic carriage stop is illustrated in figure 261. Wyssen has developed this carriage and produces them in three dimensions:

Small:	20 kN load capacity
Medium:	50 kN load capacity
Large:	100 kN load capacity

(1) Skyline (3) Haul-back line (5) Carriage (10) Tail spar (11) Tailhold (lower anchorage)

(16) A log on the skyline as carriage stop at the landing.

Figure 257. With small height difference between skyline and landing a log is useful as carriage stop.

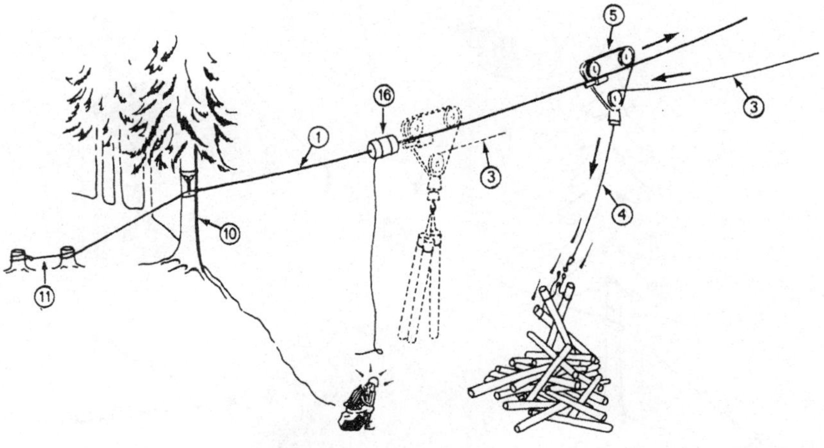

(1) Skyline (3) Haul-back line (4) Hoist line (5) Carriage (10) Tail spar (11) Tailhold (lower anchorage)

(16) A log on the skyline as carriage stop at the landing

Figure 258. With big height difference between skyline and landing the log as carriage stop leads to uncontrolled unloading.

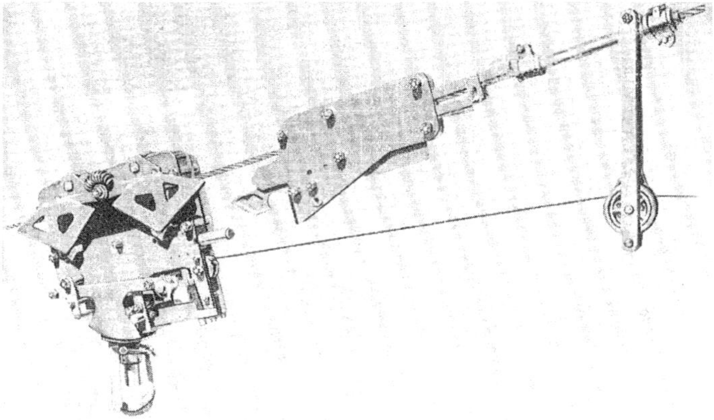

Figure 259. The Wyssen carriage with a carriage stop. (WYSSEN 1983).

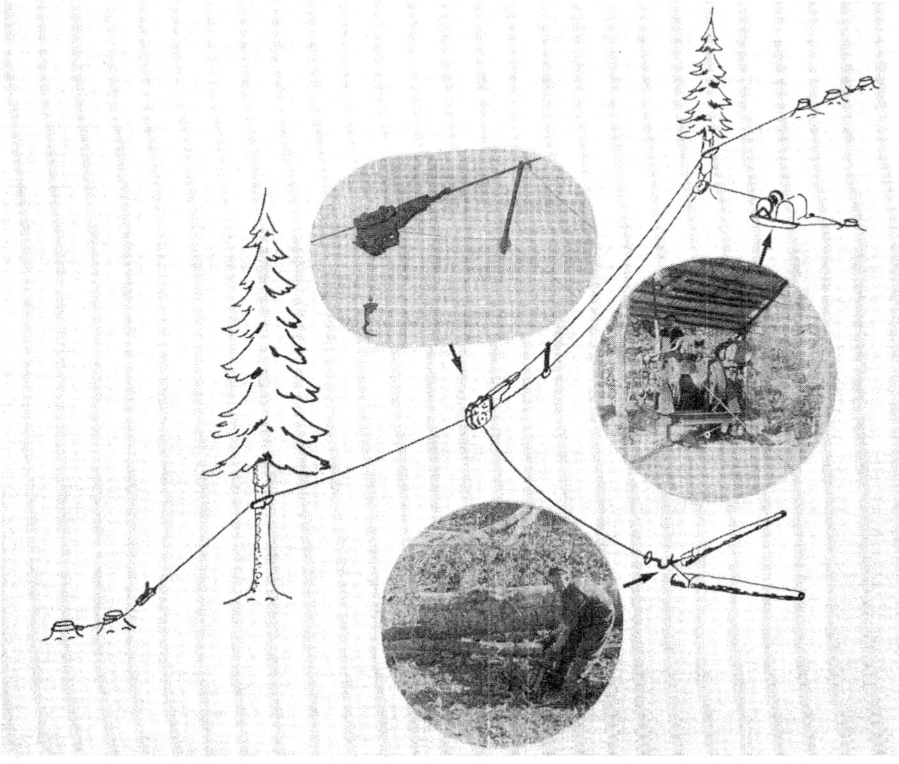

Figure 260. A Soviet Union cable crane with top mounted winch at CHADICHENSKY, Caucasus 1957.

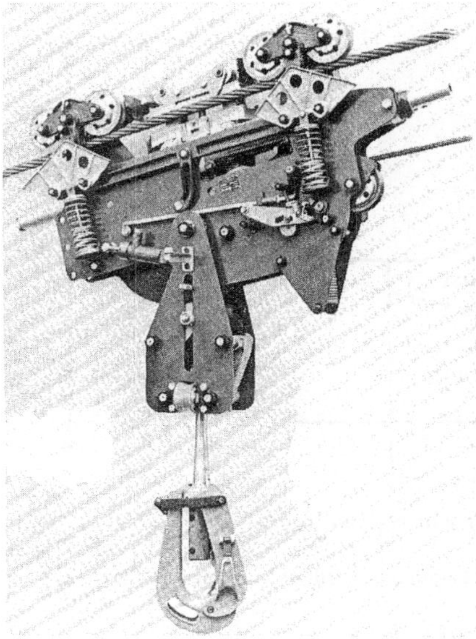

Figure 261. The Wyssen carriage with automatic carriage stop. Hydraulically operated clamp in the carriage. (WYSSEN 1983).

The carriage can pass the cable shoes in the supports at high speed. During the movement along the skyline the carriage wheels drive a hydraulic pump which transmitts oil pressure to a hydraulic accumulator. Only a few meters of movement give sufficient pressure to the accumulator.

When the carriage reaches the place were it should be stopped the winch operator changes the direction of travel. The reversing of the wheels is needed to open a valve and the oil pressure from the hydraulic accumulator is directed to the hydraulically operated clamp. The clamp fastens onto the skyline and holds the carriage.

While the carriage is being clamped to the skyline, the hook is automatically released from the carriage and the hoist line can be lowered to the ground.

The hoist line is pulled in and when the hook hits the carriage it is automatically attached to it. At the same time the oil pressure to the clamp is automatically released. A spring opens the clamp and the carriage may be pulled in one or the other direction along the skyline.

The clamp is mounted on an arm. When it is open it stands aside of the skyline. This enables the carriage to pass the cable shoe on the support. When the carriage stops and the hydraulic pressure from the accumulator activates the clamp, the arm first moves the clamp in over the skyline whereafter the clamp is pressed against the skyline.

This excellent construction makes it possible to stop the carriage anywhere along the skyline, which makes the operation flexible. There exist several types of carriages with automatic stop devices. The Wyssen construction is a good example.

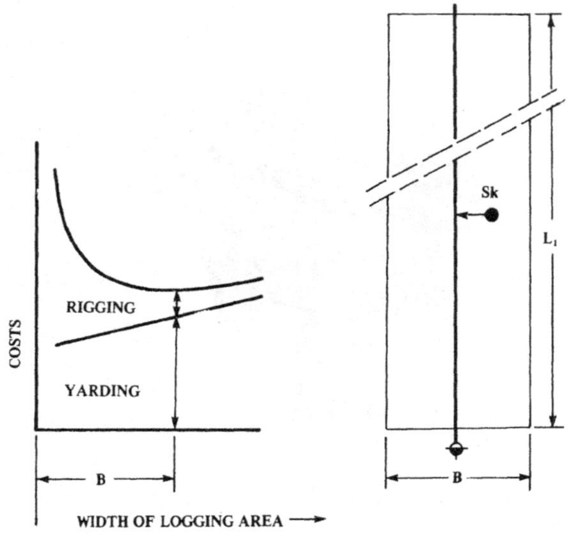

Figure 262. The optimum width of a logging area for cable crane operation.

11.3.1.2. The operational method

The set up of a cable crane with a top mounted small yarder for downhill transport of timber, is laborious. Both the rig up and rig down takes a long time. (See chapter 12.4.1.2.). This makes it desirable to yard the largest possible logging area with the same set up in order to distribute the rig up cost on a larger volume of yarded timber. On a wider logging area the skidding distance increases which also increases the logging costs. Many calculations have been made in order to find the optimum skidding distance as well as the optimum yarding area. As illustrated in figurge 262 the result depends on the cost of labour, the machine costs, the productivity of the system, the efficiency during set up, and the volume to be logged per ha. of the logging area. On the long distance cable cranes which were much in use between 1950 and 1970 the width of the logging area was often 80 to 120 meters. Increased costs of manual labour and equipment, the use of shorter cable cranes and better rig up methods have reduced the width of the logging area considerably (40 – 60 meters for a 300 – 500 meters cable crane). This trend is even more visible with other types of cable cranes and mobile yarders as will be discussed in the following chapters.

Since the lateral skidding is being done by pulling in the haul-back line with the top mounted yarder it is an advantage to pull the logs uphill towards the skyline as illustrated in figure 263. This method may even be used for selective cutting without too much damage to the remaining trees in the stand. Where the skidding distance is extra long it may be an advantage to use a corner block in order to pass obstacles and to avoid damage on standing trees.

An uphill direction of the lateral skidding reduces the forces on the equipment (chapter 5.4). It is also easier for the choker man to pull out the hoist line from the carriage to the logs.

While the system first and foremost is constructed for downhill yarding, uphill yarding may occasionally take place, for example from the lower tail end of the cable crane up to the road (figure 264).

Figure 263. A cable crane operation in selective felling (WYSSEN 1983).

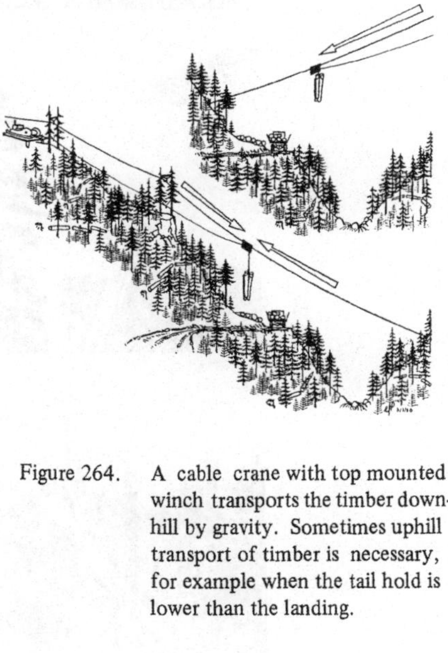

Figure 264. A cable crane with top mounted winch transports the timber downhill by gravity. Sometimes uphill transport of timber is necessary, for example when the tail hold is lower than the landing.

The crew consists of the following men:

1 yarder operator
1 chaser (unloading)
1–4 choker setters

When logging tree dimensions less than 1 cu.m. per tree the crew usually consists of three men. The choker setter pre-sets the chokers while the carriage deliver the load at the landing. Fish-hook yarding is used in small timber in order to increase the size of the load (figure 265). A double set of taglines allows the choker setter to prepare the next load while the yarder is handling the previous load. As soon as the carriage has returned to the logging area the choker setter disconnect the empty tagline with its chokers and connect the loaded tagline to the

Figure 265. The Svendsen top mounted cable crane, Heggelia, Norway 1958.

hoist line by means of a quick coupling. When the cable crane transports the timber from the top of the logging area the choker setter has plenty of time to pre-set the chokers because of the long outhaul and inhaul travel distance. Closer to the landing it may be an advantage to increase the crew to four men i.e. two choker setters.

In big timber fish-hook yarding is seldom used, because it is easier to collect sufficiently big loads. Instead, long chokers are used and the choker setter attach them directly to the hook on the hoist line (figure 266). The main problem in this case is often to wrap the choker

Figure 266. Transporting big timber in the Swiss Alps with the Wyssen cable crane. (WYSSEN 1980).

around the log. For medium sized logs a 1 meter long steel hook is a useful tool to pull the chokers under the log before wrapping.

The yarding production varies with terrain conditions, volume per hectare and volume per tree. WYSSEN (1983) has given some average figures for their cable cranes working under Central European and American conditions. Studies of Northern European cable cranes have been carried out for lighter equipment (SAMSET 1981) and the results are suprisingly similar (table 31). In both cases the average speed during downhill transport was 3 meters per second, the uphill return trip with the empty carriage is faster with the Wyssen cranes (3.3 meters per second) than with the Norwegian cranes (2.1 meters per second). The reason is that the Norwegian cranes have less engine power.

Normal production for this type of cable cranes may be based on a work place time of 12 minutes per turn (cycle) corresponding to five loads per work place hour.This is typical for a 1,000 meters cable crane with an average transport distance of L_T = 500 meters. The work place time is the effective working time plus normal delay times which according to the Norwegian studies is approximately 30 % for this type of cable crane.

Table 31. Work place time for the time elements and the total cycle of a 1,000 meters cable crane (L_T = 500 m).

Work elements	Wyssen cable cranes	Norwegian cable cranes
Return Lateral skidding Transport Loading and unloading	2.5 min. (3.3 m/s) 2.0 » 2.8 » (3.0 m/s) 4.5 »	4.0 min. (2.1 m/s) 1.4 » 2.8 » (3.0 m/s) 4.7 »
	11.8 min.	12.9 min.

Table 32. Typical production for 1,000 meters cable cranes with top mounted yarders for downhill yarding.

m^3/load	m^3/hr. (crew)	m^3/day (6.5 hrs/day)	Number of men in the crew	m^3/man/hr.	Man hr/m^3
0.5	2.5	(16)	3	0.8	1.25
1.0	5.0	(33)	3	1.7	0.59
1.5	7.5	(49)	4	1.9	0.53
2.0	10.0	(65)	5	2.0	0.50
4.0	20.0	(130)	6	3.3	0.30

Cutting is not included in these figures. The yarding production varies from 0.8–3.3 cu.m. per man per hour. The corresponding labour consumption for yarding varies from 1.2–0.3 manhours per cu.m. Time for moving and rig up time is not included in these figures.

The cable crane in Chadichensky in Caucasus was operated by three men and the production varied between 1.4 and 1.9 cu.m. per man per hour of the work place time with a labour consumption of 0.7–0.5 manhours per cu.m. The cable crane yarded oak, beach and fir trees from mature stands. Since the loads usually varied between 1.0 and 1.5 cu.m. per turn, these production figures correspond well to the figures in table 32.

11.3.2. Uphill yarding. Gravity outhaul systems

A well designed and developed road network has an access road to the steep mountain side from which feeder roads branch out. They are laid out parallel to the terrain contours and it is possible to yard the wood uphill from the area below the feeder roads (figure 267). The feeder road makes it possible to bring a mobile yarder to the top of the hillside and use it for uphill yarding.

Figure 267. Uphill yarding with a gravity outhaul cable crane.

Already in 1953 STEINLIN showed that uphill yarding with small yarders gave somewhat higher production than downhill yarding, in spite of the fact that the uphill yarding of the timber works against gravity. The reason is that this type of cable crane is simple and easy to work with. Most of the terminal work elements where production depends on man power, is in the gravity direction.

Most of the yarders and cable cranes described in the following chapters 11.4 and 11.5 may also be used for uphill transport according to the gravity outhaul system. There are many good constructions on the market. We shall limit the description to a few typical examples only, in order to give information about the principles and methods.

Figure 268. The Koller K-800 Yarder with the Koller automatic carriage. Austria 1977.

Figure 269. The Koller K-300 Yarder. Austria 1977.

11.3.2.1. Mobile yarders with tower

The mobile yarders are usually self-propelled and a system of winches and a steel spar are mounted on a truck, or tractor chassis. Figure 268 shows an example of a truck mounted yarder which is constructed for uphill yarding according to the outhaul gravity system. It is

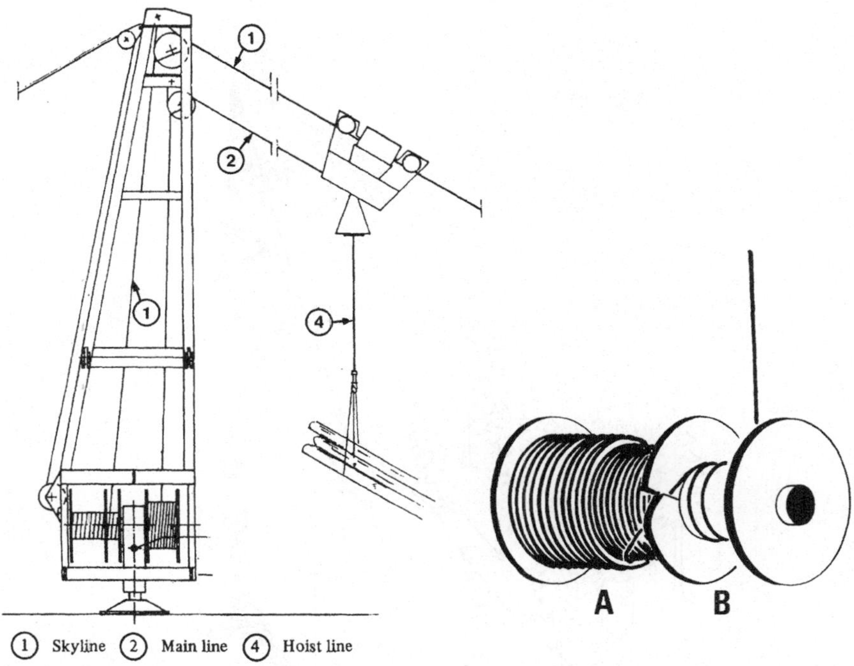

Figure 270. The Koller K-300 Yarder. To the right the skyline drum with a storage section (A) and a tightening section (B).

the Koller K-800 yarder with an 111 kW engine and a 11 meters high tower. Another example is the Koller K-300 mounted on the 3-point link of a 50 kW farm-tractor (figure 269). The winch is the Norwegian Igland winch.This machine is an example of the cooperation between the Austrian company, Koller, and the Norwegian company, Igland. The tractor unit is a cheap unit for the small contractor, and since it usually works on distances shorter than 300 meters the speed is a little more than 3 meters per second. The big unit K-800 works on distances up to 750 meters and the maximum uphill speed with load is 12.5 meters per second. The skyline drum on the tractor unit is divided into two parts, a storage drum (A) and a tension drum (B) in figure 270. The tractor unit may also be mounted on a trailer. In this case the winch is powered by a separate diesel engine of at least 50 kW.

Such yarders have usually two operating drums, the skyline drum and the main line drum (figure 271). It is a live skyline system which makes it possible to change the deflection, i.e. to lift or lower the load during transport. Table 33 shows some specifications for typical yarders for uphill yarding.

A cable crane for uphill yarding does not need a haul-back line if the slope is steeper than approximately 20 %. The carriage returns from landing to logging area by gravity. The system is sometimes called »gravity system». This term, however, does not describe the differences between downhill yarding with top mounted winches and uphill yarding with top mounted yarders. The *gravity outhaul system* uses gravity instead of a haul-back line to move the empty carriage from landing to the place where the choker setters operate.

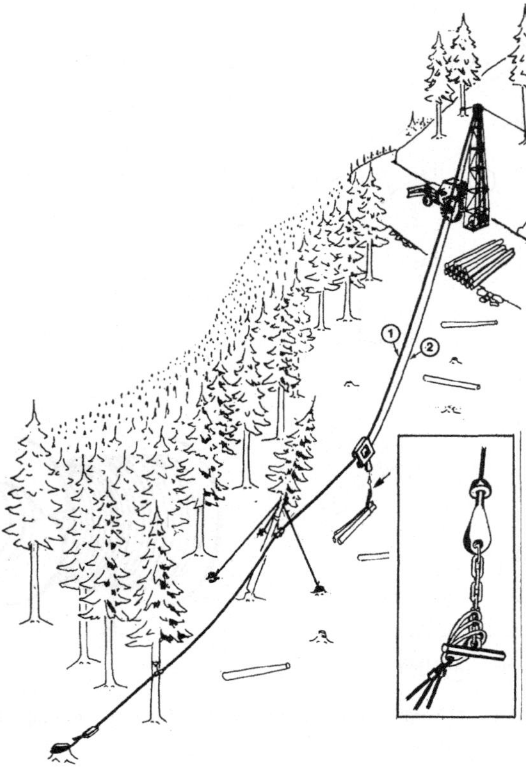

Figure 271. Gravity outhaul system with Koller K-300.

Table 33. Examples of yarder specifications for uphill yarding. Koller yarders.

Model	K-300	K-800	K-500
Tower	Tower	Tower	Swing boom
Height	7 m	11 m	10 m
Engine	50 kW	111 kW	88 kW
Skyline	350 m	750 m	500 m
	16 mm	24 mm	22 mm
Main line	350 m	750 m	500 m
	10 mm	12 mm	12 mm
Pull:	18 kN	9–16 kN	30 kN
Speed:	3.2 m/s	0–12 m/s	0–5 m/s

The big cable cranes (figure 268) often have a four man crew:

1 yarder operator
1 chaser
2 choker setters

The small cranes (figure 269) can operate more profitably with a small crew, often a two man crew:

1 yarder operator who also unhooks the chokers at landing.
1 choker setter

The cutters should be instructed in directional felling, preferably herringbonewise in relation to the skyline. It is an advantage to do the lateral skidding uphill since this direction requires less tension and minimizes the danger of delay times (see figure 263). A good coordination between the felling, cutting and yarding gives increased production. For the small yarders it is sometimes a good idea to combine felling and choker setting in one operation especially when tree dimensions are less than 1 cu.m. per tree.

When the carriage reaches to the logging area it has to be stopped. A carriage stop may be used, such as the Cristy carriage illustrated in figure 276 or the Igland carriage illustrated in figure 280. When the carriage hits the carriage stop, the hook loosens and it is easy for the crew to pull out the hoist line which is a continuation of the main line. The choker setter has set the chokers while the previous load was transported to the landing. It takes therefore, very little time for the choker setter to attach the chokers to the hook on the hoist line. When he has gone a safe distance away and given a signal to the yarder operator, the lateral skidding starts. As soon as the hook arrives the carriage it is released from the carriage stop and the uphill transport starts. There may be a second carriage stop over the landing. When the

Figure 272. Koller carriage with hydraulically operated clamp instead of carriage stop. Arrangement for fish-hook yarding. Hurdal, Norway 1983.

carriage has locked on to this carriage stop, the load is lowered to the landing and unhooked. The hook with the empty chokers is then hoisted to the carriage. This is released from the carriage stop and the return trip by gravity can start. In steep terrain the outhaul speed often is 7–10 meters per second.

Koller in Austria has developed a carriage especially designed for uphill cable crane yarding. The carriage may be used for other types of cable cranes, but has special advantages for uphill yarding (figure 272).

The carriage may be automatically clamped to the cable anywhere along the skyline. The automatic clamp is hydraulically operated. One of the carriage wheels drives a hydraulic pump which gives pressure to a hydraulic accumulator. Sufficient pressure is achieved after two or three meters movement along the skyline. When the carriage reaches the place where the yarder operator wishes to stop, he returns the carriage. This activates a valve which releases the pressure from the hydraulic accumulator and the clamp fastens onto the skyline and the hook is released from the carriage. The hoist line with the hook drops down to the ground. After the load has been skidded and lifted to the carriage, the hook is locked in the carriage while another valve releases the pressure on the clamps. These open and the transport along the skyline can start.

Another advantage of the Koller carriage is that it can pass the cable shoes on supports (figure 271). Koller produces this carriage for the following load capacities: 10 kN, 25 kN and 50 kN.

An example of an allround yarding system for uphill yarding is the Urus cable crane mounted on an Unimog tractor (figure 273 and 274). Only the skyline drum and the main line drum are used during the yarding operation. The other two drums on the yarder (the haul-back

Figure 273. Urus cable crane on an Unimog tractor. A separate winch drum set on the yarder moves the logs from landing to truck loading pile. Austria, 1978.

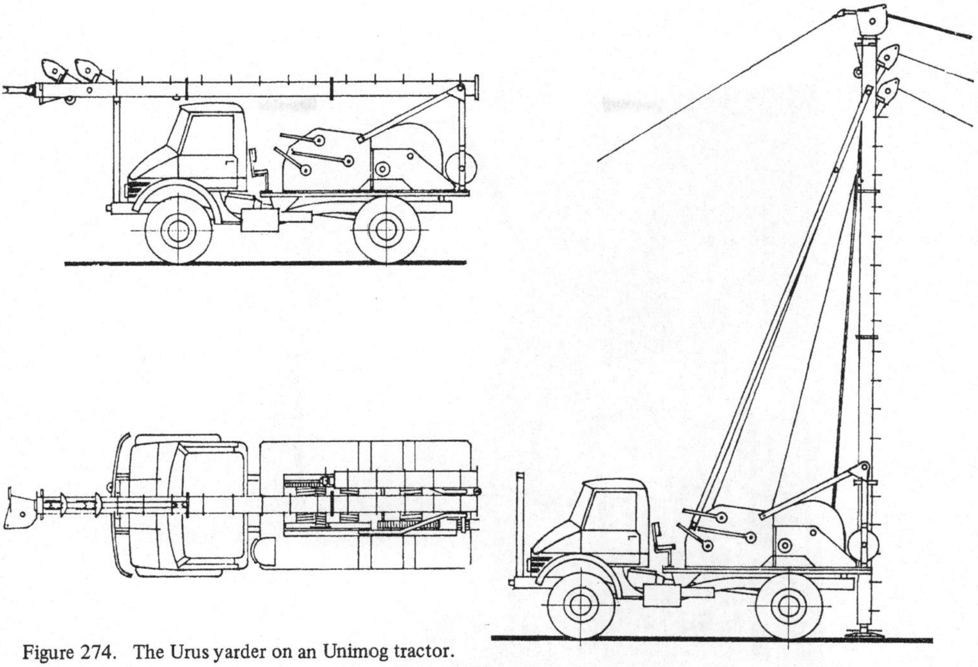

Figure 274. The Urus yarder on an Unimog tractor.

drum and the straw line drum) may be used to move the logs from the yarder to a loading place along the feeder road (figure 273). This is a great advantage in steep terrain where the feeder road often is too narrow for both the yarder and a loading place. In other cases a skidder must be used to pull the loads away from the landing (figure 268). Table 34 shows an example of the technical specifications for this type of yarder. The Urus crane has been chosen. This crane is sold in several models and we have selected a small and a big model in our example.

Table 34. Examples of specifications of a small (1.5 cu.m. load) and a big (5 cu.m. load) all round yarder. Urus cable crane, built by Hinteregger.

Model	Urus 250–1.5	Urus 500–5
Tower height	6 m	9 m
Engine	40 kW or 51 kW	66 kW
Skyline	250 m	500 m
	18 mm	30 mm
Main line	250 m	500 m
	10 mm	16 mm
Pull:	15 kN	50 kN
Speed:	0.9–4.0 m/s	0.3–9.7 m/s
Haul-back line	500 m	1,000 m
	10 mm	16 mm

In addition the yarder is equipped with drums for straw lines and guylines as well as equipment for tilting the tower.

Figure 275. Critical site yarder. The swing boom places the logs on the truck road where they are loaded directly on trucks.
Above: Uphill load transport.
Below: Gravity outhaul. Rocky Mountains, B.C. 1979.

11.3.2.2. Yarders with swing boom for uphill yarding

The swing boom yarders are usually mounted on a self-propelled chassis with wheels in order to move the machine from one logging area to another along truck roads (figure 275). The winch drums with the engine and the tower are mounted on a unit which allows the operator to swing the machine from one side to another. A vertical steel tower is placed above the point around which the winch units can be turned. The guylines are attached to the top of the vertical tower and the swing movement does not affect the guylines. The boom leans forward from the winch drums on the yarder, and the angle may be adjusted by a tackle between the vertical tower and the leaning tower.

Figure 276. Gravity outhaul system. A: Critical site yarder. B: Christy carriage. C: Linkbelt 108-C Yarder.

This type of yarder needs two operating drums only, the skyline drum and the main line drum. It is a live skyline systems, which means that the skyline deflection as well as the height of the carriage above ground can be adjusted during transport. Table 35 gives some examples of technical specifications for swing boom yarders, which have been used in the northern part of Rocky Mountains (USA and British Columbia). (Figure 276).

No carriage stop is needed at the landing for this system. When the load has been transported uphill the yarder swings the leaning boom in over the road and the yarder pulls the load off the ridge and into the road. The loading machine can take the turn and load it directly on a truck (hot logging). Usually the boom is moved in over the road with rather slack lines. Afterwards the swing unit of the yarder is locked by a brake and the load can be pulled over the ridge and into the road by the winch.

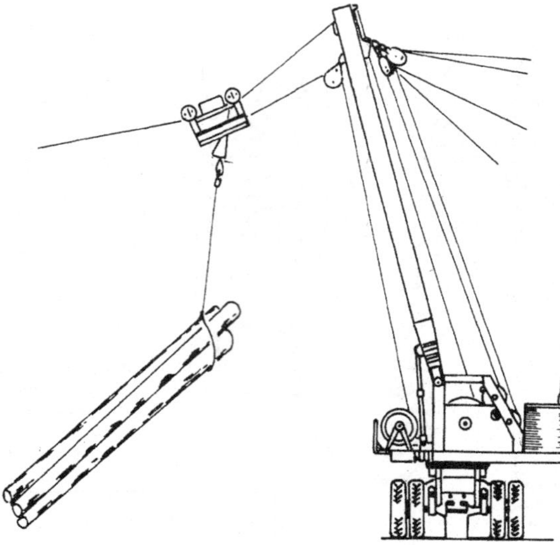

Figure 277. Koller K-500 Yarder.

Table 35. Examples of specifications for swing boom yarders.

Model	Linkbelt 108-C	Critical site yarder
Tower height	24 m	24 m
Engine power	162 kW	234 kW
Skyline	500 m	300 m
	19 mm	19 mm
Main line	500 m	300 m
	16 mm	16 mm
Carriage	Maki (carriage and carriage stop)	Cristy (carriage and carriage stop)

In the terrain there must be a carriage stop where the lateral skidding will be done. The empty carriage runs downhill by gravity. When it hits the carriage stop the hook on the main line is released from the carriage and falls down to the ground where the choker setters can pull out the main line and fasten the chokers. When they have moved a safe distance away they give a signal to the yarder operator who pulls in the load. When the hook locks onto the carriage this is released from the carriage stop and the uphill transport starts. When the load has been brought to the tower the lines are slackened and the yarder operator swings the boom in over the road so that he can pull in the load onto the road by means of the main line. Afterwards he swings the boom in the skyline direction, tightens the skyline and returns the empty carriage downhill by gravity. In some cases there is no vertical tower and the guy-lines have to be adjusted to allow for the side movement of the leaning tower (figure 277).

The critical site yarder has pneumatically operated mechanical brakes and clutches (a slipping clutch system).It is specially designed for very steep uphill yarding where it is rather small space for landings.

The Koller K-500 uses the Koller automatic carriage which may be stopped hydraulically any place along the skyline. (Figure 277).

The swing boom yarders which are examplified here are built for loads between 1 and 3 cu.meters. The Canadian and American yarders are more powerful than the European (table 33 and 35). The Canadian/American yarders are using smaller skyline dimensions and operate with higher deflection during the uphill load transport. The Central European constructions are built for terrain where less deflection can be expected and the Koller carriage may even pass supports when this is necessary.

11.3.2.3. Uphill cable crane yarding combined with tractor skidding

A double-drum winch on a skidder may be used for uphill yarding with live skyline. This system was used in the Moroto forest in Japan 1968 (figure 278). A double-drum winch was

Figure 278. Gravity outhaul system combined with tractor skidding. Moroto forests, Japan 1968.
1: Tractor at the road side. 2: Uphill transport. 3 and 5: The tractor pulls the load to landing.
4: Mitsubichi skidder with double-drum winch. 6: Measurement and bucking.

mounted on a Mitsubishi articulated frame steered tractor. The double-drum winch was equipped with 100 meters main line and 200 meters haul-back line.The diameter was D=10 mm. In the very steep terrain, the narrow feeder truck roads left little space for landings. The road spacing was less than 200 meters and highlead yarding was used for downhill yarding and cable crane yarding with gravity outhaul was used for uphill yarding. The tractor was driven up to the edge of the road near the very steep hillside. The haul-back line was pulled downhill and fastened to a big stump or to a big tail tree. A simple carriage pulled out the main line by gravity downhill to the logging area. Since the 10 mm thick haul-back line, used as skyline, had small carrying ability, the trees were limbed and transported with the top end forward. Since the lines were light the tail end of the skyline (haul-back line) was frequently moved from one stump or tail tree to another, and little lateral skidding was required.

There was only a two man crew, and when the choker setter had fastened the top end of one or two trees to the hook, the tractor operator pulled in the carriage uphill to the edge of the road. The tractor operator then released the brake on the skyline drum (haul-back drum) and applied the brake on the main drum. He then moved the tractor forward, ground skidding the load over the edge and along the road 50 to 100 meters to the landing. After unhooking he drove the tractor back again while he pulled in the skyline. He then tightened the skyline and the carriage was returned to the terrain by gravity. This flexible system gave a rather low yarding production, but a high production for the full harvesting circuit (felling and branching on the logging area and bucking and sorting on the landing). Correct bucking of the stems into assortments lead to high product prices and justified the method in this case.

This method may be used when the tree dimensions are smaller than one cu.m. per tree, and where the terrain conditions allow high deflection.

11.3.2.4. Double haul-back line as skyline

Light 2-drum winches with tower have been developed as additional equipment for farm-tractors. The winch with tower may be attached to the tractor by means of the 3-point hitch and the winch is driven by the power take-off. Some of these winches are of the standard clutch and brake type, and some of them have interlocking between the two drums. The equipment is used for cable crane yarding with running skyline and highlead yarding on distances up to 200 meters.

The equipment is cheap and useful for farmers or small contractors. The equipment available on the market is built for tree dimensions less than one cu.m. per tree (maximum 1.5 cu.m. per tree). By increasing the dimensions the same principle might be used for bigger tree dimensions. The advantage of this method is that the same equipment may be used for downhill or horizontal highlead yarding as well as uphill cable crane yarding with gravity out-haul of the carriage.

A double haul-back line is used as skyline. It is therefore, possible to use higher tension in the skyline. There can also be supports along the skyline road. Usually the maximum distance is 200 meters but by lengthening the lines it may be used on distances up to 300 meters.

The dimensions of the main line and the haul-back line is 10 mm (figure 279 and 280).

The haul-back line is pulled out downhill to the tail tree and back again to the top of the tower and anchored to a stump or a tree. A second guyline is being used in addition to this.

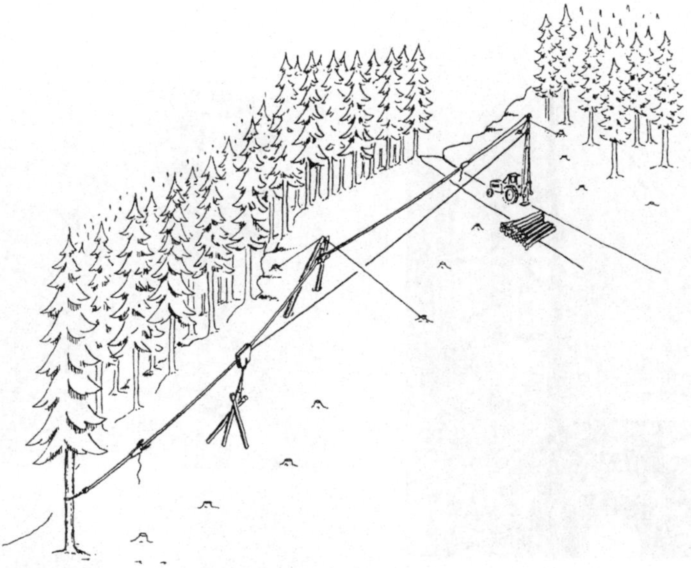

Figure 279. Gravity outhaul system. Igland double-drum winch with telescope tower on farm-tractor. The double haul-back line used as skyline. (Norway, 1979).

If supports are needed a simple twin strut support is being used (figure 280, D). These are placed sideways, parallel to the skyline and anchored with a guyline. The cable shoe has double width so that the two parts of the haul-back line have place beside each other in the shoe (figure 280, C). The cable shoe is lifted towards the top of the support by a lifting line which also serves as the second guyline on the support (figure 279). A special carriage has been developed for the purpose, the Igland carriage, for uphill cable crane yarding. The wheels in the carriage have double width in order to pass the cable shoe.

Two carriage stops are used on the skyline, one over the landing and one at the place where the lateral skidding of the pre-set chokers starts. The carriage stops are fastened to one of the double haul-back lines only, namely the part of the line which is going out from the haul-back drum on the winch. This makes it possible to adjust the tension of the skyline during operation, by using the clutch and the brake on the haul-back drum.

Usually the crew is a two man crew, one choker setter and one winch operator, who also unhooks the chokers at the landing. Three methods may be used:

1. The felling, limbing and cutting has been done by a separate crew and the two man yarding crew carries out yarding and piling only.

2. The two man crew carries out the felling, limbing and bucking simultaneously with the yarding. The feller in the terrain fells and delimbs the upper side of the logs and sets the chokers. The winch operator unhooks the chokers, returns the empty carriage to the logging area. While he is waiting for the next load he takes off the rest of the limbs with his chainsaw and makes the final bucking. This method is excellent for a father-and-son-operation in a farm forest.

Figure 280. Igland interlocked double-drum winch used as gravity outhaul cable crane. Kviteseid, Telemark, Norway 1981.

3. Full tree logging. The choker setter fells the trees and sets the chokers. The winch operator unhooks the chokers and returns fresh chokers to the logging area. He uses the time when waiting for the next load to loosen the chokers from the trees. A separate limbing machine takes care of the conversion of the trees.

The skyline is rather low above the ground, usually 5 meters except when passing cliffs, creeks etc. Usually the loads are being pulled with the front end lifted like a highlead yarding. The equipment is light and easy to work with for the crew. The change of the carriage stop from one place to another takes very little time, usually 1 – 2 minutes. When the choker man wishes to start lateral skidding higher up on the hillside, he waits until the carriage has been returned and fastened by impact to the carriage stop. On a signal the skyline is lowered and the choker setter loosens the carriage stop from the skyline. The winch operator then pulls the carriage with the carriage stop uphill 5 or 6 meters to the next point for lateral skidding, and the choker setter fastens the carriage stop to the skyline. The winch operator tightens the skyline again and the yarding continues.

Sometimes the two man crew has a surplus delimbing machine at their disposal. When the two man crew has yarded a pile of unbranched trees, for example during 1/2 – 2/3 of the day the winch operator uses the rest of the day to limb and buck the unbranched stems to logs, while the choker setter are felling the trees for the next shift. This is also an efficient organization of the work which gives rather high production and low logging costs. It is especially useful for small contractors.

11.3.2.5. The production

The production, the productivity and the labour consumption for the various cable cranes depend on the terrain conditions and the tree dimensions and big variations in productivity figures exist.

Usually the productivity for uphill cable crane yarding is 15 – 25 % higher than downhill yarding.

The daily production also depends on the length of a working day and we have referred the production to the work place hour, i.e. the effective time plus normal delay times.

Table 36 gives a summary of some typical production figures for the systems described in chapter 11.3.2.

Table 36. Typical production per work place hour (effective time and normal delay times) during uphill yarding with some cable cranes in Rocky Mountains 1979, Austria 1977 and Norway 1981.

Yarder	kW	m^3/tree	Average yarding distance m	Crew men	Production of crew m^3/hr.	Productivity m^3/man/hr	Labour consumption manhours per m^3
Koller K-800	111	1.5–2	250	4	15	3.8	0.26
Linkbelt 108-c	162	1.3	150	4	11	2.8	0.36
Critical Site yarder	234	1.3	100	5	13	2.6	0.38
Igland Telescope	50	0.3	100	2	6.2	3.2	0.31
and Koller K-300			150		5.8	2.9	0.35
			200		5.2	2.6	0.39

These figures are collected from practical operations for Koller K-800 in Austria 1977 and for Linkbelt 108-C and Critical Site yarders in Northern Rocky Mountains 1981. The small yarders are mounted on 50 kW farm-tractors. I.e. the Igland double-drum winch for highlead yarding with telescope tower but used as cable crane and the Koller K-300 cable crane. These figures are based on detailed studies over one logging season in 1981. The average slopes were in all cases 50–60 % uphill. The critical site yarder, however, worked on a very steep slope of 70 %.

Table 36 shows that the big yarders usually have a 4-5 man yarding crew and may work on a total cable crane length up to 500 meters. When the tree dimensions are between 1 and 3 cu.m. per tree the typical production for yarding only may be between 10 and 15 cu.m. per work place hour. The small yarders are usually operated by a 2-man crew and the maximum distance is 200 meters and sometimes 300 meters. A typical production would be between

5 and 6.5 cu.m. per hour for a 2-man crew. The big yarders have a productivity between 2.5 and 4 cu.m. per man per hour corresponding to 0.40 – 0.25 manhours per cu.m. The small yarders have a productivity of 2.5 – 3.5 cu.m. per man per hour corresponding to 0.4 – 0.33 manhours per cu.m. Light equipment with small crews gives excellent productivity in small and medium sized tree dimensions if a dense network of feeder roads exists.

11.4. Cable cranes with yarders at valley station

In remote mountainous terrain the first roads are laid out near the bottom of a valley, often close to a river or a lake. The first yarders developed from the donkey yarders. The main drum pulled in the load and the haul-back drum returned the rigging or the carriage to the logging area. The yarders were transported to the logging area and placed near the landing at the valley station. Such yarders remain stationary during the entire yarding operation.

Since these yarders were stationary most of the time, they were usually mounted on skids. This was also convenient for short movement, often through difficult terrain, from road-side to the yarder station. The yarder pulled itself by means of the main line and the winch.

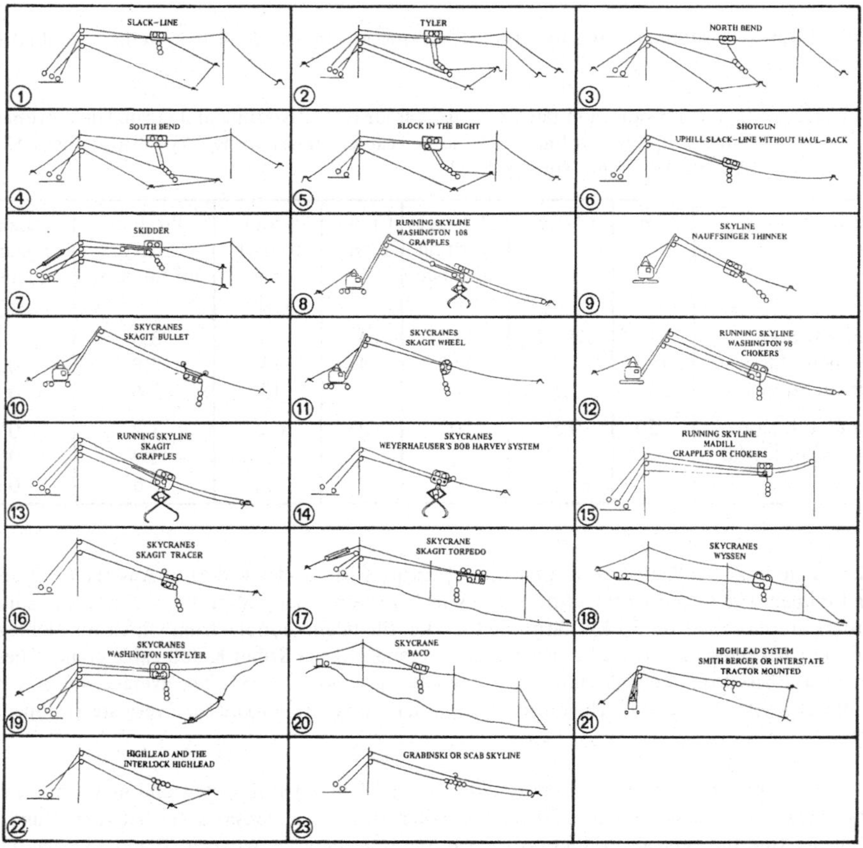

Figure 281. Some cable crane systems which have been used for logging on the American West Coast and in British Columbia (O'LEARY 1970).

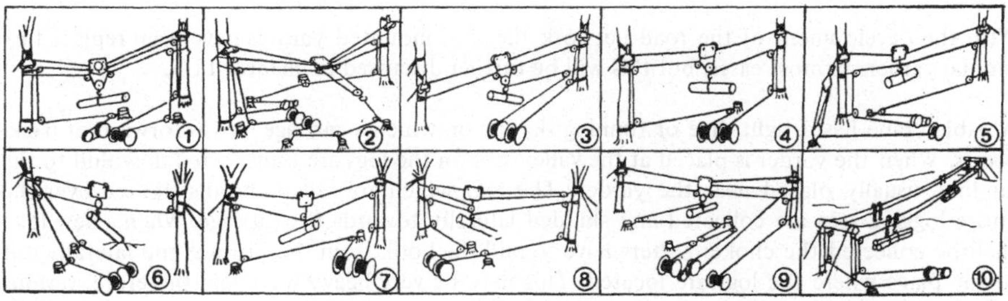

① – ③ Variants of Tyler system ④ North Bend system ⑥ – ⑧ Variants of running skyline system
⑨ Kumamoto system ⑩ Twin cable with circulating carriages

Figure 282. Some Japanese cable systems.

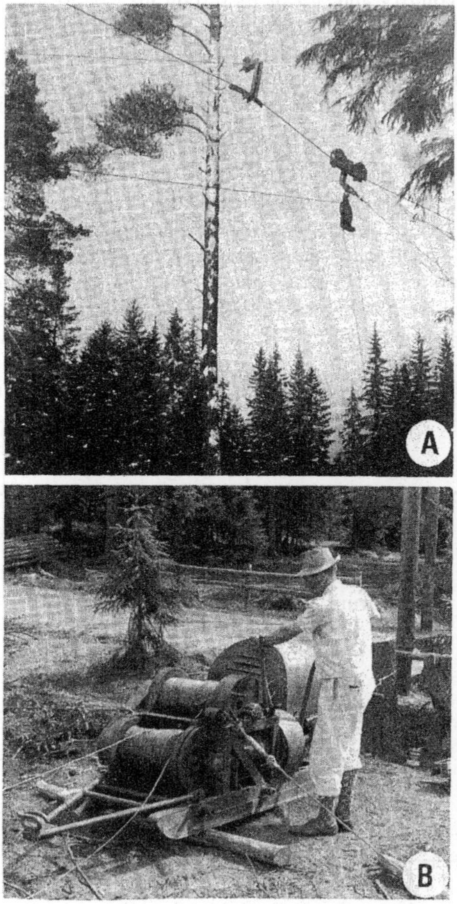

Figure 283. Cable crane yarding with the Vossa yarder. Nordmarka, Norway 1955.
A: The carriage passes a crossline support. B: The double-drum yarder powered by a Volkswagen engine.

With the development of the road network the skid mounted yarders have been replaced by mobile yarders in most cases, but this will be further discussed in chapter 11.5.

A cable crane has a tight, live or running skyline on which a carriage moves forward or backwards. When the yarder is placed at the valley station the logs are transported downhill to the landing, usually placed near the yarder. The logs which are felled and distributed over the entire logging area are collected and skidded laterally towards the carriage. When a new load shall be collected the choker setters have to pull the hook from the carriage and laterally out to the place where the logs are located. This may be very heavy work and therefore, various systems and methods have been developed in order to reduce the work load on the labour force and to increase the efficiency. Figure 281 shows some of the American and Canadian systems. The first part of the development included cable cranes with yarders at the valley station. The diagram in figure 281 also includes other systems, frequently used together with mobile yarders which will be described in chapter 11.5.

Japan is a mountainous country and 2/3 of its terrain may be logged by cabled cranes. The development of Japanese cable methods are highly advanced. It is partly based on American, Canadian or Central European methods and partly on some very interesting Japanese systems. Figure 282 illustrates a few of the commonly used cable systems in Japan and most of them are powered by a yarder mounted on runners at the valley station.

11.4.1. The yarders

Small yarders have been used for the transport of logs from 1 cu.m. per tree. The Vossa yarder was built in 1955 (figure 283). It is a reliable machine and is still in use many places. It is used for a cable crane which may yard distances up to 400 meters. It is powered by an 18 kW Volkswagen motor and has two speeds on each drum. The specifications are shown in table 37. This yarder was also used for 800 meter cable cranes. In this case the two drums were replaced by a capstan drive (figure 284).

Table 37. Examples of some stationary Norwegian yarders for cable cranes.

Yarder	Vossa yarder (figure 283)	Bjønn yarder (figure 285)	Anleggstransport (figure 286)
Engine power	18 kW	18 kW	55 kW
Yarding distance	400 m	1,200 m	800 m
Load capacity	1 m³	1.5 m³	2 m³
Winch	2 drum	2 drum	1 drum 1 capstan
Skyline	12 mm	16 mm	18 mm
Main line	7 mm	8 mm	10 mm
Max. pull	9.4 kN	28 kN	44 kN
Speeds	1.1–2.7 m/s	0.5–8 m/s	0.5–5 m/s
Haul-back line	6 mm	8 mm	
Speeds	1.4–3.7 m/s	0.5–8 m/s	
Endless line			10 mm
Speeds			0.5–5 m/s

The Bjønn yarder is especially designed for cable cranes with distances up to 1,200 meters (figure 284, A and figure 285, C). The yarder is powered by an 18 kW Volkswagen motor and there are three speeds in both directions.

Figure 284. The cable crane in Luksefjell, Norway 1961.

A: Double-drum Bjønn yarder powered by a 18 kW Volkswagen engine (constructed by Voss Sveis og Mekanikk).

B: Double capstan drive powered by a 18 kW Volkswagen engine (constructed by Hordaland Mekaniske Verkstad). The tension in the endless main/haul-back line is adjusted by a heel block behind the winch.

C: The double capstan yarder (Hordaland Mekaniske Verkstad). The wheel to the right is for speed measurements.

The Anleggstransport yarder (figure 286) has one capstan which pulls the carriage with an endless line and one main drum. It can work on distances up to 800 meters and the load capacity is 2 cu.m. The yarder is powered by a 55 kW engine and has four speeds.

None of the Norwegian yarders used a fan brake.

The Baco yarder (figure 287) has one winch drum and one groove wheel pulley. The cable carriage has an automatic stop device which allows the carriage to stop anywhere along the skyline. The Baco alround cable cranes uses an endless line to pull the carriage and when it is very steep the yarders may be equipped with a fan brake.

Figure 285. Bjønn cable crane in Ringnes, Norway 1956.
A: View from the felling site to the yarder station along the 1,000 m long skyline.
B: Prototype of the double-drum yarder.
C: The production model of the double-drum Bjønn yarder.

Table 38. Examples of Central European yarders from Baco, Switzerland. One winch drum and one groove wheel pulley.

Max. load	20 kN	30 kN	50 kN
Yarding distance	1,800 m	1,600 m	1,400 m
Skyline	20 mm	24 mm	28 mm
Main line	12 mm	13 mm	17 mm
Speed	0–8 m/s	0–8 m/s	0–10 m/s
Endless line	10 mm	11 mm	16 mm

Figure 286. The Anleggstransport cable crane with valley mounted yarder powered by a 55 kW engine. Heggelia, Norway 1951.
A: The carriage with main line, haul-back line and hoist line.
B: The yarder with a capstan for the endless main/haul-back line and a winch drum for the hoist line.

Table 39. Examples of Japanese yarders from Nansei. Two winch drums and one capstan pulley.

	Light yarders	Medium yarders	Heavy yarders
Engine power	6–7.5 kW	22–44 kW	75 kW
Load capacity	1 m^3	2 m^3	3 m^3
Yarding distance	200–400 m	300–600 m	< 800 m
Skyline	14 mm	18 mm	24–28 mm
Main line	8–10 mm	12 mm	12 mm
Pull	7.5–10 kN	20–50 kN	70–85 kN
Speed	0.5–5 m/s	0.7–8.3 m/s	0.5–6.5 m/s
Haul-back line	8 mm	10 mm	12 mm
Endless line	8 mm	10 mm	12 mm

Figure 287. Baco SWU-yarder with 1 winch drum and 1 groove wheel pulley. The Baco Universal carriage with automatic carriage stop (Baco, Switzerland 1980).

The Japanese yarders are built in many sizes and are powered with engines from 6 kW up to 100 kW. This is because of the large variety in tree dimensions in Japan from very small trees in thinnings to big logs in overmature stands, i.e. from 0.05 cu.m. per tree to 10 cu.m. per tree. Nansei is one of the biggest yarder manufacturers in Japan and examples of some of their yarders are shown in table 39 (figure 288).

The Japanese yarders are usually equipped with one or two normal winch drums and a capstan for an endless line. In some cases there are more than two winch drums.Some of the Japanese yarders are equipped with a fan brake in order to control the speed or line pull.

11.4.2. The operational methods

Some of the operational methods which are described in chapter 11.5 may also be used for downhill yarding with the yarder located in the valley. In this chapter we will concentrate on some methods which are typical for cable cranes with yarders at a valley station. These

Figure 288. A Nansei yarder with two winch drums (left) and one capstan pulley (right). (Nansei, Japan 1980).

cable cranes usually have tight skylines. In some cases the load is transported off the ground and is completely carried by the carriage and skyline. In other cases the skyline is rigged so that the carriage lifts the front end of the loads only (highlead skidding).

11.4.2.1. Carriages with hoist tackle

When the tree dimensions are less than 1 cu.m. per tree very light equipment may be used for yarding distances up to 200–400 meters. The Vossa yarder may be used as an example. The equipment is very light and flexible. It is easy to rig up and easy to handle for the crew. A 10 kN load may be transported in spite of the small cable dimensions (skyline d=12 mm, main line d=7 mm and haul-back line d=6 mm). Skill is required to rig up the equipment, so that it is loose and flexible. The elasticity in the supports and cables helps to resist high forces (figure 289 and 290). The yarding crew has usually 2 – 4 men:

1 yarder operator
0–1 chaser
1–2 choker setters

Sometimes there is only the yarder operator at the landing who also unhooks the chokers. When transporting between 200 and 400 meters one choker setter is sufficient, but if the yarding distance is less than 100 meters it may be better with two choker setters. All depends on the terrain conditions and the size of the wood. The smaller the wood, the more difficult to collect sufficiently big loads, which increases the need for manpower during choker setting. Chain chokers (figure 291) or fibre chokers are better for small sized tree dimensions while chokers of wire ropes are to be prefered for tree dimensions over 0.5 cu.m. per tree.

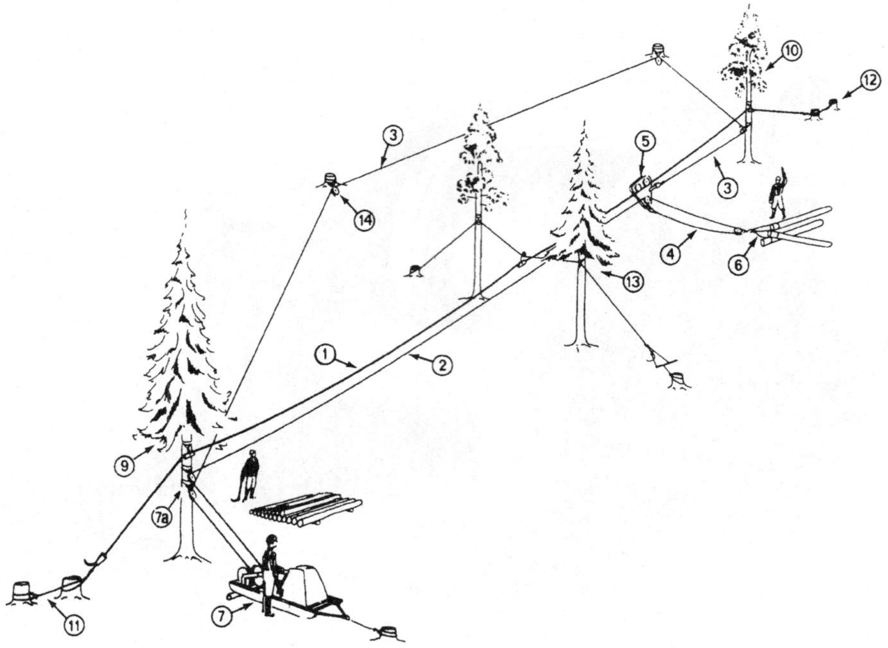

(1) Skyline (2) Main line (3) Haul-back line (4) Hoist line (double hoist line and lifting block)
(5) Carriage (6) Chokers (7) Double-drum winch (7a) Guide blocks (9) Head spar (10) Tail spar
(11) Lower anchorage (12) Tailhold (13) Support (14) Terrain blocks

Figure 289. Cable crane with valley mounted double-drum yarder and carriage with double hoist line (hoist tackle).

Figure 290. The carriage of the Vossa cable crane.

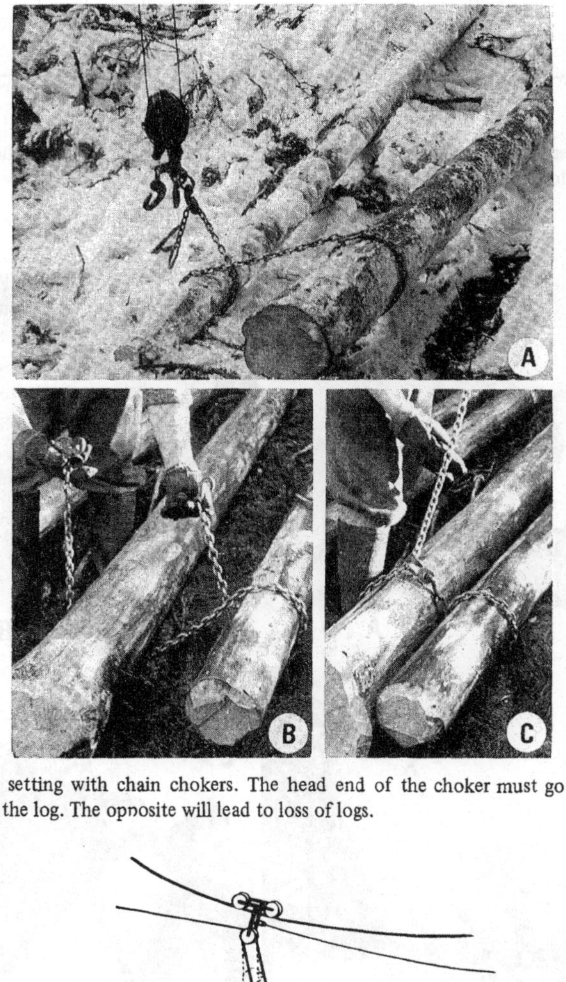

Figure 291. Choker setting with chain chokers. The head end of the choker must go out behind the sling around the log. The opnosite will lead to loss of logs.

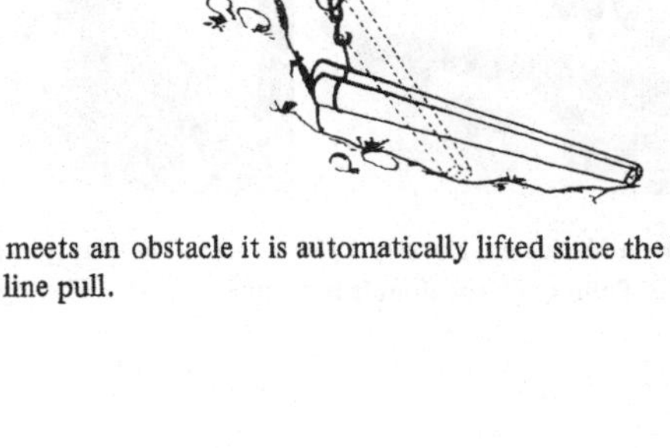

Figure 292. When the load meets an obstacle it is automatically lifted since the lifting force is approximately twice the main line pull.

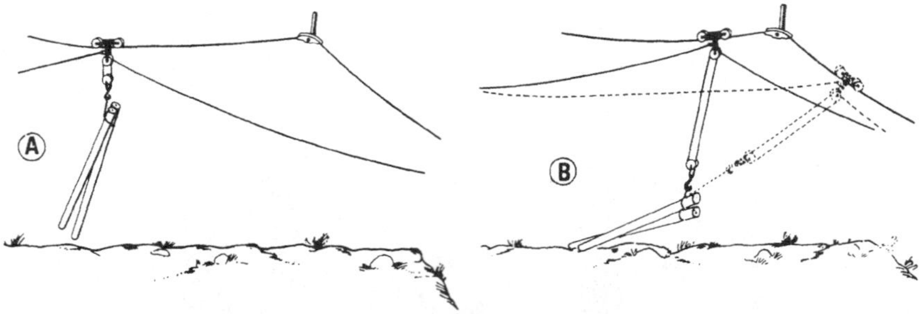

Figure 293. The Vossa cable crane is light and flexible. When the carriage passes a difficult support the load may be lowered before the carriage passes.

Figure 294. Vossa cable crane in Maridalen, Norway 1954. A: The empty carriage returns to felling site. B: Stop. C: Pulling out the double hoist line. D: The logs ready for lateral skidding.

With the light equipment the loads are highlead skidded during the yarding, in order to lift the front ends over obstacles. The hoist tackle in the carriage helps to do this (figure 292). If a load gets stuck behind an obstacle the winch operator may turn the haul-back line and brake it while pulling in the main line. The result is that the load is lifted over the obstacle.

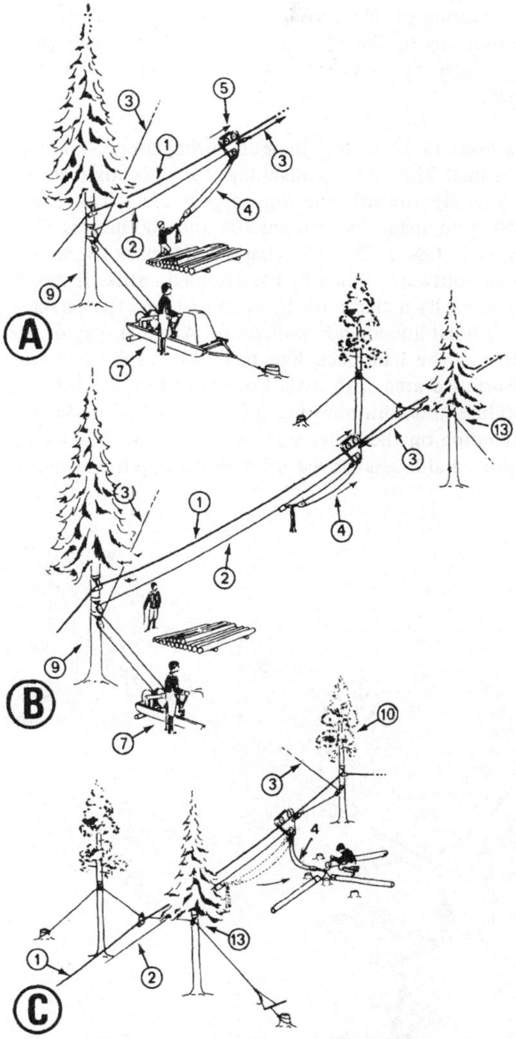

(1) Skyline (2) Main line (3) Haul-back line (4) Hoist line (double hoist line and lifting block)
(5) Carriage (7) Double-drum winch (9) Head spar (10) Tail spar (13) Support

Figure 295. It is often heavy work for the field crew to pull out the double hoist line. The work may be simplified as follows:
A: The chaser is holding back the hook.
B: The chaser attaches the hook to the main line so that the carriage returns to the felling site with the hoist line already pulled out.
C: This makes it easier for the choker setter to bring the hoist line to the logs.

Sometimes the skyline passes the cable shoe in a support at a too sharp vertical angle (figure 293). Since the skyline tension is low the main line may pull the carriage and the skyline downwards in front of the support. This creates problems when passing the cable shoe. The problem may be overcome by lowering the load in front of the cable shoe and let the carriage pass unloaded. When the carriage has passed, the load may be lifted again by the hoist tackle (figure 293). The securing of the carriage to the skyline is important with such flexible systems. Due to vibrations in the lines, the carriage may easily jump off the skyline when passing a support at high speed. The best solution is to have an individual lock under each wheel on the carriage.

The advantage of a hoist tackle is that the pull during lateral skidding is approximately twice the pull in the main line. This is a disadvantage for the choker setter when pulling the hook from the carriage laterally towards the logs (figure 294). This is especially the case during steep downhill yarding on distances between 200 and 400 meters. One way of tackling such a problem, is illustrated in figure 295. The chaser holds back the hook while the yarder operator moves the carriage outward. When he has attained a sufficient length of the hoist line the chaser attaches the hook by a clamp to the main line, and the carriage can be returned to the logging area with the hoist line already pulled out. Another way of reducing the work for the choker setters is to use the haul-back line to pull out the hook (figure 296). This method works fine on a clearfelled area with little brush and terrain obstacles. The yarding starts on one side of the skyline and continues herringbonewise to the far end of the area and back again towards the landing on the other side of the skyline. The terrain block which gives the direction of the lateral skidding is moved from one stump to another as the yarding proceeds.

(1) Skyline (2) Main line (3) Haul-back line (3a) Pullout line (4) Hoist line (double hoist line with lifting block) (5) Carriage
(7) Double-drum winch (9) Head spar (10) Tail spar (11) Lower anchorage (12) Upper anchorage (tailhold) (14) Terrain block

Figure 296. When the haul-back line is used to pull out the hoist line, the work sequence should proceed in a herringbone pattern of narrow strips, up one side of the skyline and down again on the other side of the skyline.

Figure 297. Vossa cable crane in Berby, Norway 1959.
A: The haul-back line pulls out the double hoist line.
B: The loaded carriage passes a support. The haul-back line is attached to the lifting block.

Figure 298. A. It is advantages to place the pile perpendicular to the skyline. This makes it easier to shape the height and the width of the pile on the landing.
B. When the logs rest close to the centre of gravity, an experienced operator can turn the load and place it parallel to the logs in the pile by returning the carriage on the skyline. (see figure 225).

The landing should always be placed directly underneath the skyline road and perpendicular to the skyline direction, whether yarding logs or stems in full length. When the load is being stopped close to its centre of gravity it is easy to turn the logs 90° and place them parallel to the logs in the pile. This makes it easier to shape the pile, the width of it as well as the height (figure 298).

This type of yarder may be used in any terrain, uphill and downhill. If the yarding is uphill on a slope steeper than 20% the carriage may be returned down the skyline by gravity. No

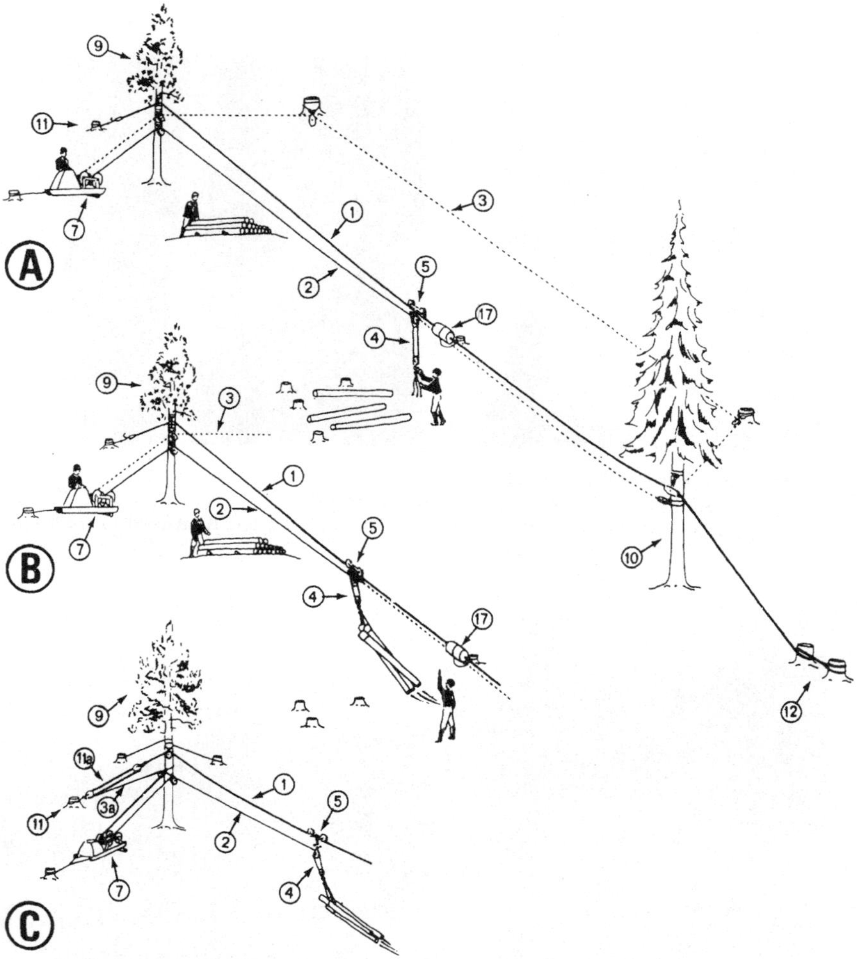

(1) Skyline (2) Main line (3) Haul-back line (not used when steeper than about 20%)

(3a) Haul-back line used to tighten the skyline (»live skyline») (4) Hoist line (double hoist line with lifting block)

(5) Carriage (7) Double-drum winch (9) Head spar (10) Tail spar (11) Inner anchorage (11a) Tackle

(12) Tailhold (17) Carriage stop (log on skyline)

Figure 299. Gravity outhaul with the Vossa cable crane.
A. and B. The haul-back line is not needed when steeper than 20%.
C. The haul-back line is used to adjust the tension in the skyline.

haul-back line is needed, instead a carriage stop must be attached to the skyline at the place where the lateral skidding is done (figure 299).

Carriages with hoist tackle have been frequently used for cable crane yarding on big timber. On the American and Canadian west coasts the following systems were used for downhill yarding with stationary yarders at the valley station (see figure 281):

Tyler system
North Bend system
South Bend system
Block in the bight

The heavy equipment and the thick wire ropes needed for the transport of big timber makes it absolutely necessary to pull out the hoist line by motorpower. This is done be means of the haul-back line which passes a terrain block at the end of the skyline (often attached to the tail tree). In other cases the terrain block is located at the end of the corridor for lateral skidding.

Some of the Japanese cable cranes are similar to the American and Canadian. The most popular is the Tyler system and the Japanese have adjusted this system to suit their conditions.

If the slope is very steep the downhill load transport is powered by gravity (figure 300 and 301). In this case a double-drum winch is used. The main line passes the head spar, the carriage and the tackle and is fixed to the tail spar. The haul-back line is attached to the hook. The hook is pulled in by the main line and pulled out by the haul-back line.

Usually the yarding crew varies between three and five men:

1 yarder operator
1–2 choker setters
1–2 chasers

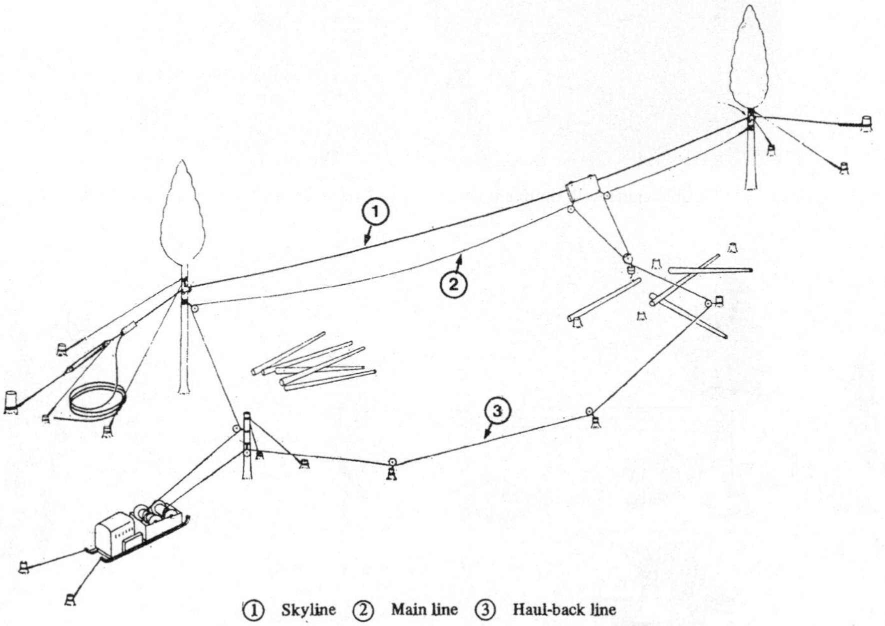

Figure 300. A Japanese cable crane arrangement of the Tyler svstem (KONUMA and SHIBATA 1976).

Figure 301. Cable crane with double hoist line. The Tyler system. Numata, Japan 1968.

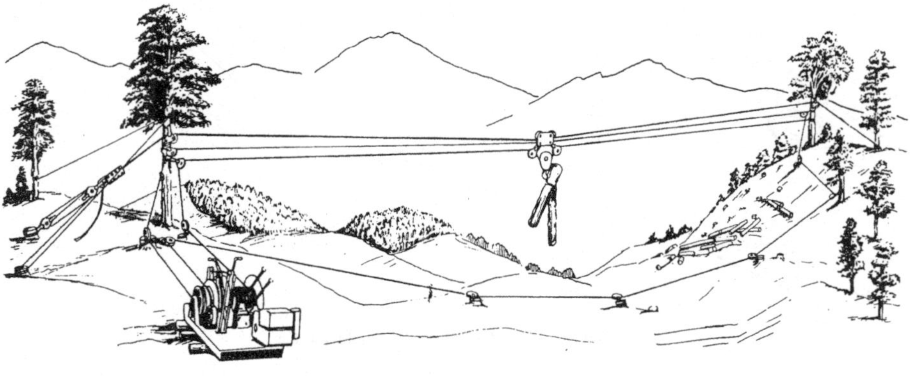

Figure 302. Endless Tyler system. Nansei, Japan 1980.

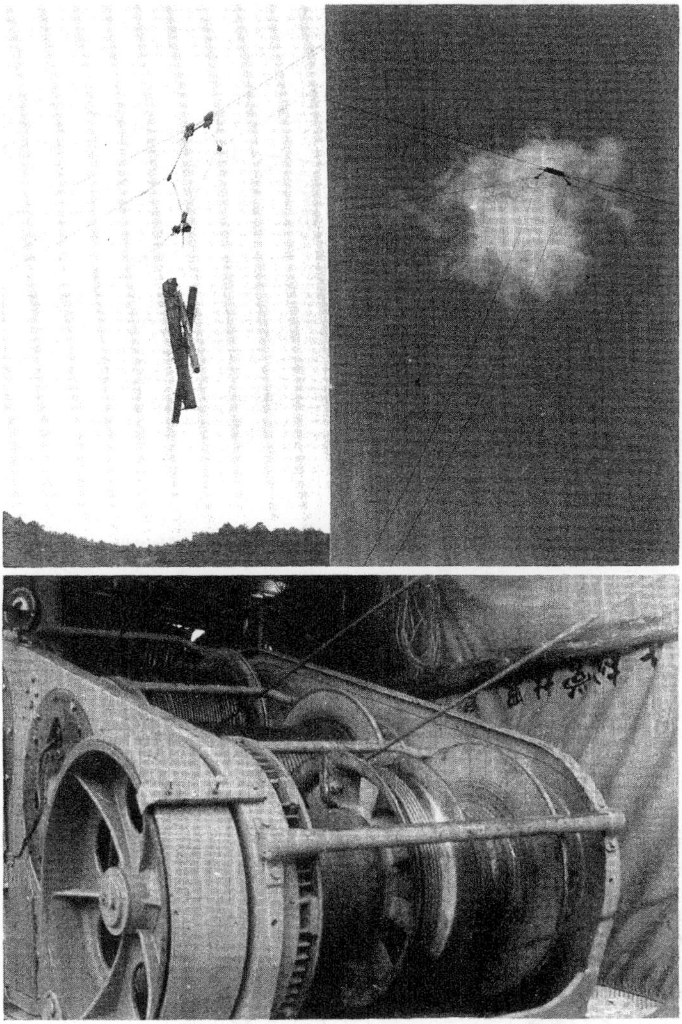

Figure 303. Endless Tyler cable crane. Numata, Japan 1968.

If the carriage (loaded or unloaded) cannot utilize the gravitational forces another variant of the Tyler system is used (figure 302 and 303). The yarder uses one winch drum for a main line and one capstan for an endless line. The endless line moves the carriage forward and backward along the skyline. The main line goes from the spar tree through the hoist tackle on the carriage and is attached to the tail tree. The hook is lowered and the load is hoisted by pulling in or out on the main line.

In some cases the haul-back line is attached to the hook on the hoist tackle instead of to the carriage (figure 304). The organization of the work is somewhat similar to the method illustrated in figure 300.

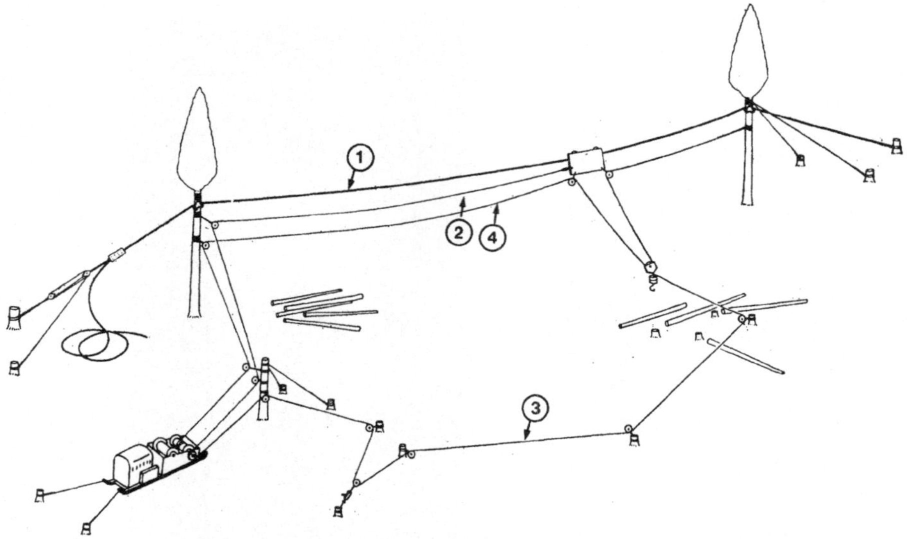

① Skyline ② Main line ③ Haul-back line ④ Hoist line

Figure 304. A Japanese modification of the endless Tyler system (KONUMA and SHIBATA 1976).

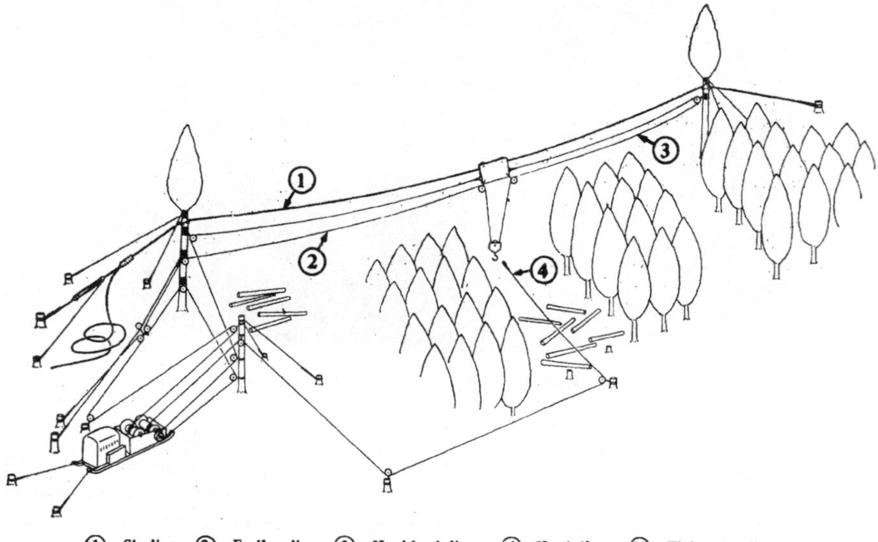

① Skyline ② Endless line ③ Haul-back line ④ Hook line ⑪ Tightening line

Figure 305. The Japanese Kumamoto cable crane (KONUMA and SHIBATA 1976).

The Kumamoto system is an interesting development of a cable crane with hoist tackle on the carriage. The Kumamoto cable crane is powered by a yarder with a double-drum winch and a capstan for an endless line. The endless line is a continuous main/haul-back line which moves the carriage along the skyline. The haul-back line runs from the rear end of the carriage through a block in the tail tree back to the carriage, through the tackle and the fall block under the carriage and the lead block in the spar tree to the capstan. The main line runs from the front end of the carriage through a lead block in the spar tree and to the capstan on the

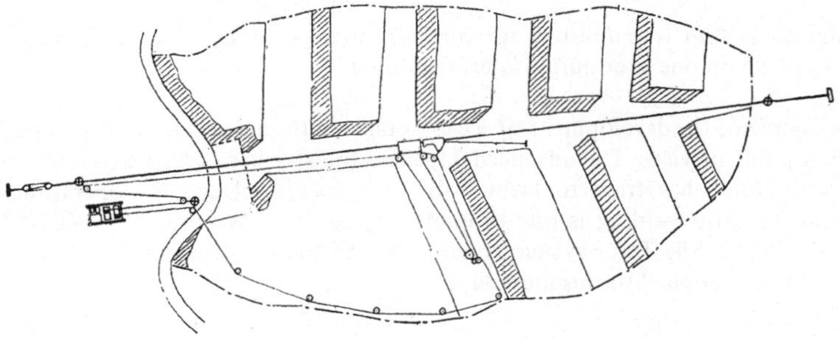

Figure 306. Branched cutting strip yarding with cable crane (KONUMA and SHIBATA 1976).

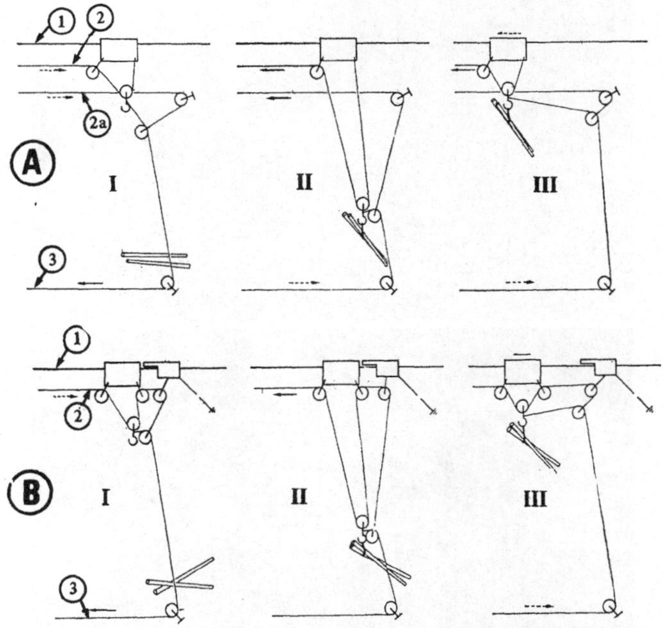

Figure 307. Pulling out the hoist line for branched cutting strip yarding (KONUMA and SHIBATA 1976).
A. Track of the haul-back line is controlled by the control line (2a) (Downhill North Bend system).
B. Track of the haul-back line is automatically regulated by the tension of the hoist line (Tyler system).
I: Pulling out the hoist line. II: Lateral skidding. III: Transport along skyline.

yarder where it is spliced together with the haul-back line, so that the main line and the haul-back line are a continuous line. One of the winch drums on the yarder pulls a tightening line. This tightens the continuous or endless main/haul-back line by means of a twin block (figure 305). By pulling in the twin block the hoist tackle with the load is being pulled towards the carriage. The hoist tackle and the fall block may be pulled laterally out from the carriage to the logs by means of a hook line. The second winch drum on the yarder is used to pull in the hook line. The choker setter may attach the hook line to the fall block or take it off again by hand.

The Kumamoto system is interesting for corridor cuttings because the yarder can keep the carriage in position on one spot during lateral skidding.

This is an example of the development of a cable crane method which is more acceptable from a silvicultural point of view. The advanced Japanese forestry has developed some cable crane methods where one has tried to meet both the silvicultural and the operational needs. Branched cutting strip yarding is one example (figure 306). According to KONUMA and SHIBATA (1976) has the Tyler system (figure 307, B) and the downhill North Bend system (figure 307, A) been applied to this method.

Figure 308. Pulling out the hoist line for lateral skidding is hard work in steep and difficult terrain (Heggelia, Norway 1951).

Figure 309. Lateral skidding and downhill transport with Anleggstransport cable crane. Heggelia, Norway 1951.

11.4.2.2. Carriages with single hoist line

In *The Anleggstransport cable crane* no arrangement is used to reduce the forces when pulling the hoist line laterally towards the logs. The system is shown in figure 164, B and the winch in figure 286. The carriage is moved forwards and backwards along the skyline with an endless line driven by a capstan. There is no stop carriage on the skyline. The hoist line is a lengthening of the main line.

After the empty carriage has returned uphill and stopped at the logging area, the hook falls down to the ground because of the weight of the hook. The choker setter has to pull the hoist line to the logs. When working high up on the hillside the choker setter has to pull the main line off the main drum and this is a very heavy work.In order to reduce the work load, the hook is sometimes anchored to a small stump in the terrain and the yarder operator moves the carriage uphill in order to pull out the hoist line. He then moves the carriage back again and this makes the work somewhat easier for the choker setter (figure 308).

The Anleggstransport cable crane is very flexible since the carriage can stop anywhere along the skyline both for loading and for unloading. The crane can operate on any terrain. This system is, however, not in use any longer due to the heavy work load on the choker setters in difficult terrain. This is especially the case during downhill yarding on very steep and long slopes.

Multiple sheave block between a front and rear carriage. The tension when pulling out the hoist line may be reduced by building the carriage in two parts, a front, and a rear carriage.

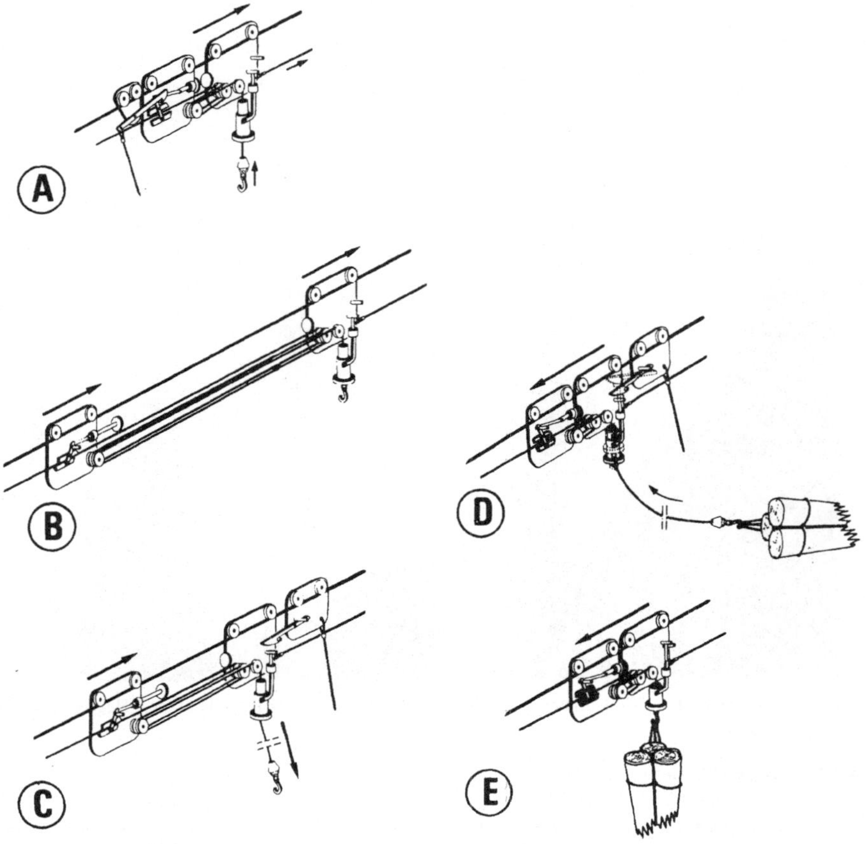

Figure 310. Multiple sheave block between a front and a rear carriage, constructed by Botolf Larsen.

A. The rear carriage is fixed to the carriage stop at the landing. The front carriage is being pulled away from the rear carriage by means of the haul-back line.
B. When there is a sufficient length of hoist line, the rear carriage is disconnected from the carriage stop. The whole system is returned to the terrain with the hoist line already pulled out.
C. With a 5-sheaved tackle between the front and the rear carriage, a force of approximately 1/5 of the tension in the main line is required to get the hoist line laterally out to the logs.
D. Lateral skidding. When the hook impacts the rear carriage it is automatically released from the carriage stop.
E. Downhill transport.

There is a multiple sheave block between them (figure 310). The main line runs from the winch drum to the carriage where it runs forward and back between the multiple sheave in the two parts of the carriage before it runs out of the front carriage as a main line. If there are six sheaves (six parts) the tension in the hoist line is reduced to 1/5 of the main line tension when pulling the hoist line out of the front carriage (see chapter 3.1, table 5).

When the carriage transports a load downhill along the skyline the front and rear carriage are pressed against each other (figure 313). When the load has been unloaded at the landing the rear carriage is fastened to the carriage stop. The yarder operator then returns the front

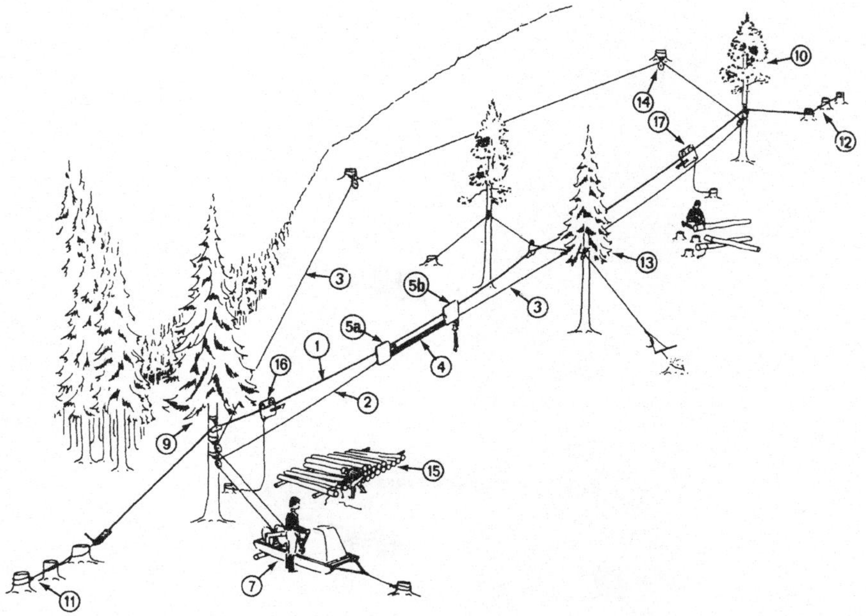

(1) Skyline (2) Main line (3) Haul-back line (4) Five sheave block between the front and the rear carriage

(5a) The rear carriage (5b) The front carriage (7) Double-drum winch (9) Head spar (10) Tail spar

(11) Lower or inner anchorage (12) Upper or outward anchorage (tailhold) (13) Support (14) Terrain block

(15) Log deck (16) Inner carriage stop (17) Outward carriage stop

Figure 311. Bjønn cable crane with valley mounted double-drum yarder. Multiple sheave block between the front and rear carriage.

carriage, which moves away from the rear carriage and pulls out the hoist line in the multiple sheave block arrangement between the front and rear carriage. When the chaser at the landing has found that a sufficient length of hoist line has been pulled out, he releases the rear carriage from the carriage stop by pulling in the unhooking line. The carriages then move uphill with the hoist line extended between the two parts of the carriage.

When the carriage train arrives at the carriage stop the choker setter pulls out the hoist line. During this work the rear carriage approaches the front carriage. When the chokers have been attached to the hook, a signal is given to the yarder operator who pulls in the hoist line. First the rest of the parts between the multiple sheave blocks are being pulled in until the rear carriage hits with the front carriage. The tension in the main line and the tension in the hoist line is practically the same (reduced by the friction in the sheaves). When the hook with the load is hoisted in, the front carriage loosens from the carriage stop and the dowhill load transport can start.

The Norwegian Bjønn cable crane works according to this principle. It is usually operated by a 4 man crew. 1 yarder operator, 1 chaser and 2 choker setters.

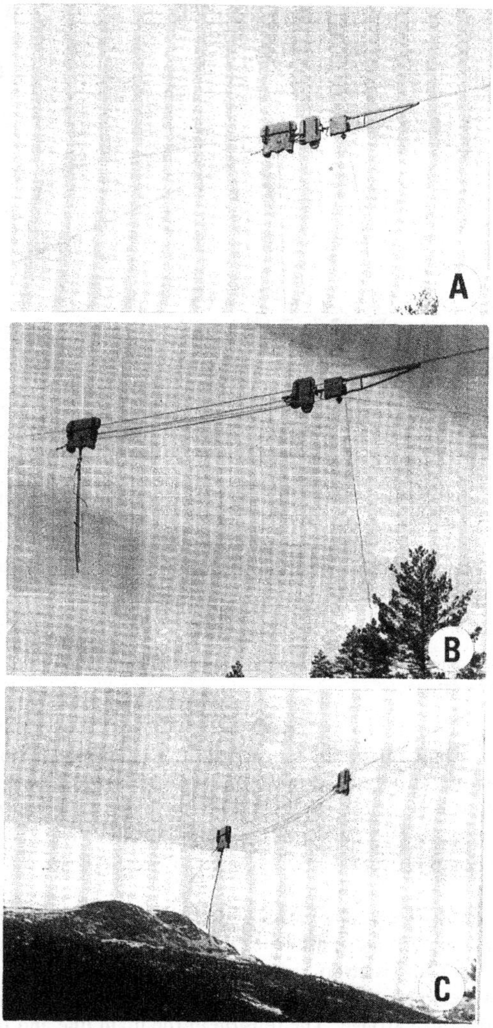

Figure 312. The Bjønn carriage. (Norway, 1956).

A. The rear carriage attached to the carriage stop at the landing. The front carriage rests against the rear carriage.

B. The front carriage is pulled away from the rear carriage which is fixed to the carriage stop at landing.

C. The front carriage with the rear carriage hanging behind during the return trip to the field.

Figure 313. The Bjønn cable crane, Ringnes, Norway 1956.

A. The front carriage with the rear carriage hanging behind impacts the carriage stop at the felling site.

B. The load passes an A-framed support during the downhill transport.

C. On the felling site the rear and front carriage is fastened to the carriage stop during lateral skidding.

D. and E. The load approaches the carriage stop at landing.

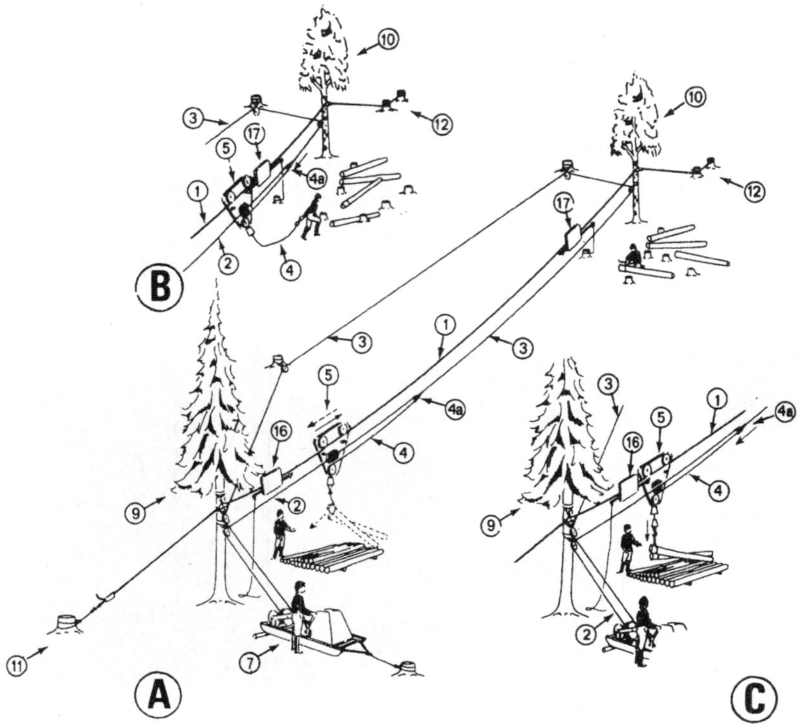

(1) Skyline (2) Main line (3) Haul-back line (4) Hoist line (4a) Three-way swivel fixing the hoist line to the main and haul-back lines (5) The automatic grip fastens the carriage to the main line (7) Double-drum winch (9) Head spar (10) Tail spar (11) Heelhold (12) Tailhold (16) Lower carriage stop (17) Outward carriage stop

Figure 314. The Sveen cable crane with valley mounted double-drum yarder.
A. The carriage is fixed by a clamp to the main line during uphill or downhill transport along the skyline.
B. When the carriage fastens to the carriage stop the clamp is automatically opened. The 3-way swivel can be pulled towards the carriage in order to pull out the hoist line.
C. The carriage is fixed to the carriage stop at the landing, and the clamp between the carriage and main line is opened. The load can be lowered to the pile by moving the 3-way swivel towards the carriage.

Clamp carriage with 3-way swivel in the haul-back line. Another way of reducing the tension when pulling out the hoist line is to use a 3-way swivel behind the carriage in the haul-back line. The distance between the swivel and the carriage corresponds to the length of the hoist line. There must be two carriage stops on the skyline, one at the landing and one at the place for lateral skidding. When the empty carriage returns uphill (figure 314) the hook is attached to the carriage and the carriage is clamped to the main line (figure 315). When the carriage arrives at the upper carriage stop it hooks onto this and the hook loosens from the carriage and the clamp is released from the main line. The yarder operator then pulls in the main line.

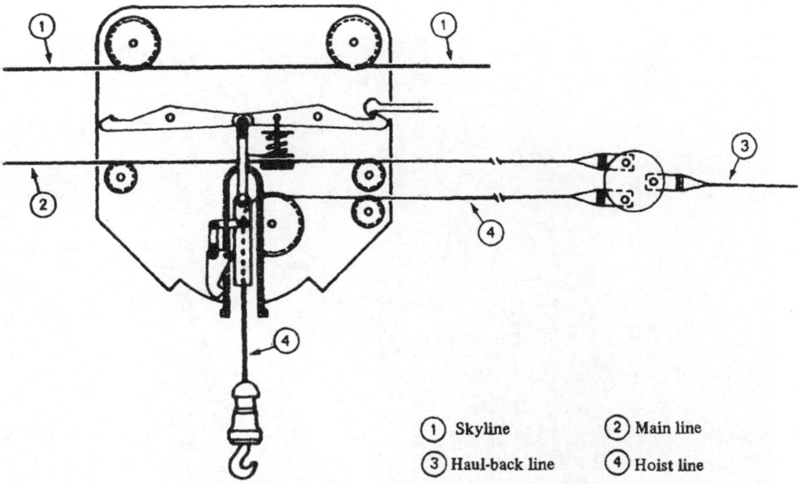

Figure 315. The Sveen carriage.

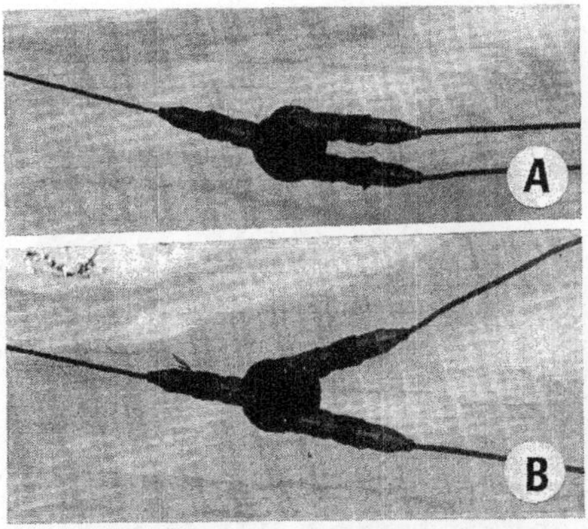

Figure 316. The 3-way swivel for the Sveen carriage. The haul-back line to the left. The main line at the upper right side and the hoist line at the lower right side.

The carriage is kept still, fastened to the carriage stop, so the 3-way swivel approaches the carriage. The hoist line drops down to the ground and it is easy for the choker setter to pull out the hook to the logs. When the chokers are hooked on, the load is laterally skidded by means of the haul-back line and hoisted until the hook fastens in the carriage. At the same time the clamp closes onto the main line and the carriage is released from the carriage stop. The downhill transport along the skyline starts.

When the carriage arrives at the carriage stop over the landing the same movements take place in the carriage until the hook has been unloaded and fastened to the carriage so that the return trip along the skyline can start.

Figure 317. Sveen cable crane at Sildrevika, Norway 1961.
A. Lateral skidding of whole trees.
B. Downhill transport of whole trees.

This system is used by the Norwegian Sveen cable crane. A double-drum winch was most commonly used (figure 285, C).

Two capstan pulleys has sometimes been used with this cable crane instead of the double-drum winch. The arrangement is shown in figure 284. The main and haul-back lines are spliced together to form an endless line. The winch use one of the capstan drives for the main line part and the other for the haul-back line part. The endless line runs around a heel block behind the winch. The heel block can be moved in order to adjust the tension in the endless line.

Normally, there is a 3 man crew, 1 yarder operator, 1 chaser and 1 choker setter.

Load carriage for pulling out the hoist line. The system is illustrated in figure 318. The cable crane uses a double-drum winch and there must be two carriage stops on the skyline, one over the landing and one over the logging area. A load carriage is placed on the skyline above the upper carriage stop. When the carriage is downhill at the landing, the load carriage rests against the carriage stop (figure 318, B). The haul-back line has a catcher attached to it at a

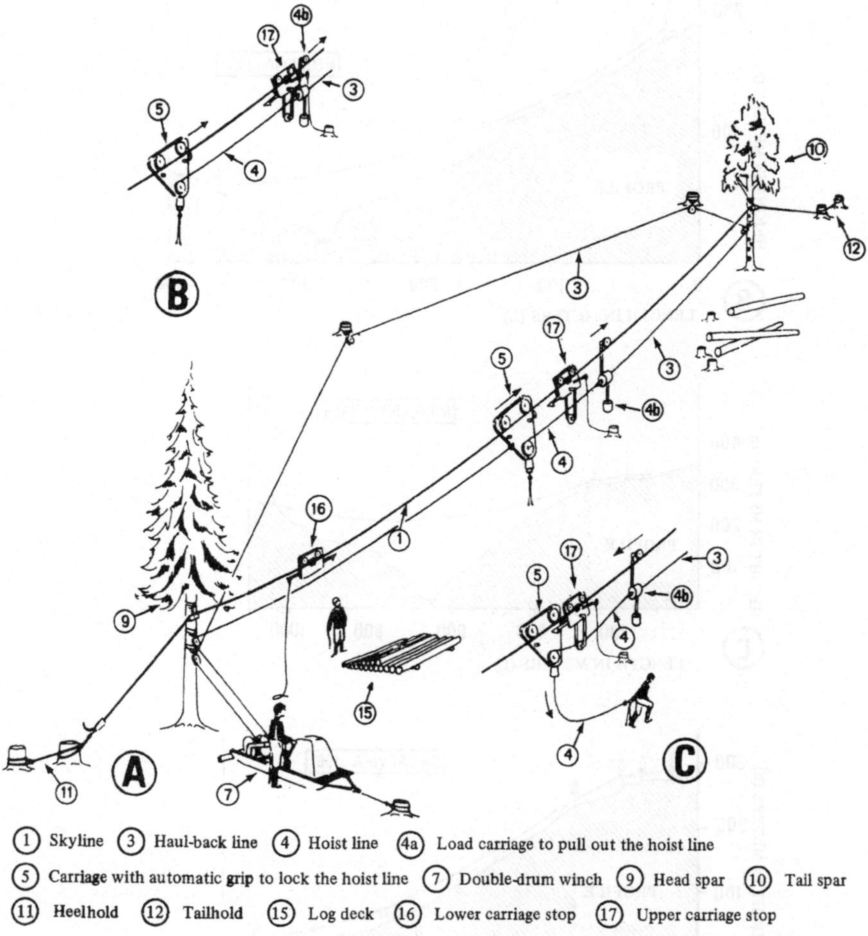

Figure 318. Svendsen cable crane with load carriage for pulling out the hoist line. The system may be used for top mounted single-drum yarder or valley mounted double-drum yarder.

A. and B. A catcher (a steel egg) on the haul-back line moves the load carriage uphill while the carriage approaches the carriage stop.

C. The load carriage helps to pull out the hoist line.

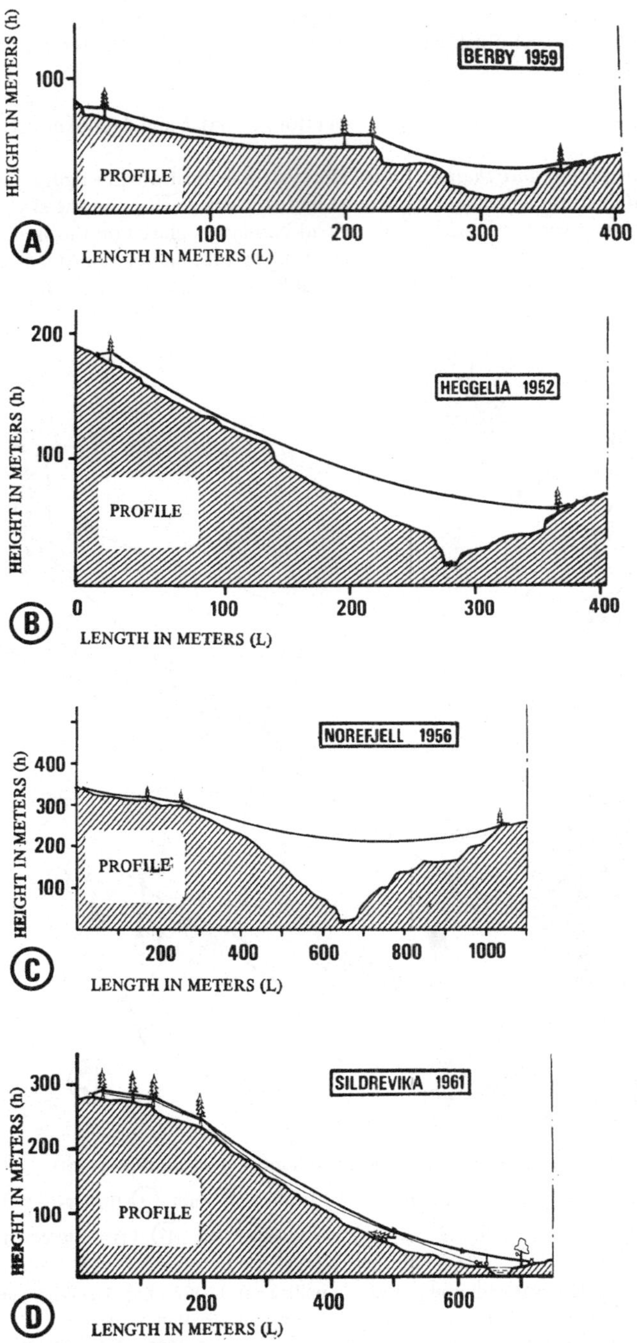

Figure 319. Typical profiles for some of the investigated Norwegian cable cranes.

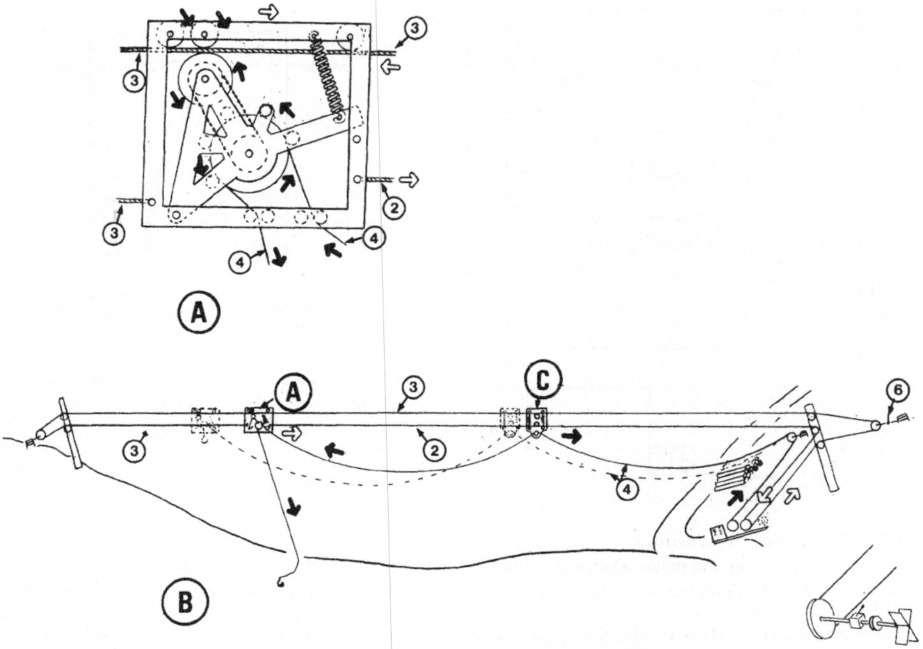

② Main line ③ Haul-back line ② and ③ Endless main/haul-back line
④ Hoist line ⑥ Tightening line

Figure 320. The Okawara system of a running skyline.
A. The carriage.
B. The running skyline.
C. The support carriage.

distance from the hook which corresponds to the length of the hoist line. The catcher is shaped like a steel egg. The haul-back line goes through the load carriage. When the carriage approaches the upper carriage stop the catcher (the steel egg) stops against the load carriage and moves it uphill. When the carriage is fastened to the carriage stop the hook is released. The load carriage is then heavy enough to pull the haul-back line downhill and simultaneously bring out the hoist line. The weight of the load carriage must be in proportion to correspond to the pre-tension in the haul-back line and must vary with the steepness and the length of the cable crane.

Carriage with friction pulley, the Okawara method. Professor Shoji Okawara, Japan, has developed an interesting cable crane where a friction pulley in the carriage pulls out the hoist line (figure 320). The cable crane has running skyline and the winch has a main drum and a capstan. The endless line has a tightening device in order to function as a running skyline. It is fastened to the carriage and moves the carriage forward and back along the skyline corridor by means of the capstan. The main line runs from the main drum in the

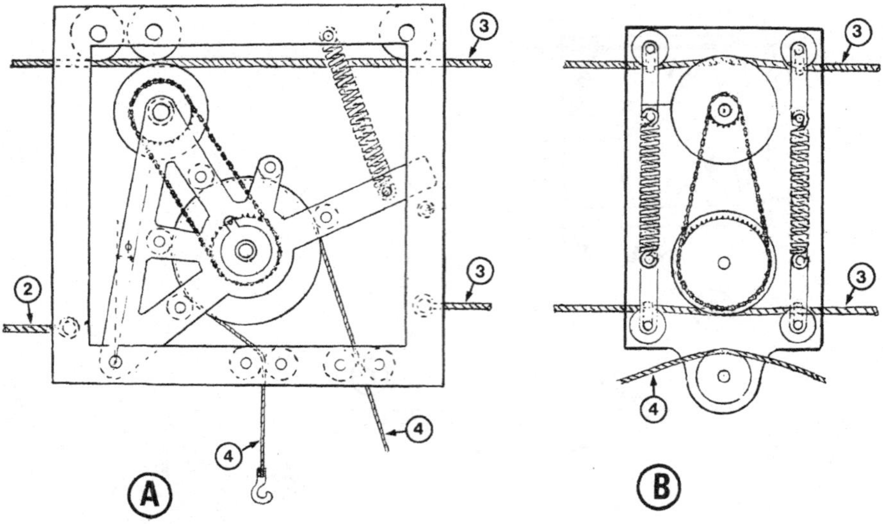

Figure 321. The Okawara carriage.
A. A friction pulley against the haul-back line (3) helps to pull out the hoist line (4).
B. A support carriage in the middle of the span helps to keep the main line off the ground.

yarder and to the carriage where it passes a friction wheel mounted on a frame. The friction wheel before the hoist line is connected to another friction wheel by means of a chain. The frame within the carriage (figure 320, A) is being pressed upward against the upper part of the running skyline by means of a spring. If there is little tension in the main line, the upper friction wheel will rotate with the movement of the carriage and pull the hoist line in or out. If there is tension in the main line, for example when pulling in the load, the frame in the carriage will be pressed down and there will be no contact between the upper wheel and the upper part of the running skyline.

When the carriage arrives at the logging area, the yarding operator returns the carriage with the main line loose. The upper friction wheel rotates and pulls out the hoist line. After choker setting the load is pulled in. The tension is high enough to keep the frame in the carriage down, so there is no contact between the upper friction wheel and the upper part of the running skyline. This is also the case during the load transport, since the load is hanging from the main line.

A support carriage in the middle of the span between the carriage and the yarder helps to keep the main line off the ground (figure 320, C). The upper and the lower wheel in the support carriage is pressed by spring tension against the upper and the lower part of the endless running skyline. Since there is a 1:3 ratio between these two wheels the support carriage will move in the same direction as the transport carriage, but with half the speed. The support carriage will therefore always be in the middle of the span between yarder and transport carriage.

The winch is equipped with a fan brake (figure 320 and 322). The speed of the fan brake may be regulated and since it is connected to the capstan drive it gives a certain pre-tension against the movement which helps to adjust the height of the load above the ground. It is for example possible to pull the load in as highlead skidding. The Okawara method is a very interesting development which is of high interest for short distance cable cranes where no supports are needed.

Figure 322. The Okawara cable crane, Takizawa, Japan 1981.

11.4.3. The production

The production of downhill cable crane yarding with stationary yarders at valley station varies with the cable system, terrain conditions, tree dimensions and volume per area. Production figures have been given in research reports or reports from practical operations, and it would be impossible to draw any sound conclusions from such figures. Instead we may give a few typical production figures, i.e. production in cu.m. per work place hour. Table 40 shows the distribution of some time elements on a yarding turn for three traditional types of Norwegian cable cranes. The Vossa cable crane transports the loads with the front end lifted only, the other two cranes carry the load off the ground on the skyline.

Table 40. Distribution of time elements for some traditional Norwegian cable cranes (SAMSET 1981).

Cable crane Length	Vossa 280 m	Anleggstransport 600 m	Bjønn 1,000 m
Return carriage	17 %	23 %	29%
Loading	41 %	29 %	26 %
Transport	24 %	41 %	30 %
Unloading	18 %	7 %	15 %
Total	100 %	100 %	100 %
Effective time	4.7 mins	10.7 mins	10.4 mins
Delay	36.3 %	17.9 %	26.4 %
Work place time	6.4 mins	12.6 mins	13.1 mins

The work place time for a rather big crane is around 12 to 13 minutes per turn, while the small yarders use a work place time of 6.4 minutes per turn. The delay time varies between 20 and 25 % for bigger cranes but is higher for the light equipment.

The productivity and the production per work place hour for the same cranes are given in table 41.

Table 41. Production per work place hour for some traditional Norwegian cable cranes with stationary yarders at valley stations (SAMSET 1981).

Cable crane	Engine kW	Length m	Yarding dist. m	m^3/hr (crew)	m^3/man/hr	Manhours per m^3
Vossa	18	200 400	100 200	8.7 5.8	2.9 1.9	0.34 0.52
Anleggs-tansport	55	200 600	100 300	11.0 5.5	3.7 1.8	0.27 0.55
Bjønn	18	200 600 1,000	100 300 500	9.5 6.8 5.2	2.3 1.8 1.3	0.43 0.59 0.76

These are typical figures but may vary a good deal depending to the conditions. The small yarders have a productivity of between 2 and 3 cu.m. per man per hour corresponding to 0.3 – 0.5 manhours per cu.m. The bigger cranes have a productivity of 1.5 – 3 cu.m. per manhour corresponding to a labour consumption of 0.3 – 0.7 manhours per cu.m. For the longer distances a labour consumption of 0.7 manhours per cu.m. would be typical for these traditional cable cranes when the average tree dimension is 0.3 cu.m. per tree.

Similar results have been shown for the Japanese cable cranes by KONUMA and SHIBATA (1976). They reported an average production of 32.3 cu.m. per day, corresponding to 1.4 cu.m. per man per hour or 0.71 manhours per cu.m. This is cable crane yarding from clear-fellings. When yarding in selective cutting, KONUMA and SHIBATA reports that the production is being reduced to 68 % as compared with the production in clearfellings.

The Norwegian investigations showed that for the traditional cable cranes the production is higher in the summer than in the winter. Table 41 shows the production and yarding on snow-free ground. The production in winter time is usually 75 % of the production in summer time.

The production increases with the tree dimensions. Very high production has been reported when harvesting very big tree dimensions on the American and Canadian West Coast using heavy equipment. The production figures in table 42 are a summary of many different reports. The figures are typical for systems like Tyler, North Bend, South Bend etc. when the yarder is stationary at the valley station. When the average tree dimensions vary between 6 and 9 cu.m. per tree, a typical productivity would be 3.5 – 4 cu.m. per man per hour corresponding to 0.25 – 0.30 manhours per cu.m. (work place time).

Table 42. Production per work place hour. Traditional standing skyline cable cranes. Downhill yarding with stationary yarders at valley stations.

m^3/tree	Production m^3/hr. (crew)	Productivity m^3/man/hr.	Labour consumption manhours/m^3
0.3	8.2	1.7	0.59
1.0	10.5	1.9	0.52
3.0	17.1	2.5	0.40
6.0	27.0	3.4	0.29
9.0	36.9	4.3	0.23

11.5. Cable cranes with all round mobile yarders

In addition to the operating drums a well designed yarder has drums and other equipment in order to make the rig up and rig down work efficient and easier for the workers. The time for movement and rigging is shorter and it is possible to move the machine faster between smaller operations. In addition the road network has a better layout and usually the yarder can be stationed on a truck road or near it.

Such yarders are mounted on mobile chassis, the socalled mobile yarders.

The faster movement from one yarding area to the next has made it necessary to equip the yarder with winch drums and other equipment which allows the yarder to power different types of cable cranes. The yarder may be placed at the bottom or at the top of a yarding area. Such allround yarders are very flexible.

The chassis on which the operating drums, the auxiliary drums, the tower etc. are mounted may vary according to the conditions. Since the movement mostly takes place on truck roads, wheeled chassis is the most commonly used. Under certain conditions, however, the yarder has to move off the truck road and a crawler construction may be necessary. In this case the yarder has to be loaded on a trailer during the transport on truck roads from one yarding area to the next (figure 323–326).

The yarder may be mounted on a trailer chassis or a semi-trailer chassis. These trailers have to be pulled by a truck during highway transport. In some cases, however, the trailer is being pulled behind a tractor and powered by the power take off on the tractor.

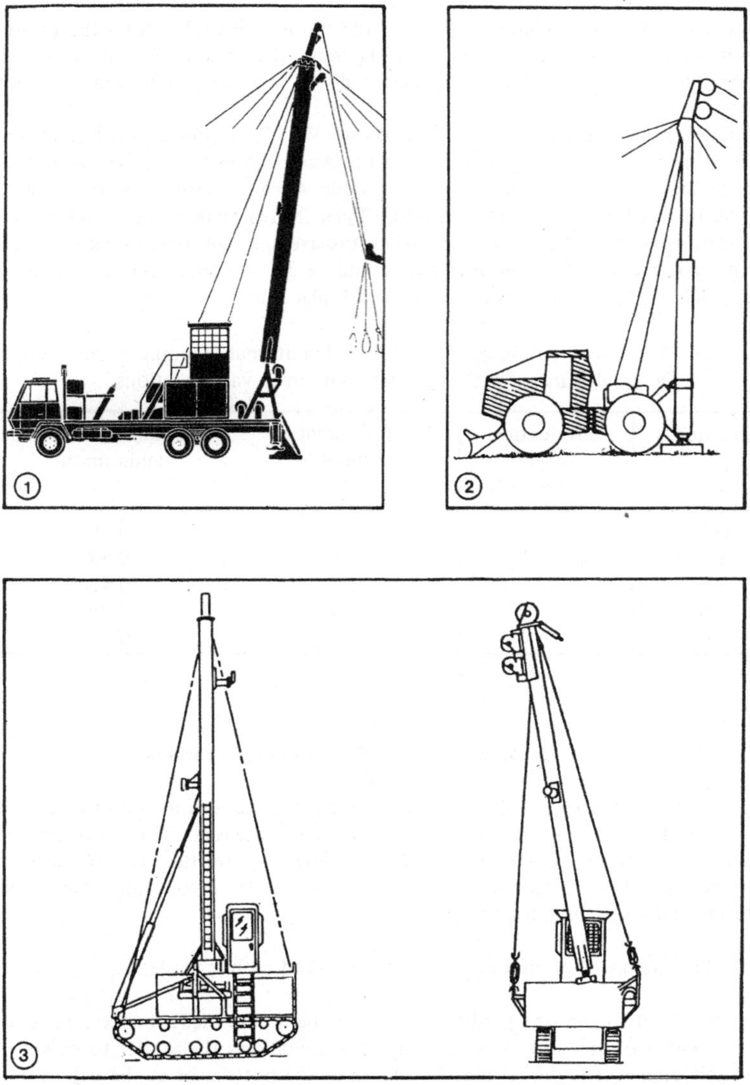

(1) Truck mounted yarder (2) Skidder mounted yarder (3) Yarder on tank type crawler chassis

Figure 323. Various types of mobile yarders (LARSEN, 1975).

Most yarders are, however, mounted on a tractor, a truck or a chassis with driven wheels or tracks. These are the self-propelled yarders. They move by their own power along the truck roads between the yarding areas. By means of hydraulic components it is quickly done to rig up the yarder after it has arrived at the logging site.

The large yarders for big timber have to be supported, usually by hydraulic feet during the operation. In other cases there is a foot under the steel spar, only, and it is kept in position by the guylines.

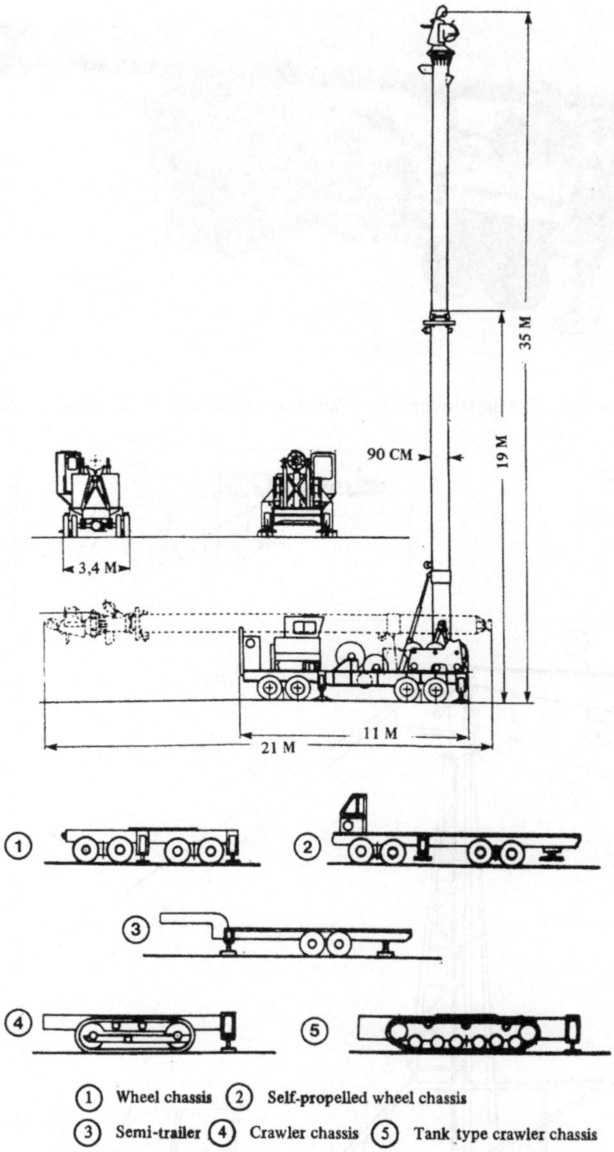

Figure 324. Various types of under-carriers for mobile yarders (LARSEN 1975)

Small yarders for small timber, however, usually have no special supports. They are supported by the vehicle itself and guylines during the yarding operation.

If the yarder has to move on public highways it is necessary to follow the regulations as far as wheel and axle pressures are concerned. This has to be taken care of by means of the number of wheels and axles which have to be used. It is also necessary to check that the weight distribution on the axles is correct in order to follow the highway regulations.

Figure 325. The tip/telescope tower laid down on the mobile yarder during transport.

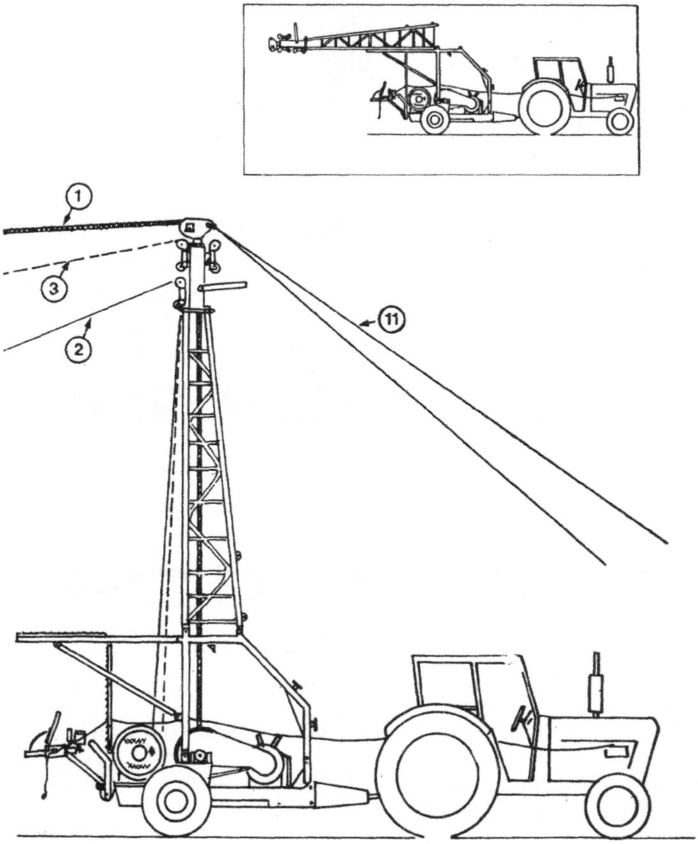

(1) Skyline (2) Main line (3) Haul-back line (11) Guylines

Figure 326. The Norwegian trailer mounted Igland yarder with tip tower for the Alp cable crane.

11.5.1. The chassis

The following chassis possibilities exists (see figure 323–326):

Mobile yarders on trailers.

- Single axle trailer
- Boggie axle trailer (semi-trailer)
- 2-axle trailer
- 4-axle trailer (2 boggies)

Self-propelled mobile yarders:

- Tractor mounted yarders
- Skidder mounted yarders
- Truck mounted yarders:
  - 4-wheel drive
  - 1 single axle and 1 boggie axle
- Self-propelled 3-axle wheeled chassis
- Self-propelled 4-axle wheeled chassis
- Crawler chassis:
  - Band tractor type
  - Tank type.

11.5.2. Tight skyline systems

These are a group of cable cranes with the skyline anchored at both ends. The skyline is tightened to a certain pre-tension, usually by a tightening device at the winch station. The tightening device might be a come-along, a winch drum on the yarder, sometimes with a multi-sheave block between the winch and the skyline. Even in the cases when the tightening device is a winch drum on the yarder the pre-tension skyline is kept constant. The carriage moves on the skyline, pulled forward or backward by means of the main and haul-back lines. A hoist line (drop line) has to be lowered from the carriage to the ground and pulled out to the logs before the lateral skidding and the hoisting of the load up to the carriage can start.

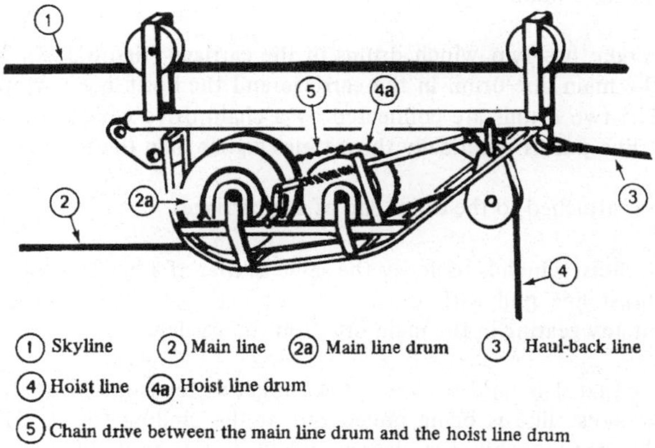

Figure 327. The Norwegian Trøim carriage.

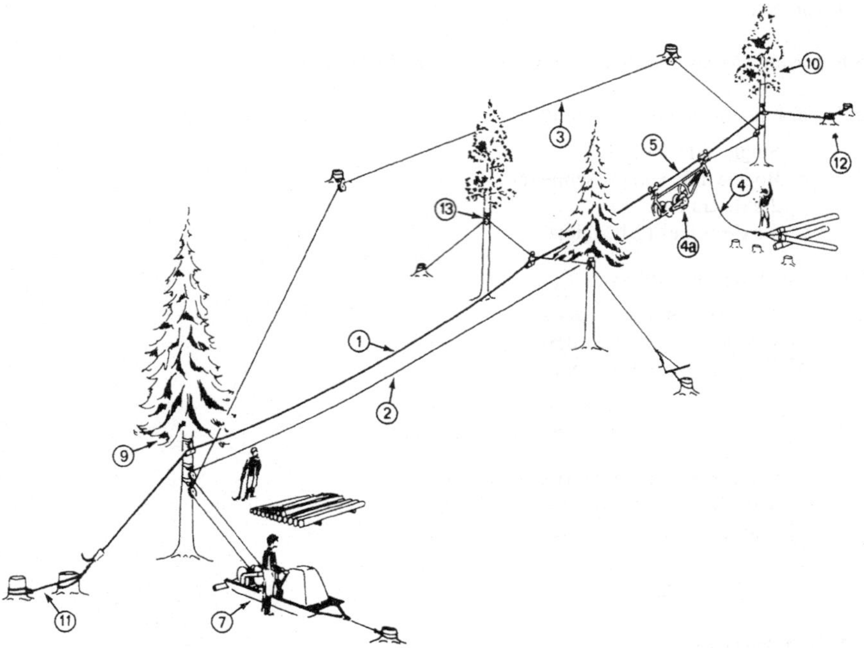

(1) Skyline (2) Main line (3) Haul-back line (4) Hoist line (4a) Hoist line drum
(5) Carriage (7) Double-drum winch (9) Head spar (10) Tail spar (11) Inner anchorage (heelhold)
(12) Outward anchorage (tailhold) (13) Crossline support

Figure 328. The Trøim cable crane with hoist winch in the carriage.

11.5.2.1. Carriages with hoist winch

The Trøim cable crane has two winch drums in the carriage (figure 327). The main line is wrapped around the main line drum in the carriage and the hoist line is wrapped around the hoist line drum. The two drums are connected by a chain drive with a gear ratio of 1 : 3. In this way the hoist line pull is practically three times higher than the tension in the main line.

The haul-back line is attached to the other end of the carriage.

The purpose of the hoist winch is basically the same as that of a hoist tackle. The object is to obtain sufficient hoist line pull with less pull in the main line. The transport speed can be kept higher without any gearing in the main drum on the yarder.

The disadvantage is that it is hard work for the choker setter to pull out the hoist line. To overcome this, the hoist line is being pulled out at the landing (figure 329). The yarder operator returns the carriage while the chaser holds back the hook. When a sufficient length of hoist line has been pulled out the hook is attached to the main line by a clamp and the carriage can return to the logging area with the hoist line already pulled out. A brake on the

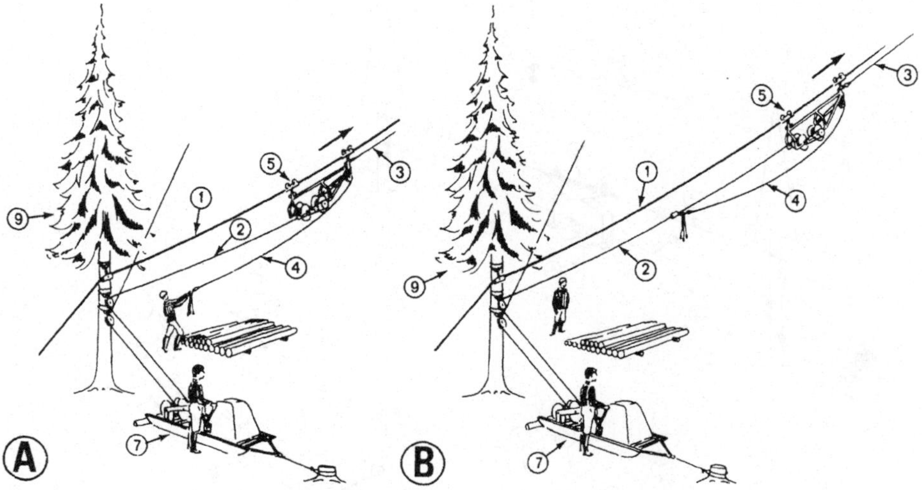

(1) Skyline (2) Main line (3) Haul-back line (4) Hoist line (5) Carriage (7) Double-drum winch
(9) Head spar

Figure 329. A. The hoist line is pulled out and fixed to the haul-back line.
B. The carriage returns to the felling site with the hoist line pulled out.

hoist line drum in the carriage prevents the hoist line from being pulled in by gravity on a steep slope when the choker setter loosens the clamp and carries the hook and the hoist line towards the logs. The brake is released by means of the tension in the hoist line during the lateral skidding.

Capstan driven hoist winch in the carriage and double main and haul-back lines.

The JoBu Combi-Cat cable crane is using a double capstan pulley (figure 330). The two capstans have the same speed, four meters per second. They are connected through a planetary drive and the speed ratio can be continuously adjusted. The capstan unit is mounted on a 50 kW farm-tractor and powered by the power take off on the tractor. A single axle trailer behind the tractor has storage drums for all the operating lines.

The carriage is hauled in or returned by means of a double set of main/haul-back lines. These are endless lines which are driven by the two capstans in the yarder. One of them, the traction line (main line), is fixed to the carriage. The other is wrapped three times around a capstan on the carriage. As long as the two capstans on the yarder rotate with the same speed, the carriage capstan does not rotate while the carriage runs along the skyline. If, however, one of the yarder capstans rotates faster or slower than the other, the carriage capstan rotates in one or the other direction. The carriage capstan is mounted on the same axle as a hoist drum (figure 331). If the carriage capstan rotates, the hoist drum rotates, and pulls in or drops out the hoist line depending on the direction of the rotation. It is a ratio of 1 : 4 between the carriage capstan and the hoist drum, the pull in the hoist line is therefore four times the pull in the hoist pulling line.

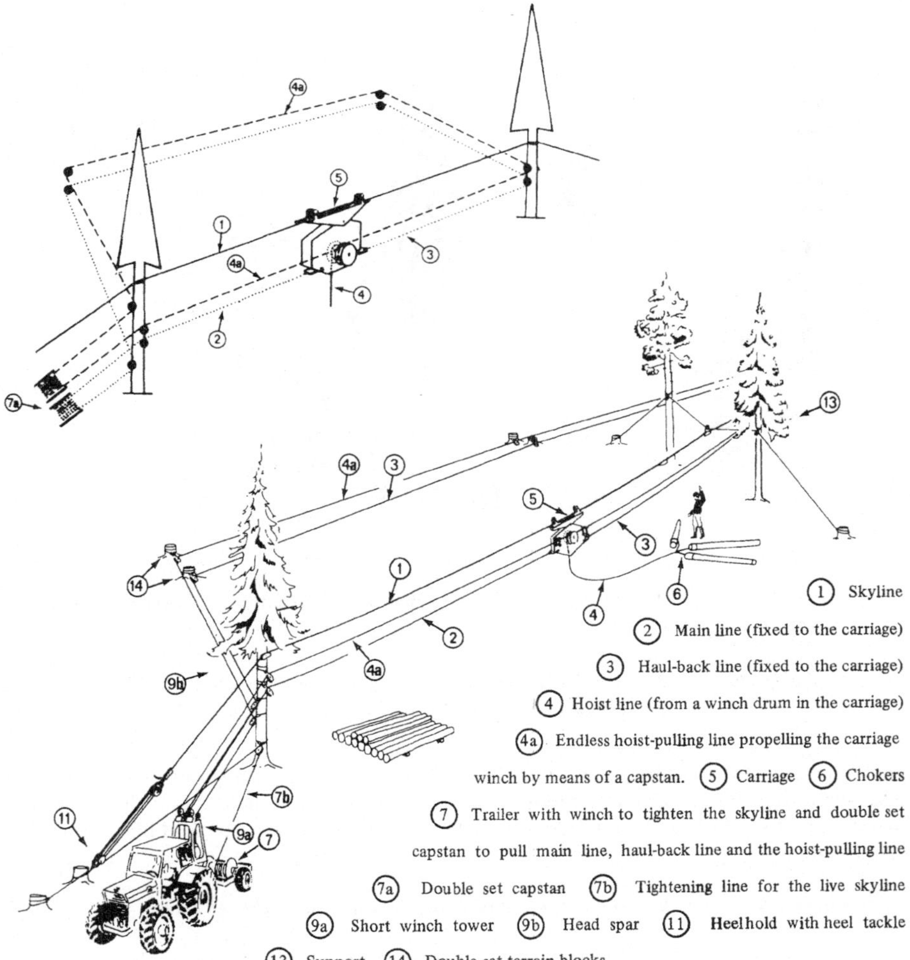

Figure 330. The Jobu cable crane with capstan driven hoist winch in the carriage (Jobu, Norway 1975).

The equipment is constructed for tree dimensions less than 1 cu. m. per tree. The standard length of the JoBu Combi-Cat cable crane is 500 meters. The carriage can pass supports and has a special arrangement to prevent the carriage jumping off the skyline (figure 331). On request the cable crane may be delivered for longer distances. The trailer has powered storage drums which simplifies the work of rigging up or down the cable crane on the yarding area. The trailer has the following drums:

Traction line	L = 1,000 m	d = 8 mm
Hoist pulling line	L = 1,000 m	d = 8 mm
Hoist line	L = 50 m	d = 12 mm
Tightening line	L = 100 m	d = 10 mm
Skyline	L = 500 m	d = 18 mm
Straw line	L = 1,000 m	d = 4 mm

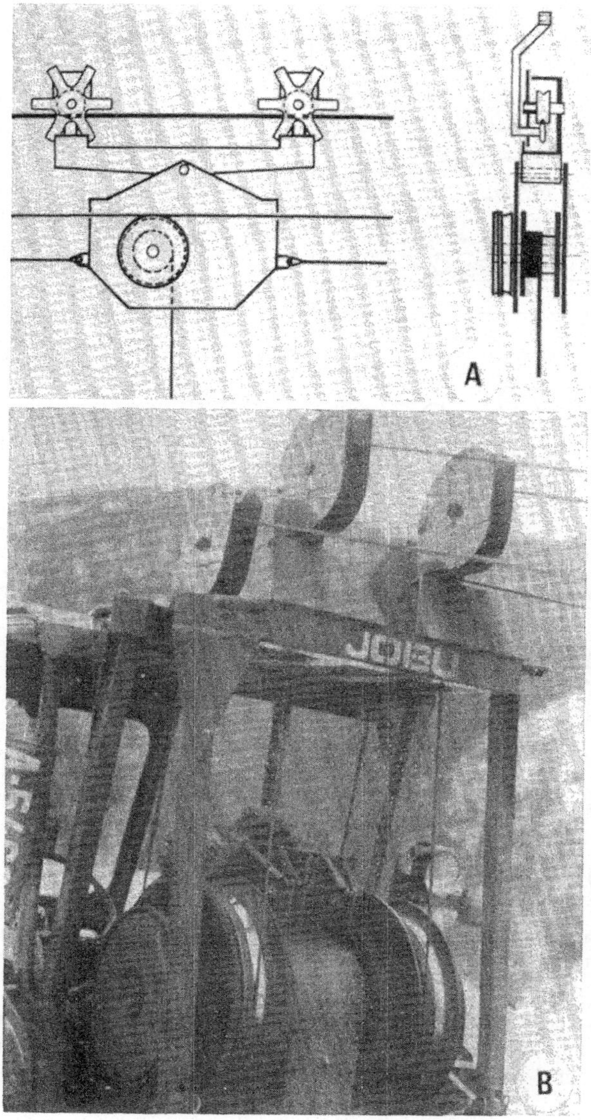

Figure 331. A. The carriage.
B. The double capstan (Jobu, Norway 1975).

The yarder is easy to operate. The carriage may stop anywhere along the skyline and the hoist line may be dropped down or pulled in when the carriage is moving or when it has been stopped. The load may therefore be high skidded along the skyline road, or lifted up to the carriage. It is a very smooth and flexible system. The capstans are pneumatically controlled. The valves are electrically operated and are driven from the battery. There is a 50 meter long electric cable and the yarder operator may stand near the landing while operating the

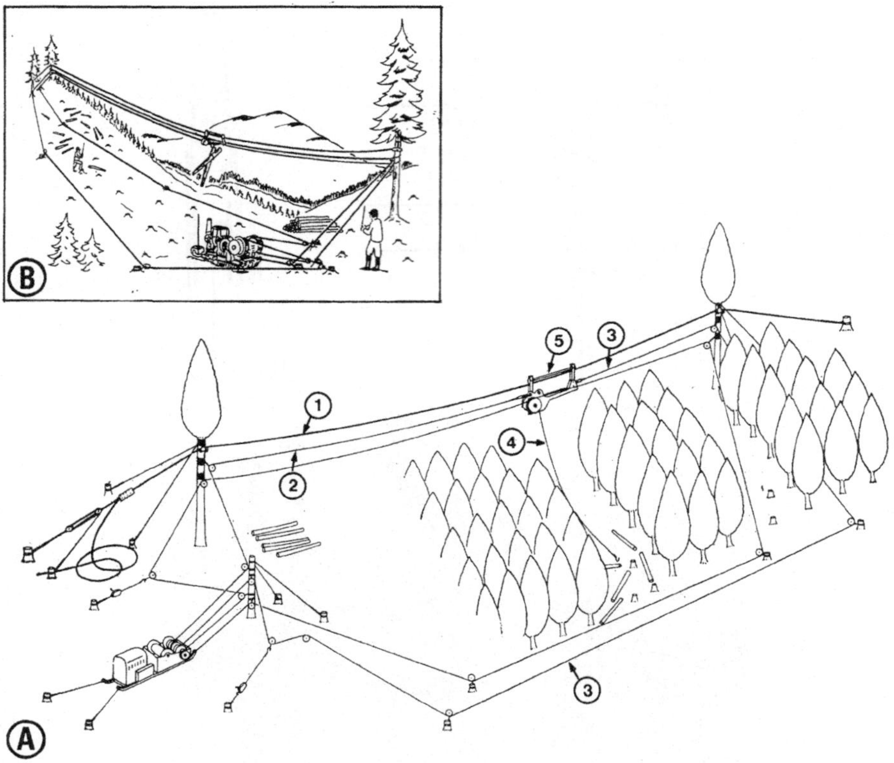

Ⓐ ① Skyline ② Main line ③ Haul-back line ④ Hoist line

⑤ Carriage with capstan driven hoist drum

Ⓑ Radio controlled yarder

Figure 332. A Japanese cable crane with hoist winch in the carriage. Nansei developed a radio controlled version of the crane (KONUMA and SHIBATA 1976).

yarder. No chaser is therefore needed. Usually the crew consists of two or three men. The yarder operator unhooks the load at the landing and works at the landing all the time. In the field there are 1 or 2 choker setters, depending on the distance and the size of the wood.

The Løland cable crane is a Norwegian cable crane which was built in 1952 (figure 42). This cable crane is basically the same type as the JoBu Combi-Cat cable crane, which first was built in 1967.

The same type of cable crane was developed in Nagano in Japan in 1957. It is now one of the most popular skyline systems in Japan (figure 332). These cable cranes are usually 200 – 600 meters long and the crew consists usually of 3, sometimes 4 men. The production is 3.0 – 4.5 cu.m. per work place hour, corresponding to 1.0 – 1.5 cu.m. per man per hour. The labour consumption is 0.7 – 1.0 manhours per cu.m. for the yarding. This productivity and labour consumption is practically the same as for the Norwegian system, when the tree dimensions are between 0.2 and 0.3 cu.m. per tree.

Figure 333. The Igland trailer mounted yarder with tip tower.
A. The tower is erected during cable yarding.
B. The tower is laid down over the tractor during transport.

Trisected hoist winch in the carriage.

The Igland Alp Cable Crane has a trailer mounted yarder (figure 333). There are five drums: main drum, haul-back drum, skyline drum with storage section and tensioning section (figure 269–270), a straw line drum and a rig up line drum. The yarder is powered by the power take off on the tractor. The minimum power required is 50 kW.

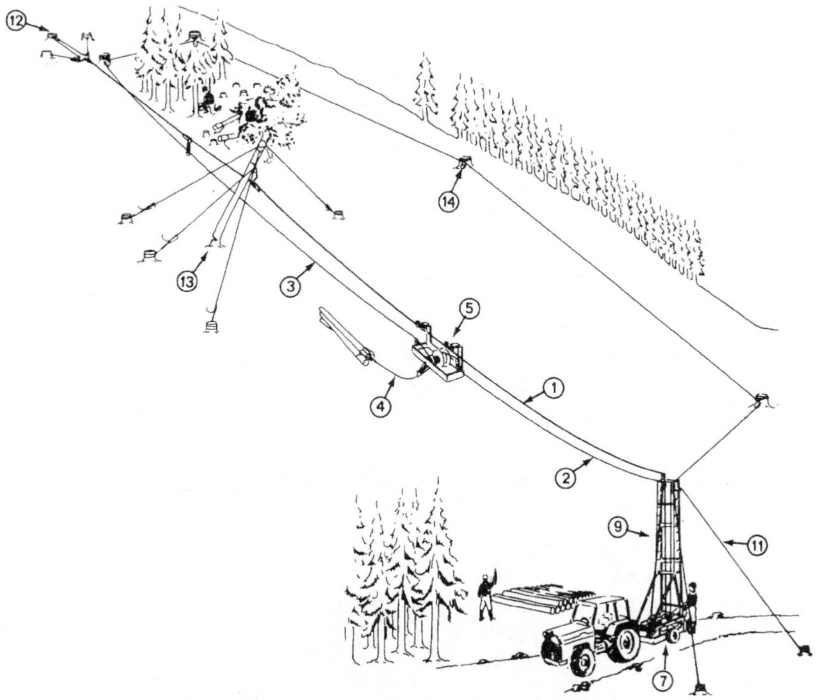

(1) Live skyline (skyline attached to a winch drum and may be lifted or lowered during operation)

(2) Main line (3) Haul-back line (4) Hoist line (5) Carriage with a three-part drum. The big diameter for main line. The small diameter for haul-back line. The mid.diameter for hoist line.

(7) Trailer with drums for skyline, main and haul-back line and straw line (9) Steel spar

(11) Guyline to the heelhold (12) Tailhold (13) Finger support (14) Terrain block

Figure 334. The Igland Alp cable crane. Norway, 1975.

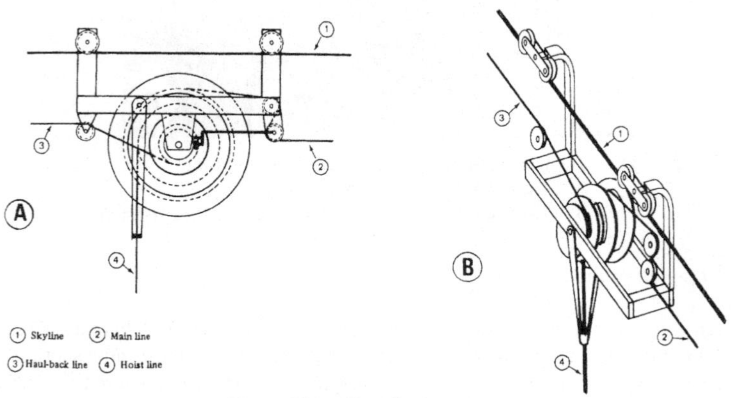

Figure 335. The Igland carriage.

Figure 336. The Igland yarder for Alp cable crane. Norway, 1975.
A. The main and haul-back drums to the right, the skyline drum with storag and pulley sections in the middle, the drum for auxiliary lines to the left.
B. The operating handles are placed in such a way that it is easy for the yarder operator to move between yarder and landing for unhooking the load.

The tower is 8.5 meters high. During transport along truck or tractor roads it is laid forward over the tractor (figure 333).

The carriage may pass supports along the skyline. It has a trisected winch drum. The three sections have different drum diameters. The hoist line is wound up on the central section of the winch drum which has the middle diameter. The main line is wound up on the drum section with the biggest diameter. Since the diameter ratio between the hoist drum section and the main drum section is 1 : 2, the pull in the hoist line is twice the pull in the main line.

The haul-back line is wound up on the drum section with the smallest diameter. Less force is needed when dropping out the hoist line, and the speed is higher than the haul-back line speed.

Figure 337. The Igland Alp cable crane in Kviteseid, Norway 1975.
A. The load arrives at the landing near the winch station.
B. The lateral skidding.
C. Downhill transport.

A brake on the winch drum helps to keep the load lifted during the transport. When pulling in the main line, the brake loosens and the trisected winch drum rotates. It is therefore, not necessary with any carriage stop at the landing. It is possible to unload the logs on several piles along the skyline road on the landing, which also makes it possible to divide the piles into different assortments.

A carriage stop is needed at the place on the skyline where the lateral skidding is done. When the empty carriage returns and impacts against the carriage stop it fastens to the carriage stop and the brake on the trisected winch drum is released. It is therefore possible to pull out the hoist line by pulling in the haul-back line.

The usual maximum length of the Igland Alp Cable Crane is 800 meters and the following wire ropes are used for tree dimensions up to 1 cu.m. per tree:

Main line	L = 800 m	d = 8 mm
Haul-back line	L = 1,600 m	d = 8 mm
Hoist line	L = 80 m	d = 10 mm
Straw line	L = 100 m	d = 8 mm
Rig up line	L = 1,600 m	d = 3.3 mm
Skyline	L = 800 m	d = 18 mm

Capstan driven hoist winch in the carriage.

Nestestog radio controlled cable crane consists of a skyline on which the carriage is being pulled in or returned out by means of an endless main/haul-back line. This endless traction line is powered by a groove wheel pulley (figure 338). There are two carriage stops, one over the landing and one in the field (figure 341). A winch in the carriage pulls the hoist line in or out. The winch is driven by a capstan and there is a chain connection between the capstan and the winch drum with a 1 : 3 ratio. The hoist line pull is 3 times as high as the main line pull (the pull in the endless line). The endless line runs from the groove wheel pulley on the yarder to the carriage, where it is wrapped three times around the capstan and continuous around the yarding area and back to the pulley on the yarder. When the carriage runs along the skyline the capstan is locked. No movement in the capstan or the hoist drum is possible so that the load can hang under the carriage.

When the carriage reaches the carriage stop at either end of the skyline it is automatically hooked onto the carriage stop and the lock on the capstan is opened. The carriage is kept in position by the carriage stop and the capstan can rotate and power the hoist winch in one or the other direction. Only two signals to the yarder is needed, inhaul and return. If the carriage is connected to the carriage stop it means that the hoist line is being pulled in or out. When the carriage is hanging in the middle of the span, the inhaul and return signal means that the carriage is being moved along the skyline, since the capstan and the hoist winch is locked.

Usually a twin corridor is rigged up which means that the skyline is rigged up in the first corridor and the haul-back line in the next (figure 338). When the area around the skyline is yarded the skyline is moved over to the haul-back line corridor.

The K-1200 cable crane has no tower, and a spar tree has to be used. The winch may be mounted on a tractor (figure 342) or a 4-wheel driven truck (figure 344). On the tractor mounted model the tractor engine drives the hydraulic system of the cable crane. The rest of the equipment is mounted on a trailer which is pulled by the tractor. On the truck mounted model an extra engine drives the hydraulic system and the truck is just for moving the equipment. All of the equipment is mounted on the truck.

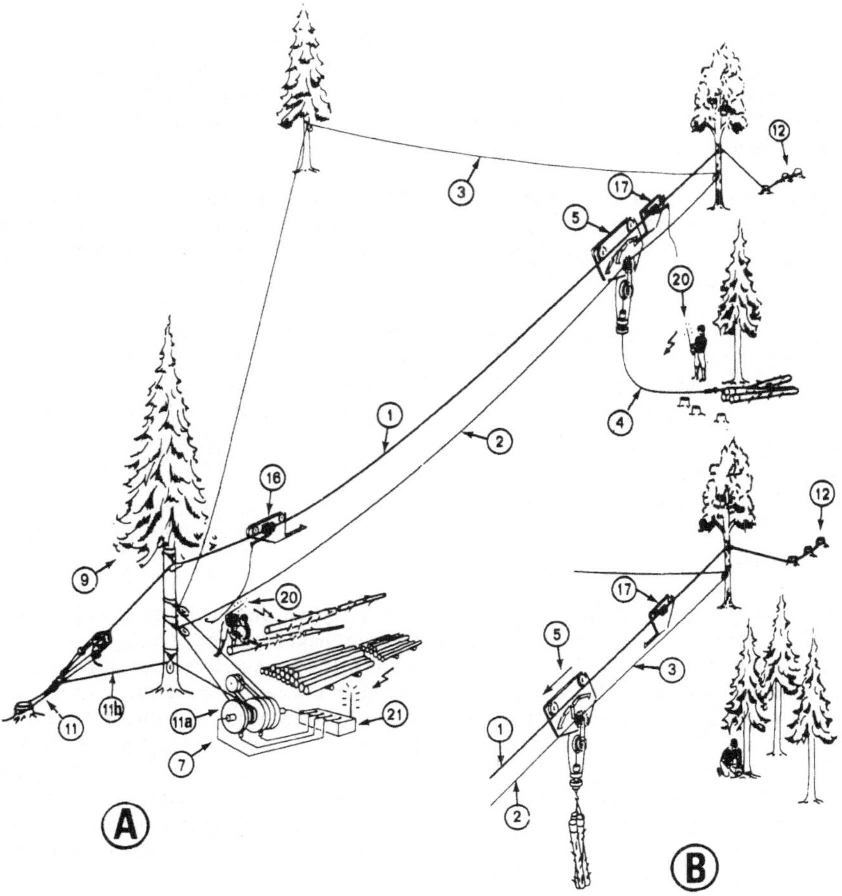

(1) Skyline (2) (3) Endless line (main and haul-back lines) pulling the capstan and winch drum in the carriage for the hoist line (4) Hoist line (5) Carriage (7) Hydraulically powered groove wheel pulling the endless line (2) and (3) (9) Head spar with guide blocks and cable shoe (11) **Heelhold** with heel **tackle** (11a) Winch drum for the heel line to tighten or lower the skyline (11b) The heel line to the tackle (12) Tailhold (16) Inner carriage stop (at the log deck) (17) Outward carriage stop (at the felling site) (20) The workers on the felling site or on the log deck have remote control of the winch by means of radio signals (21) Radio receiver to control the winch (wave, tune, rele, electricity from battery, magnet walve, hydraulic power).

Figure 338. The Nestestog radio controlled cable crane.

The power take off on the tractor or a separate engine on the truck mounted model drives a variable piston pump. This drives a slow running hydraulic motor which powers the pulley on the yarder. No brakes nor clutches are needed and the pulling direction of the endless traction line is changed by changing the direction of the flow from the pump. The system has a safety valve which stops the main engine if the feeding pressure falls below a certain limit. Connected to the timer there is a float switch which stops the engine if the oil level sinks below 90 % of the level when full.

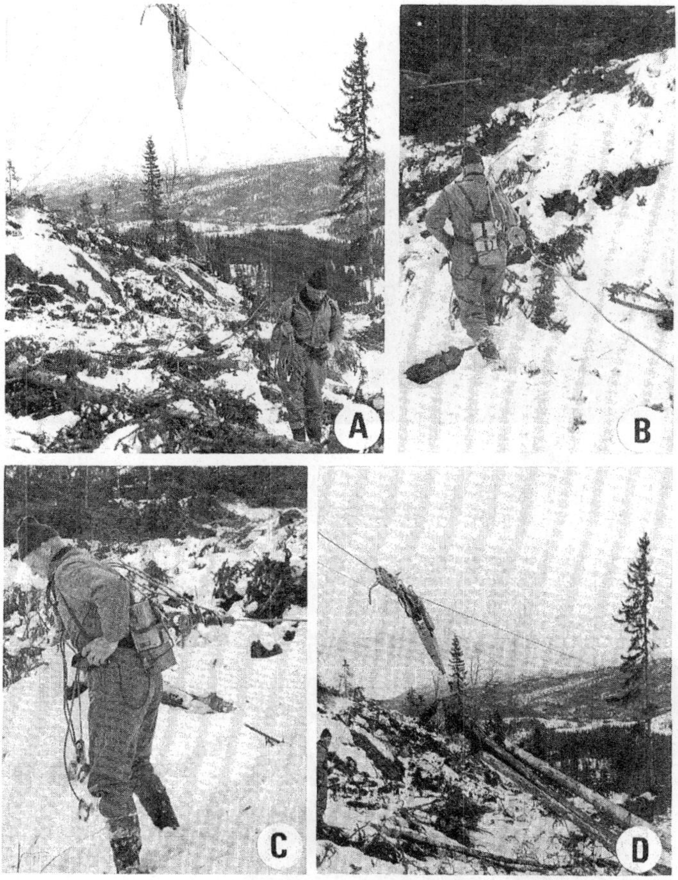

Figure 339. Nestestog radio controlled cable crane K-700, Kviteseid, Norway 1968.

A. The chokerman gives a radio signal to the yarder at the lower landing (600 m downhill) in order to pull out the hoist line from the hoist drum in the carriage.

B. and C. It is easy for the chokerman to carry out the hoist line to the logs since he simultaneously controls the yarder by means of a radio signal from his transmitter. By means of the radio signal the yarder pulls out the hoist line from the winch drum in the carriage.

D. The chokerman controls the yarder during the lateral skidding. He then turns on the timer and the carriage transports the load downhill along the skyline without his assistance (0-man operation).

Figure 340. Nestestog radio controlled cable crane in Kviteseid, Norway 1968.

A. The carriage transports the load along the skyline by means of a timer. The carriage runs automatically with no operator on the yarder (0-man operation).
B. Adjusting the timer.
C. The chaser man at the landing operates the yarder with radio signal during the transport towards the carriage stop.

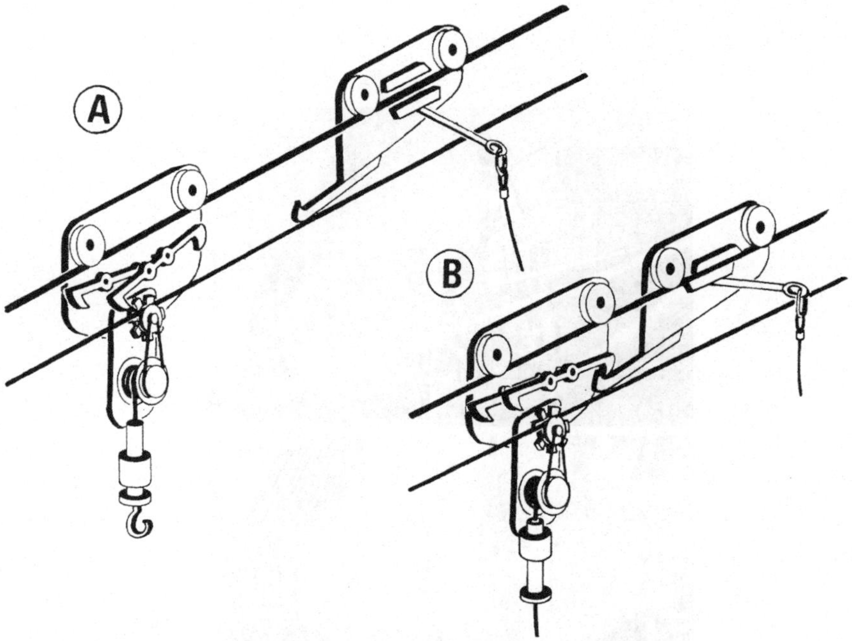

Figure 341. The Nestestog carriage for radio controlled cable crane.
A. The ratchet which locks the capstan in the carriage is closed.
B. The carriage is connected to the carriage stop and the ratchet is open. A movement in the main and haul-back line rotates the capstan. This is connected with a chain gear to the hoist drum (1–3) and pulls the hoist line out or in, depending on the direction of the movement.

The pulling capacity in the traction line is 20 kN and the cable crane is constructed for a maximum distance of 1,200 meters. The carriage may pass supports along the skyline.

The simple operation of the pulley has made it possible to use remote control. Each of the workers have a transmitter and they may send a signal to the receiver on the yarder in order to control the winch. Usually, just two signals are needed, one for pull and one for haul-back. These are linked to a trottle so the engine accelerates to a certain speed as soon as one of the signals is given. All together, there are six signals. The others are for starting and stopping the engine, starting the timer for automatic movement of the carriage along the skyline and one for stopping the timer.

The control equipment is a Phillips electronic remote control unit, frequency modulated with tone-code modulation. This unit gives 6 command signals and operates at 30 MHz. The maximum range is about 1.5 km depending on the terrain. The transmitter sends the signals which are picked up by the receiver on the yarder which via a relay unit and magnet valves operate the winch. The timer is built into the receiver on the yarder and this maintains the signals for the set time. If the timer is adjusted for 600 meters, one must first push the knob for START timer, and then again for PULL or HAUL-BACK. If the transport must be stopped before the time is up, this can be done by sending the signal, STOP.

Each of the signals consists of two tones for reasons of safety. Different tones are used for the different yarders in order to avoid the signals from one yarding crew being received by neighbouring yarder.

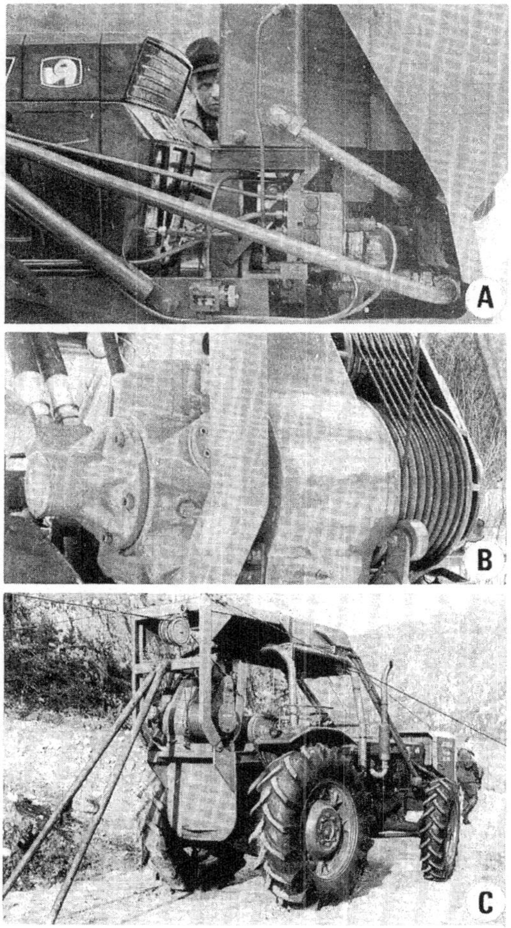

Figure 342. The Nestestog yarder for radio controlled cable crane K-1200.
A. The hydraulic pump driven by the tractor motor and the solenoid valves.
B. The groove wheel pulley powered by the hydraulic motor.
C. The anchoring of the tractor with the groove wheel pulley. The radio receiver is mounted on the roof of the cab.

If two of the transmitters try to send a signal at the same time the yarder stops. This is usually not a problem since a turn follows a certain pattern of work elements.

The Nestestog radio controlled cable crane was designed in 1965 and many units have been in operation all the time since. The radio control works excellently and with a very high availability. Figure 343 is a diagram of the winch and groove wheel drive on the Nestestog radio-controlled cable crane K-1200.

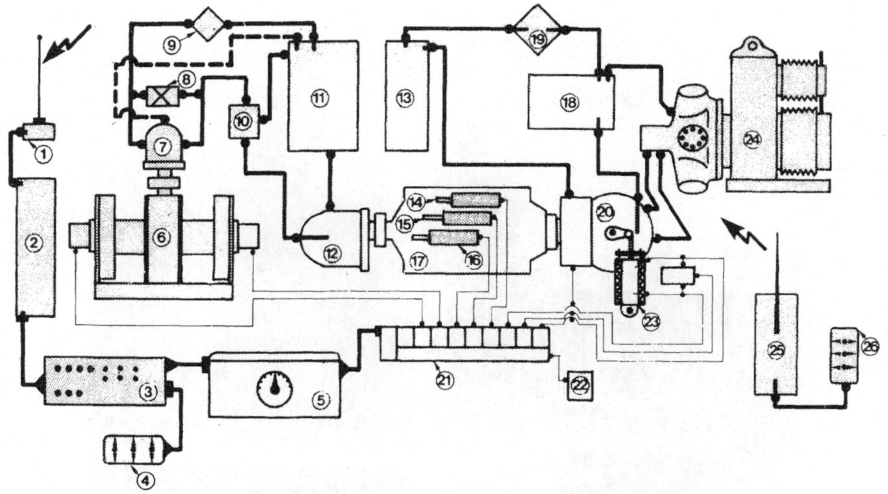

(1) Receiving antenna (2) Radio receiver (3) Relay unit (4) Control panel for remote control by cable
(5) Remote controlled timer (6) Double-drum winch (7) Hydraulic motor (8) Valve (9) Filter
(10) Valve (11) Hydraulic tank (12) Hydraulic pump (13) Oil cooler (14) Throttle control for low speed
(15) Throttle control for high speed (16) Stop switch (17) Diesel engine (18) Oil tank (19) Filter
(20) Hydraulic pump (21) Valve battery (22) Flow control (23) Displacement control cylinder
(24) Hydraulic motor with groove wheel system (25) Radio transmitter, one for each of the workers in the crew (26) Control panel

Figure 343. Diagram of the winch and the groove wheel pulley on Nestestog radio controlled yarder K-1200. (Norway, 1975).

The remote control yarder has proved that it has many advantages from a work scientific point of view. It is the worker who need the forces who also controls them. No misunderstanding between the field crew and the yarder operator is possible and the worker who sees the problems in the terrain can control the yarder himself.

The timer which controls the automatic transport movement of the carriage along the skyline has many advantages. When the carriage has stopped and is fastened to the carriage stop the choker setter lowers the hoist line with a signal from his transmitter to the yarder which is located maybe 1,000 meters away from him. He can easily carry the end of the hoist line towards the logs, hook on the logs, and then skid them laterally towards the carriage. When the load has been hoisted up to the carriage, the hook turns against the lower end of the carriage and it is released from the carriage stop. The worker then sends the signal for automatic movement of the carriage towards the landing. He can continue with his work in the field while the cable crane automatically takes care of the transport. The carriage stops automatically, approximately 20 meters from the landing. When the chaser is finished with the work on the previous load, he controls the yarder by his transmitter, takes down the load and returns the empty carriage. The empty carriage returns to the field »zero-man» operated and the chaser can do something else.

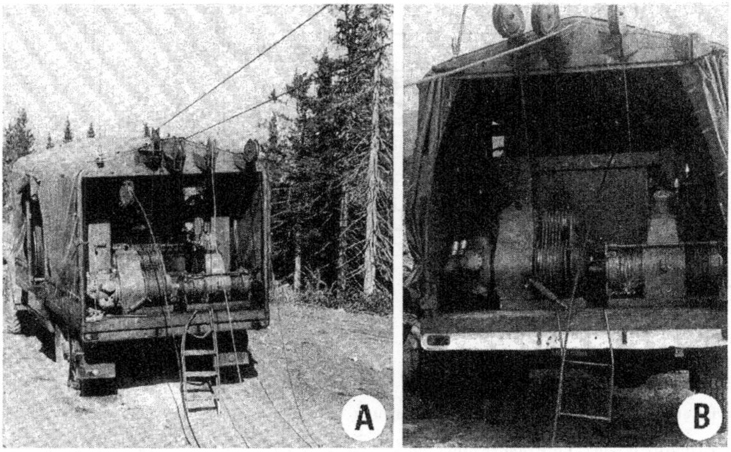

Figure 344. The Nestestog truck mounted K-1200 yarder for radio controlled cable crane. All the cables and cable equipment are placed on the truck during transport from one logging area to the next.

This has made it possible to combine felling and limbing with the yarding. While the carriage is at the landing with a load the choker setter may fell the trees for a new load and limb the upper side of the stems. When he is through with this job the carriage has already stopped 20 meters before the carriage stop and he can use his radio transmitter to run the carriage the remaining distance. He pulls out the hoist line, hooks on the trees and skids them laterally to the carriage. When the carriage has been released from the carriage stop, he sends the signal for automatic load transport and can continue felling and limbing the next load.

In the meantime, the man at the landing has finished limbing the trees from the previous load and bucked them into assortments. When he has finished the previous load, the new load is already hanging on the skyline and he can take this down to the landing and return the empty carriage automatically along the skyline. He is now ready to limb and buck the new load.

One disadvantage of the cable crane is the carriage stops. A collision between the carriage stop and the carriage causes delays. The highest percentage of delay times in the whole system was caused by this problem.

A carriage with automatic clamps may be stopped anywhere along a skyline. Figure 345, I and 346 shows one example. The top part of the Austrian Koller carriage has been built onto the Norwegian Nestestog under-carriage. The mode of operation of the Koller carriage has been discussed in chapter 11.3.2.1. (figure 272).

Another solution is illustrated in figure 345, II and figure 347. The carriage wheels drive an alternator which charges a battery. The battery drives an electric motor with a hydraulic pump which opens or closes a clamp on the skyline, and a brake on the capstan drive for the hoist winch. When the clamp is closed the hoist drum brake is deactivated. When the skyline

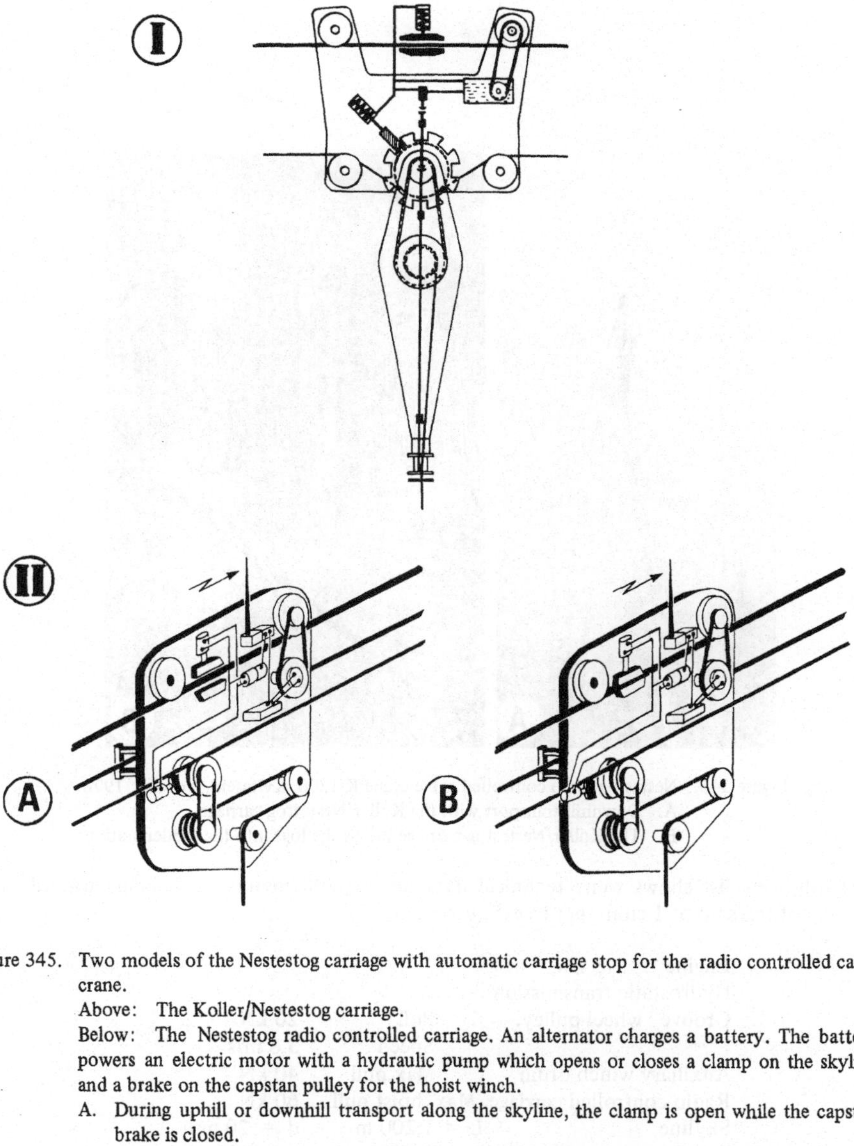

Figure 345. Two models of the Nestestog carriage with automatic carriage stop for the radio controlled cable crane.

Above: The Koller/Nestestog carriage.

Below: The Nestestog radio controlled carriage. An alternator charges a battery. The battery powers an electric motor with a hydraulic pump which opens or closes a clamp on the skyline and a brake on the capstan pulley for the hoist winch.

A. During uphill or downhill transport along the skyline, the clamp is open while the capstan brake is closed.

B. When the carriage has arrived at felling site or landing, the worker gives a radio signal to the receiver in the carriage. The clamp on the skyline closes and the capstan brake opens. A movement of the endless main/haul-back line (traction line) rotates the capstan and the hoist winch. This pulls out or pulls in the hoist line, depending on the direction of the movement.

clamp is open, the hoist drum brake is activated. There is a radio receiver in the carriage which upon signals from the worker's transmitter directs the electric power from the battery in the carriage to the hydraulic valves on the clamp or the brake cylinders. With this type of radio controlled carriage the transmitters need 8 signals, 6 to the yarder and 2 to the carriage.

Figure 346. Nestestog radio controlled cable crane K-1200, Kviteseid, Norway 1978.
A. Downhill transport with the Koller/Nestestog carriage.
B. The Koller/Nestestog carriage brings the logs into the yarder station.

The following list shows some technical data for a 1,200 meters cable crane for full tree yarding of trees up to 1 cu.m. per tree (figure 348):

Engine 75 kW		
Hydrostatic transmission		
Groove wheel pulley.	Pull	20 kN
	Speed	3.8 m/s
Auxiliary winch drum	Max.pull	40 kN
Radio controlled carriage.	Max. hoist pull	60 kN
Skyline	L = 1,200 m	d = 20 mm
Endless traction line	L = 2,500 m	d = 12 mm
Hoist line	L = 70 m	d = 12 mm
Straw line	L = 2,500 m	d = 4 mm
Rig up line	L = 2,500 m	d = 2 mm

Self-propelled hoist carriage and stationary traction line (main line).

An interesting cable crane was built in Akita in Japan 1964. The carriage is self-propelled by a 54 kW motor which powers a groove wheel pulley as well as a hoist drum (figure 349). The

Figure 347. Nestestog radio controlled cable crane, Seljord, Norway 1979.
A. Lateral skidding towards the radio controlled carriage.
B. The radio controlled carriage transports the load along the skyline.
C. The radio controlled carriage passes a crossline support.

cable crane has a fixed skyline and a stationary traction line (main line). Both the skyline and the traction line are fixed at both ends of the cable crane. The traction line is wrapped around a pulley in the carriage. The carriage moves along the skyline when the pulley rotates.

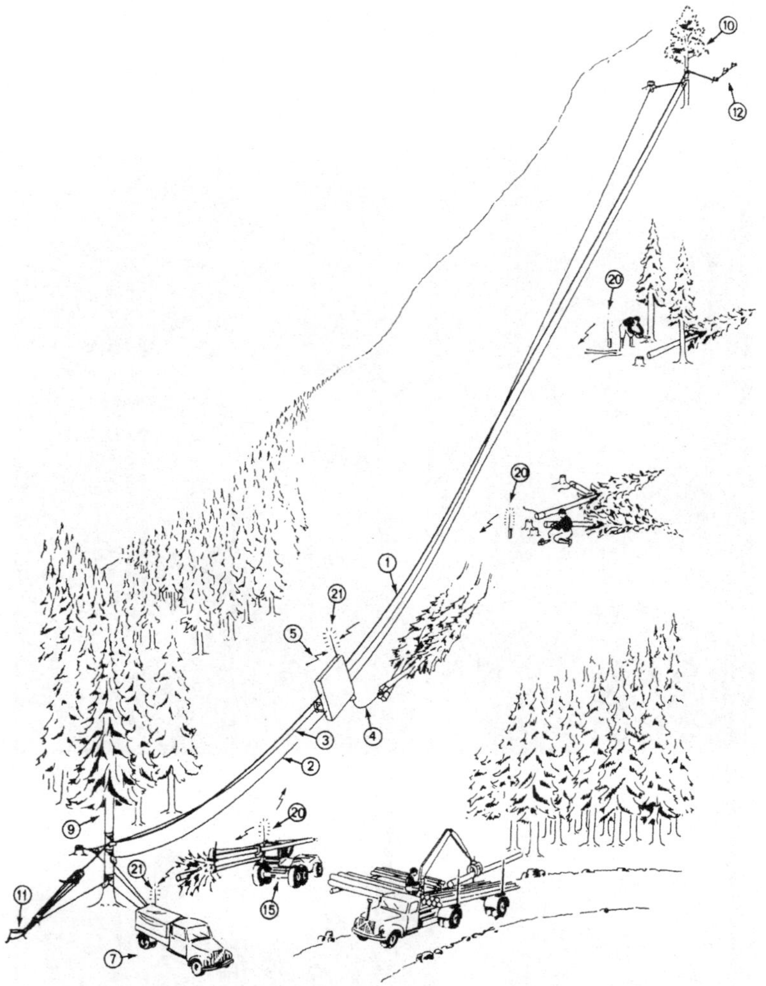

(1) Skyline (2)(3) Main and haul-back line (4) Hoist line from hoist drum in the carriage
(5) Radio controlled carriage. By radio signal the carriage is fastened to the skyline by a hydraulic grip and the capstan propelling the hoist drum is released (7) Truck mounted groove wheel and winch is propelling the cable crane (9) Head spar (10) Tail spar (11) Heel anchor with heel tackle for the skyline (12) Tailhold (15) Harvesting machine at the winch station (20) Radio senders to remote control of the truck mounted winch, groove wheel and the carriage (21) Radio receivers operating the truck mounted groove wheel, winch and the carriage

Figure 348. The Nestestog radio controlled cable crane with truck mounted yarder and radio controlled carriage. A limbing machine at landing.

The carriage is remote controlled by means of radio signals from the workers to the carriage. The choker setter, as well as the chaser has a transmitter by means of which he can control the inhaul or the return movement of the carriage. He can also start the hoist drum and feed out the hoist line or pull in the load (lateral skidding).

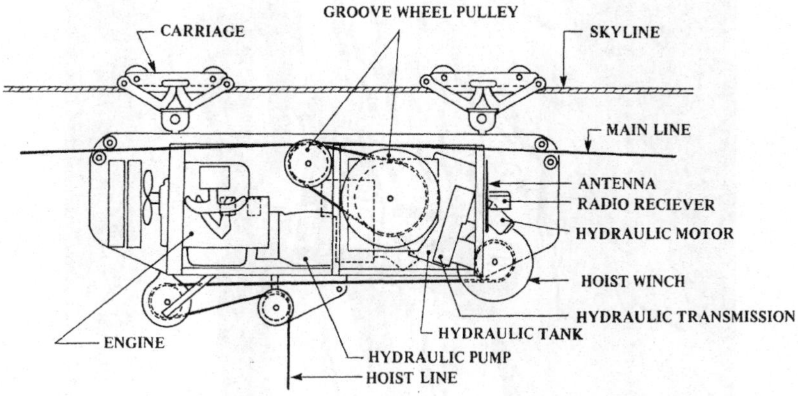

Figure 349. A Japanese self-propelled hoist carriage (KONUMA and SHIBATA 1976).

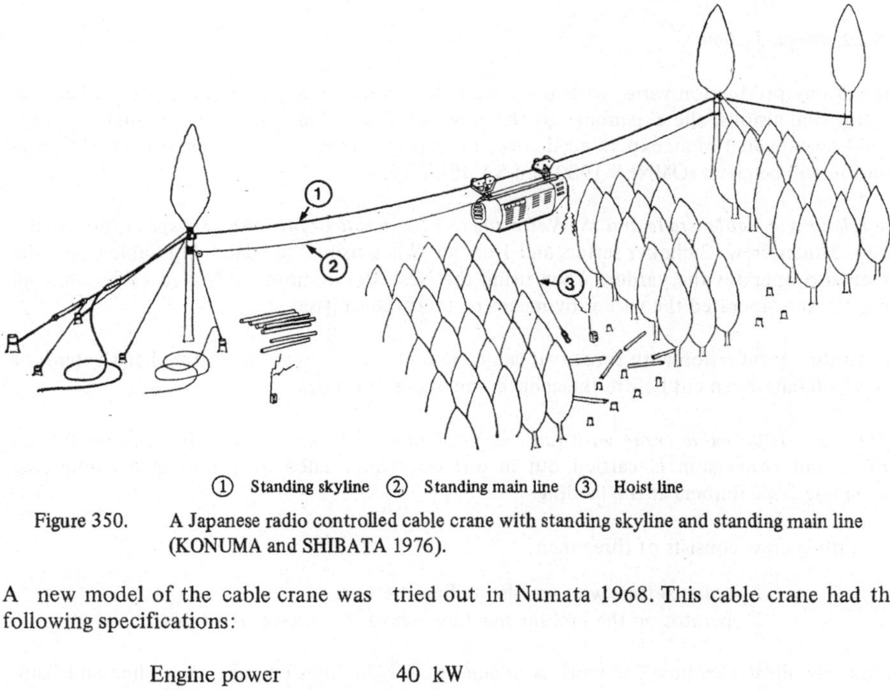

① Standing skyline ② Standing main line ③ Hoist line

Figure 350. A Japanese radio controlled cable crane with standing skyline and standing main line (KONUMA and SHIBATA 1976).

A new model of the cable crane was tried out in Numata 1968. This cable crane had the following specifications:

Engine power	40 kW	
Weight of carriage	9.8 kN	
Skyline	L = 300 m	d = 26 mm
Traction line	L = 300 m	d = 12 mm

The experiments were not promising. According to Konuma and Shibata 1976 the daily output was very low. The disadvantages were the need for another yarder to rig up the cable crane, and that the self-propelled hoist carriage had a too low availability because the mechanism of the carriage did not always function properly.

Figure 351. A radio controlled self-propelled carriage with engine powered hoist drum in Numata, Japan 1968.

11.5.2.2. The production

The yarding production varies with the system, the work conditions, the volume per hectare, the tree dimensions, the experience of the crew etc. As an example of the production which should be expected of an experienced crew, a few results from a research program of 15 years duration will be given (OMNES 1980 and SAMSET 1981).

The Igland Alp Cable Crane and the Nestestog K-1200 Cable Crane with carriage stops usually have a 3-man crew: 2 choker setters and 1 chaser. When using the Igland Alp Cable Crane the chaser also operates the yarder. When using the Nestestog remote controlled cable crane all three workers operated the yarder by means of their transmitters.

The yarding production with these two cable cranes applies to yarding of tree length stems or logs which have been cut before the cable crane operation starts.

Nestestog K-1200 cable crane with radio controlled yarder and carriage. In this case, felling, yarding and conversion is carried out in one operation. There is a limbing machine (the Logma telescope limber) at the landing.

The yarding crew consists of three men:

2 choker setters who are also felling the trees
1 operator on the limbing machine who also unhooks the trees.

Figure 348 illustrates how the work is organized. All the lines (skyline, main line and haul-back line) run along the skyline corridor and the haul-back line runs through the carriage. The felling and yarding is carried out on one side of the skyline only, in a 20 meters wide strip from the bottom to the top of the hillside. When this strip has been yarded the skyline is moved close to the edge of the standing timber. The new skyline road is parallel to the first one. The two choker setters are also felling the trees. Each of them fell the trees and set the chokers while the carriage is on its way down to the landing and back again. When the carriage returns (zero-man operated, by means of the timer), he hooks his chokers onto the hoist line, controls the yarder by means of his transmitter during the lateral skidding and sends the load down to the landing. The carriage is used alternatly by the two choker setters and this gives each of them sufficient time to fell and set the chokers on a new load.

A limbing machine converts the trees to stems or logs at the landing. Since the carriage moves automatically along the skyline the operator of the delimbing machine can use the transport time for delimbing the trees from the previous load.

Table 43 shows an example of the distribution of the work elements in a turn cycle for the three systems. The delay time is given as a percentage of the effective time. The reason for the lower delay time when using a remote control carriage is because there are no carriage stops which can cause collisions.

Table 43. Examples of the work elements distribution in a turn cycle. Length of the cable crane 600 m. L_T = 300 m. Skidding sk = 20 m. Slope 40 %. 0.3 m^3 per tree (SAMSET 1981).

Work elements	Igland Alp	Nestestog K-1200 Carriage:		
		Carriage stop	Remote control	
	Mins/turn	Mins/turn	Mins/turn	%
Carriage return	2.2	1.8	1.8	25 %
Hoist line out	0.8	0.8	0.6	8 %
Choker setting	1.3	1.3	1.3	18 %
Lateral skidding	0.8	0.8	0.7	9 %
Load transport	2.3	1.9	1.9	26%
Unloading	1.0	1.0	1.0	14 %
Effective time	8.4	7.6	7.3	100 %
Delay time %	38 %	32 %	28 %	
Work place time	11.6 mins/turn	10.0 mins/turn	9.3 mins/turn	

The productivity on these systems depends on the size of the loads. With small tree dimensions it is difficult to get sufficiently big loads on the cable crane. Table 44 shows the relationship between the tree dimension, the standing volume to be cut and yarded and the size of the load. There is no problem in getting sufficiently big loads if the tree dimensions are big enough.

Table 44. The relation between cu.m. per ha., cu.m. per tree and load size on a cable crane (Igland Alp and Nestestog K-1200) in cu.m. per turn (OMNES 1978).

m^3/tree	Standing volume	
	100 m^3/ha	200 m^3/ha.
0.1	0.53	0.73
0.2	0.66	0.88
0.3	0.82	1.01
0.4	0.95	1.15
0.5	1.08	1.27
0.6	1.19	1.38
0.7	1.29	1.45
0.8	1.40	1.54

Table 45 and table 46 give some examples of the typical production which can be expected with experienced crews operating these cable cranes. The Igland Alp cable crane yarded logs or tree length stems. The production with the Nestestog radio controlled cable crane include felling the trees, yarding and conversion into tree length stems. The remote control gives high production for the total work sequence, since the work for each man is lighter and most of the heavy work is being carried out by machine power at or near the landing.

Table 45. Typical production per work place hour for an experienced 3-man crew during yarding with Igland Alp Cable Crane. Tree length stems (SAMSET 1981).

m^3/turn	Length of cable crane								
	200 m			600 m			1,000 m		
	Yarding distance (L_T)								
	100 m			300 m			500 m		
	A	B	C	A	B	C	A	B	C
0.5	3.5	1.2	0.8	2.5	0.8	1.3	2.3	0.8	1.3
0.7	5.0	1.7	0.6	3.5	1.2	0.8	3.2	1.1	0.9
0.9	6.3	2.1	0.5	4.7	1.6	0.6	4.2	1.4	0.7
1.1	7.7	2.6	0.4	5.8	1.9	0.5	5.0	1.7	0.6

A = m^3/hr for the crew
B = m^3 per man per hour
C = manhours per m^3

Table 46. Typical production per work place hour for an experienced 4-man crew during felling, yarding and delimbing with a K-1200 radio controlled cable crane with radio controlled carriage, and a Logma T-300 delimber at the landing. (SAMSET 1981).

m^3/turn	Length of cable crane								
	200 m			600 m			1,000 m		
	Yarding distance (L_T)								
	100 m			300 m			500 m		
	A	B	C	A	B	C	A	B	C
0.7	6.0	1.5	0.7	4.1	1.0	1.0	3.2	0.8	1.3
0.9	7.7	1.9	0.5	5.3	1.3	0.8	3.9	1.0	1.0
1.1	9.3	2.3	0.4	6.5	1.6	0.6	4.8	1.2	0.8
1.3	11.2	2.8	0.4	7.7	1.9	0.5	5.8	1.5	0.7
1.5	12.8	3.2	0.3	8.8	2.2	0.5	6.7	1.7	0.6

A = m^3/hr for the crew
B = m^3/man per hour
C = manhours per m^3

11.5.3. Mobile towers and yarders

The operating winches are mounted on a mobile trailer, such as a self-propelled crawler,a self-propelled wheeled carrier, a truck or a tractor. In addition to the operating winches there are winch drums for straw lines,rig up lines etc. in order to facilitate the work with the set-up of the cable crane.

Figure 352.

A. Skagit BU16 on semitrailer.
B. and C. Self-propelled Madill 046 on crawler chassis.
D. The butt rigging with a lifting block on the haul-back line. (observe the choker setters).
E. Three section drum carriage for running skyline.
F. The cab on the yarder.
G. Self-propelled Madill 046 on a 4-axle wheel chassis and a goose-neck loader on the landing (Franklin River, Vancouver Island 1973).

Figure 353. Highlead yarding with Madill 071 on crawler chassis (Crane Brook B.C. 1977).

Figure 354. Slackline yarding with the Berger C-19 Power Yarder (Dahlgren Logging, Washington, USA 1980).

11.5.3.1. The yarders

The yarders are equipped with a tower, usually a steel spar which replaces the previous spar trees. It was often difficult to find a convenient tree near the landing or yarder station, and an artificial spar tree had sometimes to be erected. The mobile towers represented a big step forward. Because they could be rigged up anywhere the exhausting work involved in high-climbing and rigging a spar tree was avoided and the rig up time was reduced considerably.

Large towers. The size of the mobile towers and yarders vary with the tree dimensions for which they are intended. Yarding oversized tree dimensions in virgin forests such as on the coast line of British Columbia and the western United States, and the many places in the

Figure 355. The West Coast Tower (Willamette National Forest, Oregon, USA 1973).

tropics such as in the Philippines needs heavy yarders powered by 300 – 400 kW diesel engines (figure 352). The height of the steel spar is often 30–35 meters in order to give the front end of the heavy loads sufficient lift to pass big stumps and other obstacles. A slackline system or a highlead system is often used.

The maximum line pull is often 150–200 kN and sometimes as high as 400 kN, while the maximum speed often varies between 7 and 10 meters per second for the operating lines. The main line diameter is up to 35 mm and the skyline up to 42 mm in diameter. This is equipment which is difficult for the workers to handle.

Such yarders are often equipped with 5 operating drums: skyline, main line, slack-pulling line, haul-back line and hoist line (drop line, skidding line). The transmission has usually many gears in both directions. These sophisticated constructions have usually a torque converter which gives smoth running and easy adjustment of the line pull and the line speed. An example is shown in figure 356.

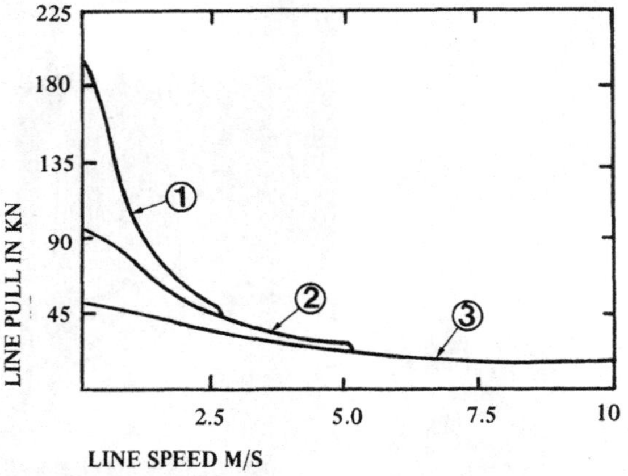

Figure 356. The relationship between line speed and line pull on various transmissions for the 176 kW West Coast Tower. ① - ② - ③ transmissions.

In the virgin forests which have not been thinned or had other silvicultural treatment there is big variation in the tree dimensions (0.5 cu.m. per tree to 20 cu. m. per tree). The yarders for these conditions are constructed for heavy timber. In spite of this, the average load is often low, because of the large amount of smaller timber in the stand. Sometimes these smaller dimensions are left on the ground. A better solution, however, is to first carry out the logging with lighter equipment and leave the big trees in the stand to the heavy mobile tower yarders (pre-logging).

A steel spar for tree dimensions up to 20 m^3 per tree need at least six guylines evenly distributed around the tower.

In second growth the need for equipment suitable for yarding tree dimensions between 1 and 3 cu.m. per tree appear. Yarders for these dimensions are powered by engines of between 100 and 200 kW and the tower is usually between 12 and 15 meters high. The towers are usually stabilized sideways by the yarder chassis and one or two back guys are usually needed to keep the tower in position. Sometimes a front guy is used for safety if the tailhold, the running lines or the skyline should break. In some countries the maximum height of a tower using one or two back guys only is 15 meters.

Side-inclined towers. These yarders are built especially for moving along feeder roads, parallel to the terrain contours. The wood is yarded uphill or downhill perpendicularly to the truck road. Highlead, slackline, or gravity outhaul systems are mostly used and the number of winchdrums for the operating lines are reduced to 3 or 4. Figure 353 shows highlead yarding with Madill 071 and figure 354 shows slackline yarding with a Berger C19 Power Yarder.

Table 47 Examples of technical data for some large yarders.

Model		Madill 071	Madill 046*	Washington 208
Weight		387 kN	650 kN	826 kN
Tower height		14.3 m	29 m	34 m
Tower type		Tower	Steel spar	Steel spar
Engine power		218 kW	355 kW	368 kW
Main line	L	960 m 665 m	700 m	610 m
	d	16 mm 19 mm	32 mm	32 mm
	Max. pull	148 kN	426 kN	165 kN
	Max. speed	7.8 m/s	8.1 m/s	7 m/s
Haul-back line	L	1,340 m	1,615 m	1,360 m
	d	16 mm	22 mm	22 mm
	Max. pull	114 kN	195 kN	110 kN
	Max. speed	7.8 m/s	17.7 m/s	10 m/s
Tagline	L	1,020 m	735 m 1,830 m	–
	d	10 mm	16 mm 10 mm	–
	Max. pull	37 kN	50 kN	–
	Max. speed	11.7 m/s	18.2 m/s	–
Straw line	L	–	1,670 m	1,650 m
	d	–	11 mm	10 mm
	Max. pull	–	159 kN	49 kN
	Max. speed	–	11.7 m/s	22.6 m/s
Skyline	L	590 m	–	670 m
	d	25 mm	–	35 mm
	Max. pull	399 kN	–	161 kN
Nationality		Canada	Canada	USA

*Main line is used as skyline.
It also has a hoist drum.

Table 47 gives some examples of technical descriptions for these types of yarders. Another interesting design is the West Coast Tower (figure 355). The 15 meters high steel spar has a rigid construction and is sideways stabilized by the crawler yarder chassis.One back guy only is needed. It is a heavy machine weighing 340 kN, and is powered by a 176 kW diesel engine. Like many other sophisticated yarders, the West Coast Tower is equipped with a torque converter which allows continuous variation of the line speed according to the load resistance during the inhaul or outhaul. The operating drums have three speeds forward and reverse, and figure 356 illustrates the relationships between the line pull and the line speeds.

Figure 357. Skidder mounted yarder and tower. (Crane Brook, B.C. 1977).

The following lines are used:

Main line	L = 850 m	d = 15 mm
Haul-back line	L = 1,700 m	d = 15 mm
Straw line	L = 1,500 m	d = 6 mm
Skyline	L = 825 m	d = 22 mm

Tractor and truck mounted yarders and towers. These are yarders built for the final cut in a sustained yield forestry (mostly clearfelling and sometimes selective cutting). Normal tree dimensions vary between 0.3 and 3 cu.m. per tree. A new generation of yarders have been developed on the American and Canadian West Coast, such as the Ecologger, the American loader, the Micro Master etc. These are yarders powered by approx. 100 kW or less and with tower heights between 9 and 13 meters. The 2 or 3-drum winches may be used for highlead, slackline and gravity outhaul systems (figure 357). The yarding distance is short, usually between 100 and 150 meters and sometimes up to 200 meters (McMORLAND and WONG 1982). Outhaul line speed is up to 4 m/s and inhaul speed is around 2 m/s.

These yarders have similar dimensions to the sophisticated yarders developed in Central Europe.

The Andritz Hydrocrane 80 may be chosen as an example (figure 358 and 359). Some technical data are shown in table 48.

The Andritz hydrocrane is hydrostatically driven. Instead of using winches, the main and haul-back lines are pulled by groove wheel pulley and a storage drum. The system is described in chapter 3.3 figure 49. This design gives good control of the relationship between the line pull and the speed and the hydrostatic drive gives a smooth running. A relief valve prevents any overloading of the system.

Table 48. Examples of technical data for Central European yarders.

		Andritz hydrocrane 80	Steyr KSK 16/20
Model		Andritz hydrocrane 80	Steyr KSK 16/20
Weight		260 kN	350 kN
Tower height		12 m	17.5 m
Engine power		235 kW	191 kW
Main line	L	700 m	650 m
	d	14 mm	16 mm
Max. pull		60 kN	41 kN
Max. speed		4.5 m/s	7.8 m/s
Haul-back line	L	1,000 m	1,300 m
	d	14 mm	14 mm
Max. pull		60 kN	41 kN
Max. speed		8 m/s	7.8 m/s
Slack-pulling line	L	1,000 m	650 m
	d	7 mm	9 mm
Max. pull			21 kN
Max. speed			18.3 m/s
Straw line	L	1,000 m	1,500 m
	d	4 mm	6,5 mm
Max. pull			21 kN
Max. speed			18.3 m/s
Skyline	L	600 m	700 m
	d	22 mm	24 mm
Max. pull		20 kN	136 kN

A slack-pulling line is running with the same speed as the main line and is used to pull out the hoist line.

The skyline drum has a maximum pull of 20 kN (pre-tension). There is a hydraulically operated clamp which holds the skyline. The clamp may be pulled in or out by the use of a tackle and a hydraulic cylinder on the yarder which allows adjustment of the skyline deflection. A relief valve prevents overload on the skyline.

Another example is the Steyr KSK 16/20 (table 48 and figure 370). It is an advanced design where standard clutch and brake controlled winch drums have been used. The brakes are water cooled. The clutches and the brakes have pneumatic controls. This also makes it possible to adjust the maximum braking forces in the lines as well as the maximum torque and line pulls. The torque converter allows smooth running and adjustment between line pull and line speeds. This construction is similar to some of the large tower yarders described in table 47.

The Steyr KSK yarder may be used for several cable systems. All the tractor and truck mounted yarders and towers may be used for uphill or downhill yarding as well as horizontal yarding for example over narrow valleys, gulleys etc. They are typical all round yarders.

The high production necessitates big landings. This problem can be avoided by using a skidder or a harvesting machine to move away the timber during the yarding. It is, however, also possible to store the wood on the feeder truck road and move the yarder sideways continuously as the yarding goes on. In such cases a swing boom is better than a fixed steel spar (figure 275). The problem of getting the load onto the feeder road can also be solved as

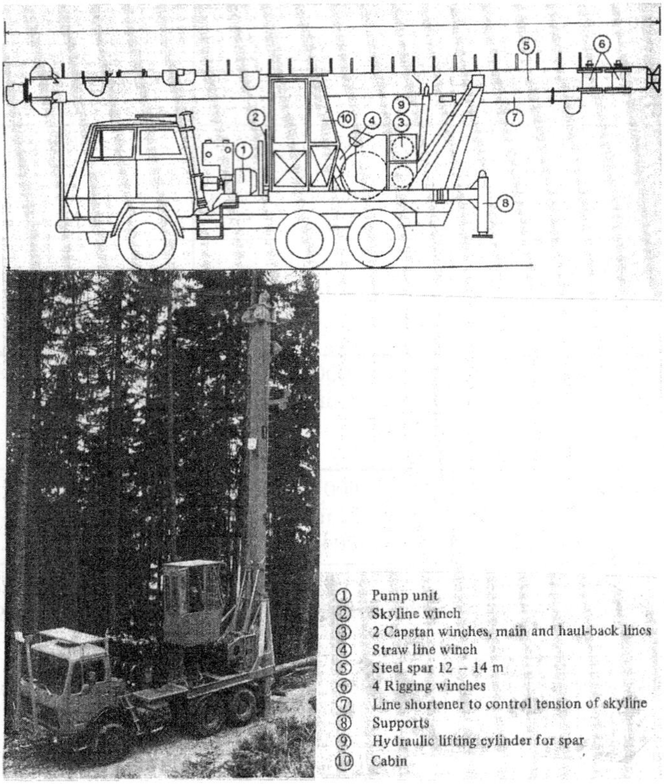

Figure 358. Andritz Hydrocrane 80 developed by Kurt Wyplel (Mayr Melnhof, Austria 1977).

illustrated in figure 360. The main line runs from the main drum via a guide block on the headspar (to obtain good spooling) and back to a terrain block mounted on the slope behind the yarder. When the load approaches the road the live skyline is slackened and the load is pulled over the edge of the road and onto the road. After unhooking, the skyline is tightened again and the carriage may return to pick up the next load. The method may be used in connection with any slackline system, during uphill or downhill yarding, with or without haul-back line.

11.5.3.2. The carriages

During highlead yarding of big timber an ordinary butt rigging is used (figure 205). In small timber it is necessary to collect a number of trees or logs in each load and a monkey block or a squirrel block allows the end of the main line to work as a hoist line (fish-hook yarding, figure 216).

Figure 359. Andritz Hydrocrane 80. A cable crane with live skyline. (Mayr Melnhof, Austria 1977).

The fixed skyline systems and the live skyline systems (slackline). The carriage runs along the skyline. The load has to be laterally skidded towards the carriage and carried by the carriage during the transport along the skyline.

It would be extremely hard work for the choker setters to pull out the hoist line from the carriages on these large cable cranes (figure 362). The hoist line (the skidding line) must be pulled out by motor power. A carriage with a 3-sectioned drum may solve the problem. The main line is wrapped around the left drum section while the slack-pulling line is wrapped around the right drum section. The hoist line is wound in on the central drum section which serves as the hoist drum. Since the main line goes in on the upper side of the drum and the slack-pulling line on the underside of the drum, they may rotate the hoist drum in one or the other direction and thus pulling in or out the hoist line. The carriage on the Andritz hydrocrane 80 uses the main line as the hoist line. The main line runs from the yarder to the carriage where it is wrapped around a pulley. This pulley can be rotated by the slack-pulling drum. The slack-pulling line runs from the yarder to the carriage, where it is coiled onto the slack-pulling drum.

As soon as the carriage has reached the place for lateral skidding, the operator pulls in the slack-pulling line which rotates the slack-pulling drum so that the main line is being fed out. The Steyr KSK cable crane uses a similar type of carriage (figure 363).

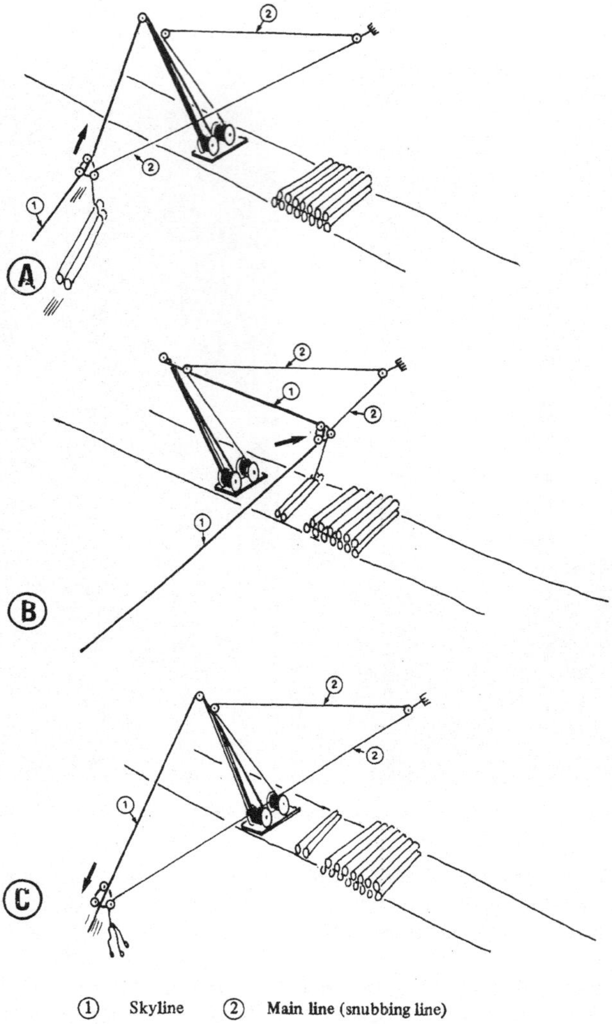

Figure 360. Uphill yarding with side inclined tower and gravity outhaul system. The heel block is rigged up behind the yarder.

A. The load approaches the edge of the feeder road.
B. While the skyline is slacked down, the main line pulls the log onto the road beside the yarder.
C. The skyline is tightened for gravity outhaul of the carriage. (Österreichische Bundesforste, Austria 1977).

Madill skyline carriage (figure 364) can stop anywhere along the skyline by means of a skyline clamp. When there is tension in the haul-back line the clamp is kept open, and the carriage can move along the skyline. As soon as the haul-back line has been slackened and the tension redused, a spring force in the carriage closed the clamp onto the skyline and keeps the carriage in position. The hoist line may be pulled out by means of a slack-pulling line. Since the clamp

Figure 361. ØBF skidder mounted yarder with side inclined tower. The heel block is mounted above the feeder road behind the yarder. By slacking the skyline, the main line pulls the load over the feeder road (Österreichische Bundesforste, Austria 1977).

Figure 362. Professor Adamowich admires the size of a three-section drum carriage for running skyline yarding in big timber. (Vancouver Island, B.C. 1973).

is constructed as an eccentric pressure grip the carriage stays in position even when there is high tension in the main line during the lateral skidding. The higher the main line pull, the higher the grip pressure of the clamp on the skyline.

The Young carriage is also a drum carriage with a 3-sectioned drum. In this case a slack-pulling line is not needed. The main line is spooled in on one side of the 3-sectioned drum while the haul-back line is wound in on the other. The hoist line (skidding line) is wound in on the central drum section. When this carriage reaches the terrain the carriage stops and a clamp fixes the carriage to the skyline. At the same time the brake on the 3-sectioned drum is opened and the haul-back line may pull out the hoist line. As soon as the load has been attached to the hoist line and lifted to the carriage, the 3-sectioned drum is locked by means of the drum brake. The skyline clamp is opened and the load may be transported by the

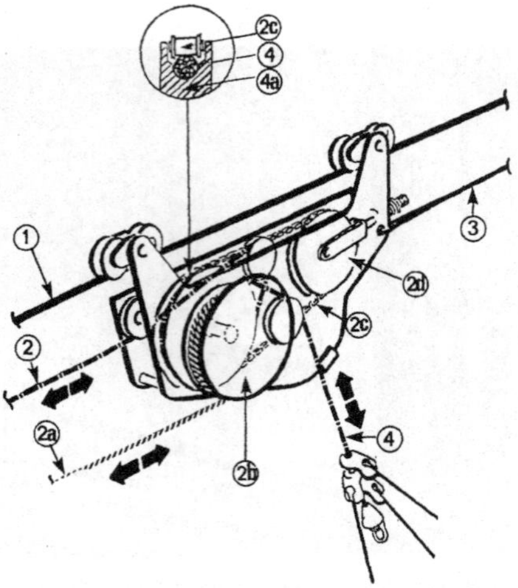

(1) Skyline (2) Main line (2a) Slack-pulling line (2b) Slack-pulling drum
(2c) Chain, pressing the main line on the pulley in order to create sufficient friction.
(2d) Tensioning wheel. (3) Haul-back line (4) Hoist line (4a) A pulley, propelled by the slack-pulling drum pulls out the main/hoistline.

Figure 363. The Steyr carriage developed by Clay Anderson. A friction chain presses the hoist line against the pulley in order to give sufficient friction to pull out the hoist line (Steyr. Austria, 1977).

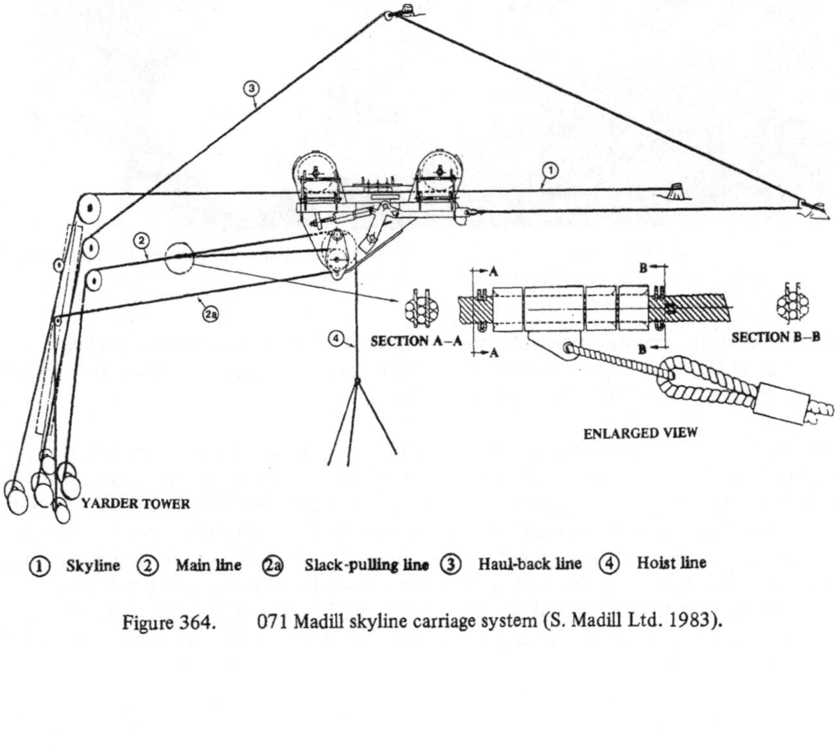

(1) Skyline (2) Main line (2a) Slack-pulling line (3) Haul-back line (4) Hoist line

Figure 364. 071 Madill skyline carriage system (S. Madill Ltd. 1983).

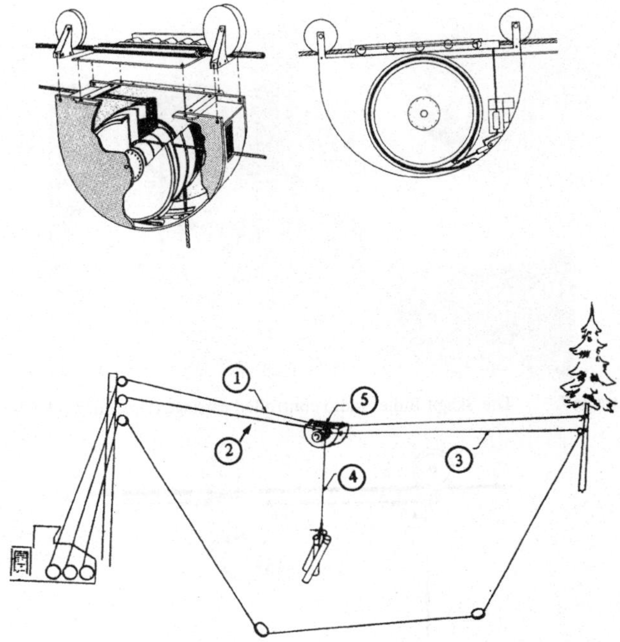

Figure 365. Live skyline cable crane with the Young 3-section drum carriage. The carriage may stop anywhere along the skyline by means of a hydraulic operated clam. The Young carriage can be radio controlled.

Figure 366. Young radio controlled carriage (Oregon, USA 1973).

carriage along the skyline. The skyline clamp and the winch drum brake is controlled by means of radio signals. Both the terrain crew and the unhooker on the landing have transmitters for this purpose (figure 365).

Figure 367. The Skagit Bullet radio controlled carriage (Oregon, USA 1973).

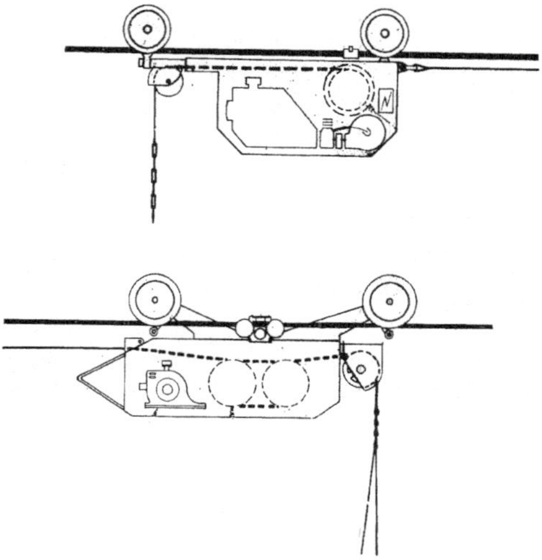

Figure 368. Above: Skagit Torpedo carriage (motor-powered winch drum).
Below: Skagit Bullet carriage (motor-powered groove wheel pulley) (STUDIER and BINKLEY 1975).

The Skagit radio controlled carriages have a motor-driven winch built into the carriage in a similar way as the Japanese radio controlled carriage (figure 351). The Skagit carriage, however, is being pulled along the skyline by means of the main and haul-back lines. (Figure 367 and 368).

The Skagit Bullet weighs 15 kN and can lift 100 kN. The main line runs from the yarder to the carriage where it is wrapped around a groove wheel pulley. The pulley is powered by a 18 kW motor. During the transport along the skyline, the groove wheel is locked by a brake. When the carriage stops for loading or unloading, the pulley brake is opened and the carriage is fastened to the skyline by means of a clamp. The pulley may pull out the main line (hoist line), using the carriage motor. There is a radio receiver in the carriage. Both the terrain crew and the workers at the landing can control the carriage by means of radio transmitters.

The Skagit Torpedo model RCC15 uses another system. It weighs 28 kN. There is a 70 kW diesel engine in the carriage which powers a hoist winch with a capacity of 120 meters hoist line (d = 22 mm). The maximum lifting capacity is 100 kN.

The workers in the terrain or at the landing can control the movement of the hoist winch in the carriage by radio signals.

11.5.3.3. The production

The large mobile yarders with towers, yarding big timber from virgin forests must be operated by a large crew. Usually there is one, and sometimes two, choker setters for each of the very heavy chokers, and the crew consists usually of seven to nine workers:

1 hooktender
1 yarder operator
1 loader operator
1 chaser
1 rigging slinger
2–4 choker setters

Table 49 gives some typical production figures when yarding virgin stands with average tree dimensions 6–9 cu.m. per tree.

Table 49. Typical yarding production for some large yarders in big timber on the West Coast of the USA (after CONWAY 1978).

		Highlead < 295 kW		Highlead > 295 kW		Slackline 368 kW		Uphill yarding 368 kW	
m^3/log	m^3/ha	A	B	A	B	A	B	A	B
0.5	300	21	0.38	23	0.34	29	0.34	33	0.30
1.0	420	26	0.31	28	0.28	34	0.29	39	0.26
2.0	600	34	0.29	37	0.22	46	0.22	50	0.20
3.0	750	39	0.20	46	0.18	59	0.17	60	0.17

A = m^3/hr for the crew
B = manhours per cu.m.

Usually the yarding distances are short, especially for highlead yarding where the production decreases considerably with increased yarding distance (figure 369). For longer distances up to 400–500 meters a skyline system has to be used.

DYKSTRA (1976) investigated the production during yarding in stands with tree dimensions 6 cu.m. per tree. (Table 50).

The production is high. This is necessary due to the high investment in equipment and the large number of men in the crew. The production per man per hour, however, is not higher than other systems where the number of men in the crew is lower even in cases where the tree dimensions are smaller, and the yarding distance bigger.

In the Inner-mountains of British Columbia the tree dimensions often vary between one and three cu. m. per tree. McMORLAND and WONG (1982) investigated the production of skidder mounted yarders (Micro-Master, Eco Logger etc.).The average yarding distance was between 70 and 75 meters (total length 140–150 meters). These yarders were operated by three to four men and the production per work place hour 10 to 15 cu.m. per hour for the crew, corresponding to 2.5 – 3.5 cu.m. per man per hour with a labour consumption 0.3 – 0.4 manhours per cu.m.

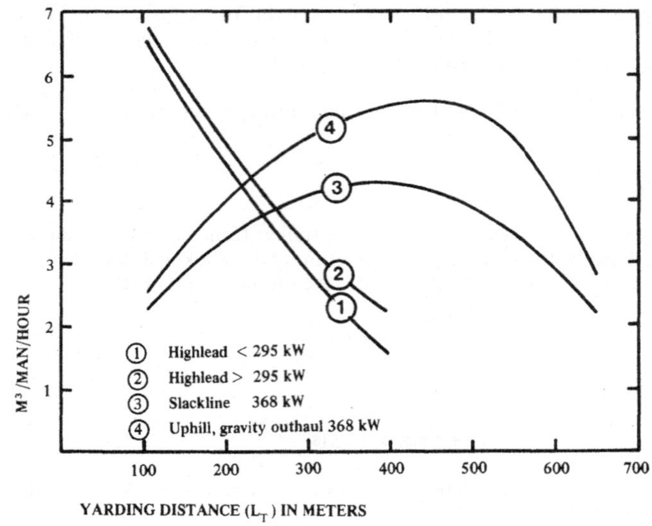

Figure 369. Relationship between production per manhour and yarding distance. (after CONWAY 1978).

Table 50. Typical yarding production for some yarders on the West Coast of the USA. 6 m^3/tree. (DYKSTRA 1976)

	Yarding distance	Logging m^3/log	m^3/hr (crew)
Berger Grabinsky	130	1.2	36
West Coast. North Bend	60	1.5	34
West Coast. Highlead	60	0.7	19
Skagit Bu 90 (gravity outhaul)	215	1.9	35
Skagit GT 3 (running skyline)	85	0.9	20

The Andritz Hydrocrane 80 is operated by a 4-man crew, and the average production during a long period when yarding tree sizes with an average of 2 cu.m. per tree, was 12.8 cu.m. per work place hour (KYLLO 1981). The yarding production was 3.2 cu.m. per man per hour corresponding to a labour consumption of 0.31 manhours per cu.m. The Mayer Melnhof operation combined felling, full tree yarding and limbing in one operation. In addition to the 4-man yarding crew, there was a feller and an operator on the delimbing machine, totalling six men in the crew. This corresponds to a total production for felling, yarding, limbing and bucking of 2.1 cu. m. per man per hour corresponding to a labour consumption of 0.48 manhours per cu.m.

The production with the Steyr KSK cable crane in Austria is similar to the Andritz Hydrocrane 80.

Tree dimensions influences the production considerably. As an example may be mentioned an experiment with the Steyr KSK cable crane in Østerdal, Norway 1982 (figure 370). The average tree dimensions were 0.3 cu.m. per tree of Norway spruce. Full tree logging was

Figure 370. Full tree yarding with the Steyr KSK 16 live skyline cable crane. A limbing/bucking machine converts the trees to logs on the landing (Østerdalen, Norway 1982).

carried out. In addition to the 4-man yarding crew (1 yarder operator, 1 chaser and 2 choker setters) there was 1 feller, 1 operator on the delimbing/bucking machine and 1 operator on the forwarder, totalling 7 men.

The production was 10 cu. m. per work place hour, corresponding to 2.5 cu.m. per man per hour with a labour consumption of 0.40 manhours per cu. m for the yarding only. The production for felling, yarding, delimbing and forwarding to truck road was 1.4 cu. m. per man per hour with a labour consumption of 0.71 manhours per cu.m. The total length of the cable crane was 350 meters.

11.5.4. Running skyline systems

All the previous cable crane systems had a standing skyline or a live skyline which carried the load over the terrain. The slackline systems have a live skyline where the deflection can be adjusted during transport. The carriage can be lifted or lowered depending on the load and the terrain conditions.

It is the running lines, the main line and the haul-back line, which carry the load in the running skyline systems. The deflection may be continuously adjusted by means of the controls on the yarder.

11.5.4.1. Tractor mounted double-drum winch with clutches and brakes

The simplest system is the tractor mounted double-drum winch. It is powered from the power take off, and each of the drums have a clutch and a brake. This kind of equipment may be used for highlead yarding.

If there is a lifting block on the haul-back line, and the main line passes a monkey block attached to the lifting block, it is possible to carry the load or lift the front end of it past obstacles. A small carriage may be used instead of the lifting block and monkey block. This becomes a cable crane with running skyline.

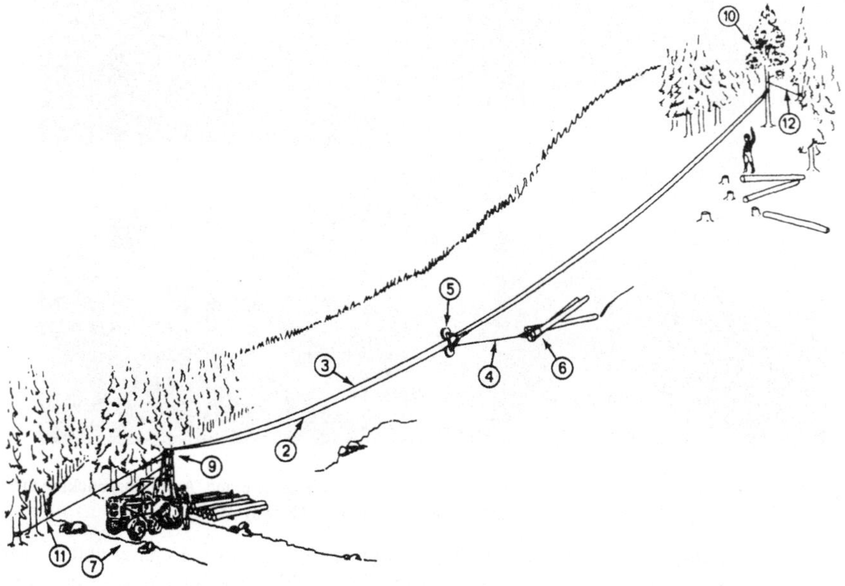

(2) Main line (3) Haul-back line (4) Hoist line (5) Carriage (6) Chokers

(7) Double-drum tractor mounted winch (9) Tilt spar

(11) Guyline for the steel spar (12) Guyline and tailhold

Figure 371. Traditional cable yarding with tractor mounted double-drum winch (The Ludvig Isachsen system, Norway 1954).

Figure 372. A Soviet Union double-drum yarder mounted on a TDT 60 crawler tractor for highlead yarding (Baikal, Sibiria, Soviet Union 1975).

There is one important disadvantage with this system. When pulling in the load, it is necessary to brake the haul-back drum in order to lift the load off the ground. The main drum must therefore be powerful enough to tighten the lines and lift the load. This is discussed in chapter 7.7. It is also difficult to control the deflection. This can cause the lines to twist around each other, especially when returning the carriage to the field. Such equipment may, however, give a good result for downhill yarding where the power requirement is less than for horizontal transport (past creeks etc.) or uphill. Figure 372 shows a Russian double-drum winch mounted on a TDT-60 crawler tractor. The rigged tower does not need to be anchored. The front blade of the tractor keeps the yarder in position during yarding. The double-drum winch has standard clutches and brakes and the yarder is partly used for highlead yarding and partly for downhill yarding as a running skyline system. The carriage construction is basically the same as the Norwegian carriage illustrated in figure 409. The yarder is powered by a 65 kW engine, the diameter of the main and haul-back lines are d=12 mm and the maximum yarding distance 300 meters.

Figure 371 shows a highlead winch for mounting on the 3-point linkage on a farm-tractor. The power requirement is 42 kW or more. The main and haul-back lines are d = 10 mm and

Figure 373. The Igland highlead winch for farm-tractor.
1: The winch drums are mounted on the lower end of the telescope tower (main drum, haul-back drum and auxiliary drum). 2: The top section of the tower. The tower blocks turn 180° in proportion to the tractor and winch. 3: Hydraulic lever controls on swing arm. Folded during transport. 4: Standard linkage adaption to farm-tractor. Four parking legs provide safe storage and quick mounting (Igland 1983).

the straw line is d = 4 mm. The height of the telescope tower is 7.5 meters. The average line speed is 2.5 m/sec. during inhaul and 3.8 m/sec. during outhaul. The maximum yarding distance is 250 meters.

The carriages illustrated in figure 406 or 410 are used. The winches are driven from the power take off on the tractor. One back guy and two side guys keep the tower in position during yarding. The farm-tractor may be disconnected from the winch, for example in the evenings. The cable crane including the anchored tower with the winch remains rigged on the logging area until the farm-tractor returns (see figure 374). It is only necessary to back up the tractor to the winch, connect the winch and tower to the three-point linkage, attach the drive shaft between the winch and the power take off and the yarding may continue.

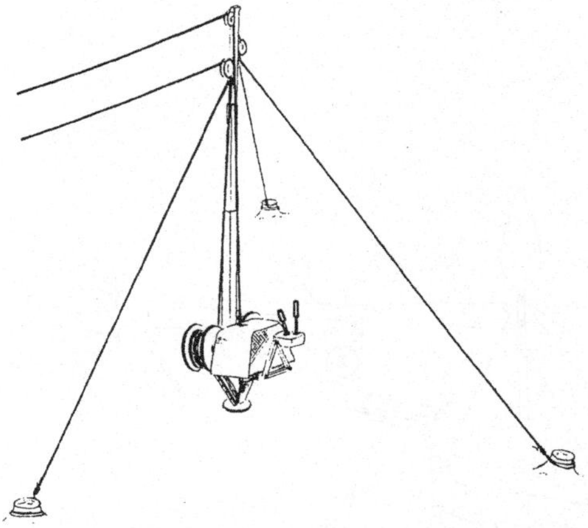

Figure 374. The farm-tractor may be disconnected from the winch and tower (3-point linkage) when the cable crane is rigged up.

11.5.4.2. Endless main and haul-back lines

The simplest way of moving a carriage along a running skyline is to use a single capstan or pulley as discussed in chapter 7.7.4. (figure 149). One disadvantage of this system is the longer rigging time because the length of the endless line has to be changed by splicing in or taking out a certain length of the line. Another disadvantage is that it is difficult to adjust the deflection and the height of the carriage during transport. The advantage of the system is its simplicity. This is the reason why a capstan drive is frequently used in Japanese running skyline systems instead of interlocked winches.KONUMA and SHIBATA (1976) have described some Japanese cable cranes which use a very interesting method. The adjustment of the tension and the deflection in the running skyline is done by a tightening line from a tightening drum on the yarder. The yarder usually has two drums and a capstan (figure 288 and 303). The capstan pulls the endless main and haul-back line. One of the winch drums is used for the tightening line and the other for the hoist or skidding line.

Figure 375, A, shows one of the simplest running skyline systems. It was in use in Hokkaido and in Akita in Japan as early as in 1959. The yarder operator may lift or lower the carriage by means of the tightening line at any time.

Another system is shown in figure 375, B. This is a Japanese variation of the Dunham system. Relatively light main and haul-back lines may be used, since there are three running lines which lift the load. The yarder may lift or lower the carriage by means of the tightening line.

In both these systems the yarder has to be equipped with at least one single-drum and a capstan. It is, however, hard work for the choker setters to move the carriage sideways. To avoid this hard work the cable cranes in figure 375 can use the second winch drum for a hoist line (skidding line). Another way of using a hoist line together with an endless main and haul-back line is shown in figure 376, A, and is called the »Obihiro system». It consists of a capstan driven running skyline and a drum driven hoist line. The main and haul-back lines as well as the hoist line are usually d = 10–12 mm, and the cable crane distance 100–300 meters. The Japanese use this cable crane to yard timber from selective cutting. The crew has usually 3 to 4 men, and a production for the crew of 2.5–3.5 cu.m. per work place hours has been reported. This corresponds to a labour consumption of 0.9–1.3 manhours per cu.m.

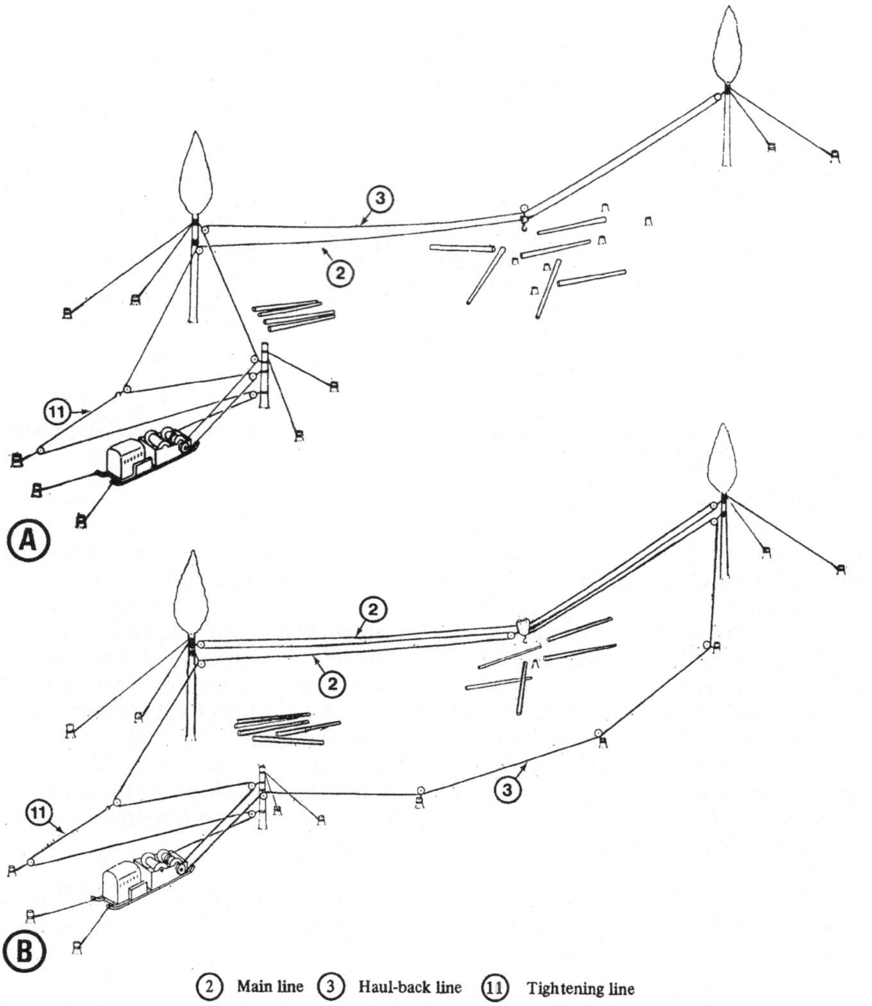

Figure 375. Two versions of Japanese running skyline system.
A. The simplest system.
B. A modified Dunham system (KONUMA and SHIBATA 1976).

Figure 376, B, shows a Japanese combination of a running skyline and the Tyler system. It is hard work to pull out the hoist line and the falling block, and this system is therefore especially suited to swing the logs from a cold deck to a landing.

Another way of using the capstan/single-drum principle is the Okawara system (figure 322), which is described in chapter 11.4.2.2.

The use of a capstan driven running skyline where the deflection can be adjusted by a tightening line arrangement is interesting and should be further developed.

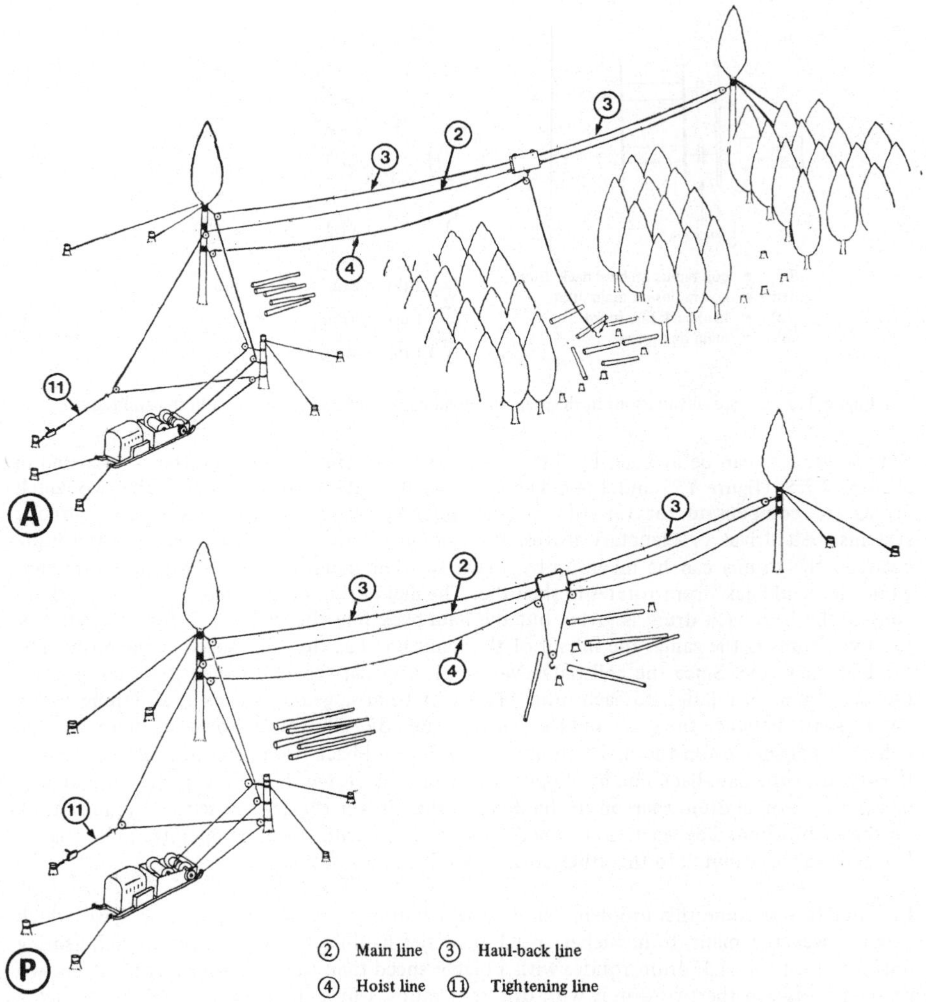

Figure 376. A. The Japanese Obihiro cable crane for selective cutting.
B. A Japanese modification of the running skyline and the Tyler system (KONUMA and SHIBATA 1976).

11.5.4.3. Interlocked winch with slipping clutch variation

If the yarder operates the running skyline by means of two drums instead of a capstan, the speed of the main and haul-back lines when carrying the load has to be adjusted. If the winch drums are driven by independent driveshafts and are controlled by clutches and brakes the main drum must be supplied with enough power to lift the load and overcome the braking forces from the haul-back drum. The brake on the haul-back drum must be able to dissipate the braking power in the form of heat. The dissipated heat is a direct power loss and can be large.

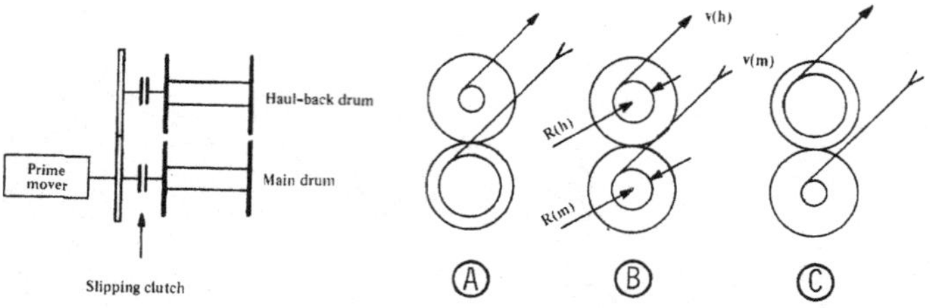

R(h) = coil radius on haul-back drum
R(m) = coil radius on main drum
v(h) = haul-back line speed
v(m) = main line speed

(A) R(h) < R(m) ; v(h) < v(m)
(B) R(h) = R(m) ; v(h) = v(m)
(C) R(h) > R(m) ; v(h) > v(m)

Figure 377. Speed variations in the main and haul-back lines due to variation in the coil radius.

This power loss can be avoided by interlocking the two winch drums and this is discussed in chapter 7.7.5. (figure 153 and 154). The adjustment of the relative speeds of the two winch drums (to compensate for variation in the spooling diameter) can be done by hydraulic systems, belt drives or planetary drives. In a slipping clutch interlocked winch, the torque between the drums can be taken up by a gear set. The main drum rotates in one direction while the haul-back drum rotates in the other.The main drum pulls in the main line simultaneous as the haul-back drum is giving out the haul-back line (figure 377). If the coil radius of the two drums is the same, the inhaul of the main line has the same speed as the outhaul of the haul-back line. Since the coil radius varies it is necessary to adjust the rotational speed of the two drums in relation to each other. This may be arranged by means of a slipping clutch on the shafts between the gears and the drums. Figure 377 illustrates, however, that when the coil radius is smaller on the main drum than on the haul-back drum, it is necessary to reduce the speed of the haul-back line by slipping the haul-back clutch. If, however, the coil radius is bigger on the main drum than on the haul-back drum, it is necessary to use the slipping clutch on the main drum. The operation of such a winch is difficult since the operator must change from slipping one clutch to the other during each inhaul or outhaul.

In order to overcome this problem, the two winch drums are driven by a gear drive which always gives the main drum higher speed than the haul-back drum during inhaul. During outhaul the haul-back drum rotates with a higher speed than the main drum. It must be two gear sets between the two drums with different ratios. During inhaul the set of gears is being used which gives the main drum so much higher rotational speeds that the main line of the empty main drum runs faster than the haul-back line on the full haul-back drum. During outhaul, the other set of gears is being used which gives the haul-back line on the empty haul-back drum higher speed than the main line on the full main drum. The result is that the operator can use the slipping clutch on the haul-back drum when pulling in a load in order to adjust the tension in the running skyline. He uses the slipping clutch on the main drum during the outhaul operation.

The slipping clutch must be cooled, for example by water-cooling, in order to take care of the heat which is being generated in the slipping clutch during the operation. This causes loss of power. Figure 378 shows that there is little slipping and little loss of power when the loaded carriage starts the inhaul movement near the tailhold of the skyline corridor. The loss of power increases as the carriage and the load approaches the landing near the yarder. The loss of power, however, is small compared with the losses when using the brake on the haul-back drum to lift the load above terrain obstacles.

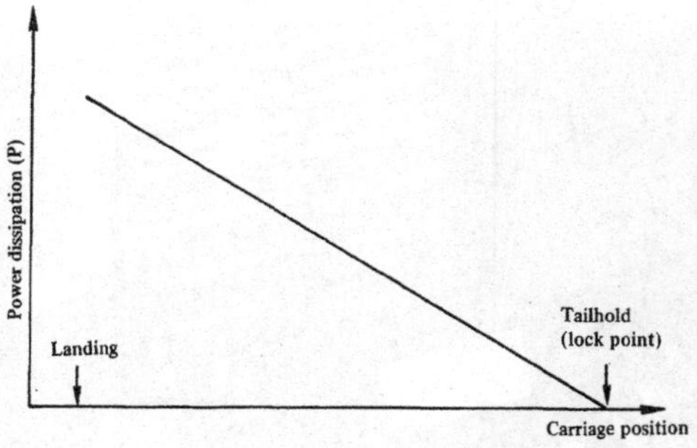

Figure 378. Power loss during inhaul in a slipping clutch interlock winch. (CARSON and JØRGENSEN 1974).

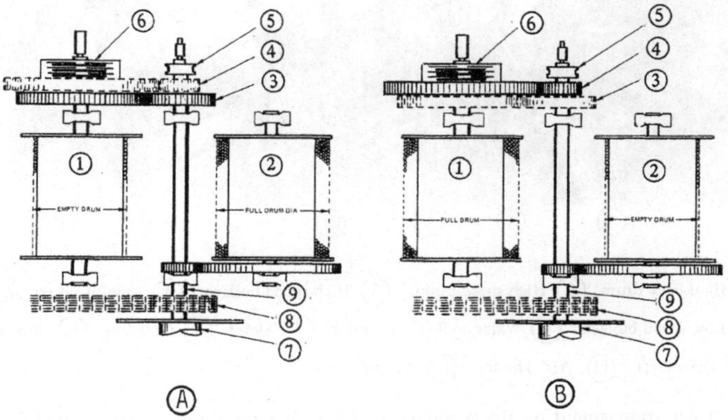

Figure 379. The two speed gearing of the haul-back drum.
A. On outhaul the haul-back line speed is faster than the main line speed.
B. On inhaul the haul-back line speed is lower than the main line speed (MUIR 1984).

This type of interlocking the main and haul-back drums, has a long tradition and was even used on some of the old steam yarders. It gives a simple and reliable construction and is being used on many types of yarders in spite of the power losses caused by the slipping clutches (Skagit, Washington, Lidgerwood, Madill, Steyr etc.).

As an example on a yarder which uses interlocked winches with slipping clutch variation, the Washington model 88 may be described (M. B. MUIR 1984).

The interlock system is illustrated in figure 379. The figure shows the main drum and the haul-back drum only, while the slack-pulling drum is not shown.

Figure 380. Gear arrangement on the Washington model 88 yarder with slipping clutch interlock (MUIR 1984).

The water-cooled slip clutch used to equalize line speeds between main drum and haul-back is identical to the brakes used on many conventional non-interlocked yarders. The shift between high and low speeds on the haul-back is made when the carriage is stopped at the end of each half-cycle. The shift mechanism is automatically protected from shifting under load. During inhaul or outhaul, the operator raises or lowers the carriage with a single lever control.

The interlock gear sets between the main, and haul-back drums allows control of line speeds and tensions. The haul-back drum has two speeds. It is geared so that on outhaul, haul-back line speed is faster than main line speed (figure 379, A). On inhaul the haul-back line speed is lower than main line speed (figure 379, B). In the figure the haul-back drum is shown as full for illustrative purposes. In reality it is half full only when the main drum is empty, since the haul-back drum also must have room for the returning length of the haul-back line.

The speeds are equalized on inhaul or outhaul by slipping a water-cooled clutch.

Shifting between high- and low-speed is done by a jaw clutch mounted between the two pinions on the driveshaft, controlled by the interlock control valve in the cab. The jaw clutch is shifted by an air cylinder mounted on the end of the drive shaft and connected to the jaw clutch slider by a piston, shift rod, and crosskey (figure 380).

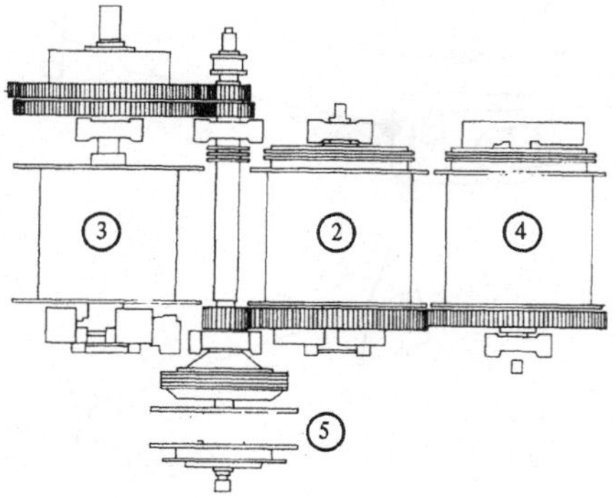

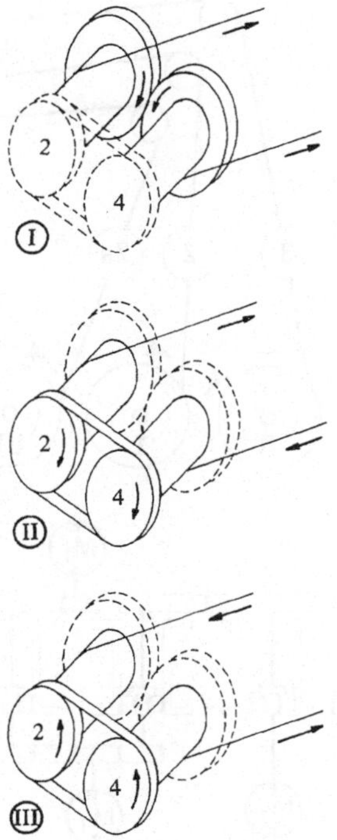

(2) Main drum (3) Haul-back drum (4) Slack-pulling drum

(I:) Gear drive between the main and slack-pulling drum. The two lines runs in the same direction.

(II) and (III): Sprocket drive between the two drums. The two lines runs in opposite directions.

(II) On outhaul, the main line runs out while the slack-pulling line runs in and pulls out the hoist line.

(III) On inhaul, the slack-pulling line runs out while the main line runs in and pulls in the hoist line.

Figure 381. The transmissions on a 3-drum slipping clutch interlock winch.

The quick drop system for slack-pulling is illustrated in figure 381.

There is both a chain drive and a gear drive system between the main drum and the slack-pulling drum on the yarder. The purpose on the system is to allow the direction of travel of the slack-pulling line to be reversed relative to the main line. The drive system provides counter slack-pulling line motion to pull slack on a hoistline carriage or to actuate a grapple.

The slack-pulling drum is identical in size to the main drum. Both drums spool identical sizes and lengths of wireropes.

The slack-pulling drum is underwound and the main drum is overwound, such that engaging the gear drive on one side of the drums moves the slackline and the main line together with the same speed during inhaul or outhaul by counterrotating the drums.

To pull slack for a hoist line or to actuate a grapple, the gear drive is declutched, and the sprocket drive on the opposite side of the drums is engaged. The result is that the slack-pulling line runs in the other direction relative to the main line. The hoist line is being pulled out when the carriage runs in the outhaul direction while the hoist line is being pulled in when the carriage runs in the inhaul direction.

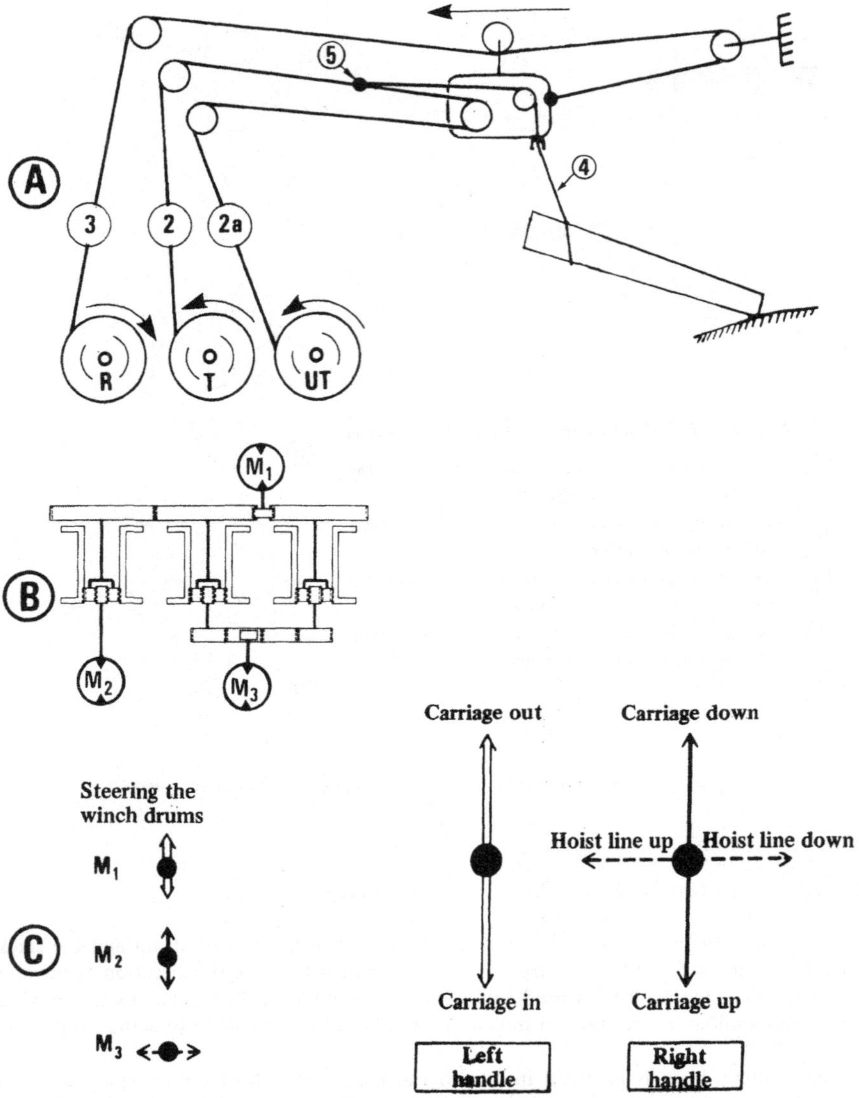

(A) Cable crane with running skyline (B) Cross-section of the 3-drum interlock winch
(C) Diagram of the winch controls (R) Haul-back drum (T) Main drum
(UT) Slack-pulling drum
(2) Main line (2a) Slack-pulling line (3) Haul-back line (4) Hoist line (5) Three-way swivel
(M_1) Primary drive motor (M_2) Interlock drive motor (M_3) Slack-pulling drive motor

Figure 382. 3-drum interlock winch with planetary drive variation.

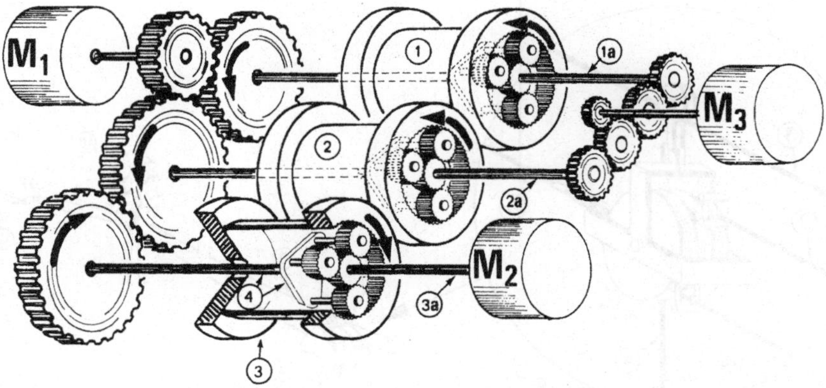

(M1) Primary drive propelling the three drums

(M2) Interlock motor connected to the sun-gear in order to adjust the RPM of the haul-back drum

(M3) Slack-pulling drive motor connected to the sun-gear in the main drum and slack-pulling drum in order to adjust their relative RPM.

(1) Slack-pulling drum (1a) Shaft to the sun-gear in the slack-pulling drum (2) Main drum

(2a) Shaft to the sun-gear in the main drum (3) Haul-back drum (3a) Shaft to the sun-gear in the haul-back drum

(4) Shaft propelling the housing of the planetary wheels.

Figure 383. The planetary drive variation of the drum speeds in a 3-drum winch.

11.5.4.4. Interlocked winches with planetary drives

Figure 382 shows an interlocked winch on a cable crane with running skylines. The winch has three operating drums in order to control the movement of the hoist line as well as the running skylines. The two inhaul drums (T and UT in figure 382, A) pull in the carriage towards the winch by means of the main line (2) and the slack-pulling line (2 A).The haul-back drum (R) rotates in the other direction and feeds out the haul-back line (3). The yarder maintains tension in the lines by transferring the torque between the haul-back drum and the main/slack-pulling drum.

As the carriage moves towards the winch the radius on the haul-back drum decreases while it is increasing on the main and slack-pulling drums. There is a planetary drive in each of the winch drums which allows the relative rotational speed of the drums to be adjusted to compensate for the variation in the diameters (figure 382, B and 383).

The hydraulic motor M1 drives the planet carriers in the three drums. As long as the sun-gears are still the planet gears will rotate and drive the drums with the same rotational speed. The haul-back drums' sun-gear is driven by the hydraulic motor M2. If this motor drives the sun gear in one or the other direction, the haul-back drum will rotate faster or slower than the two other drums. An adjustment of the rotational direction and speed of motor M2 changes the tension in the main/haul-back lines and therefore the height of the carriage above ground (figure 382, A).

The same principle is used to adjust the relative speed and direction between the main line and the slack-pulling line. When the sun-gears in the main drum and the slack-pulling drum are still the two drums rotate with the same speed. If the yarder operator adjust the direction

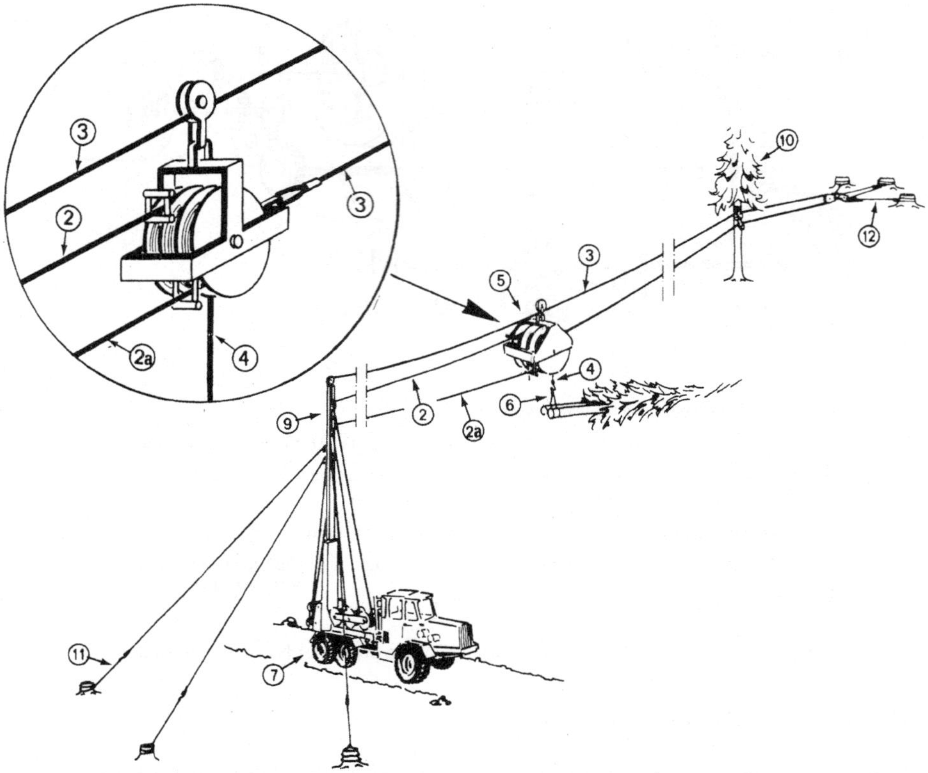

② Main line to the upper side of the carriage drum ②a Slack-pulling line to the lower side of the carriage drum
③ Haul-back line (running skyline) ④ Hoist line (from the middle drum part) ⑤ Carriage
⑥ Tagline with chokers ⑦ Three-drum interlocked winch ⑨ Tilt telescope tower
⑩ Tail spar ⑪ Straps fixed to stumps in order to anchor the guylines from the steel spar
⑫ Tailhold.

Figure 384. The Moxy cable crane with running skyline and a 3-section drum carriage.

and the speed of the motor M3, the slack-pulling line will run faster or slower than the main line. The result is that the hoist line is being pulled out or in. The continuous adjustment of the height of the carriage as well as the control of the hoist line gives a very smooth transport. It is possible to keep the load off the ground or to high skid it with the front end lifted.

During operation it is important to keep the carriage off the ground all the time.

Some carriages are illustrated in figure 384 – 386. The main line in figure 382, A is connected to the slack-pulling line. The hoist line (skidding line) runs from a swivel on the main line over a block and out in the terrain. The swivel approaches the carriage or runs out from it corresponding to the differences in the main and slack-pulling line speeds. Figure 385 shows a carriage developed by the Smith Berger Corporation and the carriage in figure 386, is developed by Washington Iron Works. These types of carriages may also be used for grapple skidding. The operator can open the grapple by feeding out the main line and pull in the slack-pulling line. Reversing this procedure closes the grapple.

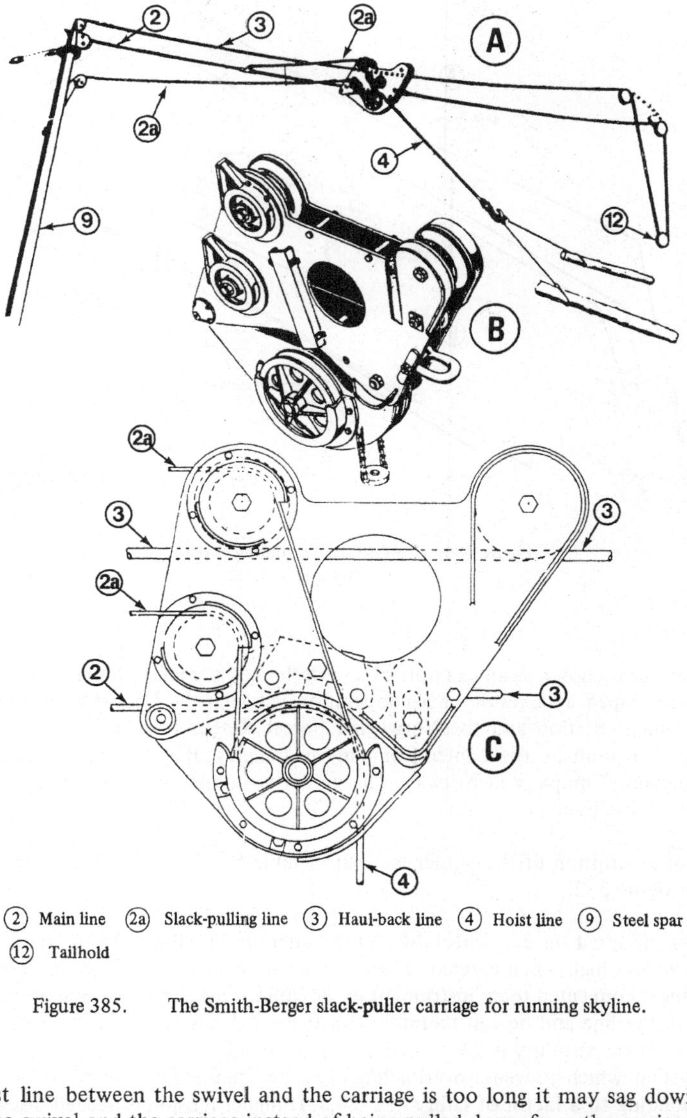

(2) Main line (2a) Slack-pulling line (3) Haul-back line (4) Hoist line (9) Steel spar
(12) Tailhold

Figure 385. The Smith-Berger slack-puller carriage for running skyline.

If the hoist line between the swivel and the carriage is too long it may sag down in a loop between the swivel and the carriage instead of being pulled down from the carriage. Different types of friction devices which help to pull out the hoist line have been developed.

The drum carriage is a simpler solution for yarding without a grapple (figure 384). The drum is divided into three sections. The main line runs in on the top of the left section, while the slack-pulling line runs in on the lower side of the right section. The hoist line is wound in on the central section of the drum. When the main line and the slack-pulling lines run with the same speed and in the same direction, the three sectioned drum does not rotate. The hoist line is pulled in or fed out by adjusting the relative speed of the main and slack-pulling lines. Figure 387 shows some constructions of slack-pulling carriages and drum carriages.

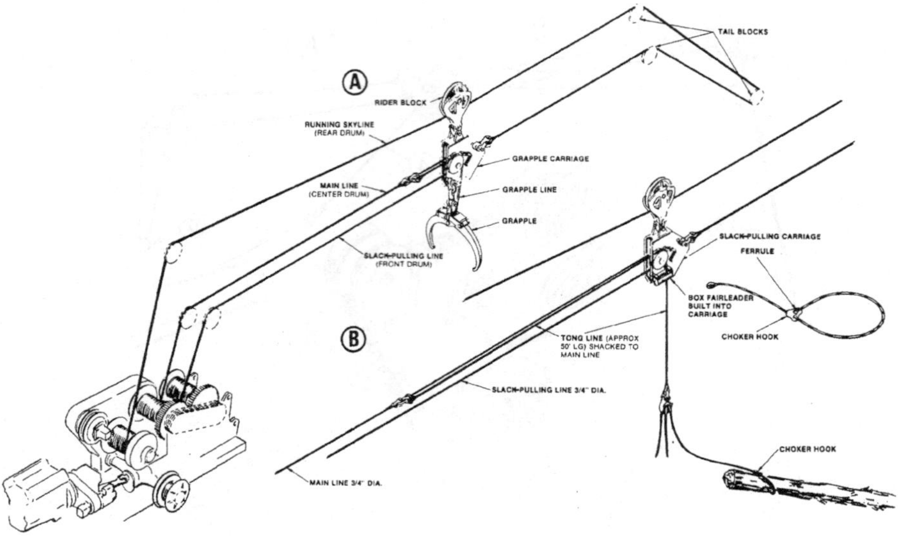

Figure 386. Washington running skyline.
A. Grapple carriage.
B. Slack-pulling carriage.

The Moxy yarder is chosen as an example of an interlocked yarder with planetary drives. This yarder was developed as a result of a cooperation between the US North West Forest and Range Experiment Station and the Norwegian Forest Research Institute. The triple drum interlock winch is built by the Lantec Company in Langley, B.C. and the yarder is built by Brown Engineering Company in Norway. The yarder is suitable for tree dimensions between 0.1 and 3 cu. m. per tree.

The technical description of the yarder is given in table 51. The triple-drum winch works as illustrated in figure 382.

The yarder is mounted on a 6-wheel driven forwarder chassis (figure 389). The tip/telescope tower is 15 meters high when erected. There is a tension gauge at the haul-back block in the tower and this is connected to an instrument in the cabin. This allows the operator to monitor the tension in the line and he can therefore run the winch faster and safer. He can also run the yarder when the visibility is low (e.g. rain or fog). In addition there is a relief valve on the hydraulic system which prevents overloading. The line tension should never be more than that corresponding to a factor of safety of s = 4.5. If higher tension is needed, for example in order to lift a load over an obstacle, the operator can adjust the relief valves for a short period of time.

The yarder is equipped with four hydraulically operated guyline drums. In addition there is a set of auxiliary lines to simplify the rigging work as well as the corridor changes. There are the two rig up lines, with diameters of 2.1 and 4 mm. The straw line has a diameter of 8 mm. During the rig up work, the foreman pulls out the thinnest rig up line. When this line has been pulled out, the second rig up line is pulled out by the yarder and this pulls out the 8 mm straw line. The method will be discussed in chapter 12.2.1.3.

Table 51. Examples of technical data for some Norwegian yarders.

Modell		Owren Vario Winch	Owren Vario Crane	Moxy Cable Crane
Weight		41 kN	150 kN	280 kN
Tower height		6.5 m	9.5 m	15 m
Engine power		40 kW	96 kW	143 kW
Main line	L	200 m	500 m	945 m
	d	10 mm	10 mm	14 mm
	Max. pull	22 kN	40 kN	33 kN
	Max. speed	4 m/s	4 m/s	11 m/s
Slack-pulling line	L	200 m	500 m	880 m
	d	10 mm	10 mm	14 mm
Haul-back line	L	400 m	1,000 m	1,490 m
	d	10 mm	10 mm	14 mm
	Max.pull	22 kN	40 kN	39 kN
	Max. speed	4 m/s	4 m/s	11 m/s
Hoist line	L	50 m	50 m	50 m
	d	10 mm	10 mm	14 mm
Straw line	L	400 m	–	1,090 m
	d	4 mm	–	8 mm
Rig up line I	L	–	2x500 m	2x760 m
	d		2.1 mm	2.1 mm
Rig up line II	L	–	1,000 m	1,370 m
	d		4 mm	4 mm
Skyline	L	–	500 m	–
	d		18 mm	
Guylines	Number	1	2	4
	L	75 m	75 m	75 m
	d	12 mm	16 mm	18 mm

11.5.4.5. Interlocked winches with hydrostatic drive

A hydraulic transmission consists usually of a hydraulic pump which powers a hydraulic motor. The pump will have variable displacement capability controlled externally by a swashplate angle, so that power flow (oil delivery) to and from this pump can be regulated. An advantage of the hydraulic drive is that it gives smooth running of the lines and smooth adjustment of the line speeds. It is rather easy for the operator to run the carriage smoothly along the lines and to place it into the correct position, for example during loading or unloading.

A running skyline cable crane has a high internal torque between main and haul-back drums. The problem is discussed in chapter 7.7.4. (table 21). The mechanical transmission such as

Figure 387. Carriages for running skyline. (Norway, 1980).
A. C. and D. 3-section drum carriage.
B. Slack-pulling carriage.

gear drive in a slipping clutch interlock or planetary drive interlock can be dimensioned for the highest torques between the drums. The maximum torque in a hydraulic drive transmission is limited by the maximum oil pressure. An oil pressure of 250 bar corresponds to 25 Mpa or 25 N per mm^2.

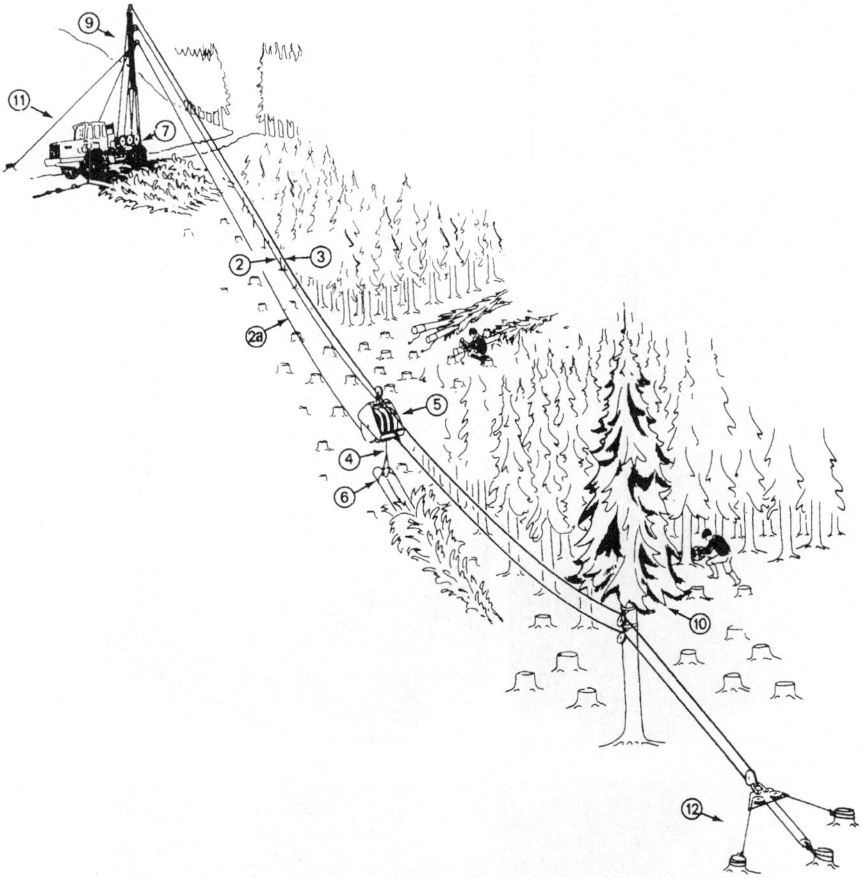

(2) Main line to the upper side of the carriage drum (2a) Slack-pulling line to the lower side of the carriage drum

(3) Haul-back line (running skyline) (4) Hoist line (from the middle drum part) (5) Carriage

(6) Tagline with chokers (7) Three-drum interlock winch (9) Tilt telescope tower

(10) Tail spar (11) Guylines from the steel spar (12) Tailhold

Figure 388. Uphill yarding with the Moxy running skyline cable crane.

When the line tensions give the hydraulic motor a higher torque then corresponding to the maximum oil pressure, the relief valve in the circuit will open. It happens that this limits the possibility of lifting a log or a load over an obstacle. This is also an advantage since the relief valve prevent overloading the transmissions and the wire ropes as well as the other cable equipment.

The illustration in figure 391 shows the hydraulic interlock variation of line speeds during inhaul CARSON and JØRGENSEN (1974). The pump has variable displacement capability so that the powerflow to and from the pump can be regulated regardless of the speed of the drive shaft coming from the haul-back drum.

Figure 389. The Moxy yarder with a 3-drum Lantec interlock winch. (Norway, 1980).

During inhaul the main drum is driven by the driving shaft from the engine (the prime mover). The haul-back drum is declutched from the driving shaft. The haul-back drum is connected to a hydraulic pump. The haul-back line tension, caused by lifting the load above the ground, results in a torque which powers the pump. The pump activate a hydraulic motor connected

Figure 390. The Moxy yarder. (Norway, 1980).
A. Top section of the tip/telescope tower.
B. The hydraulically driven auxiliary drum sets at the rear of the yarder.
Left: The spools for a 2.1 mm and the 4 mm rig up lines.
Right: The drum with the 8 mm straw line.
C. Two of the guyline drums and side view of the Lantec 3-drum winch.
D. Drums for the main line, the slack-pulling line and the haul-back line.

to the main drum. Consequently, the braking power in the haul-back drum is regenerated, and used to power the main drum. During inhaul the main drum is partly driven by the driving shaft from the engine (the prime mover) and partly by the hydraulic motor.

The prime mover supply the power to yard the logs and to overcome losses at other points in the system. The hydraulic transmission supply the torque needed to give the necessary tension in the haul-back lines to carry the carriage and the load.

The haul-back line tension is proportional to the pressure in the hydraulic unit and the relative drum speeds are proportional to the fluid flow controlled by the pump displacement.

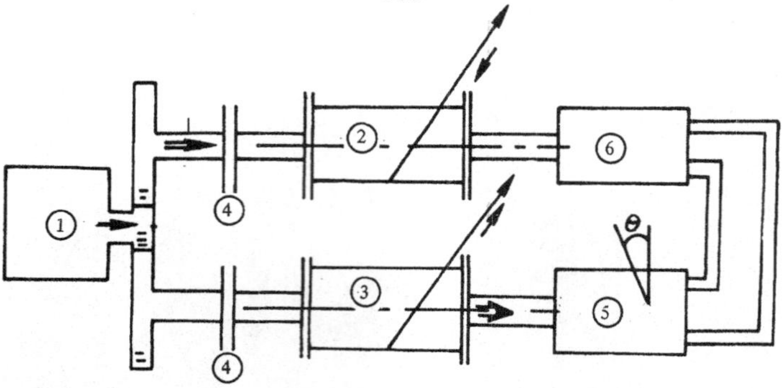

Figure 391. The system of a double-drum winch with direct drive, hydraulic interlock, and hydraulic speed variation between the drums (CARSON and JØRGENSEN 1974).

During outhaul, the prime mover powers the haul-back drum while the clutch on the main drum is disconnected. In this case the hydraulic pump is driven by the shaft from the main drum, and powers the hydraulic motor connected to the haul-back drum.

Another method of hydrostatic drive variation is to use one hydraulic motor on each operating drum. If there are three drums, main drum, slack-pulling drum and haul-back drum, the engine powers three variable displacement hydraulic pumps. Each of these pumps powers one hydrostatic motor on each of the drums and the line speeds are adjusted by varying the oil flow capacity from the variable displacement hydraulic pumps (figure 392).

The Owren Vario Winch may be chosen as an example of a yarder for a cable crane with a running skyline which has a hydrostatic drive, (figure 392). It is a 40 kW yarder with a 6.5 meter high tip tower, suitable for uphill or downhill yarding with a maximum yarding distance of 200 meters. Table 51 gives some technical specifications for the yarder.

The yarder is mounted on a small hydrostatically driven crawler tractor. The same pumps which power the crawlers also powers the winches in the yarder. The rigid tip tower is usually anchored with one 12 mm guyline and two hydraulic feet are used for lateral support. The total weight of the yarder is 41 kN and the yarder is suitable for tree dimensions up to 1 cu.m. per tree.

It is quick to move the yarder from one skyline road to the next, and the yarder is an interesting machine for thinnings. (It takes approximately 5 minutes to take down the yarder, move it 20 meters to the next thinning corridor, and rig it up again. This includes the yarder and the anchoring, but not the moving of the lines). This yarder can also be mounted on a skidder (figure 393). The weight of the winch and tower unit only is approximately 18 kN.

The HPC yarder has a similar type of a hydrostatic driven interlock winch (KAIERLE and HÖFFLER 1980). The Bavarian Forest Research Institute, in cooperation with the Institute of Forest Operations at Munic University and with the HPC-manufacturing company, developed the yarder. It is a trailer mounted yarder, powered by a 52 kW diesel engine. The 3-drum interlock winch is driven by three hydrostatic motors powered by three variable pumps (figure 394). The yarder has remote control by means of a 40 m long electrical cable, which reduces the crew to 2 men only, the choker setter and the yarder operator who also unhooks the logs.

Figure 392. Owren Vario Winch, Hurdal, Norway 1980.
A. and C. The tip tower is raised hydraulically.
C. Two hydraulic feet and one guyline support the tower. From the top of the tower and downwards: haul-back line, main line and slack-pulling line.

The yarder has a 10 m high tip tower, and the maximum yarding distance is 300 m (running skyline). The maximum speed with a 20 kN load is 2.1 m/s, and, the maximum tension in the main line is 35 kN, in the slack-pulling line 6,5 kN, and in the haul-back line 25 kN.

Figure 393. The Owren Vario Winch mounted on a skidder.

The Owren Vario Crane is a larger hydrostatically driven yarder suitable for cable crane yarding of trees between 0.1 and 2.0 cu.m. per tree on distances up to 500 meters. It has been developed under a cooperation between the Norwegian Forest Research Institute and Owren Mechanical Workshop. Table 51 gives some technical specifications for the yarder.

The yarder has been built on a crawler chassis. The winch sets and the tower are built on a frame and may be placed on a forwarder or a truck chassis, according to the needs.

There are three drums used in this running skyline system: main drum, slack-pulling drum and haul-back drum. In addition there is a skyline drum. This all-round yarder may, therefore, be used with different skyline configurations (figure 397) i.e. live skylines for uphill or downhill yarding or running skylines. The yarder may therefore be used on any kind of terrain. There are auxiliary drums in order to facilitate the rigging work. There are two hydraulicly driven guyline drums.

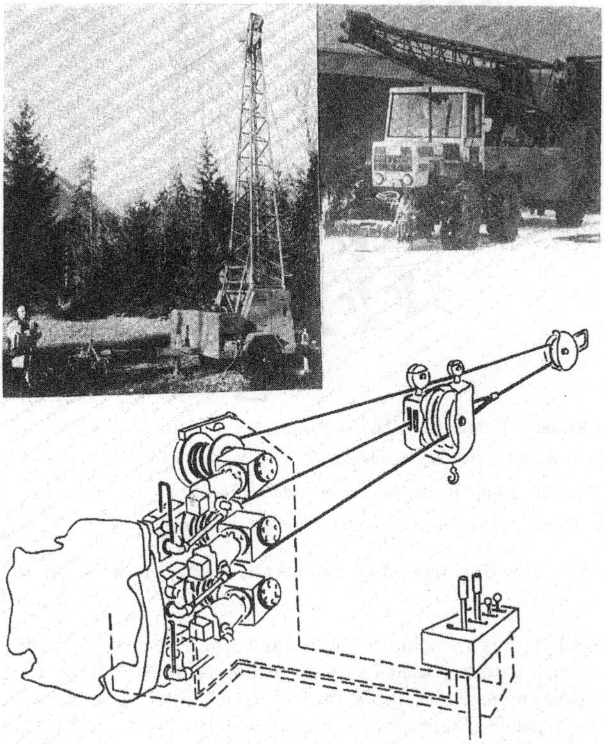

Figure 394. The hydraulic interlock system on the HPC cable crane with running skyline. (KAIERLE and LÖFFLER 1980).

The yarder can also be used on a running skyline cable crane with endless main and haul-back lines (figure 400). In this case, a capstan must be mounted on the main drum (figure 401). The haul-back drum is used to pull the tightening line in order to control the height of the carriage and the load.

The rigid tower has a special construction which makes it very fast and easy to rig up or take down (figure 396).

The two guylines are anchored opposite to the yarding direction. The sideways stabilization of the tower is done with two hydraulic feet.

Figure 398 shows a hydraulic diagram for the four pumps and the hydraulic circuits on the yarder.

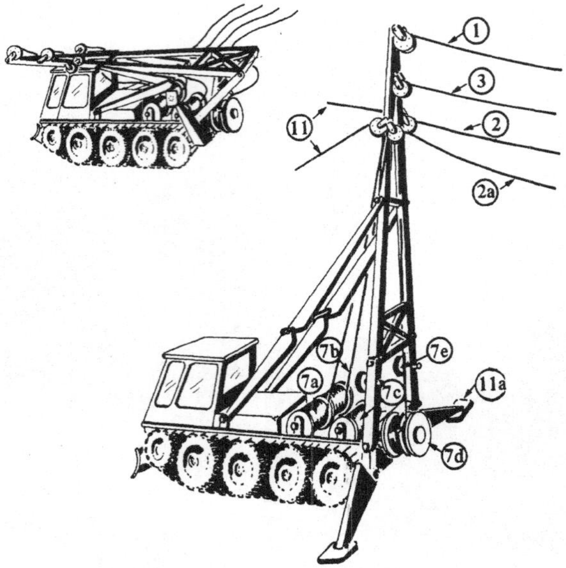

(1) Skyline (2) Main line (2a) Slack-pulling line (3) Haul-back line

(7a) Main drum (7b) Slack-pulling drum (7c) Haul-back drum

(7d) Skyline winch with storage drum and tightening drum (7e) Guyline drum

(11) Guyline (11a) Hydraulic feets for sideways stabilization

Figure 395. The Owren Vario Crane with the tip tower rigged up and laid down.

The above mentioned yarders are equipped with three drums. A three-sectioned drum carriage is suitable when such a yarder is used to power a running skyline cable crane (figure 401). This carriage has also been used for uphill yarding. Under difficult terrain conditions one or several supports are needed. There exist a support which may be used for running skylines which will be discussed in chapter 11.5.4.6. Under extreme terrain difficulties it is better to use a live skyline system. In this case, the main line and slack-pulling line are connected and wrapped around a capstan in the carriage. This capstan powers the hoist drum in the carriage by means of a sprocket drive.

If the yarding direction is uphill, a gravity outhaul system may be used. No haul-back line is needed since the carriage returns to the logging area by gravity. The skyline drum operates the live skyline, and the main line drum the main line. The Koller automatic carriage is excellent for this type of yarding (figure 402).

The two other operating drums on the yarder may be used for other purposes, for example to move away the loads of unbranched trees from the yarder towards a delimbing machine.

11.5.4.6. Interlocked winches with belt variators

Yarders with interlocked winches for running skyline cable cranes are sophisticated and expensive. This is especially the case when planetary drive or hydrostatic drive is used to achieve the necessary speed variations of the drums.

Figure 396. The tip tower on the Owren yarder.

Experiments have been carried out in order to find a cheap interlocked winch for yarders for the small contractors or the farmers. These are yarders for 200 – 300 meters cable cranes and for tree dimensions up to 1 cu.m. per tree (unbranched). If the trees are limbed and bucked into logs, even bigger tree dimensions may be handled.

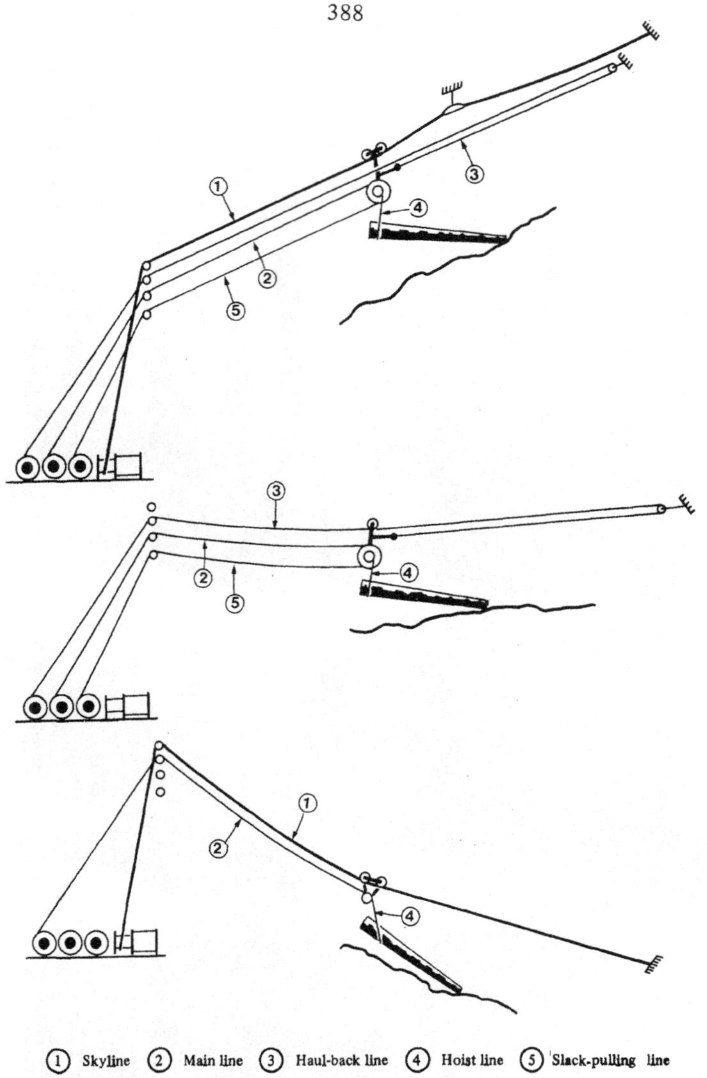

Figure 397. Some cable crane systems with the Owren yarder: live skyline, running skyline and gravity outhaul system.

Belt variators are widely used for example in harvesting machines for agricultural crops or in small automobiles.

A prototype (figure 403) was used to try out a belt variator with four belts. The number of belts were reduced and it was found that one set of belts was sufficient to transfer the torques (figure 404). The system consists of a double pulley which pulls both the main drum (T) and the haul-back drum (R). The central flange in the driving pulley is loose and can move sideways on the axle. A hydraulic cylinder can move the driving pulley sideways and shorten the distance to one of the drums and lengthens the distance to the other. This changes the radius in the driving pulley continuously, which results in a continuous variation of one drum speed in relation to the other.

VARIO CRANE 6912

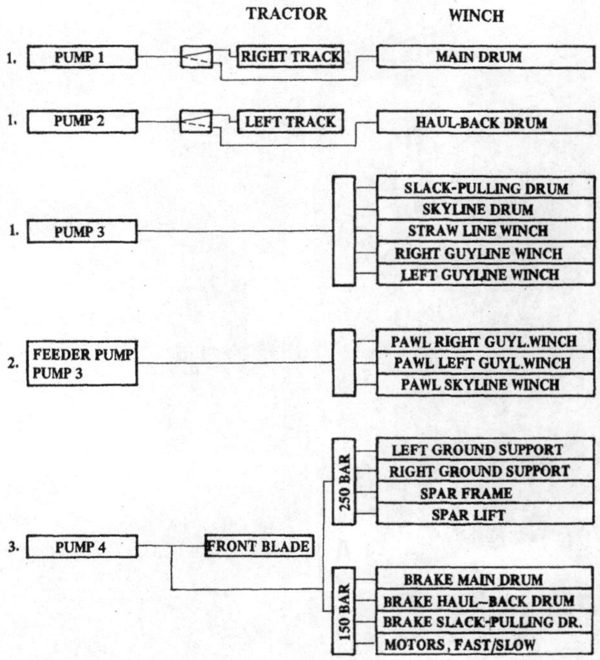

1. OIL DELIV.VARIABLE 0–3,3 l/sec. MAX. PRESSURE 350 BAR
2. PRESSURE 16 BAR
3. OIL DELIV.VARIABLE 0–0,67 L/SEC. PRESSURE, MAX. 250 BAR

Figure 398. Diagram of the hydraulic system used on the Owren Vario Crane.

Table 52. Some spesifications of the Igland Interlocked winch with belt variator.

Weight		12 kN
Height of telescope tower		6.7 m
Main line	L_1	200 m
	d	10 mm
Haul-back line	L_1	400 m
	d	10 mm
Straw line	L_1	400 m
	d	4 mm
Main and haul-back lines		
	Max. pull	41 kN
	Belt variator pull	15–5 kN
	Belt variator speed	2.5–8.1 m/s

Figure 399. Yarding with the Owren running skyline in Hurdal, Norway 1981.
A. Pulling out the hoist line.
B. Fibre rope chokers.
C. Setting the chokers.
D. Signalling.
E. Lateral skidding.

The belt variator is mounted on the Igland highlead winch shown in figure 373. A device for changing the direction of rotation of the drums is required. The interlocked winch is shown in figure 405. A two year test period has shown that the belt variator is reliable, and well suited for the smaller yarders.

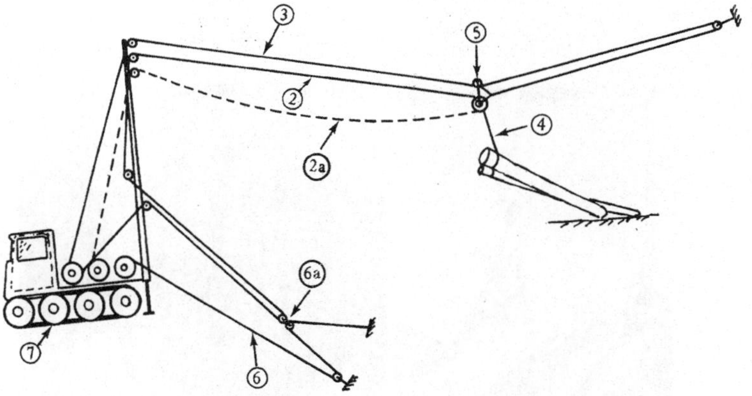

(2) and (3) Endless main and haul-back lines. (2a) Slack-pulling line
(4) Hoist line (skidding line) from the hoist drum in the carriage.
(5) Carriage with hoist drum, powered by the main line (2) and the slack-pulling line (2a) (7) Tractor with capstan and 2-winch drums.
(6) Tightening line. (6a) Twin block.

Figure 400. Owren Vario Crane used as running skyline with endless main/haul-back line, and tightening line. (Norway, 1984).

The belt variator gives a smooth operation of the running skylines and it is easy to operate the yarder. The limitation, however, is the same as for some of the hydrostatic drives. When the internal torque between the main and haul-back drums reaches a certain limit, the belts will slip in the pulleys. This limitation is also an advantage since it helps the operator not to overload the winch, the ropes or the other equipment.

This type of yarder has two operating drums only, the main drum and the haul-back drum. It would be more complicated and expensive to include a slack-pulling drum. It is, therefore, necessary to pull out the hoist line in order to facilitate the work for the choker setters. Two different methods shall be described.

Carriage with slack-pulling block on the haul-back line.

The movement of the haul-back line can be used to pull out the hoist line. When the carriage reaches the point for lateral skidding a catcher fastened to the haul-back line reaches the carriage. The catcher is formed like a steel egg (figure 406). If the yarder operator continue to haul out the carriage, the steel egg moves inward with the haul-back line. The steel egg (3 a) is small enough to pass the carriage block, but there is a slack-pulling block immediately behind the carriage which is too narrow to let the steel egg pass. The steel egg catches the slack-pulling block and pulls it along the haul-back line in the direction of the yarder.

The hoist line is fastened to the main line by means of a three-way swivel (2 c). The slack-pulling line runs from the swivel, past a block on the carriage, through the slack-pulling block and back again to the carriage where it is fastened. When the steel egg pushes the slack-pulling block away from the carriage, it also pulls in the three-way swivel towards the carriage and the hoist line is fed out.

A and B: Owren Vario Winch used as ordinary running skyline.

C, D and E: Owren Vario Winch used as running skyline with endless main, and haul-back line.

C: A capstan mounted on one of the winch drums (left) pulls the endless main/haul-back line. The other drum pulls the tightening line. D: The twin block with the tightening line.

E: The carriage on the running skyline to the left. The bight of the endless main/haul-back line to the right.

Figure 401. Running skyline yarding with Owren Vario Winch, Hurdal, Norway 1981.

Figure 402. Cable crane yarding with the Owren Vario Crane.

When all the logs have been skidded in one place, the carriage and the steel egg have to be moved to the next place for lateral skidding. The distance between one skidding place and the next is usually 10–15 meters. The choker setter moves to the next place for lateral skidding. By signalling to the yarder operator he stops the lines when the steel egg arrives at his place. The choker setter loosens the steel egg on the haul-back line and holds it still while the yarder operator moves the carriage towards the new skidding place. The choker setter fastens the steel egg by means of two socket head cap screws, whereafter the skidding and transport can continue from the new place. The change of skidding road takes on average 2.4 minutes.

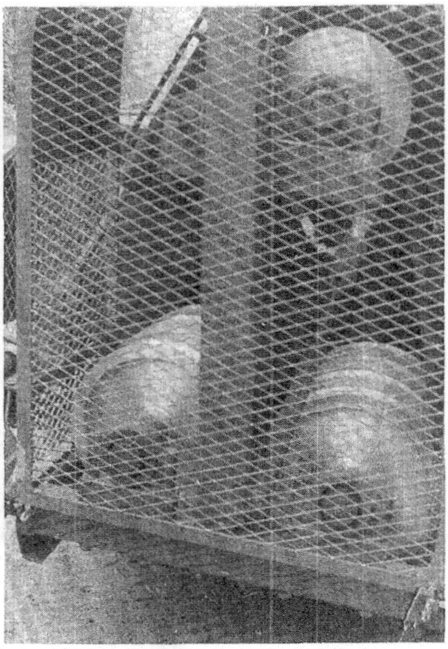

Figure 403. Adjustment of the main and haul-back line speeds by means of a belt variator used in the Igland interlocked winch.

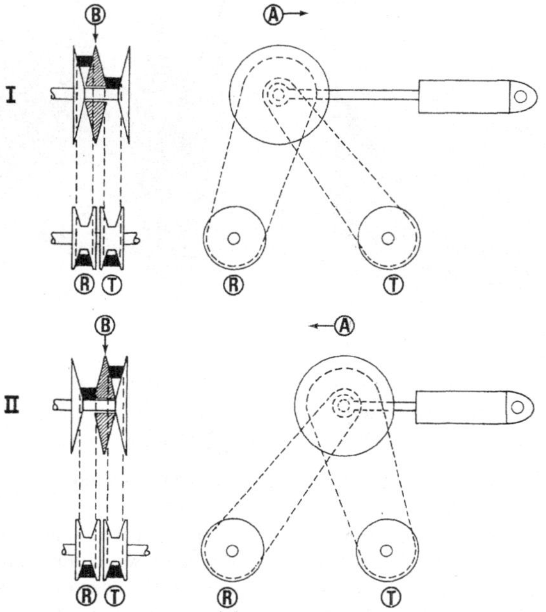

Figure 404. The belt variator system. A. The driving double pulley pulls the main drum (T) and the haul-back drum (R). The central flange in the driving pulley (B) can move sideways on the axle.

I. Left position of the pulley. Low main drum speed and high haul-back speed.

II. Right position of the driving pulley. High speed of the main drum and low speed of the haul-back drum.

Figure 405. The Igland 2-drum interlock winch with telescope tower connected by means of the 3-point linkage to a farm-tractor. The carriage for running skyline.

In the hands of an experienced crew the method of pulling out the hoist line by means of a slack-pulling block is efficient and easy to work with. It can also be used with chokers for automatic unhooking.

Twist between the hoist line and the slack-pulling line can sometimes cause problems. It is an advantage to use torsion-free rope constructions in the slack-pulling line and in the hoist line.

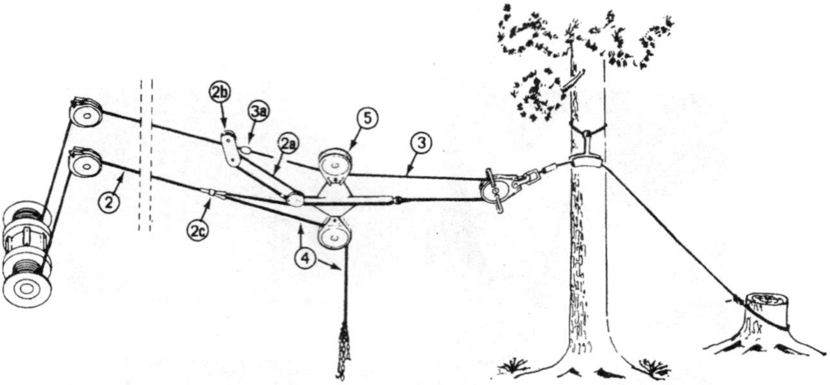

(2) Main line (2a) Slack-pulling line (2b) Slack-pulling block

(2c) Three-way swivel between the main line and the slack-pulling, and hoist lines (3) Haul-back line

(3a) Catcher fixed to the haul-back line which can pass the carrying block (5) but brings with it the slack-pulling block (2b)

(4) Hoist line (5) Carriage

Figure 406. The Igland carriage with slack-pulling block for pulling out the hoist line on a running skyline with 2-drum winch.

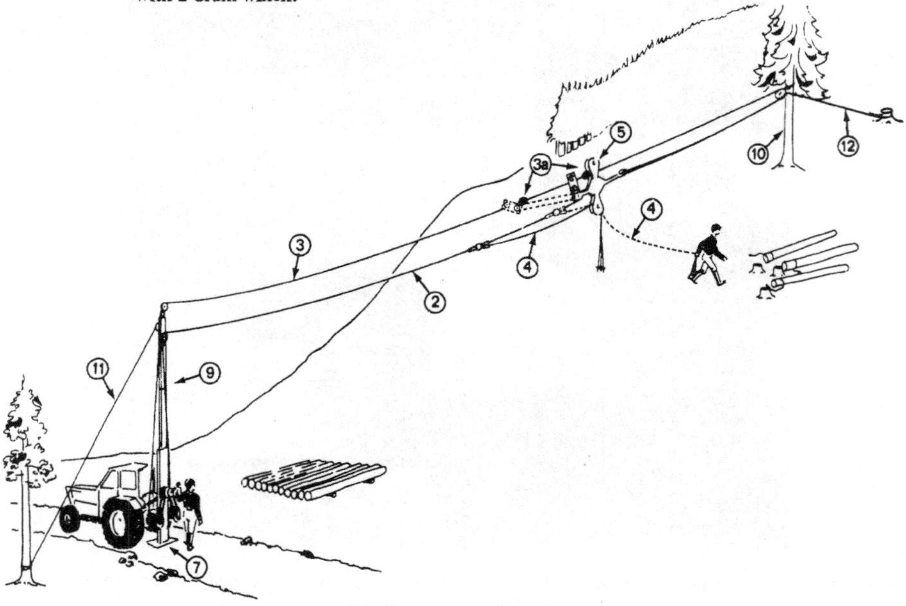

(2) Main line (3) Haul-back line (running skyline) (3a) A catcher fixed to the haul-back line brings the slack-pulling block in the direction of the winch and pulls out the hoist line

(7) Tractor mounted double-drum interlock winch with belt variator (9) Tractor mounted telescope tower with one back guy from the guyline drum

(10) Tail spar (11) Guyline (12) Tailhold

Figure 407. The Igland running skyline cable crane with two winch drums on the yarder and slack-pulling block to pull out the hoist line.

Figure 408. Running skyline cable crane with a 2-drum winch on the yarder. (Telemark, Norway, 1978).
A. Pulling out the hoist line with the slack-pulling block.
B. Transport along the skyline.
C. The yarder with interlock winch and belt variator mounted on a 4-wheel driven tractor.

Carriage with fixed hoist line.

A very simple solution to the problem is to pull out the hoist line on the landing and fasten it in the carriage (figure 409). The front end of the main line is used as hoist line. A catcher shaped as a steel egg is fastened to the main line at a point where there is a sufficient hoist line length.

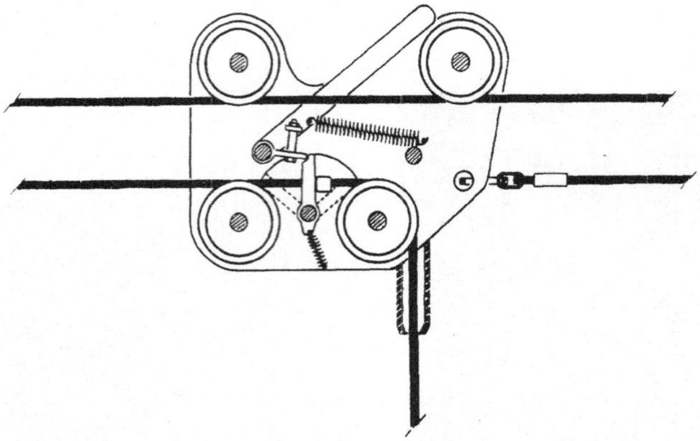

Figure 409. The Igland carriage for fixed hoist line.

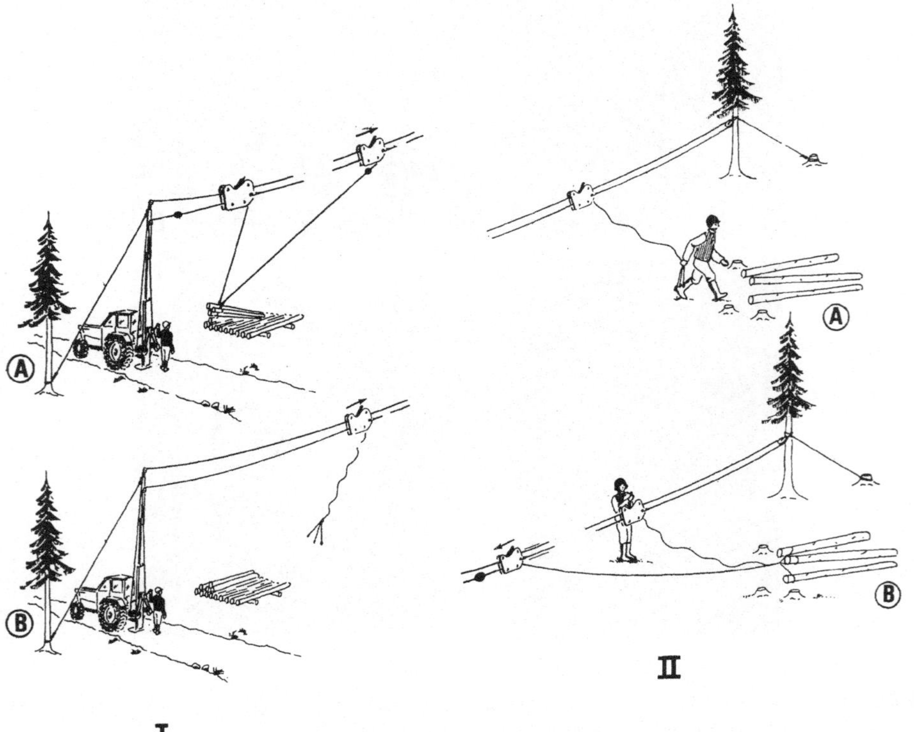

Figure 410. The Igland cable crane with 2-drum interlock winch and carriage with a fixed hoist line.
I. At the landing.
II. In the field.
I A. The carriage is being returned towards the field until the catcher (the steel egg) is fastened in the carriage. I B. After the chokers have been unhooked the carriage returns to the field with the hoist line pulled out.
II A. Carrying out the hoist line to the logs. II B. Loosening the catcher from the carriage before lateral skidding.

Figure 411. The Igland carriage for fixed hoist line. Telemark, Norway 1980.

Before unhooking the chokers, the yarder operator hauls out the carriage until the steel egg has passed the carriage (figure 410, I A). There is a locking device shaped like a fork in the carriage which holds the steel egg. When the chokers have been unhooked the carriage can return to the logging area with the hoist line already pulled out. The choker setter may carry the loose end of the hoist line to the logs. He then walks behind the carriage and releases the steel egg with a handle. The lateral skidding and transport can start.

Figure 412. Cable crane yarding with Igland interlock winch, telescope tower, and carriage for fixed hoist line. Telemark, Norway 1980.
A. The yarder at the landing.
B. and C. The carriage with the fibre rope chokers.
D. Unhooking the chokers.

A better solution is shown in figure 413. There are two catchers on the main line with a 1 to 2 meters spacing between them. The first catcher locks the hoist line to the carriage and the second catcher releases it again. On the landing the yarder operator hauls out the carriage until the first catcher (steel egg) is fastened in the carriage. He then unhooks the chokers and returns the carriage to the logging area. When the choker setter has attached the chokers to the hoist line, the yarder operator continues to haul out the carriage until the second catcher has passed it. This is watched by the choker setter who gives a new signal in order to start the lateral skidding and the transport.

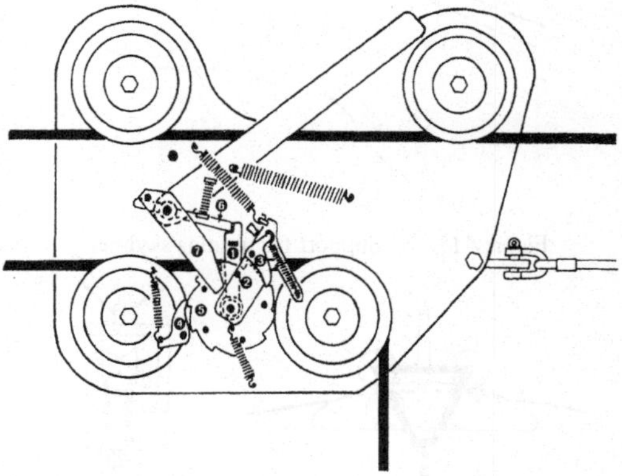

Figure 413. The Igland carriage with automatic locking and loosening of the hoist line. There are two catchers on the main line: the first catcher locks the hoist line, and the second releasees it again.

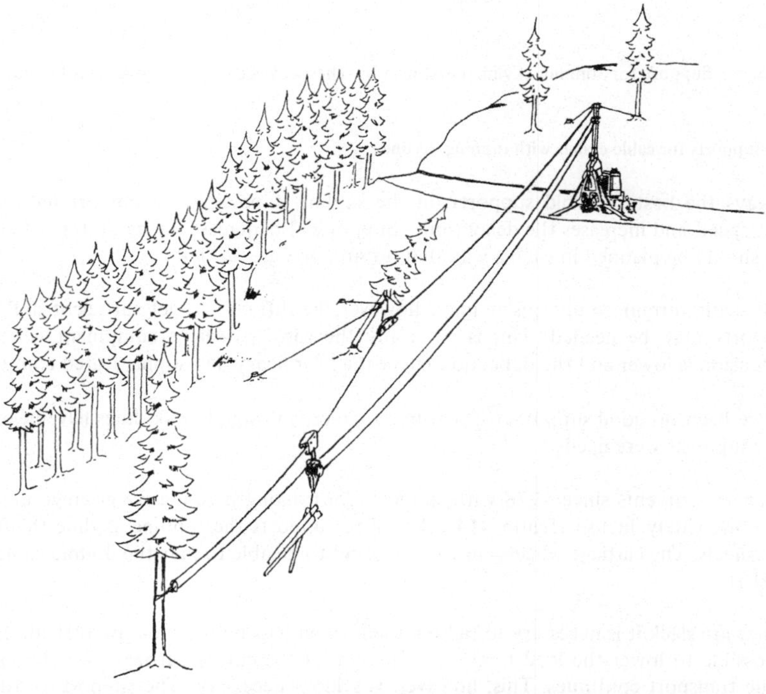

Figure 414. Uphill cable crane yarding with Owren Vario Winch. A support on the running skyline developed by The Norwegian Forest Research Institute, 1979.

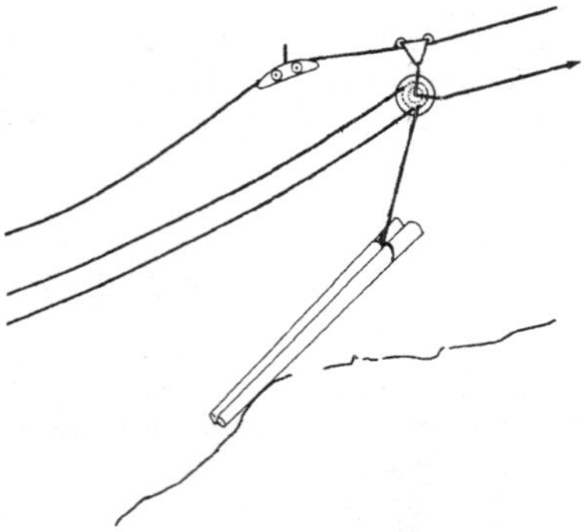

Figure 415. Support for running skyline.

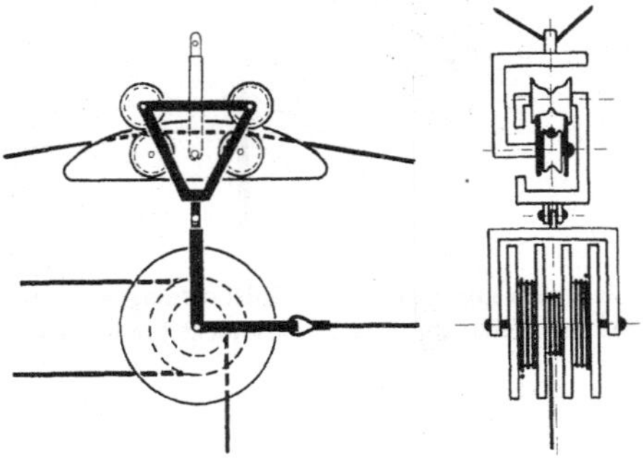

Figure 416. Support for running skyline developed by The Norwegian Forest Research Institute, 1979.

11.5.4.7. Supports for cable cranes with running skylines

It is always the best to avoid supports on the skyline if possible. A support reduces the transport speed and increases the delay times. Supports also increase the rig up time. The road network should be planned in such a way that no supports are needed.

Under difficult terrain conditions it may, however, be difficult to get sufficient deflection and supports may be needed. This is especially the case with light equipment where the skyline tension is lower and the deflection larger than for heavy and sophisticated equipment.

There have been no good supports for running skylines, and standing skylines have been used whenever supports were needed.

Norwegian experiments since 1978 with supports for running skyline have given good results and are now widely in use (figure 414). The shoe supports the running skyline (haul-back line) by wheels. The carriage wheels are wide in order to be able to pass the double cable shoe (figure 416).

If the lines are slack it is necessary to tighten them when passing the shoe. In difficult cases it is also possible to lower the load to the ground and let the empty carriage pass the support before the transport continues. This, however, is seldom necessary. The support for running skylines has been in operation during a five year period, both with heavy equipment (the Moxy yarder) and with lighter equipment (Igland interlocked winch, Owren Vario Winch).

Figure 417. Support for running skyline (Kviteseid, Telemark 1980).
A. A-framed support with lifting line to the left and guyline to the right.
B. The wide carriage wheel for a running skyline support.

Figure 418. The Igland carriage passes a running skyline support.

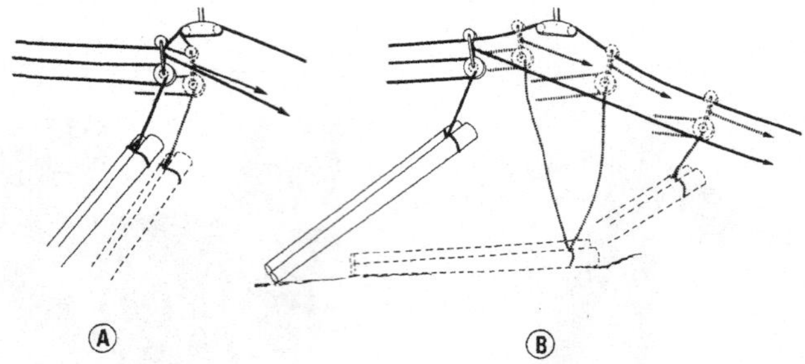

Figure 419. The haul-back line should be tightened when the carriage passes the support. In difficult cases it is possible to lower the load before the carriage passes the support. (This situation has never happened during 3 years of operation with experienced operator).

11.5.4.8. Yarding cranes

Yarding cranes may be used with live skyline system for slackline yarding. Yarders with swing boom for uphill yarding were discussed in chapter 11.3.2.2. (the Linkbelt and Critical Site yarders). Another example is the Schield Bantam which is powered by a 75 kW engine (figure 420):

Skyline	L	300 m
	d	19 mm
Main line	L	300 m
	d	16 mm
Haul-back line	L	500 m
	d	12 mm

A yarding crane is mounted on a wheeled or a crawler chassis (figure 421). The yarder unit is mounted on a turn table. The tower is placed over the center of the turn table and is anchored with a back guy. The winch set with the engine and the operator's cabin may be turned around the center of the turn table. The swing boom is anchored to the tower by a tackle which makes it possible to change the slope of the swing boom. The swing boom may be turned to each side and cover a triangular area as illustrated in figure 424. For an experienced operator it is possible to swing the carriage sideways in relation to the skyline corridor.

When the yarder pulls in a load, the swing boom is positioned in the direction of the skyline corridor. When the load approaches the landing it is possible to swing the boom sideways and place the load, for example beside the yarder. If the load is too heavy the operator releases the tension in the main and haul-back lines, he then swing the crane into side position and brake the turn table, whereafter he can pull in the load on the landing.

The swing boom yarder can operate a cable crane with a hoist line. In big timber it is better to use grapple yarding (figure 423). This reduces the yarding crew. If the average tree dimensions are between 6 and 9 cu.m. per tree the yarding crew may be reduced from 8 to 3 or 4 men by using a grapple instead of a hoist line. Another advantage is the reduction of the time per turn, since both the loading and the unloading time is much shorter than when chokers are being used.

Figure 420. Schield Bantam yarder with swing boom. Live skyline cable crane with gravity outhaul system. Oregon, USA 1977.

It is a disadvantage that one log only can be yarded in each turn. The grapple yarder needs particularily good deflection, and more deflection than an ordinary cable crane with hoist line. Supports can not be used in the span. The method is only suitable for big tree dimensions on logging areas where there is sufficient deflection.

The grapple should be positioned in the same direction as the main and haul-back lines. The timber shoud be directionally felled, perpendicular to the direction of the skyline (figure

425). The choker setter communicates with the yarder operator by radio and tells the operator where to position the grapple.

Table 53 gives some examples of specifications of yarding cranes.

Table 53. Examples on specifications of yarding cranes.

Modell		Madill 044		Washington 118
Weight		–		510 kN
Tower height		18.3 m		16.4 m
Tower type		Yarding crane		Yarding crane
Engine power		355 kW		316 kW
Main line	L	535 m	345 m	495 m
	d	25 mm	32 mm	22 mm
	Max. pull	255 kN		390 kN
	Max. speed	10.7 m/s		9.0 m/s
Tagline	L	455 m	1,310 m	495 kN
	d	16 mm	10 mm	22 mm
	Max. pull	55 kN		180 kN
	Max. speed	12.7 m/s		9 m/s
Haul-back line	L	700 m	1,370 m	1,000 m
	d	22 mm	16 mm	22 mm
	Max. pull	277 kN		102 kN
	Max. speed	10.7 m/s		10.5 m/s
Straw line	L	719 m	2,194 m	1,350 m
	d	11 mm	6 mm	10 mm
	Max. pull	230 kN		42 kN
	Max. speed	8.5 m/s		24.3 m/s

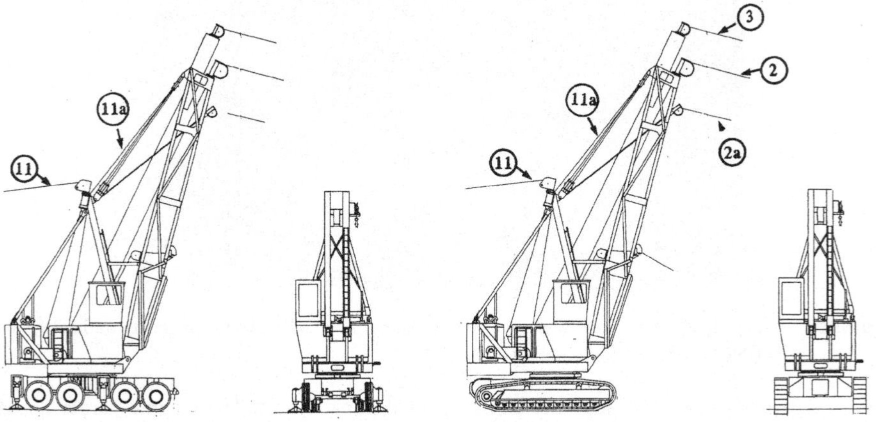

(2) Main line (2a) Slack-pulling line (3) Haul-back line (11) Back guy

(11a) Multiple sheave blocks to lift or lower the boom

Figure 421. Mobile swing boom yarders.
Left: Wheeled undercarrier (4-axle).
Right: Tracked undercarrier (tank type).

Figure 422. Grapple yarding with swing boom yarders. The terminal operations (loading and unloading) are reduced to a minimum. In the field a tongue operator helps the yarder operator by radio communication to direct the tongue on the log. No chaser man is needed at landing. The swing boom directs the log to correct place and the unloading time is reduced to open the tongue (Oregon, USA 1977).
Right: A wheeled undercarrier.
Left: A tracked undercarrier.

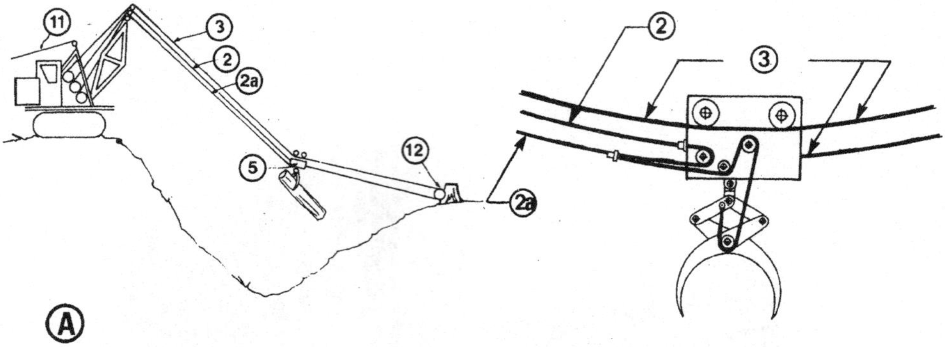

(2) Main line (2a) Slack-pulling line (3) Haul-back line (5) Carriage (11) Back guy
(12) Tailhold

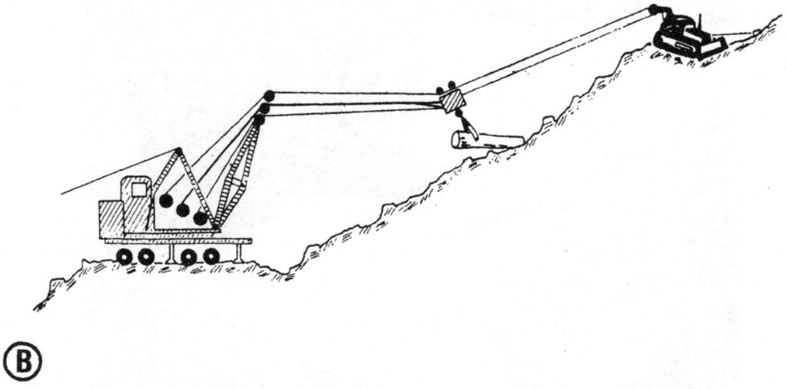

Figure 423. Grapple yarding needs good deflection. No supports are possible in the span. The rig up time is short.
A. Stump as a tailhold anchor.
B. Tailhold tractor is convenient under certain terrain conditions.

Grapple yarding with a yarding crane has a high production and it is necessary to carry out hot logging. A loading machine must be placed at the landing and transfer the logs directly to a truck. The landing should have sufficient space for the yarder, the log pile, the loading machine and the truck. A secondary road may be placed above the landing in steep terrain (figure 426). This gives good height difference between log pile and skyline, and good space for the loading operation.

Moving the yarding crane from one position to the next is done quickly since it is a well stabilized machine, anchored with one back guy. If the terrain allow the use of a tail tractor at the tail end, moving the crane from one corridor to the next can be done even more quickly. Since it is easy to move the crane from one position to the next it is often preferable to move the crane instead of the tail anchor (figure 427). The result is a shorter yarding distance which increases the yarding production.

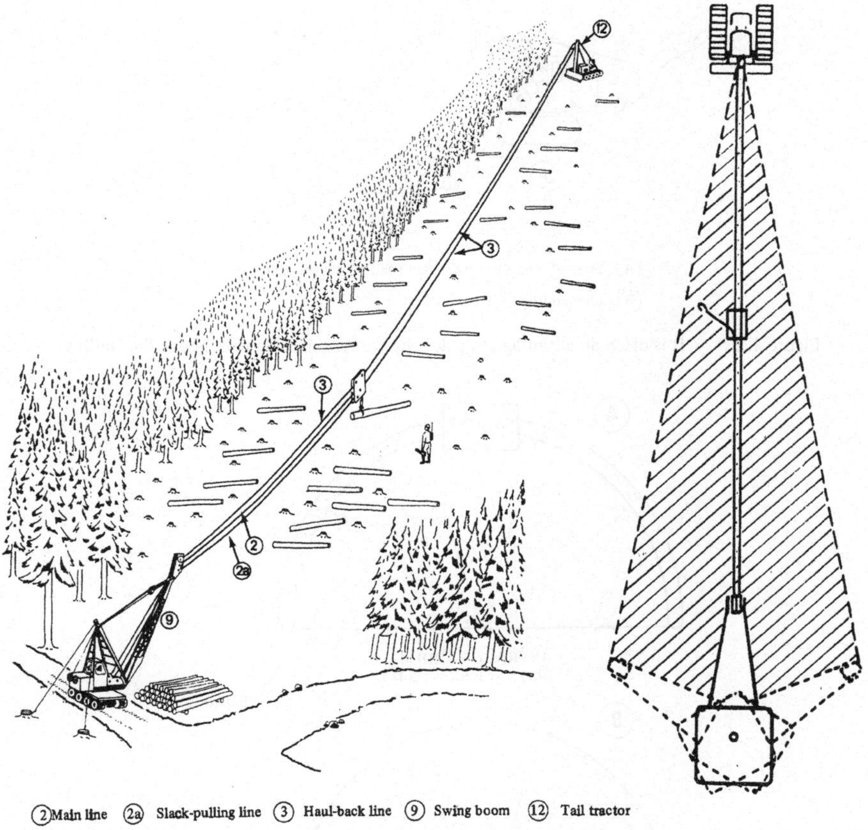

Figure 424. A swing boom yarder in one end and a tailhold tractor in the other give a high degree of flexibility. The swing boom can direct the grapple and the carriage sideways to the logs with an experienced operator on the yarder.

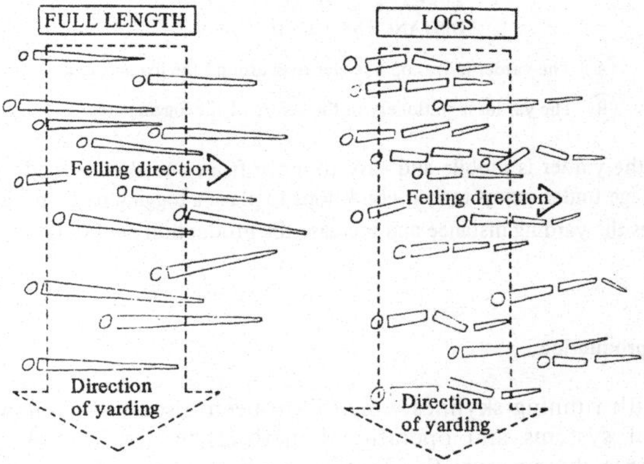

Figure 425. The direction of felling for grapple yarding.

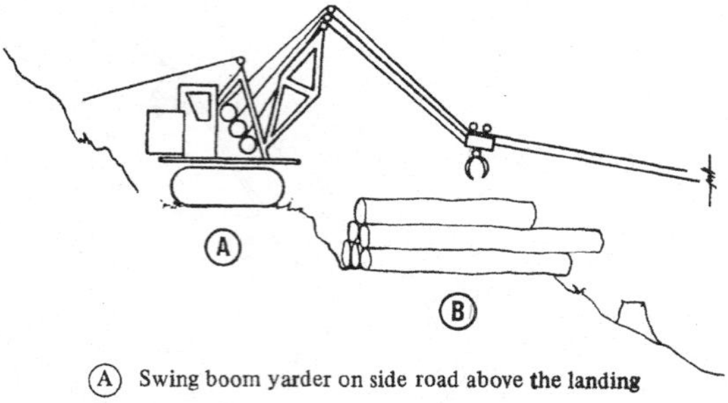

(A) Swing boom yarder on side road above **the landing**

(B) The landing

Figure 426. It is often an advantage to place the yarder on a side road above the landing.

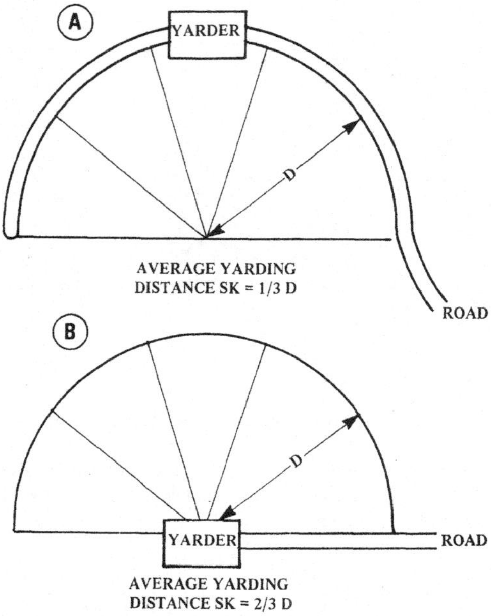

(A) The yarder moves on a feeder road around the logging area

(B) The yarder is stationary in the centre of the logging area

Figure 427. Since the yarder is mobile and easy to move from one skyline road to the next, it may be an advantage under many terrain conditions to place a logging road around the logging area. This reduces the yarding distance and increase the production per hour.

11.5.4.9. The yarding production

The cable cranes with running skylines which have been discussed in chapter 11.5.4. include a great number of systems and operational methods in use throughout the world. The tree species, the tree dimensions, the terrain conditions, the climatic conditions etc. are different and so is the production.

This chapter will give a few examples on typical production and productivities which might be expected for todays running skyline cable cranes when yarding coniferous forests. The production figures is limited to yarding only, of trees, stems or logs which have been cut on the yarding area beforehand. Time for rig up and rig down of the cable crane is not included.

Light yarders for tree dimensions up to 1 cu.m. per tree.

These are light yarders powered by 40 to 50 kW engines, such as for example the Owren Vario Winch, the Igland interlocked double-drum winch etc. The maximum yarding distance is usually 200 meters, which sometimes may be extended to 300 meters.

The best economic result has been achieved with a 2-man crew, 1 yarder operator and 1 choker setter. In some cases they are yarding only, in other cases they undertake both felling limbing, bucking and yarding. In the latter case they sometimes take the cutting first and the yarding afterwards, for example half a day on each operation and sometimes they take felling, yarding and conversion of the trees into assortments in one continuous transport sequence. This is always the case in winter time since the snow fall may cover the logs and reduce production.

Table 54. Distribution of the time elements per turn (OMNES 1984).

Crane Yarding distance	Moxy 270 m	Owren Vario Winch 70 m
Return carriage	16 %	18 %
Loading	41 %	30 %
Transport	25 %	15 %
Unloading	18 %	37 %
Total	100 %	100 %
Effective time	3.54 min./turn	3.08 min./turn
Road change and delay	1.91 min./turn	3.14 min./turn
Work place time	5.45 min./turn	6.22 min./turn

Table 55. Production per work place hour (effective time + normal delay time). 0.1 m^3/tree – 1.5 m^3/tree. Owren Vario Winch. 2-man crew (OMNES 1984).

m^3/load	m^3/hr (crew)		Manhour per m^3	
	Yarding distance, m.			
	50	150	50	150
0.6	5.1	3.5	0.39	0.57
0.8	5.9	4.3	0.34	0.46
1.0	6.5	5.1	0.30	0.39

Table 54 is an example of the distribution of the time elements in a turn, and table 55 shows typical production for 100 meters and 300 meters cable cranes (the Owren or the Igland systems). Such cranes may handle logs from trees of up to approximately 1.5 cu. m. per tree.

The production varies normally between 2 and 3 cu.m per man per work place hour with a labour consumption between 0.3 and 0.5 manhours per cu. m. for yarding only.

Medium sized yarders for tree dimensions up to 3 cu. m. per tree.

These are yarders powered by 140–150 kW engines which may be used for running skyline cranes up to 600 meters. Time studies have shown that longer cable cranes than 400–500 meters should be avoided. It happens, however, that the crane has to take some longer spans due to specific terrain conditions and the crane should be constructed for at least 700 meters.

The Norwegian Moxy yarder which has been investigated by detailed studies over a long period of time is chosen as an example for some typical production figures.

A 4-man crew operates the cable crane:

1 yarder operator
1 chaser
2 choker setters.

Table 56. Production per work place hour (effective time + normal delay time). 1 m^3/tree – 3.0 m^3/tree. Moxy cable crane. 4-man crew. (OMNES 1984).

m^3/load	m^3/hr. crew			Manhour per m^3		
	Yarding distance, meters					
	100	200	300	100	200	300
0.8	9.1	7.4	6.2	0.44	0.54	0.64
1.0	9.7	8.3	7.2	0.41	0.48	0.56
1.2	10.5	8.8	7.8	0.38	0.45	0.51
1.4	10.8	9.5	8.3	0.37	0.42	0.48

Table 56 shows figures for yarding only. Rig up time and rig down time are not included. Normally the productivity is between 2 and 3 cu.m per man per work place hour with a labour consumption of 0.3–0.5 manhours per cu. m.

The productivity is approximately the same for yarding of unbranched trees. If the felling is combined with choker setting the crew has to be increased by one man. The cable crane carries out full tree yarding and a delimbing machine at the landing takes care of the conversion from unbranched trees to logs.

Large yarders for tree dimensions between 4 cu. m. per tree and 20 cu. m. per tree.

The yarders for such cable cranes are usually powered by engines with more than 300 kW. The lengths of the cable cranes are seldom more than 300 meters and the average yarding distance is usually 70–150 meters.

The yarding crew are usually somewhat smaller than for the cable cranes with standing skylines. The yarding crew for a cable crane with running skyline is usually 5 men for tree-dimensions less than 2 cu. m. per tree, 6 men if the tree dimensions are smaller than 6 cu.m. per tree and 7 to 8 men for large sized tree dimensions. When the cable crane has a carriage with a tong for grapple yarding there are usually four men in the crew: yarder operator, loader operator, one tong operator and one foreman.

The trees must be felled and bucked into logs before the yarding starts. The yarding production is very high and it is necessary to carry out hot yarding. There must be a loading machine at the landing in order to move away the logs simultaneously with the yarding. A loader operator is therefore included in the yarding crew both for the cable cranes with running skylines and for the grapple yarding.

Table 57. Production per work place hour (effective time + normal delay time) for running skyline cable cranes (Run) and for grapple yarders (Grap). (Compare with table 42 page 315).

m^3/tree	m^3/hr. crew		m^3/man/hr.		Manhours pr. m^3	
(average)	Run	Grap.	Run	Grap.	Run	Grap.
0.3	10.6	9.3	2.3	2.4	0.43	0.42
1.0	12.8	13.4	2.6	3.4	0.38	0.29
3.0	19.3	25.0	3.7	6.3	0.27	0.16
6.0	29.0	42.5	5.2	10.6	0.19	0.09
9.0	38.7	60.0	6.8	14.9	0.15	0.07

When the tree dimensions are over 3 cu.m. per tree, the production is normally between 20 and 40 cu. m per work place hour for the running skylines and between 25 and 60 cu. m. per work place hour for grapple yarding.The production per man per work place hour is not extremely high compared with the production for yarding smaller tree dimensions. The reason is the high number of men in the crew. This problem is even more visible for cable cranes with standing skyline (see table 42). The running skyline systems has usually a lower labour consumption than the standing skyline cable cranes. This is very noticeable with grapple yarding, where the labour consumption varies between 0.07 and 0.16 manhours per cu. m.

When yarding small timber a main problem is to collect sufficient number of logs in order to get acceptable size of the loads in the turns. Pre-bundling should be considered.

When yarding big timber a main problem is the large number of men in the crew. Yarding cranes with grapple yarding is one way to solve the problem.

12.0. THE RIG UP WORK OF CABLE CRANES

The potential productivity of a cable crane may or may not be utilized by the crew. The resulting production depends on the knowledge, experience and training of the men in the crew and their leaders. In no other part of the job, is this more visible than during the rig up and rig down of the cable cranes.

The organization of the work as well as the working technique have changed with time. The previous stationary cable cranes, sometimes with two to four weeks rig up time, have been replaced by mobile yarders and the yarder mounted auxiliary equipment has resulted in a rig up time of only a few hours.

The previous 1,000–1,500 meters cable cranes are seldom used any more. Todays cable cranes are usually 300–500 meters long, because of the denser network of feeder roads in difficult terrain.

The communication between the workers has been made simpler and better by the radio which has reduced waiting time and other delay times during rig up work.

The previous more stationary cable cranes consisted of parts and equipment which were built together on the yarding area.

Todays cable cranes have most of the equipment mounted on the yarders, and the rig up work has been reduced to stringing out the cables and fastening them to anchoring points in the terrain.

In regions where a sustained yield forestry has been carried out, sometimes through several tree generations, the equipment has been smaller than in regions with virgin forests and oversized trees.

With time areas with over-mature virgin forests are being reduced and replaced by second growths. The cable operation under these conditions are carried out by smaller equipment. A world wide exchange of experience and designs is taking place. This also applies to the rigging techniques. The previous large crews have been replaced by smaller crews with specialized training. They are also trained to do special work elements during the rig up. The foreman is usually a working foreman. He is not any longer the leader only, he also carries out a certain job during the operation. He does the pre-rigging on the yarding area before the yarder and the cable crew arrives and does the preparatory work for roadchanges while the rest of the crew carries out the yarding on the previous road.

12.1. The planning

Before any cable operation starts, the operational activities on the entire area should be included in a plan of operation. In connection with the survey in the forest the terrain must be divided into terrain sections, and the terrain conditions must be classified (terrain classification).

The next step is to lay out the access roads in order to connect the terrain sections with the road network for long distance transport.

The feeder roads branches out from the access road and divides the terrain sections into operational units. The layout of the cable cranes has to be placed in relation to the feeder roads, and the planning of the cable cranes is a deciding factor for the layout of the feeder roads, determining the spacing between them, their location etc. The winch and cable systems are an integrated part of the transport network in a forest, and are usually a feeder road of second order.

The planning is important and has been discussed and analyzed in other publications. In this book we shall limit the description to some of the problems only which is of interest in connection with planning the use of the cable cranes.

12.1.1. Layout of the road network

Sometimes the road network consists of logging roads only. These are built to take sufficient payload and have to allow for the length of the loaded trucks.

In a sustained yield forestry the roads shall serve other functions besides the logging, such as cultivation, thinning, forest fire protection etc.

The road network consists of the following:

The access roads
The feeder roads
The terminals.

12.1.1.1. The access roads

These roads are the main roads in the forest. In steep terrain the access roads have to overcome height differences and steep slopes may be needed. Norwegian experiments have shown that such roads may be built with a slope of up to 20 % (FRØNSDAL 1980). The driving conditions on the road depends on the slope as well as the road width and the curvature. If the slope is increased the curve radius and the road width should be increased accordingly.If the single lane width is 3.5 meters on a horizontal road, a road with 20 % steepness should have a width of 5 meters. The road should have no curves on the steep slope and the slope should be reduced at the bottom so that there is a flat stretch of road for stopping at the ende of the steepest slopes.

Climatic erosion can create problems on steep slopes. It is an advantage to spray an asphalt emulsion on the road surface in order to reduce the erosion (SOLBRAA and SKAAR 1969). The erosion is also reduced if the speed of the water from rain or melting snow is slowed down by open surface culverts. Professor Franz Hafner suggested the following spacing between the surface culverts:

Table 58. Distance between open surface culverts on steep forest truck roads (access roads). (After HAFNER and HEDENIGG 1959).

Slope in percentage	Distance between the culverts, m
< 10	36
12	32
14	28
16	26
18	23
20	21

12.1.1.2. The feeder roads

The feeder roads branches out from the access roads and are usually parallel to the terrain contours. It is an advantage to have at least a 1 % slope, since it reduces the climatic erosion on the road surface. If the feeder road passes a small creek, a surface culvert may be built. With high water discharge a closed culvert should always be used. In this case the feeder road should have at least 1 % slope downhill towards the culvert (in both directions). The average slope of a feeder road should usually be 3 – 5 % in the transport direction.

It is an advantage to build the feeder roads as truck roads, in order to avoid double handling during the transport sequence.

The location of a feeder road in a steep hillside is important (figure 428). Supports along the skyline should be avoided. It is also important to place the feeder road near the edge on the upper end of a slope. This reduces the problems during uphill transport of loads.

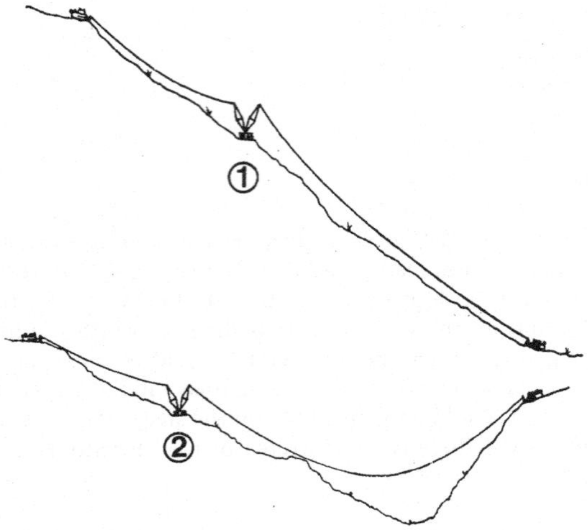

Figure 428. Cable crane yarding needs deflection in order to avoid supports in the span. Special care should be taken for the loaction of the feeder roads in relation to the terrain conditions.

The feeder roads are laid out parallel to each other and the spacing will depend on the cable system to be used. A 300 – 400 meters spacing allows the use of short distance cable cranes with their high production.

In rocky and very difficult terrain the road building is costly and a more open network has to be laid out. A spacing of 800 – 900 meters makes it possible to use cable cranes shorter than 500 meters which is an advantage both from an ergonomic as well as from a production point of view.

12.1.1.3. The terminals

The terminals are an important part of the transport network. They should be placed where it is possible to yard the timber from the surrounding area without the use of supports. The terminals must have sufficient place for the yarder, the landing, the loading machine and the truck. This again, depends on the size of the yarder and the size of the timber, and whether hot or cold yarding is being carried out.

In many places it is an advantage to use the feeder road as a terminal. With full tree yarding the cable crane make a continuous pile or deck of trees on one side of the road, the delimbing machine converts the trees to logs and place them on the other side of the road. In these cases one should make sure that there is sufficient space on both sides of the feeder roads for the trees and the logs.

12.1.2. The location of the cable cranes

The cable cranes are feeder roads of second order. They should be placed so that they cover most of the area with as few moves as possible. The advantage of avoiding supports has already been discussed.

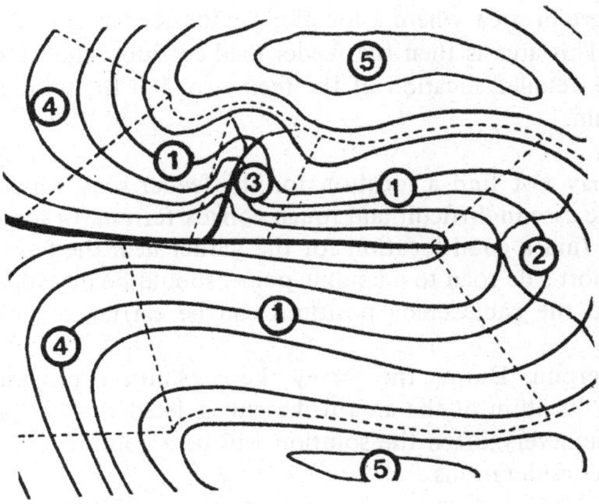

Figure 429. 1. Uniform hillsides. 2. Basin-shaped hillsides. 3. Closed valleys. 4. Cone-shaped hillsides. 5. Plateaus.

The layout of the cable cranes vary with the terrain conditions (figure 429). In uniform hillsides where the terrain contours are approximately parallel to each other, the cable crane may be moved sideways in such a way that one skyline corridor is parallel to the next. If the cable crane skids the logs laterally from both sides of the skyline, the skyline corridor should be approximately 90° in relation to the terrain contours. If, however, the yarder skids the logs laterally from one side of the skyline only, the angle between the terrain contours and the skyline should be less. In this way the load is kept away from the area where the choker setters work when it is transported along the skyline. This makes the work for the terrain crew safer.

In a basin-shaped hillside, the skyline corridors branches fan-shaped out from a landing at the yarder station. A tailhold must be moved during the road change while the yarder stays at the same location. If a feeder road is placed high up in a basin-shaped hillside, the timber may be yarded uphill and the yarder has to be moved during the road change from one corridor to the next. In this case the tailhold is the same during the whole operation.

In a cone-shaped hillside the opposite is the case.

Both in the case of a basin-shaped hillside and a cone-shaped hillside, it is of importance to find a good terminal place since a large amount of timber will be handled at this point. It may be worthwhile to build a short road from the feeder road in order to get a good location for the yarder and sufficient space for the timber as well as the loading machine or the delimbing machine.

12.1.3. The planning method

The previous laborious levelling has been replaced by the use of aerial photographs and/or good maps together with computer planning (SKRÅMO 1971).

When it is decided which cable crane method is going to be used as well as the approximate spacing between the feeder roads, the aerial photographs or the map are used in order to find out where the road has to be laid in order to avoid any supports along the skyline. In a similar way the approximate location of a terminal can be found.

Usually, there is a certain area where a location of the feeder road will give a support-free span on the skyline. This area is then the feeder road corridor, and should be marked on the map. The correct and detailed location of the feeder road or the terminal is selected after an inspection in the terrain.

In other cases one may not find a location for the feeder road which gives a support-free span. This is often the case in difficult and rough convex terrain. In some instances, however, it may be possible to find a good location for the yarder near the feeder road and the possibility of building a short side road to a landing point, should be investigated. The planning of the road network and the cable crane positions can be carried out in the following steps:

1. Survey in the terrain. During the survey the silvicultural programme is planned and the size and the location of the area to be cut is decided. The terrain conditions may give limitations and very often the solution will be a compromise between operational and silvicultural considerations.

2. Planning on maps, aerial photographs and computer. The location of feeder road corridors and terminals is being investigated by the use of computer programmes. The places and the corridors are marked on the maps.

3. Survey with detailed planning. The correct location of the feeder road is decided and marked on the maps. The same is done for the yarder stations and the possible skyline corridors.

The maps have to be updated after the road has been built. Before the yarder operation starts, the foreman should be supplied with the updated maps where the skyline corridors, the location of tail trees and yarder station etc. is marked.A longitudinal profile of terrain and skyline is a good help for a foreman when he starts the pre-rigging work, the rig up work, and the operation itself. Good planning is one of the secrets behind a high production.

12.2. Rig up tools

Some of the equipment and tools have been described in chapter 6.0. where anchors, spars and intermediate supports were discussed. Standard wire rope equipment including wedge grip, eccentric pressure grip, tightening devices such as small hand winches, come-along, etc. are useful tools used for the rig up or rig down work. Generally, cable clips or double base clamps which have to be tightened by screws are avoided, partly because it is laborious to deal with and partly because they can be dangerous. The diameter of the wire rope gets smaller as the rope gets older and the clip-screws have to be tightened after a while. Spliced eye or pressed metal sleeve eye with shackle is safer and quicker to use.

12.2.1. Yarder mounted auxiliary winches

Traditionally the yarders were equipped with drum set for all the operating lines. In addition todays yarders are equipped with winch drums for different types of auxiliary lines (figure 390 B and C). The auxiliary winches replaces the previous laborious manual operated hand-winches or other tightening devices (fig. 430).

12.2.1.1. Guyline drums

The guyline drums are usually powered from the main yarder engine, for example by hydraulic

Figure 430. Tightening device (»come-along») for manual tightening of the skyline or guyline tension. (Norway, 1954).

drives. The power should be sufficient to give the guylines equal tension. For medium or large yarders 20 kN is usually sufficient. Each guyline drum should have a stop device, for example a hydraulically operated pawl and ratchet wheel which can withstand the forces when the guylines are stretched during yarding. The guyline runs from the guyline drums over the steering blocks (guy blocks) near the top end of the steel spar and outwards to the anchoring points (the anchoring straps). The large yarders with 25–35 meters high steel spars need 6 to 8 guy drums. The medium sized yarders with 15 meters high steel spars have 2 to 4 guyline drums and the light yarders with 6 to 9 meters high steel spars have 1 to 2 guyline drums which sometimes are motorpowered and sometimes hand operated (hand winches).

12.2.1.2. Straw line drums

These are powered by the engine. The drum capacity must be sufficient to pull in a length of straw line which is twice the length of the main line pluss 20 %. The large yarders usually have a straw line diameter of d = 10–12 mm. On the medium sized yarders the diameter of the straw line is d = 8 mm and on the light yarders the diameter is d = 4–6 mm.

Figure 431. Two men are needed to pull out the 6 mm straw line around a 300 m long yarding area.

Figure 432. 2 back-pack spools with 1,000 m of 2.1 mm rig up line weighs less than 20 kgs. If the main and haul-back lines shall run along the same corriodor the worker walks downhill and rolls out a double rig up line. These are connected to each other and placed around a block at the top of the hillside (Norway, 1968).

12.2.1.3. Rig up drums - rig up spools

It is a heavy work to pull out the straw line by hand. Two men are needed to pull a 6 mm straw line around the corridors for the main and haul-back lines on a 300 meters long cable crane (figure 431).

The use of thinner rig up lines reduces the work load on the men.Based on Norwegian experiences the following dimensions may be suggested:

Nylon line (Harpoon line)		S_{Br} = 1.5 kN
Rig up line I :	d = 2.1 mm	S_{Br} = 3.0 kN
Rig up line II :	d = 4.0 mm	S_{Br} = 8.4 kN

The nylon line is used together with a harpoon gun.

The rig up line (I) may be mounted on back-pack spools which can be carried by one man (figure 432). A length of 1.000 meters of the 2.1 mm thick rig up line weighs 150 N and the weight including the ruck sack frame is less than 200 N.

The yarder should have standardized driven shafts where the spools may be mounted. When the rig up line (I) has been pulled around the yarding area, the other rig up line (II) may be pulled around the area by the driven spools and can then be used to pull the straw line around.

If the terrain is very steep with boulders and cliffs, it can be difficult to walk and even more difficult to pull a rig up line. The same is the case when the cables shall cross water such as a river, or a lake, or when the lines has to cross a narrow valley with steep hillsides. A harpoon gun may be of help in such cases (figure 433).

The harpoon gun M-52 is being used by the Life-saving service at sea and the coastguard. The gun has a 12 mm diameter bore. An arrow is put into the barrel and against the cartridge. A 150 meters long nylon line is coiled up in a box which is fastened to the gun barrel. The way the line is coiled up leads to very little resistance when the line is being pulled out of the box by the arrow. If the arrow head is made of steel or rubber, the line can be shot out over a horizontal distance of 120–150 meters.

The steel or rubber head may be replaced by a rocket. In this case the horizontal shot distance is approximately 300 meters.

The recoil from the gun is hard. It is recommended that the operator places the gun against a stump before discharge. An elevation of 25–30° gives the best result when a rocket arrow head is used.

The operator usually carries the gun to the top of the hillside and shoots downhill. It is easy in this case to shoot out the line over a distance 500–700 meters.

When shooting over such distances the nylon line should be coiled up as illustrated on the left in figure 433.

Two shots may be fired after each other. The two parallel lines are tied together by a knot and the bight is laid over a block. The line may be used to pull the rig up line (II) around the corridor.

① The harpoon gun with a rocket as arrow head ② The rocket is automatically ignited when the arrow is 50 meters away from the gun

③ A method of winding up the rig up line manually ④ When the line is wound up it is placed in a box

⑤ The harpoon gun. The box with rig up line ready for operation. ⑥ Rocket head ⑦ Steel head

Figure 433. A harpoon gun is a handy rig up tool under difficult terrain conditions. It can shoot a rig up line (nylon) 300 m horizontally and up to 700 m downhill from the top of a hillside. (Telemark, Norway 1962).

Normally, it is sufficient, however, to use the back-pack spools (figure 432). Each of the spools has 500 meters and if the main line and haul-back line follow different roads, the worker has to walk around the main and haul-back line corridors while the rig up line is being rolled out behind him. If, however, the main and haul-back lines follows the same corridor as in many of today's cable systems, the two lines, one from each of the spools may be rolled out simultaneously. The worker walks to the top of the hillside, with the entire length on the back-pack. The two lines are rolled out parallel to each other as he walks downhill. Before he starts he fastens the end of the two lines to each other and lay the bight over a block at the top of the hillside. When he has walked down the hill to the landing, the first rig up line (I) may be used to pull out the second rig up line (II). The second rig up line (II) is used to pull out the straw line and the rest of the rig up work can start.

12.2.1.4. The rig up line method

The back-pack spools with the rig up line may be too heavy to carry when walking on uneven surface or through thick brush. It is sometimes easier to pull out the 4 mm thick rig up line (II) directly. This will depend on the terrain conditions and the cable crane method (see figure 434). Figure A illustrates gentle terrain. If the length of the cable crane is shorter than about 250 meters it is better to pull out the rig up line (II) directly. One man is sufficient for that job. Over longer distances it is better to use the back-pack spools and roll out the rig up line (I) while walking around the main and haul-back line corridors.

If the slope is steeper than 20% and the length of the cable crane is more than 200–250 meters, the back-pack spools should be used. If the main and haul-back lines run in different corridors (figure 434, B I) the single rig up line is rolled out. When the main and haul-back lines run along the same corridor, it is better to roll out two parallel rig up lines, while walking downhill from the top towards the landing (figure 434, B II).

Sometimes the cable crane crosses a deep valley (figure 434, C). In this case it is usually better to pull the haul-back line and the rig up line (II) together down to the bottom of the valley. When the crew has reached the bottom of the valley, the rig up line (II) is pulled uphill to the tail end of the cable crane, laid around a pass block, and pulled down to the bottom of the valley where it is connected to the haul-back line. The haul-back line is then pulled around the corridor and back to the yarder by means of the rig up line (II). The haul-back line may now be used to pull out the tail block and the tail tackle as well as the main line and the skyline.

When yarding uphill (figure 434, D) the steepness and the roughness of the ground decides the method. On gentle slopes it is easiest to use the back-pack spools and lay out two rig up lines parallel to each other while walking downhill. At the tail end the two parts of the rig up line are attached to each other and the bight is laid around a pass block. The rig up line (I) can then pull out the rig up line (II) using the motor and this in turn pulls out the straw line.

On steep slopes with a smooth surface the skyline can be pulled downhill directly and anchored to the tail anchoring. If it is a running skyline it is better to pull down the haul-back line and the rig up line (II) parallel to each other. When the two lines have reached the tail end the worker connect them with each other, lays the bight around a pass block and pulls the haul-back line up again with the rig up line (II).

The rig up lines reduce the physical stress on the worker and saves manpower. The heavy work load is transfered to the engine on the yarder.

12.2.1.5. Separate rig up winches

These light portable winches are handy tools for the rig up work. The chainsaw winch can be

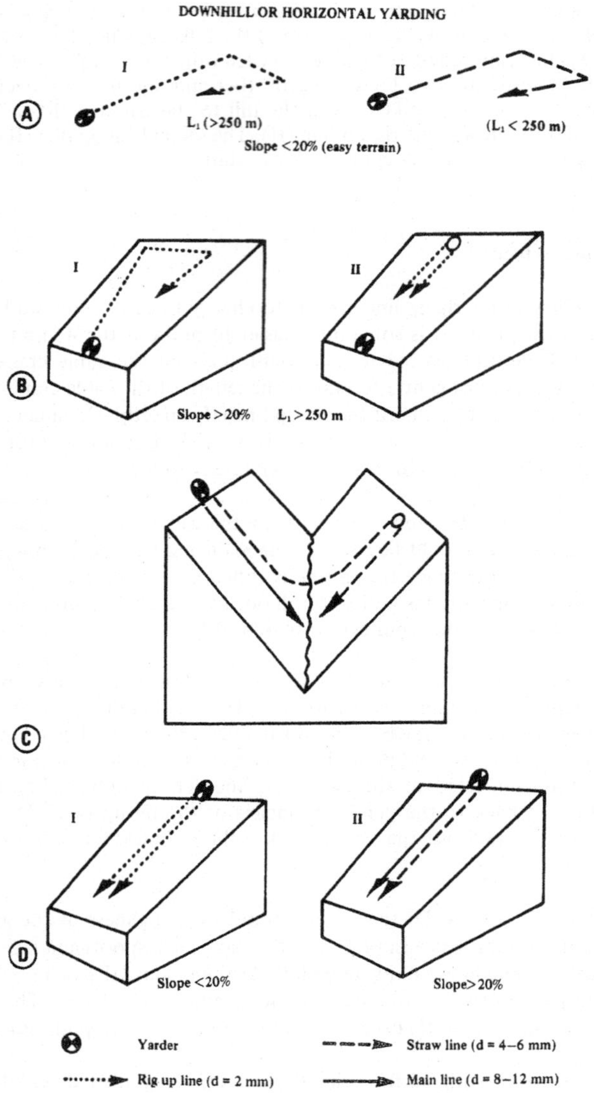

Figure 434. A. In easy terrain, shorter than 250 m, a 4–6 mm straw line may be pulled out directly. On longer distances it pays to use the 2 mm rig up line on back-pack spools.

B. The rig up line method is usually advantages in very steep terrain.

I. If the main and haul-back lines run in different corridors, the single rig up line is being rolled out.

II. If they run parallel in the same corridor, the double rig up line is being rolled out.

C. Yarding over a small valley with steep hillsides. The main line and straw line are being pulled downhill to the bottom of the valley. The straw line is being pulled uphill and downhill again on the other valley side.

D. Uphill yarding. When the slope is gentle the double rig up line is being rolled out. In steep slopes the main line and the straw line are pulled out directly (parallel to each other).

Figure 435. Left: The Jobu chainsaw winch for rig up work.
Right: The Austrian Ackja rig up winch.

carried by two men. It is an additional equipment for the chainsaw. The chainsaw blade is replaced by a gear and a winch. The winch positions itself in the direction of the pulling line like a motor-driven block (figure 435). The operator controls the spooling on the drum by moving the motor sideways. The power capacity is 2.5 kW. The average speed is 0.2 m/sec. and the average pull is 9 kN. The main line diameter is d = 6 mm and the length up to 100 meters. Sometimes it may be necessary to use a block to increase the pulling force.

Figure 436 shows one use of the chainsaw winch when rigging up a light cable crane for downhill transport with a top mounted winch. The yarder and the drum with the skyline are placed beside each other at the landing in the valley. The skyline and haul-back line are connected by means of a 3-way swivel to the pulling line of the chainsaw winch. The two workers walk uphill, approximately 50 meters, anchor the winch to a stump and pull up the skyline and the haul-back line. They continue in a similar way uphill, step by step, to the top. It is important to make sure that there is no twist between the haul-back line and the skyline. One of the two workers places the two lines on oposite sides of stumps, brush etc. when moving uphill.

When the chainsaw winch has reached the top of the hillside, the skyline and the haul-back line lies parallel with each other along the entire length of the cable road.

The chainsaw winch can now be used to pull the cable crane winch from the valley and uphill to the top of the hillside.

First, the small yarder is attached to the skyline. The skyline is anchored in both ends, and tightened until the yarder hangs clear off the ground on the skyline by means of two blocks. Since the haul-back line has been pulled out from the drum on the yarder to the top of the hillside, it can be used to pull the yarder uphill along the skyline.

The chainsaw winch is the power-source for the uphill transport of the yarder. As soon as the 2-man crew has anchored the chainsaw winch, they pull the pulling line downhill approximately 50 meters (the length of the pulling line). At this point it is attached to the haul-back line by means of a wedge grip and the chainsaw winch pulls in the pulling line. The result is that the yarder is pulled 50 meters uphill along the skyline. The haul-back line is then anchored to a stump or a tree at the top of the hillside by means of a strap with a wedge grip. The pulling line from the chainsaw winch can now be pulled 50 meters downhill again by the

Figure 436. The chainsaw winch used for rigging up a top mounted cable crane, Heggelia, Norway 1957. A., B. and C. The skyline and the haul-back line are being pulled uphill by the chainsaw winch in 50 m steps (the helper must walk outside and not between the lines).
D. and E. The chainsaw winch pulls the yarder uphill on the slack skyline in 50 m steps.

crew, and attached by the wedge grip to the haul-back line. The chainsaw winch can now pull the haul-back line with the yarder a new 50 meter distance uphill along the skyline.

The crew continues to pull the yarder, one 50 meter step after the other, until the yarder has reached the top of the hillside. The yarder can then be rigged up at the top end of the cable crane.

Another rig up winch is the Austrian Ackja winch (figure 435):

Weight		0.75	kN
Power		4.8	kW
Average pulling force		8	kN
Average speed		0.8	m/s
Main line	L =	100	m
	d =	6.5	mm

The motor construction allows the winch to operate in any position. It moves very well over terrain obstacles. It can therefore be used to pull out the straw line from the yarder 100 meters at the time. It is also a useful tool at the tail end of a cable crane for moving the tail block and the tail anchor during road changes. It is used with medium sized and smaller yarders. It may, however, be necessary to use a block in order to double the pulling forces. These types of rig up winches would be too light for the heavy equipment used with large yarders in virgin forests. In this case the rig up winch may be used to pull out the straw line only.

Figure 437. The small Vossa yarder with its winch trailer pulls itself forward in the terrain by means of the main line.

Figure 438. Correct installation of the Vossa yarder. The log keeps the drums horizontal. The rear anchoring is in the centre of the drums. The two side guys are adjustable to place the yarder correctly in relation to the guide block in the spar tree.

Small yarders are sometimes used for thinnings or for final cut with tree dimensions less than 1 cu. m. per tree (figure 437 and 438). The yarder and trailer are pulled by the yarder itself step by step to the yarding position. The yarder is light and must be well anchored. It should be placed with the front end horizontally on a log. The winch is anchored by a strap to a stump or a tree. The yarder's construction allows the tension to go in a straight line from the strap through the centre of the main drum and out along the main line. The small yarder is kept in position by two side guys. A tension rod on each of the guylines makes it possible to adjust the sideways position in order to obtain good spooling on the winch drums. The thinnest line, the haul-back line, can be used as a straw line during the rig up work.

12.2.1.6. The straw line method

Sometimes the straw line is used to tighten the standing skyline by using a combination of an

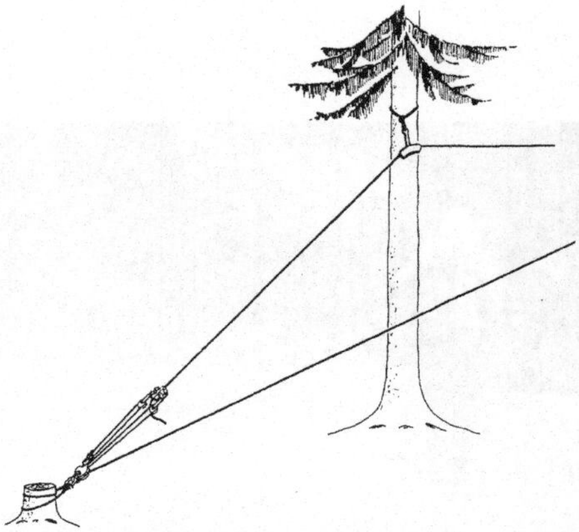

Figure 439. Tightening the skyline by means of eccentric grip, tackle and tightening line.

eccentric grip and a tackle (figure 439). It is necessary to make sure that the distance between the anchoring stump and the spar tree is long enough to be able to tighten the skyline without having to change the grip. The eccentric grip should be fixed to the loose skyline as near the spar tree as possible. Sometimes the yarder has a special winch drum which is used when tightening the skyline. When the skyline is tight the end of the wire rope in the tackle is fastened, for example by means of a wedge grip. This is sufficient to keep the skyline in position during the yarding. It is recommended to anchor the loose end of the skyline to another anchoring point as a safety precaution.

It is better to use two straw lines operated by two drums, 1 main straw line drum and 1 haul-back straw line drum. The two straw lines enclose the yarding area and are connected to each other with a three-way swivel. The third connecting point on the swivel is used to connect whatever line or equipment that has to be pulled or moved. The Nestestog radio controlled cable crane uses an endless straw line pulled by a groove wheel pulley. The straw line arrangement may also be used to rig up a support (figure 441). The pulling line from the 3-way swivel can be fixed to the crossline, by means of a wedge grip. Since the groove wheel pulley on the yarder is radio controlled, the worker can control the yarder which pulls the 3-way swivel uphill and tightens the crossline in the crossline support. A similar working method may be used to erect an A-framed support as illustrated in figure 442. This system using main and haul-back straw lines and a radio controlled yarder has been successfully used in Norway for 20 years. The system does not need a radio controlled yarder. The 3-way swivel may be moved in a similar way by a double-drum straw line winch on the yarder without radio control, but the yarder operator then has to follow the instructions from the terrain crew.

Moving from one road to the next in steep terrain can be a heavy work, and a good method for using the straw line is necessary. There should always be a double set of anchoring tackles, and the foreman transports the tackle to the new yarding road by means of the straw line. This can be done while the yarding crew undertakes the yarding on the previous road. As soon as the yarding is finished along one skyline corridor the tail block is loosened from the anchoring tackle and transported to the next road by means of the straw line. The straw line lead block should be placed behind the anchoring tackle in order to be able to work with slack lines when the tail block is fastened to the tail tackle.

Figure 440. A. and D. Rolling out the rig up line from the back-pack spools.
B. Powered drums for rig up lines, straw line, and skylines on the trailer.
C. Eccentric grip and tackle for tightening the skyline by means of the tightening line. (Telemark, Norway 1968).

Since the straw line drums are mounted on the yarder the operator has to control both the operating drums and the straw line drums simultaneously if part of the road change work is done simultaneous with the yarding. It may therefore be an advantage to use a separate rig up winch when the foreman does the preparatory work for the road change (figure 435).

Figure 441. The 3-way swivel between the main and haul-back lines tightens the crossline in a cross-line support.

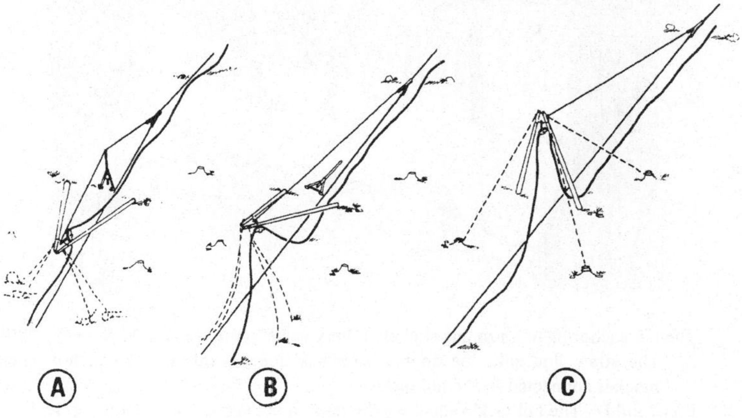

Figure 442. Rigging up an A-framed support by means of radio signals to the yarder and a 3-way swivel between the main and haul-back lines.

A. and B. The straw line from the 3-way swivel lifts the support .

C. The erected support is being anchored by means of guylines.

Figure 443. The tail anchoring of a running skyline. (Moxy cable crane, Telemark, Norway 1980).
A. The straw line pulls the double haul-back line towards the tail anchor where the tail block is connected to the tail tackle.
B., C. and D. The tail tackle equalizes the tension between 3 or 4 anchoring trees.

In some of the cable systems the main and haul-back lines run along the same corridor. During the road change the two parallel lines may become twisted, and it is heavy work to take out the twist after the tail block has been anchored and there is tension in the lines. *A tail block swivel* makes this work easier (figure 445). This swivel has a bolt which can lock the

Figure 444. Moving the tail block of a running skyline during road change. (Moxy cable crane, Telemark, Norway 1980).

A. The straw line or the rig up winch pulls the double haul-back line into the new position.
B. The tail block is connected to the tail anchor (sometimes an anchoring line and sometimes directly to the anchoring tackle).
C. The tail block connected to the anchoring rope which is lifted by the tail tree.

swivel in four positions. When the cable crane has been rigged up in a new corridor the yarder operator moves the carriage slowly outwards. When it approaches the tail end, the twist between the lines is easy to see. The foreman removes the bolt in the swivel and the tail block can easily be turned until the twist has been removed. He then locks the swivel in order to prevent the tail block from turning.

Figure 445. A tail block swivel (for a light running skyline). The swivel can be fixed in one of four positions by means of a steel pin. (Hurdal, Norway 1980).

Figure 446. A universal anchoring strap for light cable cranes (Hordaland Mekaniske Verkstad). The strap may be used to hang up blocks, for anchoring a light yarder, or as a cable shoe for crosslines in supports, etc. (Ås, Norway 1956).

12.2.1.7. The spar tree

Modern yarders are usually equipped with a steel spar and it is usually unnecessary to rig up a spar tree. It still happens, however, that a spar tree is needed and has to be rigged up. Before cutting the area a spar tree with a suitable location has to be selected and the location of the landing depends much on where a good and stable spar tree can be found. The light and medium sized cable cranes need a spar tree which gives the skyline sufficient height in order to transport the load off the ground. It is not usually necessary to top the trees and the anchoring of the skyline may serve as one of the guylines (figure 102). The size of the spar tree should as a rule be as big as the biggest trees to be cut during the logging operation.

Sometimes climbing shoes were used to rig up the spar trees (figure 103). It is, however, better to use a rigging ladder (figure 104), which is a ladder with 3 meter sections fastened to the tree by small chains.

When yarding virgin timber with tree dimensions up to 20 cu. m. per tree and average tree dimensions for example between 6 and 9 cu. m. per tree the choise and the rigging of a spar tree are an important part of the rig up work. A good spar tree should be selected and marked before the cutting. The preparation and rigging of a spar tree can be done before the yarder and the yarding crew arrives on the area. A rigging tractor is needed in order to pull out the guylines etc.

When a good spar tree has been selected, the cutters fell the logs on the landing and around the guyline circle. The high climber walks up the tree and cuts the branches. He tops the tree at 34 meters height where the top diameter must be at least 40 cm. As soon as he has cut off the top, he fastens a pass block by a strap immediately beneath the top. Then he descends approximately 5 meters and remove the bark from a 1 meter high strip around the tree. This is for the tree plates. The tree plates are being lifted up to the high climber by means of the pass line which is a fibre rope with the pass chain at the end. The pass chain is used as a mean of attaching the pass line to the various pieces which must be lifted to the top of the tree. It also serves as a seat for the climber. When the tree plates have been nailed to the tree with railroad spikes, each of the guylines are lifted up and wrapped around the tree. Each guyline is fastened by means of an eye splice and and a shackle (figure 447).

The first guyline is wrapped clock-wise around the tree, the second is wrapped counter-clock-wise etc. If all the guylines were wrapped the same way they would twist the tree. By alternating the directions twisting is avoided. If the spar tree shall be used for highlead yarding, 6 guylines are sufficient. When the spar tree is used with live or tight skyline systems 8 guylines have to be used.

After the top guylines have been fastened to the tree the high climber fastens the buckle guys to the tree, approximately 8 meters below the top. Three buckle guylines are sufficient with highlead yarding, but there has to be five buckle guys with live or tight skylines.

The anchoring points should be evenly distributed around the spar tree at a distance of at least 60 meters from the tree. If six guylines should be used a watch may be used to find the location of the best positioned anchoring stumps. They are chosen so that they correspond to 2-4-6 o'clock etc. on the watch. The stumps must be fresh and solid and have a dimension similar to the maximum size of the trees to be yarded. The rigging tractor (usually a crawler tractor) is used to tighten the guylines. The tail of each guyline is pulled out to the stump by the straw line and the line is wrapped around the stump. Before doing so a notch around the stump has to be made by an axe or a chainsaw. When the line is tightened it is fastened with at least 8 railroad spikes to the stump (figure 447). The next wrap of the guyline is placed around the stump and fastened with 2 or 3 spikes. The third and final wrap is fastened with 6 spikes or more. Before each of the wraps is fastened they are tightened by the rigging tractor. The spikes keep the guyline wraps tight to the stump. It is, however, the friction between the guyline wraps and the stump which takes the main part of the tension in the guyline.

When all the guylines are tightened the high climber can contiue to rig the tree. He first hangs up the bull block (the main line lead block). The block and the strap are sent up to the tree using the pass line, the two ends of the bull block strap have a steel eye and are attached to the block with a shackle. The bull block can weigh between 0.2 and 0.7 kN. A second strap is therefore hung between the block and one of the top guys as a safety precaution.

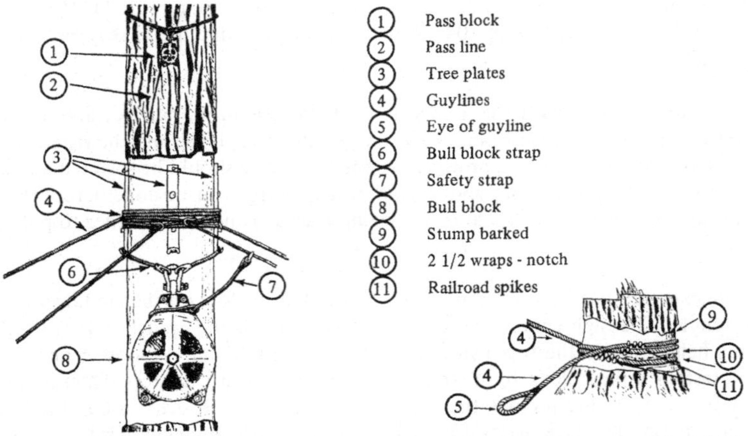

Figure 447. The arrangement on a large spar tree for cable yarding of big timber (CONWAY 1978).
Left: Top section of the spar tree.
Right: A stump for anchoring one of the 6–8 guylines.

The haul-back block is somewhat lighter. It is mounted 1 – 1.5 meters below the bull block.

When both blocks are placed, a section of the straw line is threaded through each of the blocks and will later be used to pull the main and the haul-back lines through the blocks.

Rigging up a spar tree is a difficult, heavy and time consuming work. It has been replaced by the modern yarders with steel spars. The heavy work is done on the ground. The steel spar is erected by hydraulic cylinders and supported on the ground by hydraulic feet. Before erecting the steel spar the foreman must inspect the lead blocks and the other equipment on the top of the spar. The spar is then erected with all the guylines and the operating lines already mounted. Afterwards the guylines have to be distributed around the steel spar in a similar way as with the spar tree. A 35 meters high steel spar needs 6 guylines while a 15 meters spar may have 2 or 3 guylines and a 6 to 9 meters high spar needs 1 to 2 guylines.

12.3. Pre-rigging

One has to admire our ancestors, these giants of work: the lumber jacks, the teamsters. Some of these mountain forest workers used their hands, their own strength and their axes to send down valuable timber from the most difficult mountain terrain through dry shoots to the valley. Some of them took out the biggest trees from the difficult terrain to the railroad or the sawmill with their donkey yarders. These gentlemen of work build our nations. You asked them, the hooktender, the superintendent, whose education had been hard practical work: how do you manage? It's just to go to the job and do it!

Much water has flowed down the rivers since that time. Appropriate tools, sophisticated yarders, better working techniques, more advanced operational methods and organization of the work have led to reduced stress on the worker, better accident prevention and more efficient systems. We have learned to specialize the work within a working crew, each member has a better knowledge about his own part of the total work sequence. We have learned to plan the work by means of modern planning methods and as a consequence we may divide the jobs between the men, not limited by the character of the job only, but even as far as the period of time is concerned.

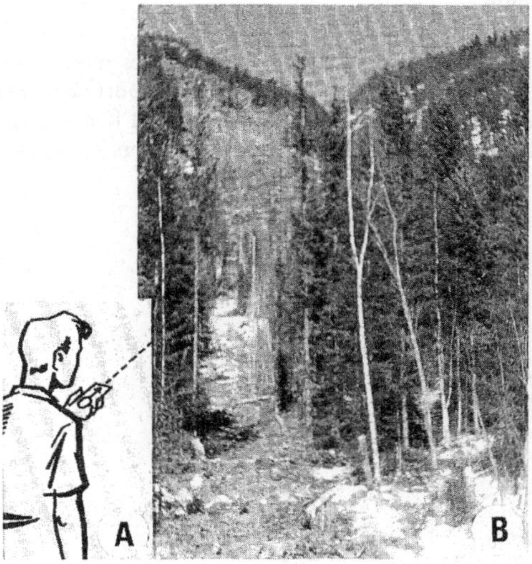

Figure 448. The compass is of good help when alingning the corridors for the skyline or the haul-back line.

While some of the crew members take care of the yarding, others do part of the rigging or the road change. It is not necessary for the yarder and the yarding crew to arrive at the logging area while the preparatory work is being carried out. This may lead to down time such as waiting time for the crew and the expensive machinery. It is better to have the yarding area, the landing, the anchoring points etc. prepared before the yarder and the yarder crew arrives. When the plan of operation has been made the roads have been built and the landing site has been prepared the foreman gets the map where the cable crane roads and the profiles have been marked. He may need a worker with him to cut the skyline corridor, and if necessary the corridor for the haul-back line. If brush or standing trees blocks the view he may use a compass to take out the direction of the corridor (figure 448). The guyline anchors may be choosen and the guyline straps mounted on the anchors. These may be big stumps. In other cases they may be »dead man anchors» or bolts in rocks.The same is the case with the tail anchors including the tail tackle. The foreman and if necessary, his helpers, may also roll out or string out the rig up lines before the yarder and the yarding crew arrive. If a spar tree is needed it is sometimes an advantage to rig it up beforehand.

Such pre-rigging makes the work efficient. When the yarder and the crew arrives, the rig up time is short before the yarding starts.

The yarding crew have become specialists on operating the yarder efficiently. When they have started their part of the operation, the foreman can continue his preparatory work. Pre-rigging is an important means to increase the work efficiency in cable crane yarding, especially for large and medium sized cable cranes operated by large yarding crew.

The pre-rigging for light cable cranes with 2 or 3-man crew is different.

If the plan is to cut the timber on the logging area before yarding, the responsible person in the yarding crew makes an inspection before the cutting starts. He has received a map or aerial photograph from the forest administration where the logging area as well as the location of the cable crane is marked. He then selects the spar trees, tail trees, support trees etc. as well as the anchoring points.

Many times, however, especially with the light cable cranes mounted on farm-tractors the 2-man crew carries out the felling and yarding in one transport sequence. The leader selects anchoring points for the steel spar as well as tail anchors and if necessary tail trees while the other man cleans the skyline corridor. This is a 3 to 5 meters opening in the stand from the landing to the tail anchor.

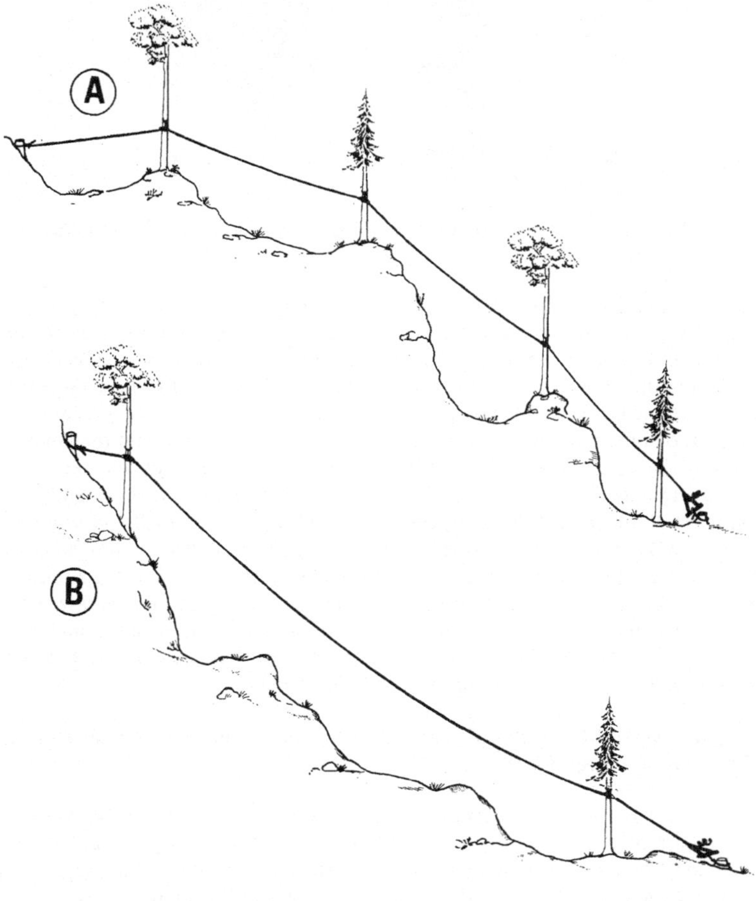

Figure 449. A. Convex profile leads to many supports.
B. Concave profile simplifies the rig up work and the yarding operation.

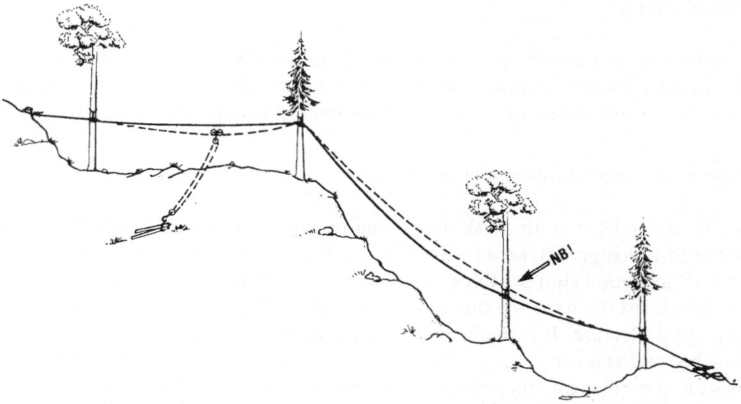

Figure 450. Wrongly placed support. The lateral skidding in upper span increases the skyline tension and can lift the cable shoe in the lower span.

12.4. The rig up work

The qualifications needed for rigging up a cable crane are different from the qualifications needed for keeping a high yarding production. During yarding the crew consentrate on getting big loads in the shortest possible turn time and smallest possible down-time. The work of rigging up a cable crane requires knowledge and experience in finding good locations for the cables, the anchoring points, the landings etc. in relation to the area and the timber to be yarded. The ability to find a good layout and work strategy is important.

The Norwegian Forest Research Institute has carried out cable crane yarding with several crews per year and many different cable crane systems throughout a 30 year period. These experiments have led to some general plans of operation for the rig up work. The resulting rig up methods have also been tried out in practical operations in other places. The results have influenced the following chapter and the methods used in Northern Europe are used as examples of different types of rig up methods. Examples from other forest types and cable systems used elsewhere are based on experiences from study trips and reports from other countries.

The rig up work has been divided into rig up steps. Within each of the steps the job to be undertaken by each man in the crew has been described.

The selection of cable systems has been made in order to show different rigging methods which have been tried out.The rigging steps have been described under normal or usual conditions and with a certain number of men in the crew for each system. The description therefore gives the main course of action during the rig up work only, which is useful under many conditions in normal terrain. Differences in the terrain conditions or changes in the number of crew members may make it necessary to adjust the rig up steps accordingly.

12.4.1. Light cable cranes

These are cable cranes powered by one or two drum winches mounted on runners (skids) or light cable cranes powered by double-drum winches mounted on the 3-point linkage of farm-tractors. The light cable cranes are useful for tree dimensions up to 1 cu.m. per tree.

12.4.1.1. Light cable cranes for ground lead yarding

Normally the 10 – 12 mm thick skyline is mounted 5–10 meters above the ground. The spacing between the supports varies with the terrain conditions and can be 50 to 150 meters. It is an art to select good support trees. If possible, the skyline should be located in a concave terrain which reduces the need for supports (figure 449). The light equipment has to be rigged up with a good deflection. It is therefore necessary to ensure that the location of a support is not too low because this can cause problems during lateral skidding (figure 450). The support is often placed at a hill crest where the angle of the slope changes. For downhill transport the support tree should be located below the crest in order to avoid that the load gets stuck when passing the support (figure 451).

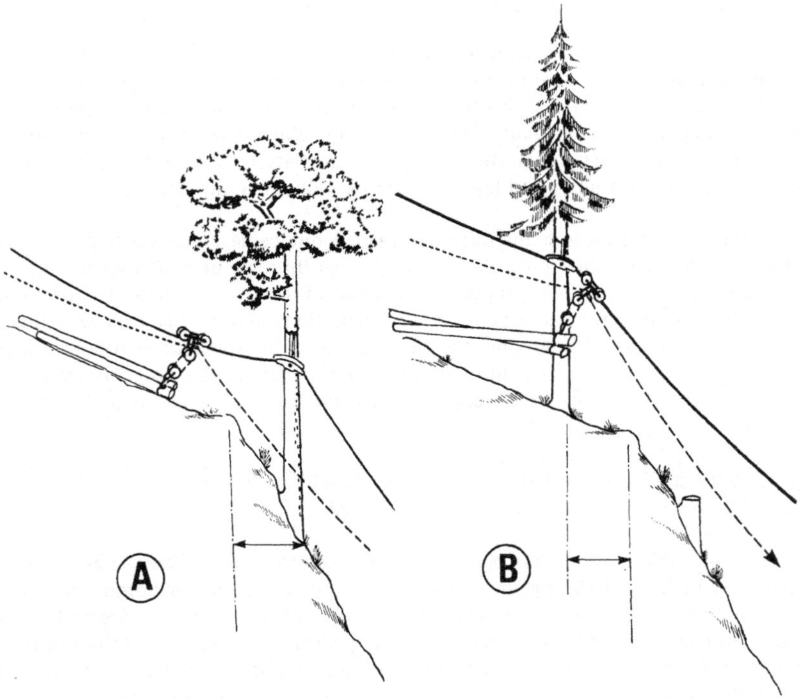

Figure 451. The location of a support for downhill yarding.
A. A good location.
B. Wrongly placed support. The load may get stuck in the terrain.

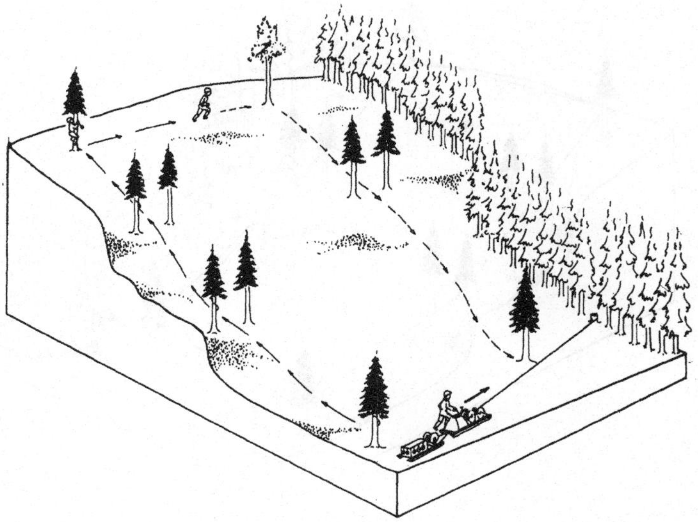

Figure 452. Step I.
The yarder operator pulls the yarder with the winch trailer towards the winch station by means of the haul-back line (figure 437). The choker setter and the chaser carry the blocks around the skyline and haul-back line corridors and fasten them to suitable trees. The cross-line with the cable shoe and the tightening device (come-along) is placed at the support trees.

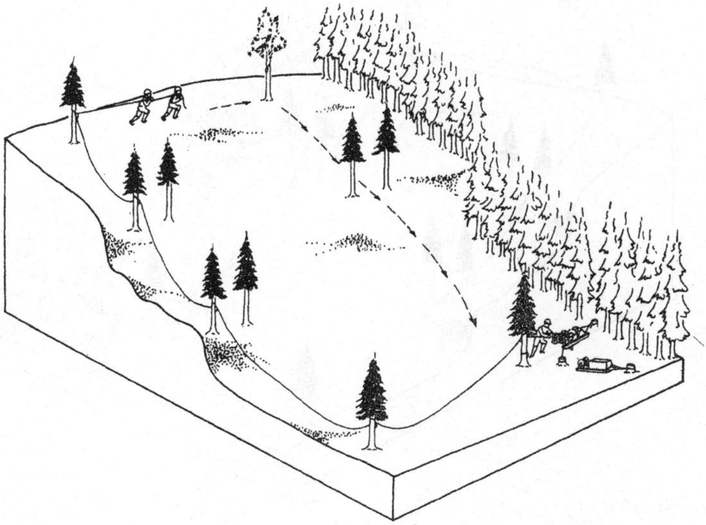

Figure 453. Step II.
The choker setter and the chaser pull out the 6 mm haul-back line along the haul-back corridor and the skyline corridor. Underway they position the haul-back line in the terrain blocks. The yarder operator riggs up the yarder. He also checks that the haul-back line runs freely out from the haul-back drum.

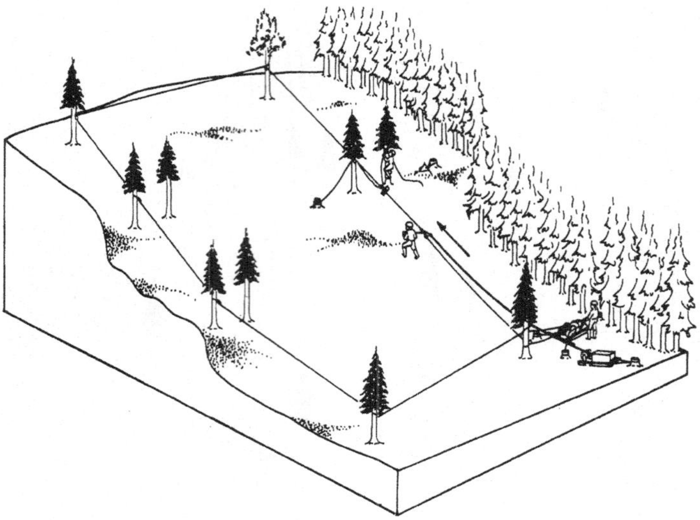

Figure 454. Step III.
The haul-back line is connected to the main line and the skyline by means of a 3-way swivel. The main line and the skyline are pulled out to the tail end of the skyline corridor by means of the haul-back line. The yarder operator controls the winch and the chaser follows the 3-way swivel and ensures that there is no twist between the main line and the skyline. The choker setter starts to rig up the crossline support.

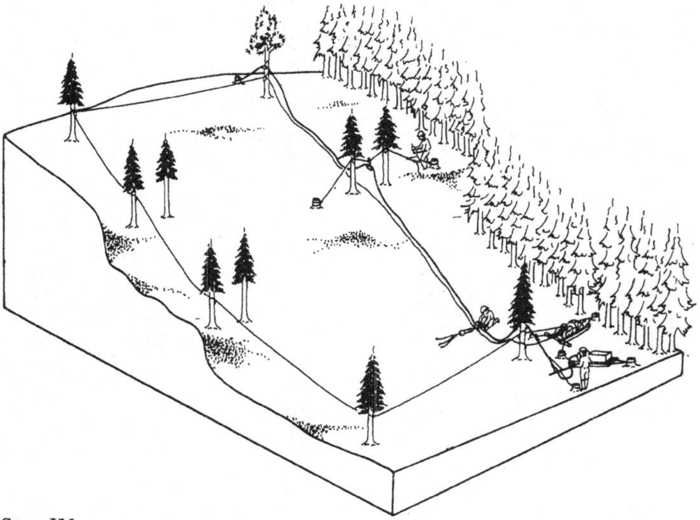

Figure 455. Step IV.
The chaser walks down the skyline corridor and checks the haul-back line and the skyline. They should not be tangled. He then places the carriage on the skyline and fastens the main line and the haul-back line to the carriage. The choker setter attaches the tail end of the skyline to the tail anchoring. He places the skyline in the cable shoe at the tail tree. The same is done at the support. Afterwards he tightens the crossline in the support. The yarder operator can start to tighten the skyline

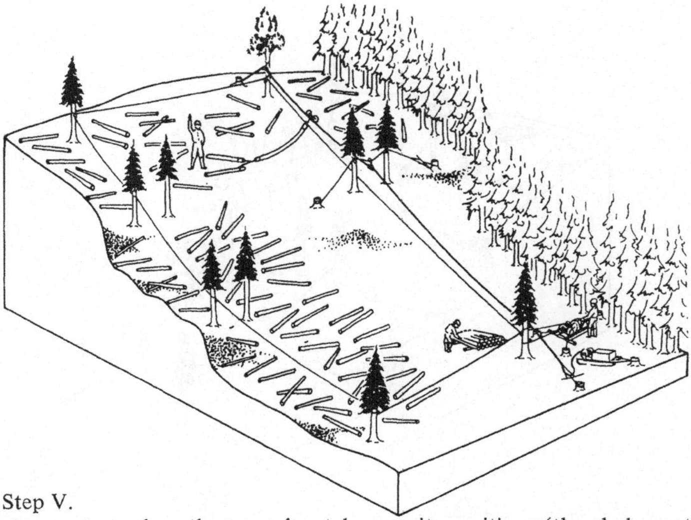

Figure 456. Step V.
The transport can start when the crew has taken up its positions (the choker setter in the terrain, the chaser at the landing and the yarder operator on the winch). After a few loads the terrain blocks, the cable shoes etc. are inspected and the skyline tension adjusted, whereafter the transport can continue. When the area around the skyline corridor has been yarded, the skyline is pulled around and down the haul-back corridor. The rest of the rigging follows the same procedure as above.

The described rig up method can be used for dowhill or horizontal yarding, and for uphill yarding when the slope is less than 20 %. If the slope is steeper than 20 % the carriage can return to the terrain by gravity. No haul-back line is needed. Instead a short log is fastened to the skyline at the point where lateral skidding shall take place. When the carriage returns and hits the log the falling block comes down, and it is easy to pull out the main line to the logs. For a 2-man crew the rig up work may be carried out as follows:

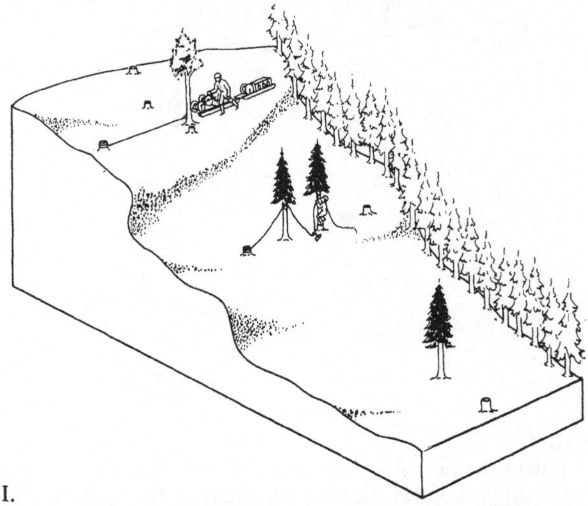

Figure 457. Step I.
The yarder operator pulls the yarder with the winch trailer to the yarder station by means of the haul-back line. The choker setter riggs up the support.

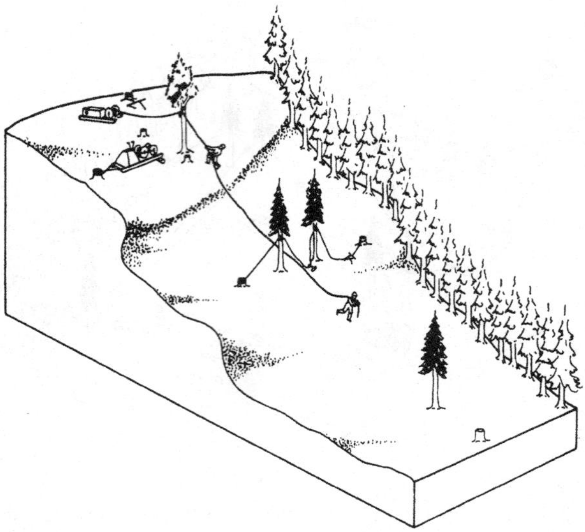

Figure 458. Step II.
The choker setter pulls the skyline downhill and the yarder operator helps him with the work. Two men is usually sufficient since they work with the gravitiational forces. The choker setter hangs the skyline in a cable shoe in the tail tree and fastens the end of the skyline to the tail anchor (usually 1 or 2 rigid fresh stumps). The yarder operator starts to rig the yarder.

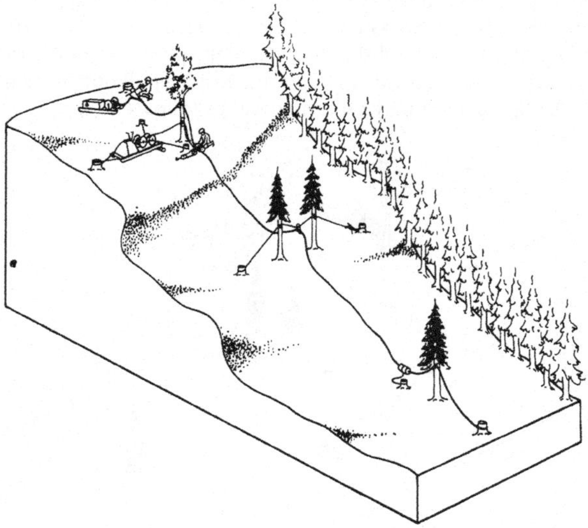

Figure 459. Step III.
The choker setter fastens the carriage stop (a log) to the skyline. He then places the skyline in the cable shoe at the support and tightens the crossline in the support. He walks up to the winch station and positions the carriage on the skyline. In the meantime the yarder operator finishes the rigging of the yarder and he then tightens the skyline.

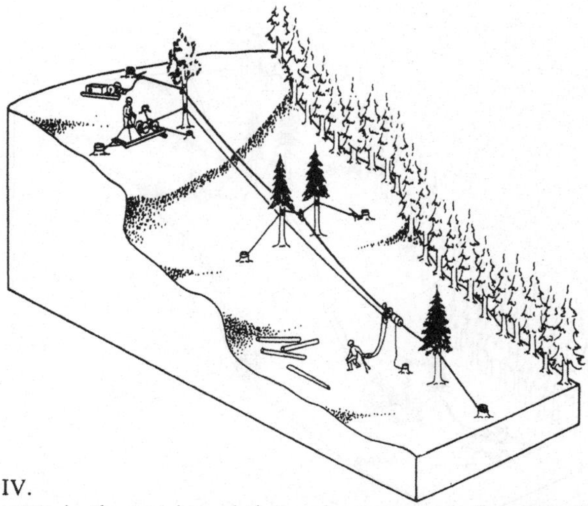

Figure 460. Step IV.
With the choker setter in the terrain and the yarder operator at the yarder the transport can start. The yarder operator unhooks the chokers at the landing during the yarding operation. After a few loads the crew inspects the anchors, the supports etc. before the yarding continues.

12.4.1.2. Light cable cranes for downhill yarding with top mounted winch

As an example a cable crane with a 16 mm thick skyline and 8–10 mm thick haul-back line is described. The single-drum winch powered by an 18 kW motor is mounted at the top end of the logging area. The maximum length of the skyline is 1,200 meters but is usually 500–700 meters. The cable crane is operated by a 3-man crew who also carries out the rig up work.

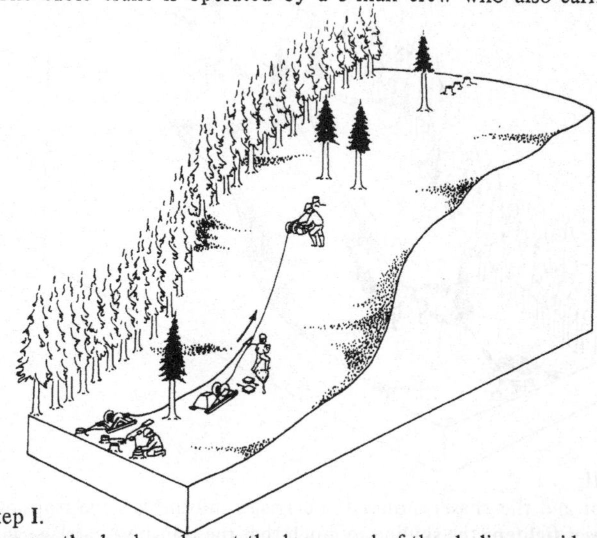

Figure 461. Step I.
The chaser riggs up the heel anchor at the lower end of the skyline corridor. The yarder operator uses a chainsaw winch to pull a swivel with the skyline and the haul-back line uphill. The choker setter follows the swivel in order to avoid that the haul-back line and the skyline gets twisted. The uphill pulling is done in steps, 50 meters at the time.

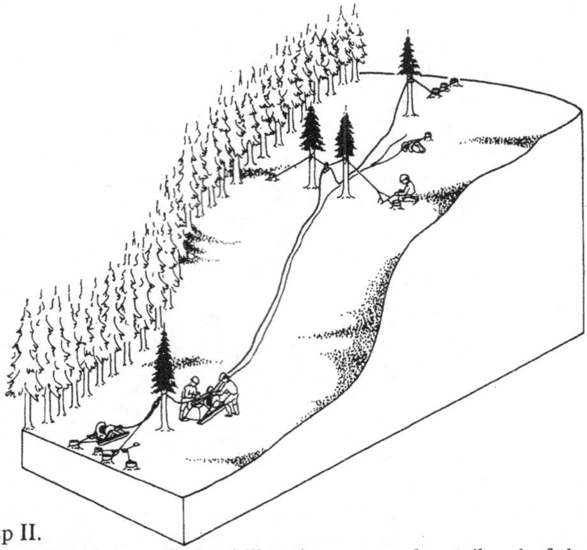

Figure 462. Step II.
When the two lines have been pulled uphill to the upper end or tail end of the skyline corridor, the yarder operator and the choker setter fasten the tail end of the skyline to the tail anchoring. The yarder operator walks downhill while he is inspecting that the haul-back line and the skyline are not twisted and that no brush will be in the way when moving or lifting any of the lines. The choker setter riggs up first the tail tree and then the support, whereafter he tightens the crossline. In the meantime the yarder operator and the chaser fasten the yarder to a transport carriage on the skyline.

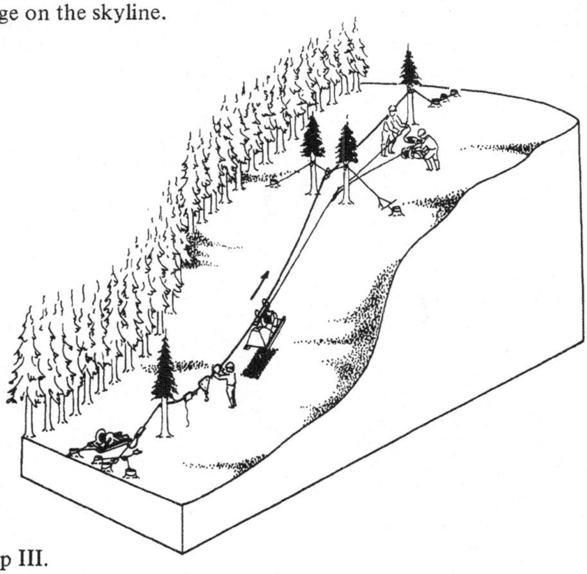

Figure 463. Step III.
The yarder operator and the chaser mount the carriage and the carriage stops on the skyline. Afterwards the chaser tightens the skyline so much that the transport carriage with the yarder is hanging low but clear off the ground. The yarder operator walks uphill. Together with the choker setter he pulls the yarder uphill step by step (50 meters at a time), using the chainsaw winch.

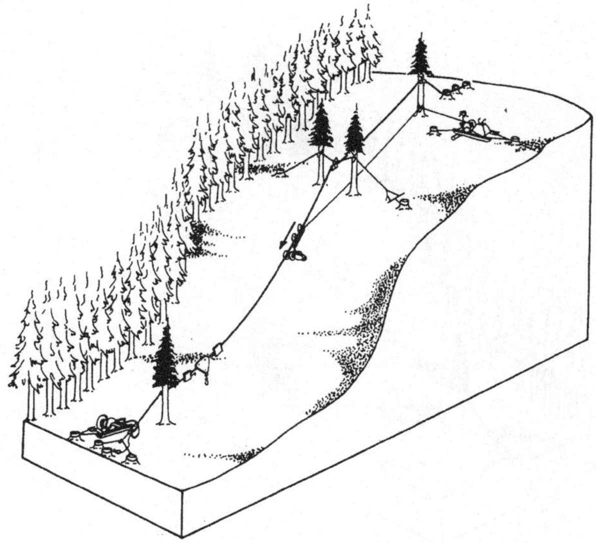

Figure 464. Step IV.
The yarder operator and the choker setter rig up the yarder at the top end of the cable crane. Afterwards they send the transport carriage downhill by gravity. The chaser riggs up the landing.

Figure 465. Step V.
The chaser fastens the skyline to the heel anchor and tightens the skyline until it has the necessary pre-tension. The choker setter takes chokers, chainsaw etc. and moves to the point where the lateral skidding shall start. The yarder operator pulls the carriage and the upper carriage stop uphill by means of the haul-back line to the point where the lateral skidding shall start.

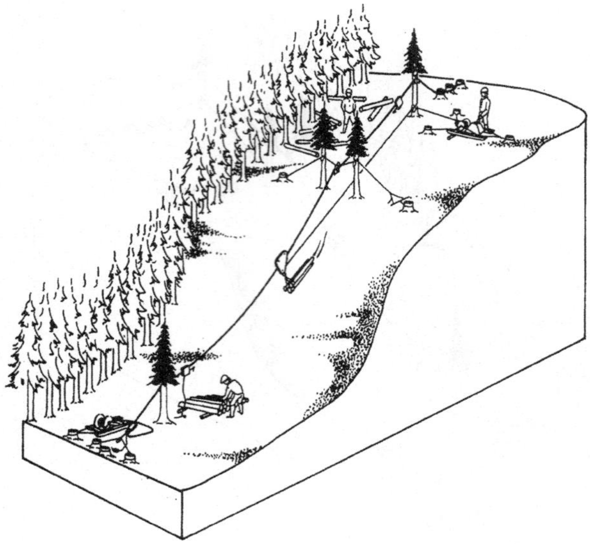

Figure 466. Step VI.
When a barrel of fuel has been transported to the top of the cable crane, the crew can start yarding with the yarder operator at the yarder, the choker setter in the terrain and the chaser at the landing. After the first day's yarding, the whole cable crane is inspected to check that all anchoring points, blocks etc. are safely fastened. With a top mounted cable crane the yarding starts from the top where there is good contact between the choker setter and the yarder operator. When the lateral skidding and yarding is finished at one lateral skidding location, the carriage stop is moved 15–20 meters downhill. To do this, the yarder operator pulls the carriage uphill until it hits the carriage stop. The choker setter loosens the carriage stop by means of a releasing rope. The yarder operator can now move the carriage with the carriage stop downhill to the new lateral skidding location. At this point the choker setter fastens the clamp in the carriage stop to the skyline by means of the releasing rope and the cable yarding can continue.

12.4.1.3. Light farm-tractor mounted cable cranes for downhill or uphill yarding

These are small yarders such as the Igland interlock yarder or the Owren Vario Winch mounted on a farm-tractor's 3-point linkage or on a skidder. The running skyline system is used and the diameter of the main line, haul-back line and sometimes the slack-pulling line is d = 10 mm. The maximum yarding distance is 200–300 meters. The cable crane is operated by a 2-man crew, one yarder operator who also unhooks the chokers and one choker setter. The steel tower is kept in position by guylines. When the back guy has been attached to a stump the tractor is moved ahead in order to tighten the back guy. On the Igland cable crane the side stability is kept by two side guys which are fastened by a double wedge grip (figure 93). Yarders of these types are illustrated in figure 373 and 393.

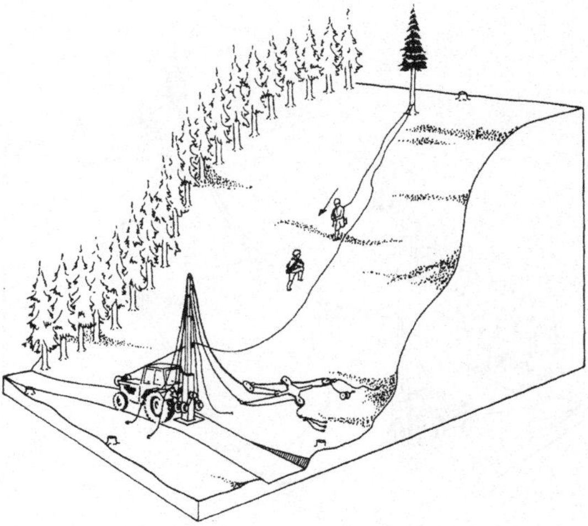

Figure 467. Step I.
The choker setter walks out with a 5 mm thick straw line, lays the bight in a pass block fastened to the tail tree, and starts walking downhill again. The yarder operator clarifies the main and haul-back lines as well as the tail block. He then walks uphill to meet the choker setter bringing with him a guyline. The choker setter takes over the guyline and walks back to the tail end of the cable crane while the yarder operator brings the straw line downwards to the yarder.

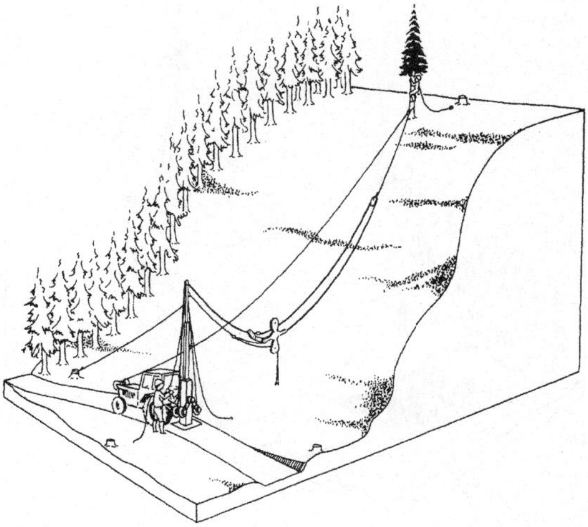

Figure 468. Step II.
The yarder operator controls the yarder and pulls the tail block towards the tail tree by means of the straw line. An anchoring strap and a tensioning device is sent with the tail block. In the meantime the choker setter riggs up the guyline in the tail tree.

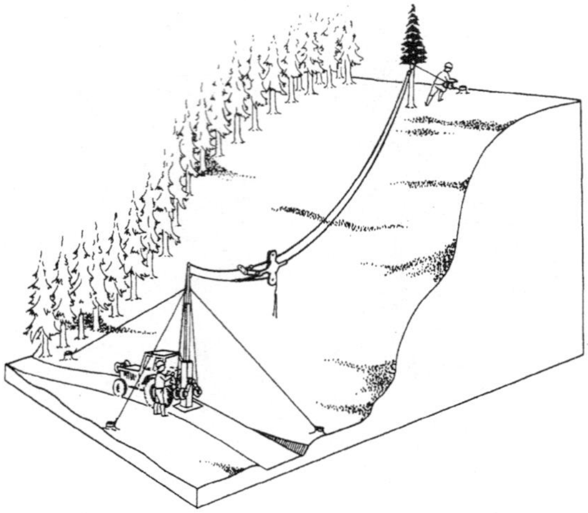

Figure 469. Step III.
The choker setter tightens the guyline by means of the tensioning device. The yarder operator secures the steel spar by means of the back guy and and the side guys (or the side feet) whereafter he hauls out the carriage towards the terrain.

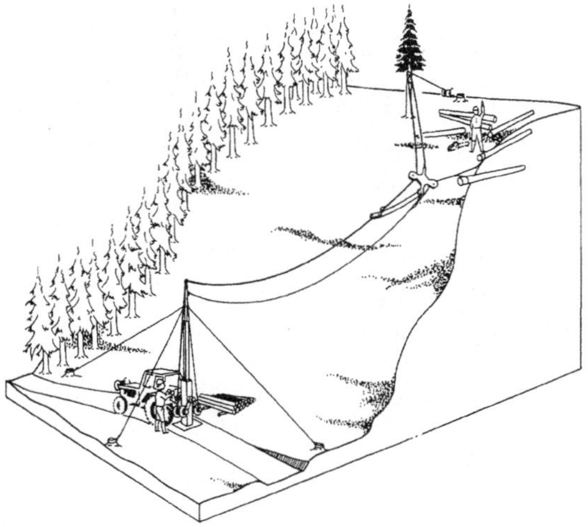

Figure 470. Step IV.
The yarding starts with the choker setter in the terrain and the yarder operator at the landing. When the visibility is good, arm signals may be used. If not, sound signals are preferred.

12.4.1.4. Light farm-tractor mounted cable cranes for uphill yarding and gravity outhaul

The same equipment as described in the previous chapter can be used for uphill yarding. The double haul-back line is used as skyline. When the terrain is steeper than 20 % the carriage is returned by gravity.

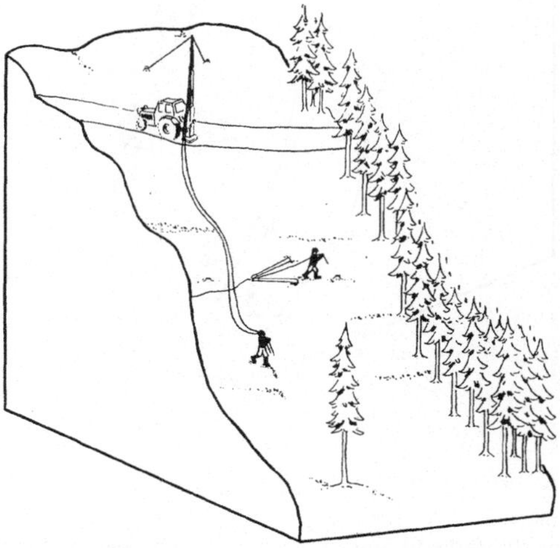

Figure 471. Step I.
The yarder operator riggs up the yarder. He then starts to rig up the support if needed. In the meantime the choker setter pulls the haul-back line and the 5 mm thick straw line downhill to the tail tree. He fastens the tail block to the tree, fastens the straw line to the haul-back line and places the bight in the block. If necessary, he stabilizes the tail tree with a guy-line.

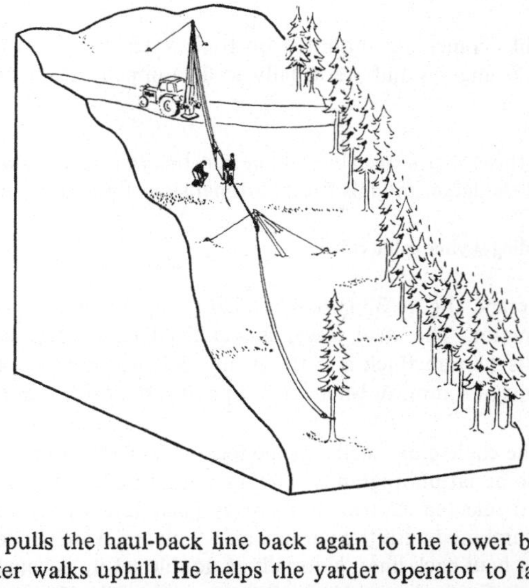

Figure 472. Step II.
The yarder operator pulls the haul-back line back again to the tower by means of the straw line. The choker setter walks uphill. He helps the yarder operator to fasten the head end of the haul-back line in the terrain. Afterwards they both put the carriage and the carriage stop on the skyline (the double haul-back line) and finish the rigging of the support if necessary.

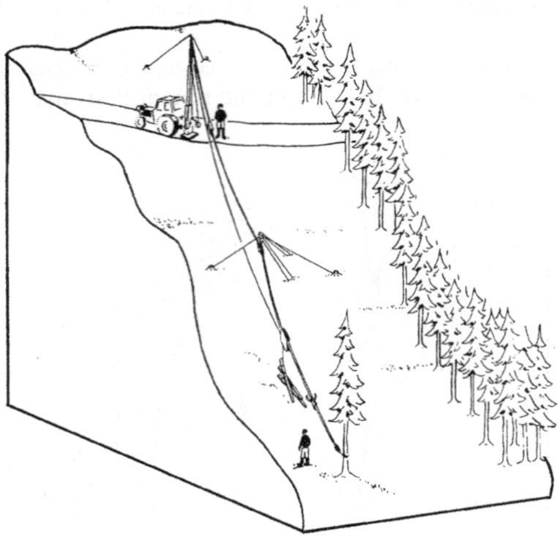

Figure 473. Step III.
The yarding starts from the tail end and moves uphill. The best is to yard a 20 meters wide belt on one side of the skyline. When the yarding on this area is finished the cable crane is moved parallel into the edge of the forest where it is rigged up by the two workers for felling and yarding a new 20 meters wide strip.

12.4.2. Medium sized cable cranes

The medium sized cable cranes are mounted on trailers, tractors or trucks. They are often powered by 60–110 kW engines and are usually well equipped with auxiliary equipment for rigging purposes.

The lighter models of these types of cable cranes can be used for logging tree dimensions up to 2 cu.m. per tree and the larger models for tree dimensions up to 3 cu.m. per tree.

12.4.2.1. Traditional standing skyline cable crane

The Igland Alp cable crane (figure 334) has a trailer mounted yarder, powered by the power take off on a farm-tractor. The steel tower is stabilized by a back guyline and two side guylines. The skyline is 18 mm thick and the main and the haul-back lines are 8 mm. The straw line has a diameter of 4 mm. A 3-man crew operates the cable crane.

Before the crew and the cable crane arrive at the logging area the cable crane setup has been planned and located on aerial photographs. The anchoring points, tail trees and if necessary, support trees have been selected. Cutting has already been done by separate wood cutters. In cases where felling is combined with the yarding a 3–5 meters wide corridor has been cleared for the skyline and the haul-back lines. Normally, a double set up is being used, which means that when the yarding is finished around the first corridor of the skyline, it is moved over to the haul-back line corridor. Such a layout saves time and work for the road change.

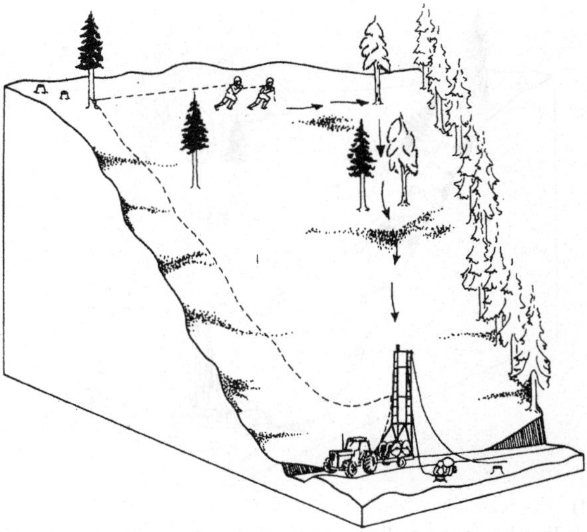

Figure 474. Step I.
The yarder operator pulls out the back guy and fastens it to a rigid anchoring point, usually a big stump. He tightens the guyline by hand and afterwards he moves the tractor a little until the back guy is tight. He then riggs up one or two side guylines. While he does this he also ensures that the 4 mm thick straw line is running freely out from the straw line drum. In steep terrain where the terrain is longer than 250 meters two men are needed to pull out the straw line. They hang up the terrain blocks (the corner blocks) and the tail block.

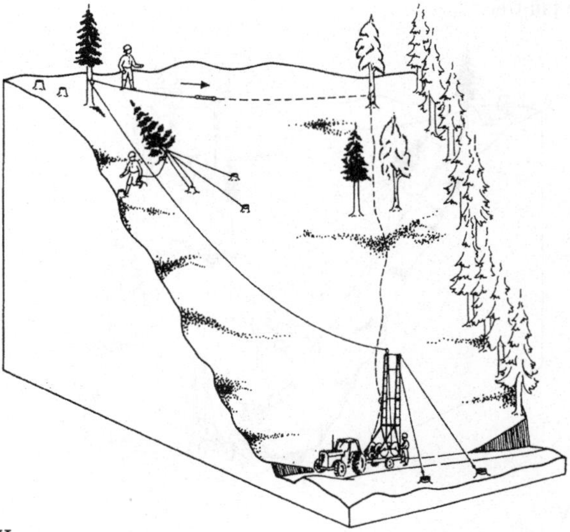

Figure 475. Step II.
The yarder operator pulls out the 8 mm thick haul-back line by means of the straw line. The two other men in the crew walk uphill and carry with them the equipment for the support. One of the workers checks that the connection between the straw line and the haul-back line passes smoothly through the blocks. The other starts to rig up supports.

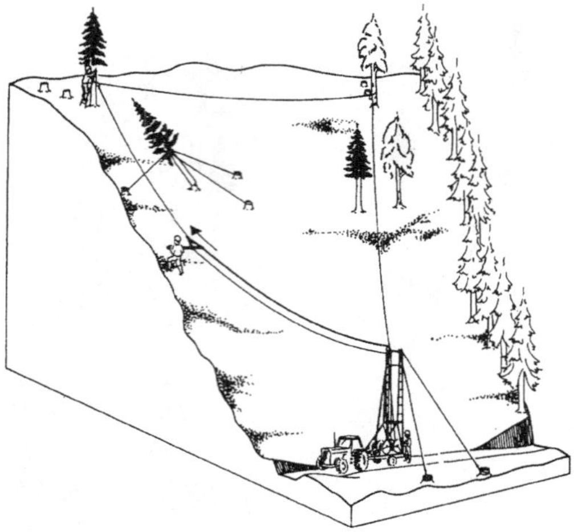

Figure 476. Step III.
When the haul-back line has been pulled out the head end is attached to a 3-way swivel. The main line is connected to the other end of the swivel. The 3-way swivel is a transport device for equipment needed in the terrain such as guylines and other equipment for supports and for the tail tree and the tail anchoring.When this transport is finished the 3-way swivel is moved back to the yarder and the skyline is fastened to it. The 3-way swivel has a certain width in order to keep the main line and the skyline away from each other in order to avoid twisting between them. The yarder operator controls the yarder while the skyline is being pulled up to the tail tree.One of the men follows the 3-way swivel during the uphill pull in order to undo any twist. The other man hangs up the cable shoe for the skyline and fastens the guylines to the tail tree.

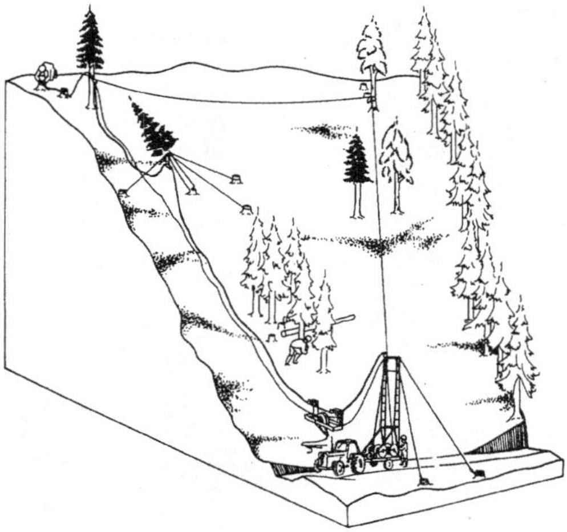

Figure 477. Step IV.
While the yarder operator places the skyline in the carriage and fastens the haul-back line and the main line to the drum in the carriage, one of the terrain workers prepares the tail anchoring for the skyline. The other terrain worker places the skyline in the cable shoe at the support. He then hoist the cable shoe up to the support and fastens the hoist line to a stump. When he has finished this work he starts the felling.

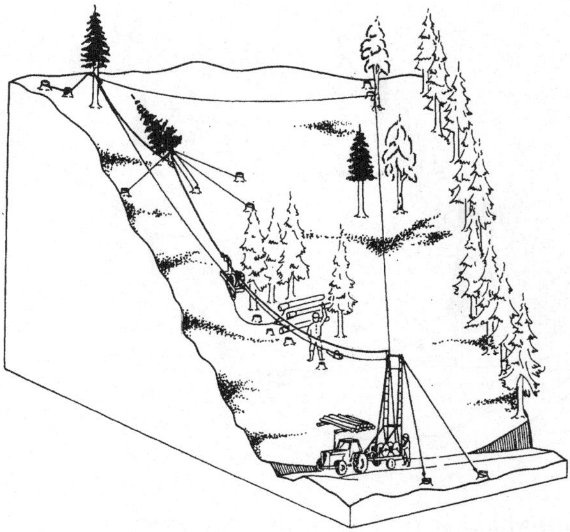

Figure 478. Step V.
The skyline is tightened by the skyline drum and then the carriage is hauled out past the supports to check that everything functions. The two terrain workers start the felling and cutting at the lower end of the logging area. The yarding and transport can start. The two cutters fell and limb the trees into logs in the terrain and set the chokers. The yarder operator skids the loads laterally to the skyline and transports them downhill to the landing. He unhooks the chokers and returns them to the terrain. After the first days yarding the whole cable crane is inspected to see that everything is right.

12.4.2.2. Radio controlled standing skyline cable cranes

The Nestestog radio controlled cable crane has a tractor mounted or a truck mounted yarder. In the case of the tractor mounted yarder all the storage drums and other auxiliary equipment are placed on a trailer. They are placed on the truck if the yarder is truck mounted.

The felling is usually combined with the yarding. A two or three men crew operates the yarder, one or two men in the terrain, and one at the landing. The latter one, combines operating the yarder with operating the delimbing machine.

The endless main/haul-back line is 10–12 mm thick and the skyline 20 mm. The endless line serves as a straw line during the rig up work. In addition there are two rig up lines, one which is 2.1 mm thick and the other 4 mm thick. We describe the rig up method for a double set up similar to the set up described in chapter 12.4.2.1. With a modern radio controlled cable crane, even the carriage is radio controlled and the haul-back line runs along the same corridor as the skyline. In this case the rig up method is similar to the rig up work for a running skyline which is described in chapter 12.4.2.3.

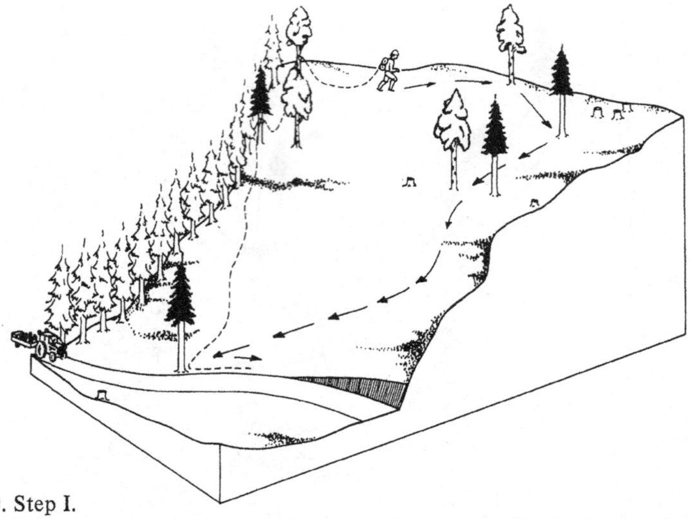

Figure 479. Step I.
The pre-rigging starts with clearing a 3–5 meters wide corridor for the haul-back line and the skyline. This is carried out by the foreman, and sometimes he is assisted by one of the cable crew members. The foreman selects the spar tree, the support trees and the tail trees and he selects the anchoring points and fastens the anchoring straps to the stumps. They fasten corner blocks and tail blocks on suitable trees and the foreman rolls out the rig up line I. This is the only line which has to be transported manually. When the yarder arrives at the site, the preparatory work should be practically finished.

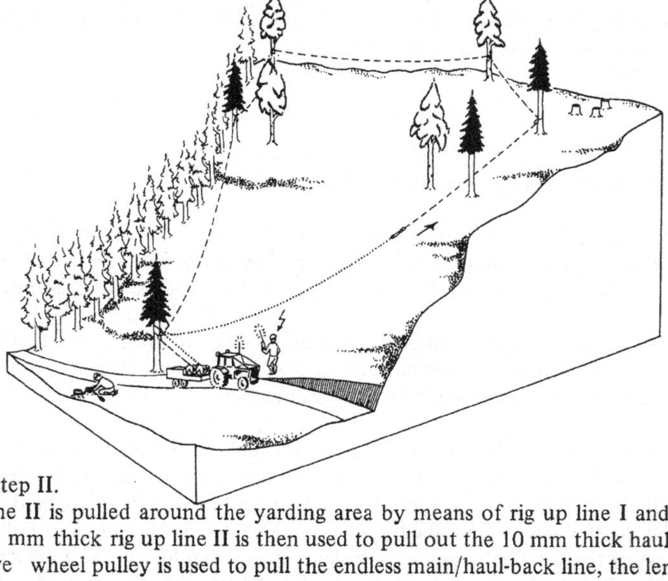

Figure 480. Step II.
The rig up line II is pulled around the yarding area by means of rig up line I and the rig up winch. The 4 mm thick rig up line II is then used to pull out the 10 mm thick haul-back line. Since a groove wheel pulley is used to pull the endless main/haul-back line, the length has to be adjusted by splicing in or taking out a section of the line. A 3-way swivel is attached to the head end of the main and the tail end of the haul-back part of the endless line. The 3-way swivel connects in other words the two lines to each other. The 3-way swivel serves as a transport means during the rig up work. While one of the workers carries out these tasks, the other starts to rig up the heel anchor.

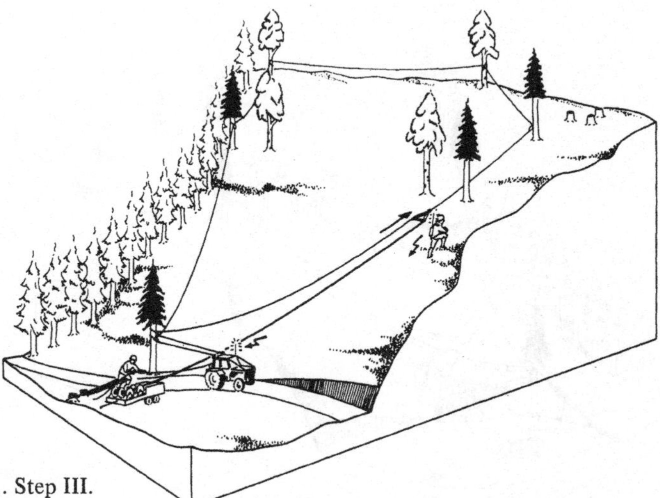

Figure 481. Step III.
The yarder operator moves the winch trailer towards the heel anchoring point. He wraps the endless line around the groove wheel pulley on the yarder. He then moves the tractor forward until the endless line is tight, and anchors the tractor at this point.
The other worker continues to rig up the heel tackle. When this is finished both workers cooperate by attaching the skyline to the 3-way swivel. The skyline is then pulled uphill by the endless line. Since the yarder is radio controlled it is the worker who follows the 3-way swivel uphill who controls the yarder with his transmitter. At the tail end he anchors the skyline to the anchoring point. The other worker continues to prepare the landing and to arrange the heel anchor.

Figure 482. Step IV.
When the skyline has been fastened to stumps or rock bolts at the tail end of the cable crane, one of the workers riggs up the tail tree and then prepares the supports if necessary. He uses his transmitter to control the yarder when he wishes to move the 3-way swivel up or downhill. The 3-way swivel is being used to transport all equipment necessary for the tail tree and the supports. The other worker at the landing attaches this equipment to the swivel. When the support has been rigged up the worker fastens the crossline to the 3-way swivel and tightens the crossline by moving the 3-way swivel.
In the meantime the other worker rigs up the spar tree and prepares an extra straw line from a straw line drum which is used to tighten the skyline.

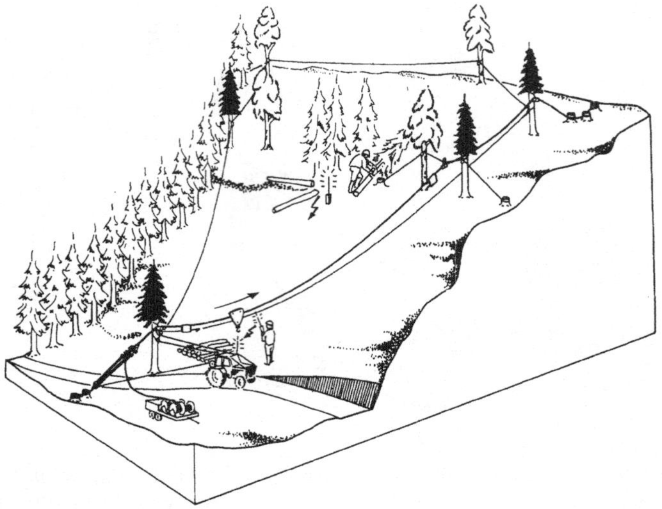

Figure 483. Step V.
When the tail tree, the spar tree and the support have been rigged up and the carriage has been placed on the slack skyline, the skyline is being tightened by means of the straw line and the tackle with the eccentric grip (figure 439). The tension may be controlled by means of an extensometer or by means of stop watch (chapter 7.8.1. and 7.8.2).When the skyline has the correct pre-tension, the yarding can start. After a few turns, all the blocks and anchors should be inspected before the yarding continues.

12.4.2.3. Running skyline cable cranes

The Moxy yarder (figure 384) is an example. It can be used on distances up to 600 meters, but is usually used on distances between 300 and 500 meters. The main line, slack-pulling line and haul-back line are 14 mm thick. There are not usually any supports in the span. It is, however, possible to use a support on the running skyline system.

There are 4–5 men in the crew: One working foreman, one yarder operator, one chaser and one or two choker setters. The felling is usually combined with the yarding. The choker setters fell the trees, set the chokers and hook them on the hoist line. A limbing machine on the landing does the rest of the conversion.

The felling and yarding is carried out in 20 meters wide strips. One starts in one end of the area to be logged and rig up the first cable crane perpendicular to the feeder road and the terrain contours. A 20 meters wide strip on one side of the skyline is felled and yarded, whereafter the skyline is moved close to the edge of the forest and parallel to the first set up.

The yarder is equipped with a tip/telescope tower which is supported by two or three guylines (figure 384). The powered guyline drums tighten the guylines. There are also powered spools or drums for the 2.1 mm thick and the 4 mm thick rig up lines as well as the 8 mm thick straw line.

When yarding downhill, horizontally or uphill on slopes gentler than 20 % the following rig up method may be used:

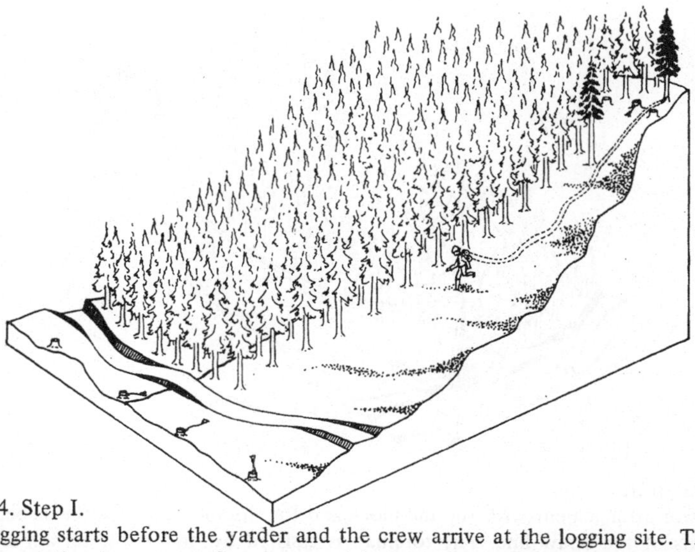

Figure 484. Step I.
The pre-rigging starts before the yarder and the crew arrive at the logging site. The foreman plans the operation by means of maps or aerial photographs where the cable crane location has been marked. He may use one of the crew as a helper. They clear the skyline corridor for the first set-up and place a pass block on a tree behind the tail anchor. The two head ends of the rig up line I are connected to each other and laid around the pass block.The foreman, or his helper, walks downhill while the two rig up lines are being rolled out (figure 432). Under certain terrain conditions it saves time to string out the 4 mm thick rig up line II as shown in figure 434.
The foreman also attaches anchor straps to good and rigid stumps so they are ready for anchoring the guylines.

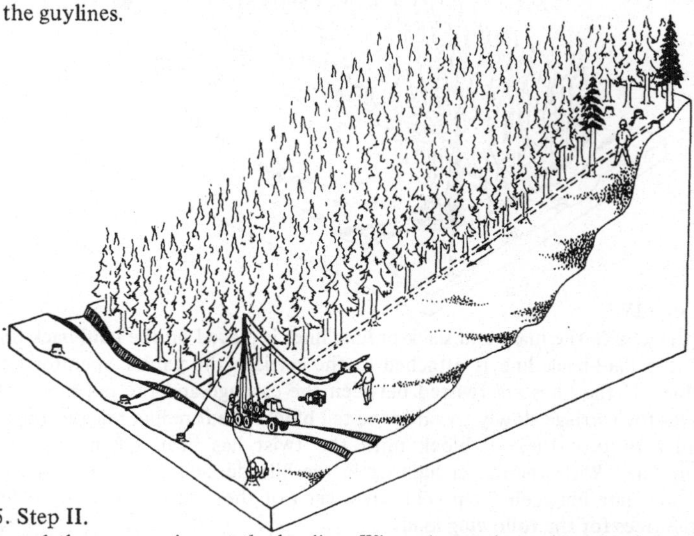

Figure 485. Step II.
The yarder and the crew arrives at the landing. When the steel spar is erected the terrain crew string out the guylines and fasten them to the anchor straps. Afterwards they clarify the tail block with the haul-back line and the tail tackle.
The yarder operator pulls out the 4 mm thick rig up line and then the 8 mm straw line. The foreman is at the tail end of the skyline and has contact with the yarder operator by radio.

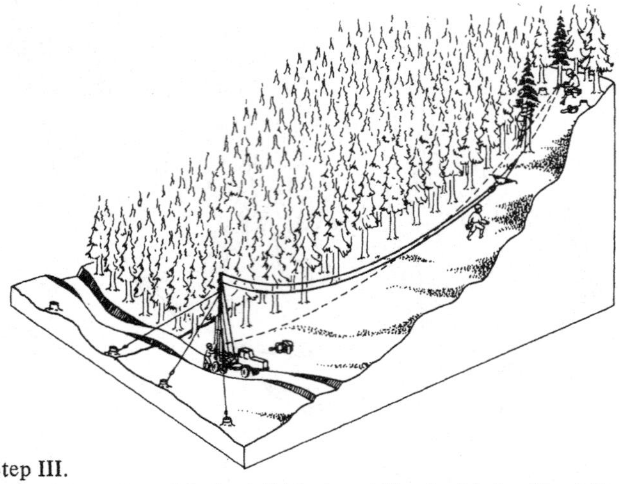

Figure 486. Step III.
The tail anchor arrangement with the tail block and the double haul-back line is being pulled out by means of the straw line. The foreman is assisted by one of the terrain crew while fastening the tail anchor to the anchoring stumps and also when rigging up the tail tree. Afterwards the worker walks to the place high up on the hillside where he starts to fell trees.

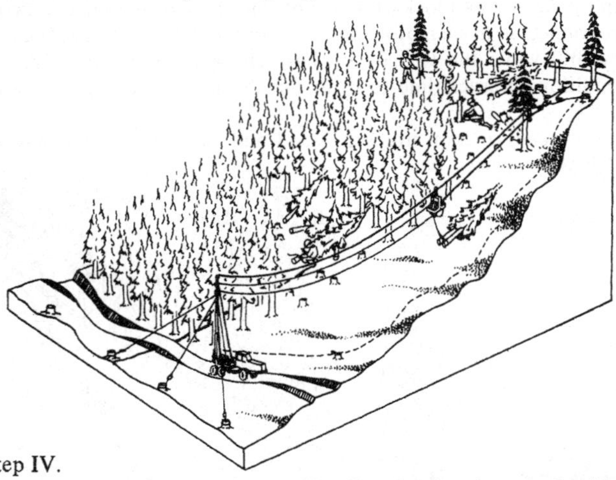

Figure 487. Step IV.
When the carriage with the main and slack-pulling lines is placed on the haul-back line and the head end of the haul-back line is attached to the carriage the yarder operator tightens the running skyline. If the lines are twisted between the carriage and the tail block, the yarder operator moves the carriage slowly towards the tail block.The foreman releases the bolt in the swivel in order to turn the tail block until the twist has been taken out. the yarding operation can start. With one cutter high up in the hillside, and the other lower down, the carriage can alternate between them. This gives each of them sufficient time to fell and set chokers on the trees for the following load.
During the yarding operation the foreman starts to prepare for the next road change. For that purpose, he uses the straw line by signalling to the yarder operator. Alternatively he may use a separate rigging winch for this purpose. The foreman moves an extra set of anchoring blocks and ropes to the top of the next skyline corridor. He prepares the next tail tree. He also has to select and set anchor straps on stumps for the new guyline anchors if necessary.

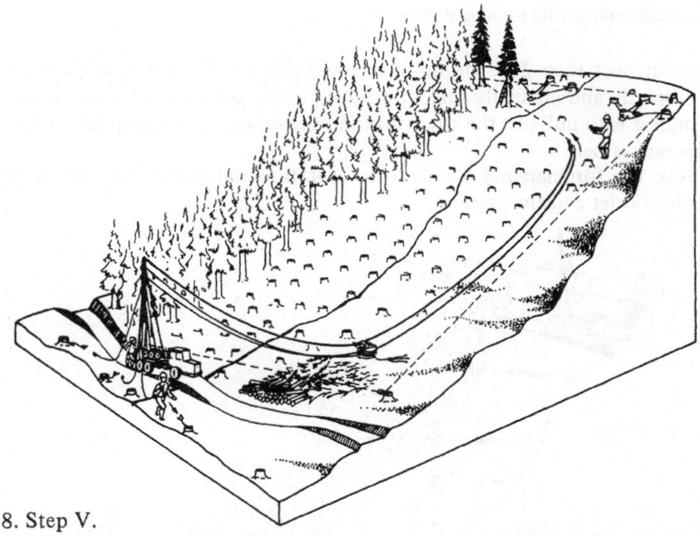

Figure 488. Step V.
When the first strip has been logged and yarded, the yarder operator lowers all lines to the ground. The foreman moves the tail block with the haul-back line from the first anchoring arrangement to the next one by means of the straw line or the separate rig up winch.

The yarder operator slackens the guylines and the yarder is moved 20 meters along the feeder road to the next skyline corridor. Very often the same guyline anchors may be used for two set ups. Sometimes, however, the guylines must be released from the anchors and moved to the new anchoring points.

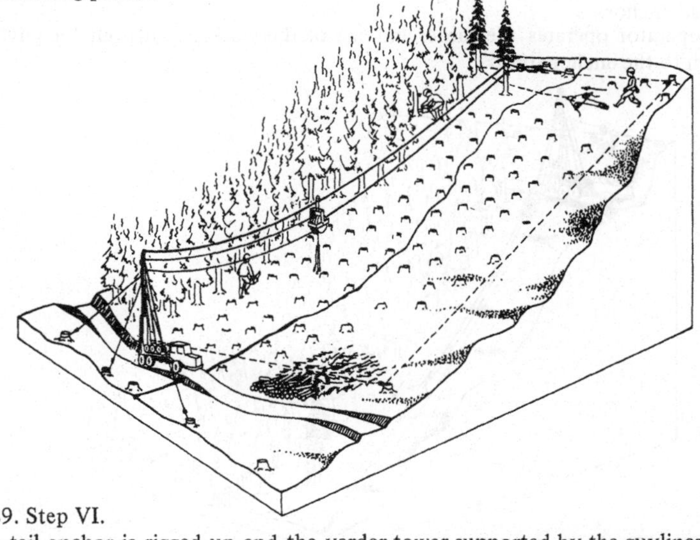

Figure 489. Step VI.
When the tail anchor is rigged up and the yarder tower supported by the guylines the yarder operator tightens the main and haul-back lines and lift them to the new skyline corridor. If necessary, he takes out the twist between the lines with help from the foreman. Before the yarding can continue on the new strip the tail anchor arrangements have to be pulled from the previous skyline corridor to the next by means of the straw line or the rig up winch.

12.4.2.4. Steep uphill yarding with running skyline

If the terrain is steeper than 20% the rig up work is simpler. The 8 mm thick straw line may be pulled out directly and manually since it is pulled in the same direction as the gravitational force. The simplest is to pull out the straw line and the 4 mm thick rig up line simultaneously and parallel to each other.
In this case too, the foreman has planned the operation and made some pre-rigging before the arrival of the yarder and the crew.

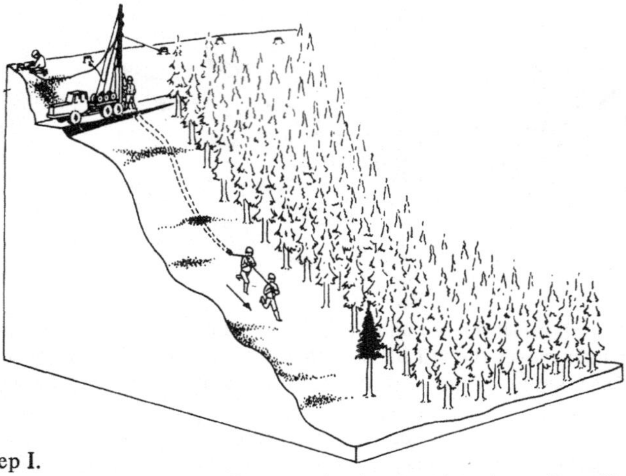

Figure 490. Step I.
One or two men pull out the straw line and the 4 mm thick rig up line. The two lines are connected and laid around a pass block which the workers attach to a stump or a tree behind or near the tail anchor.
The yarder operator operates the yarder and one of the workers pulls out the guylines and attaches them to the anchor straps.

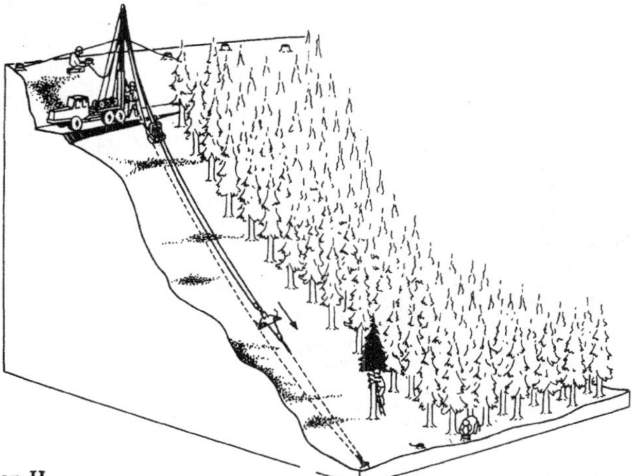

Figure 491. Step II.
The foreman and one of the cutters prepares the tail anchor and the tail tree. The other cutter continues the work with the guylines. The yarder operator pulls up the end of the straw line with the rig up line. He then pulls out the tail anchor arrangement with the tail block and the double haul-back line.

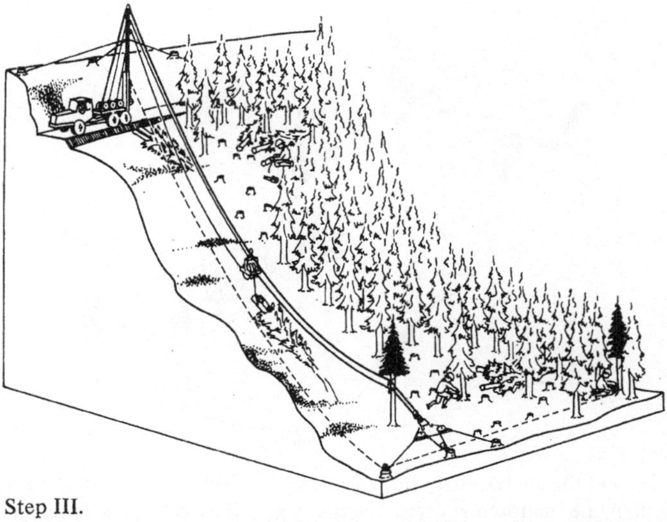

Figure 492. Step III.
The yarder operator supports the steel spar by tightening the guylines with the guyline drums. The foreman helped by one of the cutters fixes the tail anchor tackle to rigid stumps and hang up the double lifting blocks in the tail tree. The yarder operator tightens the main and haul-back lines with the ready mounted carriage and the yarding can start. One of the cutters works low down in the terrain and the other higher up in the terrain. They are felling and setting the chokers and the trees are transported unbranched to a landing near the feeder road. They must be placed in such a way that the limbing machine can reach them and convert them to assortments. During the felling and yarding, the foreman prepares for the road change to the next 20 meters wide felling and yarding strip.

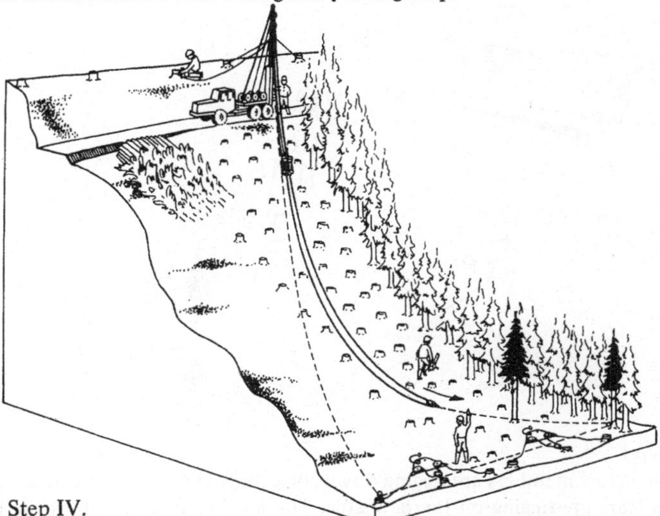

Figure 493. Step IV.
When the first yarding strip has been logged the yarder is moved approximately 20 meters along the feeder road to the next skyline corridor. The steel spar is anchored by the guylines to new anchoring points. The foreman pulls the tail block from the previous anchoring tackle to the anchoring tackle at the end of the new skyline corridor. He uses the straw line or a rig up winch to move the tail block.

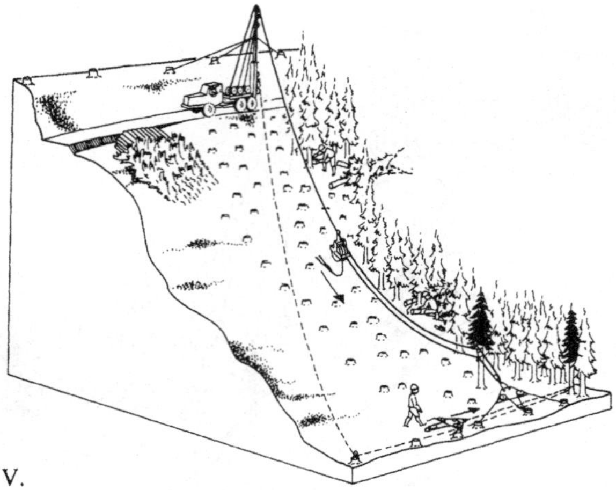

Figure 494. Step V.
When the yarder and the anchors have been arranged the felling and yarding continues on the new yarding strip while the foreman starts the pre-rigging for road change number 3.

12.4.2.5. Steep uphill yarding with standing skyline (gravity outhaul)

These are truck, or forwarder mounted yarders with 10–15 meters high tower, and 100–130 kW engines. The yarder and the cable crane are constructed for yarding tree dimensions up to 3 cu.m. per tree. The diameter of the skyline is 24 mm and main line 12 mm.No haul-back line is needed (chapter 11.3.2.1. and figure 268). The crew has usually 4 men: one yarder operator, one choker, one chaser who also operates a tractor to pull away the stems and one bucker.

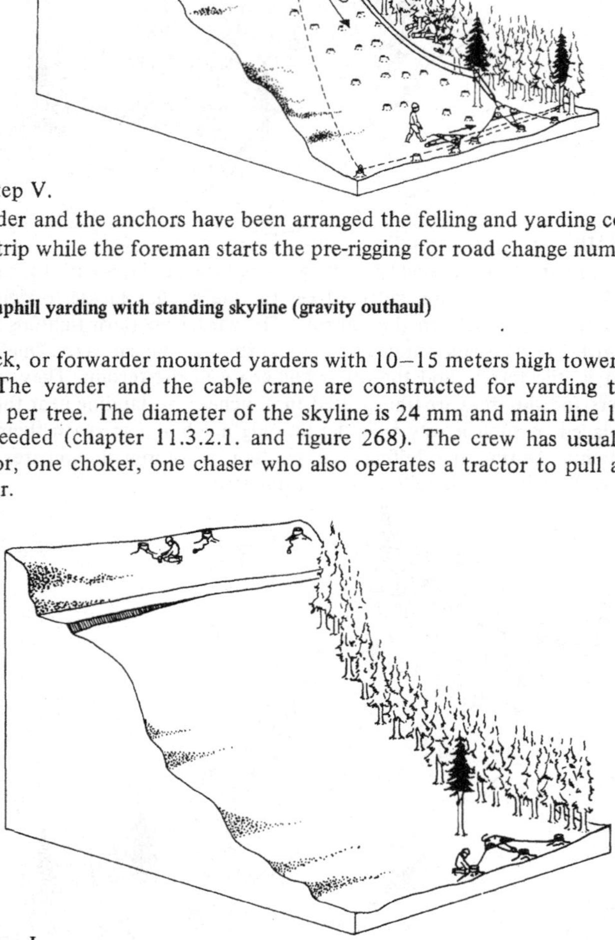

Figure 495. Step I.
While two of the crew members are rigging down the cable crane on the previous yarding area the two others start pre-rigging on the new area. The logging area as well as the location of the cable crane have been marked on a map or aerial photograph, and the foreman selects stumps for guyline anchors as well as the tail anchor. They fasten anchoring straps on the stumps and carry a preliminary tail anchor to the tail end of the cable crane. The tail anchor tackle is too heavy to carry, and will be transported out on the skyline as soon as it is fastened to the preliminary anchor (see step II).

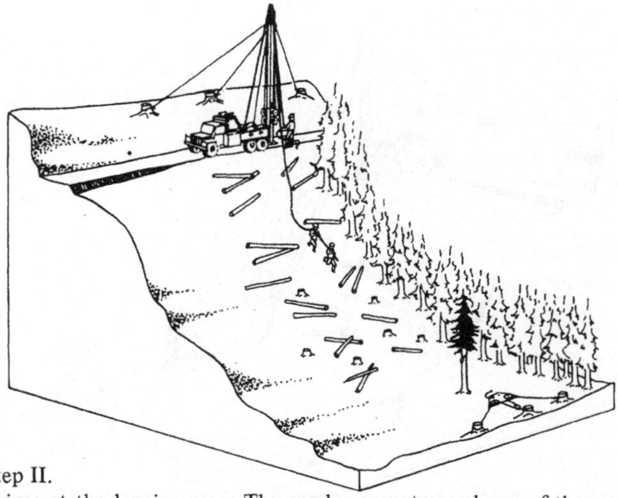

Figure 496. Step II.
The yarder arrives at the logging area. The yarder operator and one of the workers erect the tower, pull out the guylines and attach them to the anchoring straps and tighten the guylines. The two other workers pulls out the skyline. In steep terrain two workers are sufficient for that job. In cases where the terrain is difficult it may be better to pull out the straw line and use this to pull out the skyline afterwards. This is especially the case if the skyline has to pass a creek and has to be pulled uphill at the tail end (figure 434). The two workers fasten the tail end of the skyline to a preliminary anchor in order to use the skyline for transport purposes.

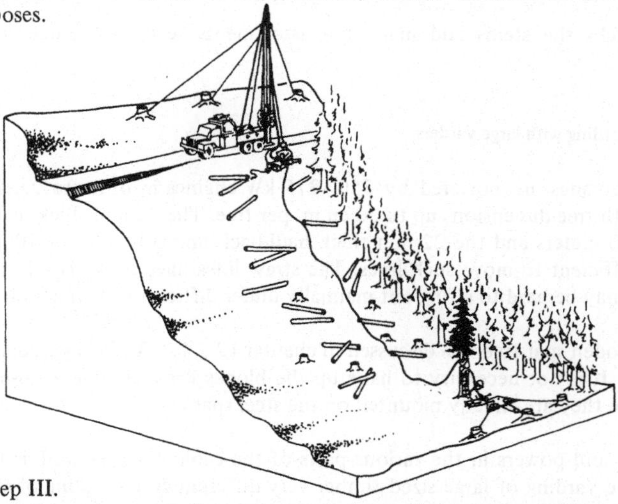

Figure 497. Step III.
The yarder operator and one of the workers put the carriage on the skyline and give the skyline sufficient pre-tension for transport of equipment such as tackle for the tailhold, rig up ladder and cable shoe with strap for the tail tree etc. The two workers at the tail end finishes the preparation of the tailhold and the tail tree.
In some cases there are large enough tail tree and stumps at the tail end and no special tailhold tackle is needed. In this case it is easier to carry downhill the anchoring strap as well as the cable shoe for the tail tree. This simplifies the work and no extra equipment transport is needed. The two workers fasten the tail end of the skyline with the anchoring strap to the anchoring stumps.

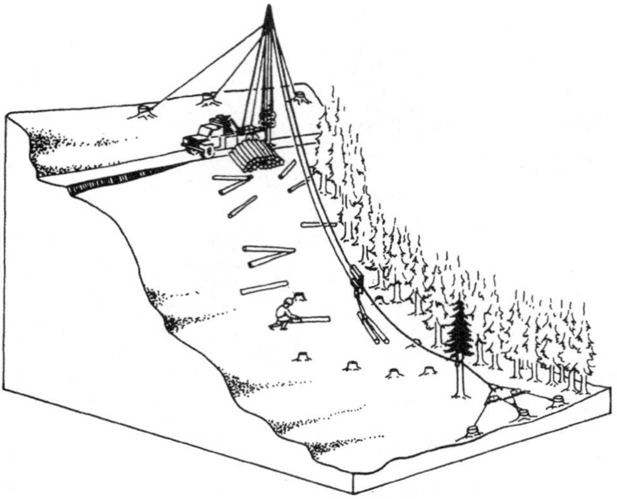

Figure 498. Step IV.
When the rig up work is finished the skyline is tightened and the yarding can start. The choker setter works in the terrain and the chaser unhooks the chokers. He uses a landing tractor to pull the stems away from the yarder along the feeder road.

One worker bucks the stems and make the assortments ready for truck transportation.

12.4.3. Highlead yarding with large yarders

The large cable cranes are powered by 350–375 kW engines in order to yard timber from virgin forests with tree dimensions up to 20 cu.m. per tree. The 32 mm thick main line weighs 0.38 kN per 100 meters and the 22 mm thick haul-back line 0.16 kN per 100 meters. Manpower is not sufficient to move such lines. The straw lines have to be 10–11 mm thick and even such lines may be hard to string out manually under difficult terrain conditions.

Rigging up a wooden spar tree was discussed in chapter 12.2.1.7. Todays yarders are equipped with steel spars. It is not necessary to hang up the blocks for main line, haul-back line, guy-lines etc. because they are already mounted on the steel spar.

The high forces and powers in the various parts of the cable crane as well as the anchoring points makes the yarding of large sized timber very different from yarding of second growth and smaller tree dimensions. The experienced foreman, the hooktender, has to select fresh stumps as anchoring points which can withstand the high tensions. Normally, these should have the same order of dimensions as the biggest trees to be yarded during the operation. If such stumps are not available several smaller stumps may be used and the tension from the guyline or the skyline has to be distributed between them using equalizing blocks.

Sometimes it is necessary to use a dead man anchor (chapter 6.1.2.). Since the skyline tension may be as high as 250–300 kN more than one dead man may be needed. If so, the tension has to be distributed between them by equalizing blocks.

The yarding crew is usually 7–8 men, the foreman (hooktender) included. The rig up work of cable cranes with large yarders differ much from rigging up cable cranes for smaller tree dimensions with 2–4 man crews (BINKLEY and STUDIER 1975, CONWAY 1978, LILEY 1983). In this connection we shall limit the discussion to the rig up work for a highlead operation with a large yarder.

The foreman has received the maps from the administration where the logging area as well as the location of the cable crane are marked. The pre-rigging consists of preparing the landing, felling the area for the guyline circle, and selecting the guyline anchors as well as the anchors for the corner block and the tail block. The foreman (the hooktender) has selected these anchoring points, the fellers have felled the trees and the straps have been attached to the anchoring stumps.

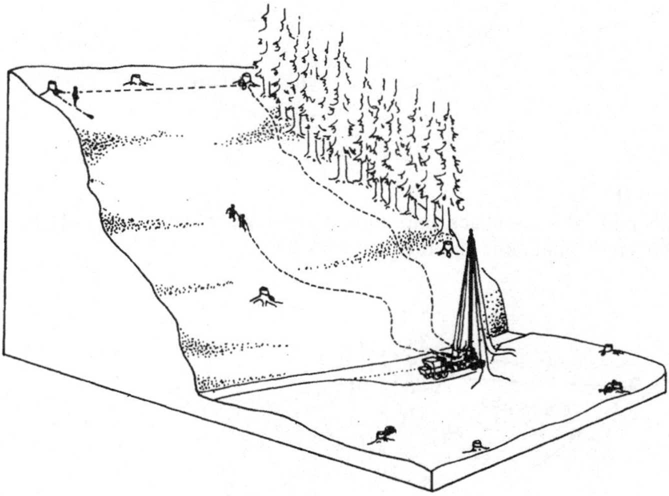

Figure 499. Step I.
The yarder operator checks that the top of the steel spar (the blocks, the guylines etc.) functions satisfactorily. He then erects the steel spar usually by hydraulic power. Two of the workers pull out the guylines and attach them to the straps on the anchoring stumps. Afterwards the guylines are tightened in order to stabilize the steel spar.

In the meantime two workers pull out the straw line in sections, one section along the waste line (the tail part of the haul-back line), one section along the main line corridor and one section along the haul-back line corridor. The rest of the crew carry out the corner block and the tail block and fasten them to their respective anchoring stumps.

This is the traditional layout of the work. It may be preferable to equip the yarder with spools or drums for 2 mm and 4 mm thick rig up lines. The rig up line I may be rolled out along the main line and the haul-back line corridors during the pre-rigging, the 4 mm rig up line may be pulled around by means of the 2 mm thick rig up line and the straw line may be pulled around the area by means of the 4 mm rig up line. In this case pass blocks are temporarily fastened to the corner stump and the tail stump. The straw line is then used to pull out the heavy corner block and tail block.

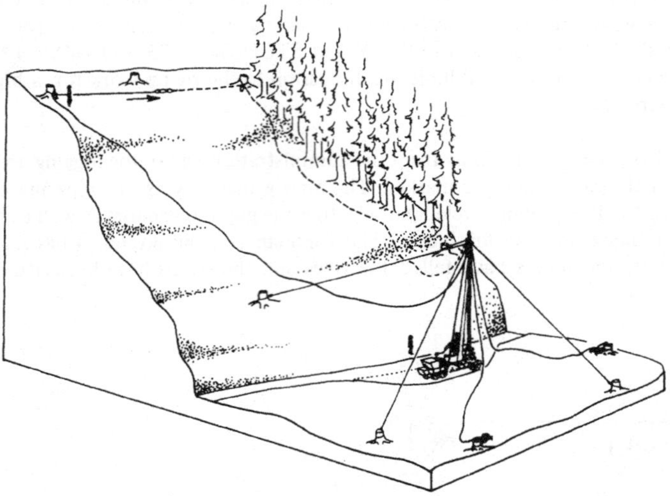

Figure 500. Step II.
The straw line pulls the haul-back line around the logging area. The butt rigging is then connected between the main line and the haul-back line.

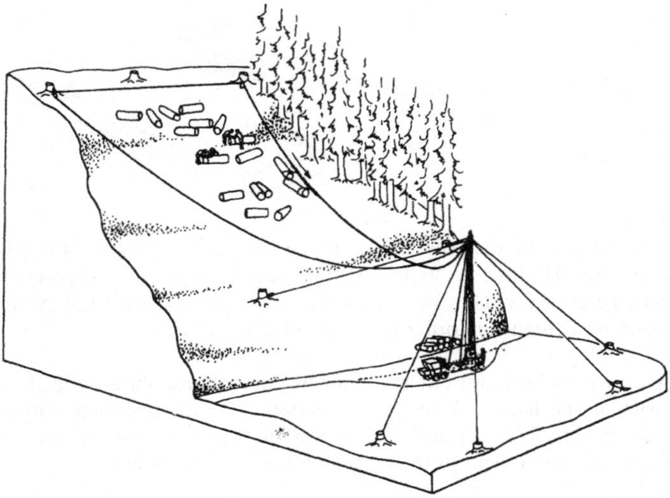

Figure 501. Step III.
The yarding starts. All timber along the main line corridor, and within the reach of the butt rigging is yarded to the landing near the yarder, where they are continously loaded on trucks for the long distance transport.

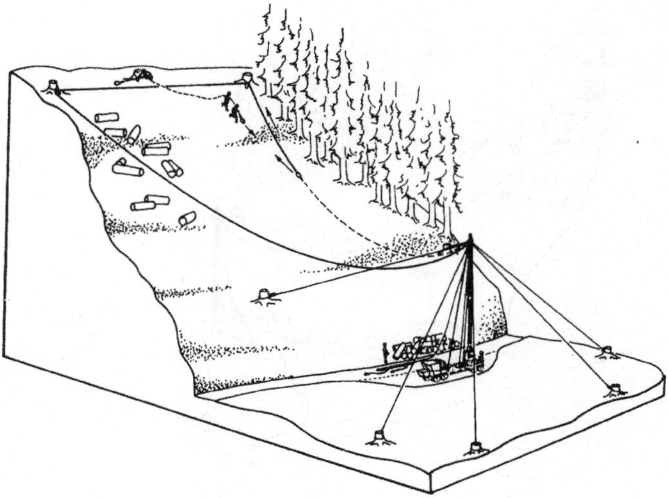

Figure 502. Step IV. (Road change).
If the logging area is located between the main and haul-back lines, the next main line corridor is being moved in the direction of the haul-back line corridor.A new tail block is placed at the tail end of the new main line corridor. The crew carries out one or two sections of the straw line. Each section is 60 m long and coiled up. The hooktender strings out the straw line section between the haul-back line and the new main line.

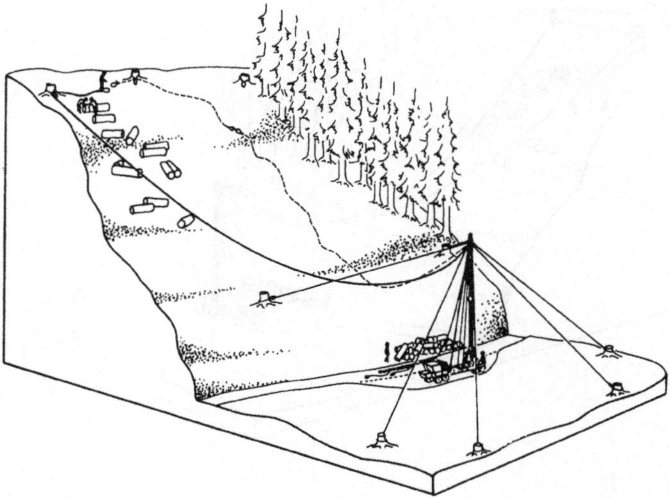

Figure 503. Step V.
The yarder operator has pulled in the rig, unhooked it from the haul-back line, and attached the yarder mounted straw line to it. The haul-back line pulls out the straw line to the head end of the straw line section in the terrain. The head end of the straw line section is connected to the straw line, and the tail end of the section to the haul-back line.

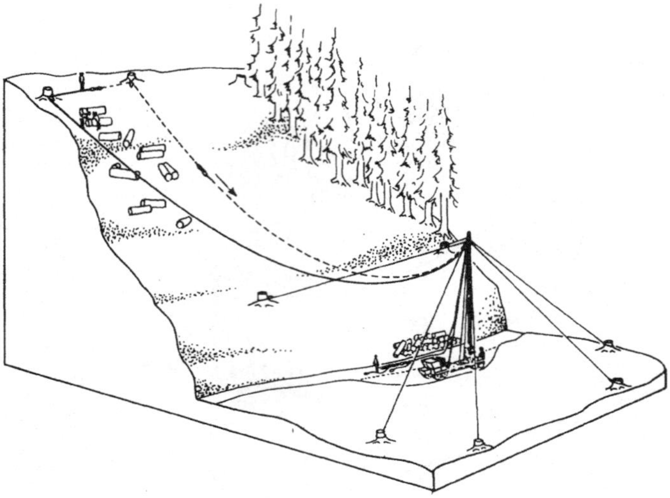

Figure 504. Step VI.
The yarder operator pulls in the haul-back line by means of the straw line, first through the tail block in the new corridor and then along the new main line corridor. The straw line is then disconnected from the haul-back line.

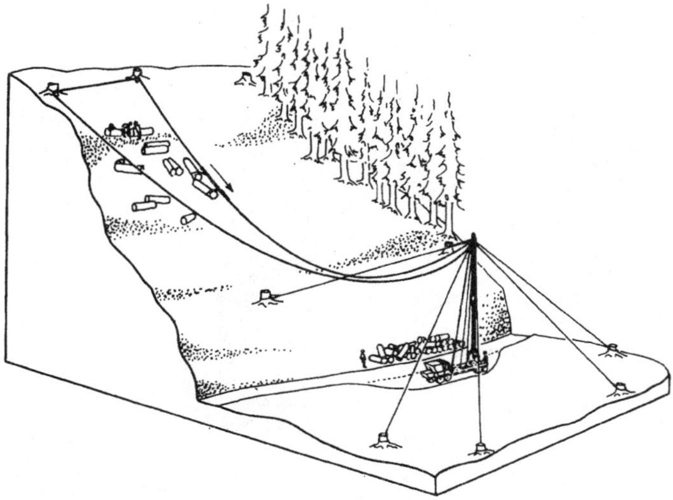

Figure 505. Step VII.
The head end of the haul-back line is attached to the butt rigging and the yarding can continue along the new main line corridor.

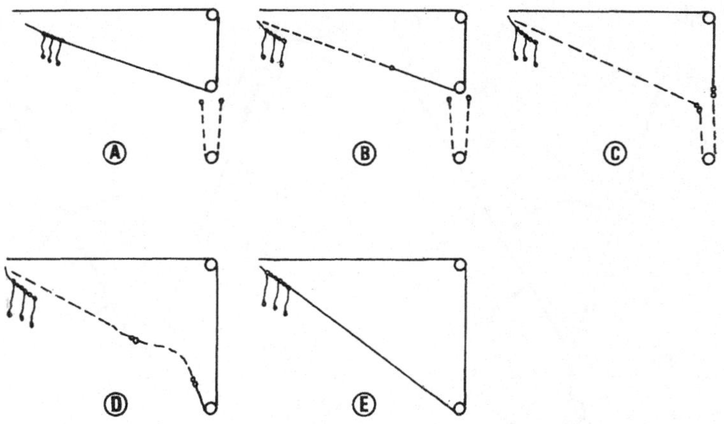

Figure 506. A New Zealand method for road change during highlead yarding (LILEY 1983).

The road change method described in figure 502–505 is commonly used on the American West Coast (CONWAY 1978). LILEY (1983) has described another method which has been used in New Zealand (figure 506). The highlead yarding starts with short spacing between the main line and the haul-back line corridors. During the road change the main line is moved away from the haul-back line.

A: The tail block in the new corridor is mounted on a stump and a section of the straw line is laid out double between the first and the second tailhold.

B: The butt rigging is returned to the landing. The haul-back line is disconnected from the butt rigging and hooked onto the straw line. The straw line is then pulled out towards the tailhold where it is disconnected from the haul-back line.

C: One end of the straw line section is attached to the haul-back line and the other to the straw line from the yarder.

D: When the yarder operator pulls in the straw line he first pulls the haul-back line through the tail block in the new main line corridor and then inwards along the new main line corridor.

E: The straw line is removed from the head end of the haul-back line. The haul-back line is now attached to the butt rigging and the yarding can continue along the new main line corridor.

The rig up work which is described above (figure 499–506) may be used when the large yarder shall operate according to the highlead system.

In some cases the terrain conditions necessitate the use of a skyline in order to carry or lift the load over terrain obstacles. This is usually a live skyline. Figure 507 illustrates two different methods of moving the skyline from one skyline corridor to the next.

A: The straw line is connected directly to the skyline by means of a catch chackle before the skyline is released from the tailhold in the first skyline corridor. The straw line pulls the skyline to the other skyline corridor.

B: The skyline is anchored to a stump in the middle of the logging area by means of a strap and a catch chackle, before the skyline is released from the tailhold. Afterwards the skyline is picked up by the straw line and pulled over to the next skyline corridor.

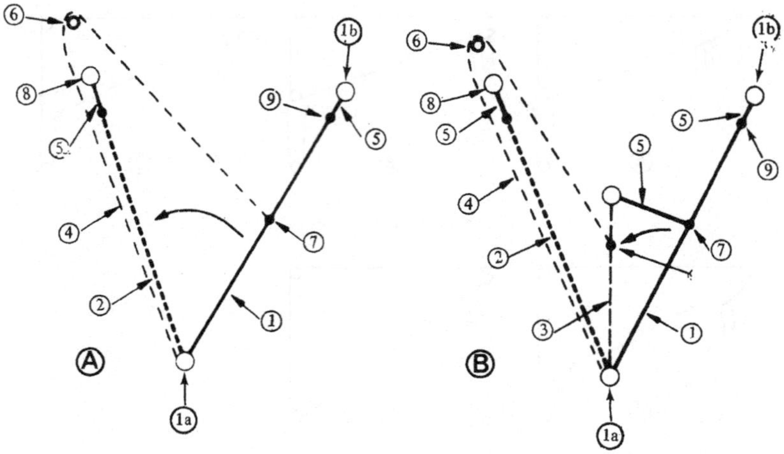

(1) Skyline (1a) Head tree (1b) Old anchor (2) New skyline road
(3) Skyline before the straw line pulls it into the new anchor (8)
(4) Haul-back line (5) Strap (6) Block (7) Catch chackle
(8) The new anchor (9) Knockout chackle

Figure 507. Two methods of road change for heavy cable cranes, used on the American West Coast (CONWAY 1978).

12.5. Time consumption for rig up work

The time needed to rig up a cable crane depends on the terrain conditions, the length of the cable crane and the system. It was a tiring and time consuming job to set up these old stationary cable ways or cable cranes, often more than a 1,000 meters long with many supports. Even when the men were experienced, they often used two to four weeks to construct such a cable way which sometimes remained stationary for several years.

Todays mobile yarders, where auxiliary equipment is of great help for the men when rigging up the cable crane, can be ready for yarding in a few hours. The layout of a good road network has helped to reduce the yarding distance and to avoid supports. The time consumption during the rig up work depends a lot on whether a good road network has been laid out on the difficult terrain and which cable system is being used.

First and foremost the variation in time consumption depends on the training and experience of the men in the crew. One crew may use 1–3 days to rig up a modern 400 meters long cable crane while another crew uses 3–4 hours only.

Average production under various conditions will therefore not be discussed in this chapter since so much depends on the local conditions, the forest type, the terrain, the road network etc. as well as the qualification of the men.Instead a few typical examples of time consumptions which could be expected from good experienced men under normal conditions will be given. These are based on statistics from a long series of cable systems and operations throughout a 30 year period in Norway as well as experiences from Central Europe and the American West Coast.

The time consumption depends also on the length of the working day. It is, therefore, given in number of ***work place hours*** which is the effective time pluss the normal delay times. (see chapter 14, table 75).

As an example of the development one may compare the rig up time for the stationary long cable cranes which were in use in the 1950ies with the modern cable cranes of today (table 59).

Table 59. Examples of the rig up and down times for some Norwegian cable systems during the period 1951–1980.

Cable system	Year 19–	Length m	Supports	Crew	Hrs/day (x	Number of days Rigging Up	Down
Twin cable system	53	1,100	7	4.1	7.5	26.6	–
Stationary pendulum cable way	51	1,200	2	3.0	7.5	12.0	–
Movable pendulum cable way	51/53	448	1.5	3.0	7.5	4.2	0.6
Cable crane. Top mounted winch	55–59	354	1.4	3.0	7.0	2.8	0.8
Cable crane. Valley » »	55–58	581	1.1	3.0	7.0	3.4	0.6
Radio controlled cable crane	67–80	480	1.0	3.0	6.0	2.2	0.7
Moxy running skyline	78–80	398	0	5.0	6.0	0.46	0.20
Moxy road change						0.10	
Light running skyline	78–80	107	0	2.0	6.0	0.27	0.07
Light running skyline, road change						0.10	

(x Work place hours.

12.5.1. Light cable ways and cable cranes

The mobile light pendulum cable ways are driven by the gravitational force (chapter 10.3). The 3-man crew had no engine power to help them during the rig up work. It took them 3 1/2 days to rig up and rig down again a 300 meter long cable way and nearly 8 days to rig up and down a 700 meters cable way.

Table 60. Smoothed times for the installation and dismantling of light pendulum cable ways. Wo = 6.0 work place hours/day. Manhours and days per crew.

Cable length (L_1), m	300		500		700	
Number of supports	1		2		3	
Work operations	Man-hours	Days (crew)	Man-hours	Days (crew)	Man-hours	Days (crew)
Rig up	55.5	3.1	81.9	4.6	114.9	6.4
Rig down	8.7	0.5	16.2	0.9	24.9	1.4
Total	64.2	3.6	98.1	5.5	139.8	7.8

The small yarders for light cable cranes were different. The crew could use the haul-back line as a straw line (figure 438). Table 61 gives an example of the time consumption when three experienced men are rigging up and rigging down a light cable crane. They needed 1–2 hours to rig up a 100 meters long cable crane and 4–6 hours to rig up a 400 meters long cable crane. The rig down time varied between 1/2–3 hours.

Table 61. Smoothed times for installation and dismantling of cable cranes for yarding with the Vossa-winch. Work place time (Wo) in hours for a 3-man crew.

Cable length (L_1), m	<100	100–200	200–300	300–400
Number of supports	0	1	2	3
First installation				
Rig up	2.1	3.1	4.3	5.7
Rig down	0.9	1.4	2.1	2.8
Total hours/crew	3.0	4.5	6.4	8.5
Road change				
Rig up	1.1	1.9	2.9	3.8
Rig down	0.4	0.7	1.1	1.4
Total hours/crew	1.5	2.6	4.0	5.2

The rig up method for downhill yarding using cable crane with top mounted winch was described in chapter 12.4.1.2. (figure 461 – 466). Table 62 gives an example of the time consumption for a 3-man crew when rigging up and rigging down a cable crane. It took them 2–3 days to rig up a 300 meters cable crane and 4–7 days if there is 3 supports and the length of the cable crane is between 700 and 1,000 meters. The rig down time varies between 1/2–2 days.

Table 62. Smoothed times for a 3-man crew installing and dismantling cable cranes for cable transport with top mounted single-drum winch. Wo = 6.0 work place hours/day. Manhours and days per crew.

L_1 average	300		550		850	
Number of supports	1		2		3	
Work operation	Man-hours	Days (crew)	Man-hours	Days (crew)	Man-hours	Days (crew)
			First installation			
Rig up	57.0	3.2	88.5	4.9	121.5	6.8
Rig down	20.1	1.1	30.0	1.7	42.0	2.3
Total	77.1	4.3	118.5	6.6	163.5	9.1
			Road change			
Rig up	37.5	2.1	53.4	3.0	75.0	4.2
Rig down	11.1	0.6	18.3	1.0	27.6	1.5
Total	48.6	2.7	71.7	4.0	102.6	5.7

The light farm-tractor or skidder mounted cable cranes have usually a 2-man crew (figure 467 – 470). The method was described in chapter 12.4.1.3. and 12.4.1.4. The 2-man crew needs 2 hours to rig up a 100 meters cable crane with no supports and approximately 4 hours to rig up a 200 meters cable crane with one support. To rig up an extra support takes usually 1–2 hours depending on the conditions. The average time of a road change is 0.6 hours for the crew.

12.5.2. Medium sized cable cranes

Table 63. Smoothed times (in manhours and hours/crew) for the installation, dismantling and for the clearing of the cable corridor for the radio controlled **K**-1200 cable crane and for the Alp cable crane.

Cable length (L_1), m	100	300	500	100	300	500
	Manhours			Crew hours		
Rig up and down	7.9	34.2	60.5	2.6	11.4	20.2
Clearing the cable corridor	7.2	11.3	15.4	2.4	3.8	5.1

The Norwegian radio controlled cable crane is an example of a cable crane without a steel spar. Both spar tree and tail tree are usually needed. When the cable length is 100 meters, supports are not usually needed. It is common to have one support with a 300 meters long cable crane and two supports with 500 meters length. An extra support takes usually one man 2–3 hours to rig. When cutting is combined with yarding it is necessary to clear the cable corridor. The time consumption for rig up and rig down is 2.5–20 hours for a 3-man crew. 70 % of this is rig up time. The time consumption for rigging up and down the Igland Alp cable crane is approximately the same as for the radio controlled cable crane.

Table 64. Smoothed times (in manhours and in hours/crew) for rig up and rig down work and road change of Moxy cable crane with running skyline.

Cable length (L_1), m	100	300	500	100	300	500
	Manhours			Crew hours		
Rig up	3.5	10.5	17.5	0.70	2.10	3.50
Rig down	0.7	2.5	7.5	0.14	0.50	1.50
Total	4.2	13.0	25.0	0.84	2.60	5.00
Road change	1.2	2.1	3.0	0.24	0.42	0.60

Table 65. Examples on time consumption for rig up work and road change with some cable systems in British Columbia (COTTELL, McMORLAND and WELLBURN 1976).

Yarder	(1	Tower height m (2	Power kW	System (3	Horizontal yarding dist. m	Crew	Rig up time Crew hours	Road change Crew hours
Ecologger	(Ti)	T 13	97	H R	75	4–5	3.3	0,2
Madill 70'	(W)	T 21	134	H	105	5–6	–	0.5
Madill 71	(Cr)	T 15	164	H R	57	4	4.3	0.6
Madill 70'	(Cr)	T 21	116	H	135	4	4.5	0.5
Wash. 078 skylock	(Cr)	B 14	138	Gr	100	3	8.0	0.6
Skagit GT3	(Cr)	B 14	164	Gr	99	3	2.5	0.1
Koering Bantam	(W)	B 9	67	H R	91	4	1.2	0.3
Madill 71	(Cr)	T 15	164	H	135	4	1.3	0.6

(1 Ti = Timberjack
W = Wheel undercarrier
Cr = Crawler undercarrier
(2 T = Tower
B = Swing boom

(3 H = Highlead
R = Running skyline
Gr = Grapple yarding

The time consumption for rigging up and down a running skyline may be examplified with the Moxy yarder (table 64). This is a cable crane suitable for yarding tree dimensions up to 3 cu.m. per tree.

In this case the foreman of the crew has carried out pre-rigging which means that anchors for guylines, tail anchor etc. are prepared before the yarder arrives. In most cases the rig up line has been strung out also. The rig up and rig down time varies from 1–5 hours for a 5-man crew and length of cable cranes between 100 and 500 meters. There are no supports on the skyline. The time consumption for road change is normally less than half an hour for the crew.

COTTELL, McMORLAND and WELLBURN (1976) made an investigation of the productivity and the time consumption for rig up work and road change with various medium sized cable cranes in British Columbia. The mobile yarders were mounted on wheeled chassis or crawler chassis. The length of the cable cranes were defined by the average horizontal yarding distances. This shows that the length of the cable cranes investigated usually varied between 150 and 225 meters and there were no supports in the span.

With the mobile yarders and 3–5-man crews the usual rig up time varied between 1 and 4 hours per set-up.

The road change for the highlead systems and running skylines varied between 10 and 35 minutes. The mobile yarding cranes with swing booms used 7 to 33 minutes per road change. The equipment and the methods are described in chapter 11.5.4.8. (see table 53). The time consumption for road changes shown in table 65 corresponds well with the findings of SAUDER (1980 and 1981). With a Madill 044 or an American 7250 yarding crane and a crawler tractor as tailhold, the road change represented around 10 % of the work place time. Each road change took a few minutes only, and with 100–140 meters yarding distances the average yarding time between two road changes was 35–40 minutes only. The grapple yarding is efficient and with a tail tractor the set-up time is short. It can only be used, however, under very good and smooth terrain conditions where there is sufficient deflection for the running skyline and the grapple.

12.5.3. Large cable cranes with mobile towers

It is hard work to rig up a 35 meters high spar tree (chapter 12.2.1.7.). The experience of the crew is the most critical factor. An experienced crew can rig a standing tree in 8–12 hours, while an inexperienced crew may use 2–4 days (CONWAY 1978). A yarder with a mobile tower is much easier to rig up and it takes an experienced crew less than 3 hours to do it. If the crew has no experience the same work may take as much as 3 days. SAUDER (1982) investigated the productivity of the Madill 046 yarder equipped with a 29 meters high tower (chapter 11.5.3.1., table 47).

Table 66. The yarding crew with Madill 046. (SAUDER 1982).

	Cable system	
	Highlead	Slackline
Landing crew:		
Yarder operator	1	1
Bucker	1	1
Chaser	1	1
Yarding crew:		
Rigging slinger	1	1
Choker men	1	2
Back rigging crew:		
Back rigger	1	1
Back rigger helper	1	1
Hook and rig	1	1
Total	8	9

The total crew was 8 men during highlead yarding and 9 men during slackline yarding. The average yarding distance in both cases was 185 meters. The average volume was 1.21 cu.m. per log and the daily production 184 cu.m. per day with highlead yarding and 208 cu.m. per day with slackline yarding. The high production leads to frequent road changes. Accordingly, 2 men of the yarding crew, the back-rigger and his helper were continuously occupied with the preparation of the next road change.

Table 67. Rig up time and road change in minutes for the crew with Madill 046. (Average yarding distance 185 m). (SAUDER 1982).

Cable system	Pre-rigging (preparation) mins	Rig up time mins	Total rig up time mins
Highlead	3	12	15
Slackline:			
Skyline	9	69	76
Haul-back line	8	22	33
Yarder	–	–	175

The average time of rigging up the yarder was 2 hours and 55 minutes. With highlead yarding the time consumption averaged 15 minutes per road change. During slackline yarding, it took 33 minutes to change the location of the haul-back line, but 1 hour 16 minutes averagely to change the location of the skyline.

13.0. BALLOON LOGGING

Many of the cable crane systems lift the logs above the ground and transport them along the skyline above the terrain obstacles. This is sometimes called »to fly the timber». Other transport means such as aeroplanes, helicopters, airships and balloons do the same. These are vehicles which carry the load by means of aerostatic or aerodynamic lift. They fall outside the framework of this book on winch and cable systems.

The standing skyline lifts the load on a cable way or a cable crane while the running lines (main lines, haul-back lines etc.) move the load along the skyline corridor. The skyline may be replaced by a balloon which lifts the load above the ground by means of aerostatic lift instead of the tension in the skyline. Also in this case the load is being moved along the cable corridor by means of a winch and wire ropes.

13.1. Examples on aerostatic and aerodynamic lift

13.1.1. Helicopter

Helicopters have been used for timber transport since the early 1950ies. The load is lifted and transported by means of aerodynamic lift.

Table 68. Examples of some helicopters used for timber transport.

Make and type	Power kW	Empty weight		Lifting capacity		Total kN
		kN	%	kN	%	
Bell 204 B	810	19.6	51	19.0	49	38.6
Bell 205 A	1,030	20.0	53	18.0	47	38.0
MI-4	1,250	40.0	55	32.1	45	72.1
MI-6	6,470	285.0	70	120.0	30	405.0
Sikorsky S-61N	1,100	55.6	65	30.6	35	86.2
Sikorsky S-64	8,080	78.3	45	94.3	55	172.6

Figure 508. A. Full tree yarding with Bell 205. A limbing/bucking machine at landing converts the trees into logs (Drangedal, Norway 1981).

B. Sikorsky S61N yarding 3 m^3 bundles of logs to floatway (Kviteseid, Norway 1970).
C. Bell 204B yarding 2.4 m^3 pre-bundled logs to floatway (Kviteseid, Norway 1963).
D. The USSR helicopter MI6 yarding 8 m^3 bundles of logs to floatway (Kviteseid, Norway 1967).
E. Sikorsky S64 (flying crane) yarding logs from selective felling (Washington, USA 1972)
F. The Sikorsky S64 helikopter at bunkering area.

The helicopters used for timber transports has a payload of between 30 and 50 % of the total weight which can be lifted by the rotor blades (table 68). Some helicopters are specially constructed for carrying heavy loads such as Sikorsky S-64, »The flying crane», where the lifting capacity is 55 % of the gross weight.

Different types of helicopters have been used for timber transport in mountainous regions of Norway during the period 1964 to 1981 (SAMSET 1964, 1967, 1970 and FRØNSDAL 1981). These transports have been followed up by statistical analyses and detailed time studies and in the reports there are reference lists of reports from other logging operations with helicopters throughout the world. Table 69 summarizes a few production figures from the Norwegian tests. They refer to coniferous forests averaging 0.3 m^3 per tree. The first three operations in table 69 are based on pre-bundled barked logs which were air-dried before helicopter transportation in order to obtain the most effective use of the loading capacity. The average loads utilized 85 % of the helicopters lifting capacity.

Table 69. The production with helicopter transport of timber in Norway 1964–1981.

Make	Average payload m^3/trip	Average speed. Km/hr.		m^3/hr
		Load direction	Return	
Bell 204 B	2.3	67	89	28
Sikorsky S-61N	3.5	67	102	56
MI-6	12.8	80	133	218
Bell 205 A	1.5	77	81	20

The last operation summarized in table 69 was full tree yarding with the helicopter and a limbing/bucking machine at the landing. In this case the average loads only utilized 65 % of the helicopters lifting capacity. It was difficult for the crew to get sufficiently big loads and the helicopter had to transport fresh wood including the branches. This reduced the effective use of the helicopters transport capacity.

The three first methods in table 69 gave a high utilization of the transport capacity. The preparatory work with delimbing, barking, bucking and pre-bundling had to be carried out by hand and was too costly. In the last operation (Bell 205 A) the timber fellers were transported to the logging area by the helicopter each morning. They felled and set the chokers. The trees were transported downhill to the landing where a harvesting machine branched and bucked the trees into assortments. As a whole it was an efficient operation.

These examples from Norwegian tests show that it is difficult to get big enough loads. Reports from many operations throughout the world shows that it is difficult to utilize more than 70–80 % of the lifting capacity of the helicopter.

The weather factor is also of importance. Investigations on the American West Coast shows that the flight availability for helicopter logging varies between 30 and 75 % throughout the year. The lowest availability is in the winter month and the highest at mid-summer. It is difficult to get an average higher than approximately 50 %. During the rest of the time the helicopter logging has to be stopped due to difficult weather conditions.

The helicopter logging may be of interest under extremely difficult terrain conditions where road building is too costly or has to be avoided for other reasons. The logging costs are always very high. The Norwegian experiences agree with experiences from other countries. The helicopter logging usually costs twice as much as cutting and yarding using cable cranes. The helicopter is a practical machine for many other purposes in forestry apart from logging, such as spraying of pesticides or herbicides, transport of equipment for constructional purposes, or transport of equipment when rigging up cable ways or cable cranes in extremely difficult terrain.

13.1.2. Airships

Airship transportation carries the load by means of aerostatic lift. An envelope filled with lighter-than-air gas carries the load. The airship is moved forward by the thrust from a propeller which also adds some aerodynamic lift to the aerostatic lifting capacity.

The first experiments with airships took place at the end of 19th century. Graph Zeppelin constructed the first Zeppeliner in 1898–1900. It was 160 meters long and the diameter was 12 meters. The cigar shaped balloon envelope contained 11,000 m³ of hydrogen gas. Two engines, 12 kW each, were mounted on gondolas which gave the airship a speed of 30 km/h.

Many new and better constructions appeared and the airship Graph Zeppelin carried out a circum navigation of the globe in 1929 (20 days and 4 hours). This airship had a regular route over the Atlantic during the period 1930–1936, when the airship »Hindenburg» replaced »Graph Zeppelin». The airship »Hindenburg» was 245 meters long and the envelope contained 190,000 m³ hydrogen gas. The weight of the airship was 195 ton and the propeller drive was powered by 4 diesel engines with 880 kW each.

The airship transport came to a dramatic end when »Hindenburg» exploded over Lakehurst airport outside Philadelphia, May 6, 1937. The reason may have been an electrostatic charge in the airship. The hydrogen gas was ignited by the discharge of electrons from the airship to the earth immediately before landing.

During World War II airships were used for military purposes. The »Zeppeliners» were rigid airships. The shape of the balloon envelope was kept by an aluminium framework (figure 509). Modern airships are non-rigid (figure 510). The shape is kept by the cut of the balloon envelope but only as long as it is inflated. The modern airships use helium instead of hydrogen.

Figure 509. Graph Zeppelin airship with rigid construction.

Figure 510. **Non-rigid** airship (used for seeking after submarines).

Table 70. Specifications of some non-rigid airships from Goodyear.

Type	GX-20	ZPG-2W	ZPG-3W	ZW-1
Length	59	104	123	142
Diameter (m)	14	23	26	34
Weight of the envelope (kN)	17	70	93	121
Area of the envelope (m^2)	2,010	5,870	7,890	10,250
Volume (Helium) (m^3)	5,740	27,620	42,210	79,320
Max. aerostatic lift (kN)	55	276	421	791
The empty weight of the airship (kN)	40	166	212	402
Payload (according to aerostatic lift) (kN)	15	112	213	397
Addition for aerodynamic lift due to motor/speed (kN)	4	27	48	68
Number of motors	2	2	2	2
Power of each motor (kW)	155	590	940	1,120
Normal speed (km/h)	56	74	75	64
Max. speed (km/h)	80	130	151	134
Normal height (meter)	300–900	300	300	300
Max. height (meter)	2,600	2,900	2,900	3,100
Max. flight distance (km)	800	3,900	5,900	7,700

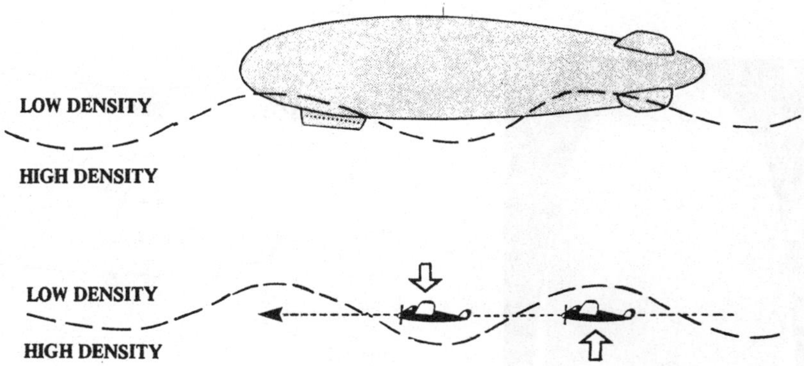

Figure 511. The movement of an airship versus a small airplane through air turbulences.

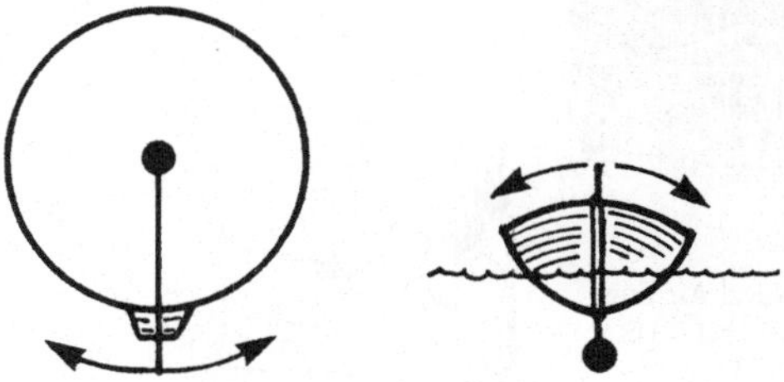

Figure 512.

Goodyear Aerospace has developed different types of non-rigid and manned airships. The first model contained 9,900 m³ gas. The largest was constructed in 1930 and contained 196,000 m³ gas. The existing models contains 42,000 helium gas (figure 510). Helium gas is non-combustible and has completely replaced hydrogen gas in airships or in balloons. Some examples on specifications for non-rigid airships are given in table 70.

Airships have some advantages, which seems to be of interest in connection with relatively slow transport of heavy loads, such as timber transport. They are rather stable in air turbulences, due to their large size (figure 511). The side movement caused by air turbulences is »positive» like a pendulum, since the load hangs under the balloon envelope during transport. This is opposite to ship-transportation, where the side movement caused by waves is »negative» since the load rests on the ship on the top of the water (figure 512).

13.1.3. Hybrid vehicles

Helicopter transport is technically a good solution under extremely difficult terrain conditions where road building is impossible. The disadvantage is that the helicopter itself is very expensive and that it has a very high fuel consumption. Experiments are being carried out in order to combine the advantages of aerostatic lift with the advantages of aerodynamic manoeuvring. Two of these experiments may be mentioned:

Figure 513 a.
The Helistat in erection at Naval Air Engineering Center, Lakehurst, NJ, USA. November 1984.

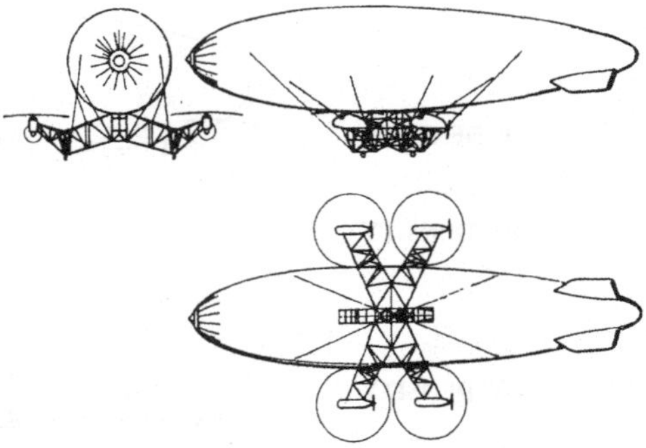

Figure 513 b.
The Helistat consists of a non-rigid airship, and four helicopters equipped with rotors for aerodynamic lift and propeller for forward movement (PIASECKI 1980).

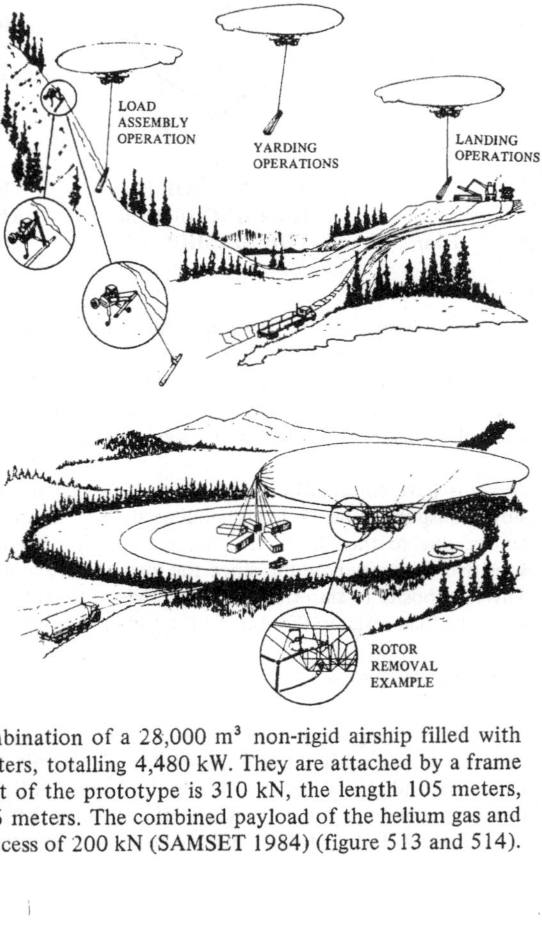

Figure 514.
Hypothetical method for Helistat yarding (US Forest Service 1980).

13.1.3.1. The Piasecki Helistat

The prototype of this vehicle is a combination of a 28,000 m^3 non-rigid airship filled with helium gas and four 1,120 kW helicopters, totalling 4,480 kW. They are attached by a frame construction to the airship. The weight of the prototype is 310 kN, the length 105 meters, the height 34 meters and the width 46 meters. The combined payload of the helium gas and the helicopters is estimated to be in excess of 200 kN (SAMSET 1984) (figure 513 and 514).

Figure 515. The Cyclo-Crane anchored to the airship tower on the Tillamook Airport before it was destroyed by a hurricane in October 1982 (notice the people on the ground).

13.1.3.2. The Cyclo-Crane

Another experiment is the Cylo-Crane. The balloon envelope has a diameter of 21 meters and the length of the Cyclo-Crane is 56 meters. The envelope contains 9,350 m^3 helium gas (figure 515 and 516). The aerostatic lift from the helium gas is 91 kN on the prototype. Four wing/blade assemblies generate the hydrodynamic lift. Each of the four wing/blade assemblies are equipped with a propeller drive powered by a 68 kW engine. The total power for the four propeller drives is 272 kW. The Cyclo-Crane is hovering when the wings are parallel to the horizontal axle of rotation. The propeller drives rotate the balloon envelope with the wings. The wings provide a lift via cyclic control. As soon as the payload has been lifted from the ground the blade airfoils are used to generate force in the line with a horizontal axis for forward (or reverse) movement. This is accomplished by turning the entire wing/blade assembly. When the wing/blade assemblies are directly aligned with the direction of flight, the vehicle has reached a non-rotating condition, and the maximum forward speed.

The weight of the prototype of the Cyclo-Crane is 82 kN and lighter than the aerostatic lift of 91 kN. The Cyclo-Crane must therefore use the motorpower by means of the wing/blade assembly in order to descend. The wing/blade assembly is therefore used, both for hovering and descending the Cyclo-Crane. The total payload of the prototype is estimated to 18 kN.

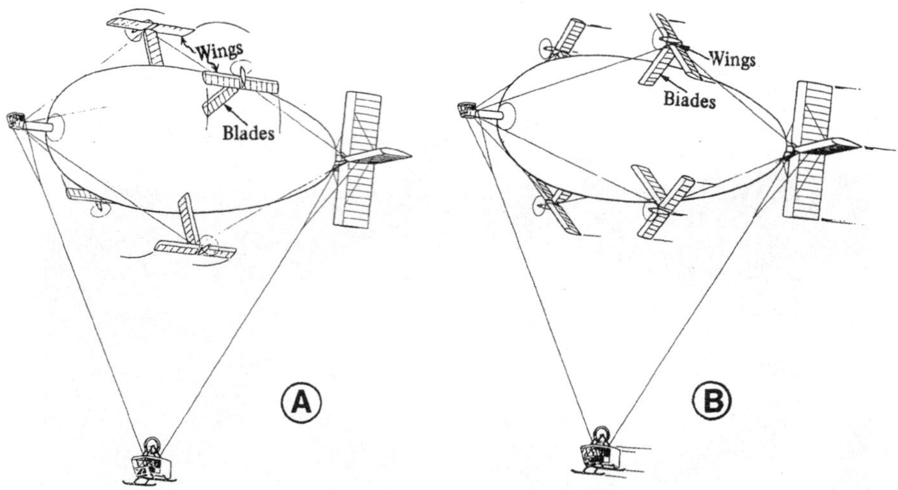

Figure 516. A. The Cyclo-Crane is hovering when the wings are parallel to the horizontal axis of rotation. The cab (and the sling load attached below the cab) is supported via bearing assemblies mounted for and aft of the main horizontal structure. The wings provide a lift via cyclic control.

B. The wing/blade airfoil assemblies continue to turn as the vehicle accelerates until the maximum forward speed is reached. At this point all airfoils (and engines) are directly aligned with the direction of flight and the vehicle has reached a non-rotating condition.

13.2. The history of balloon logging

The first experiment with balloon logging was carried out in 1956 (SUNDBERG 1962). A Norwegian 2-drum Vossa vinsj was used to pull the main and haul-back line and a barrage balloon from World War II was used to lift the front end of the logs during the skidding. The 500 m^3 cigar-shaped balloon envelope could lift 5.6 kN. When the weight of rigging and the envelope was subtracted, the effective lift was between 2 and 2.5 kN, sufficient to lift the front end of the loads above the terrain obstacles (figure 517).

A new experiment was carried out on Vancouver Island 1963. Two airship-shaped balloons with 2,380 m^3 helium gas were used during the experiment (MATHESON 1964).

In 1963 Faye STEWART, Bohemia Lumber Company and Goodyear Aerospace Corporation did an experiment with a V-shaped twin hull balloon (figure 518). The practical operations were followed by teorethical studies at the Pacific North West Forest and Range Experiment Station in Portland (MANN 1965) and at Vancouver Forest Products Laboratory in British Columbia (MacINTOSH 1968).

The V-shaped balloon gave an aerodynamic lift in addition to the aerostatic lift during the load transport. When the balloon was pulled forward at high speed, the aerodynamic lift could be significant (figure 519). The V-shaped balloon, however, was difficult to control. As long as the V-shaped balloon is being pulled forward it produced an aerodynamic lift. When the balloon is being pulled backward the opposite happens. When the empty balloon returns to the logging area, one often had to pass quite a distance beyond the point for choker setting in order to have room to turn the balloon around to position the balloon correctly for the load transport. This reduced production.

Figure 517. Balloon logging in Sweden, 1957.
1. The balloon is filled with 500 cu.m. of Hydrogen.
2. Transport to the logging site.
3. The Norwegian Vossa yarder was used during the balloon logging.
4. Balloon logging.
5. The landing.

The difficulties in controlling the V-shaped balloon resulted in quite a few accidents with total balloon failure. A broken balloon envelope and the loss of helium gas is costly to replace. From 1966 the natural shaped balloon has been used in balloon logging (figure 520).

Figure 518. V-shaped balloon logging in Oregon, USA 1963.

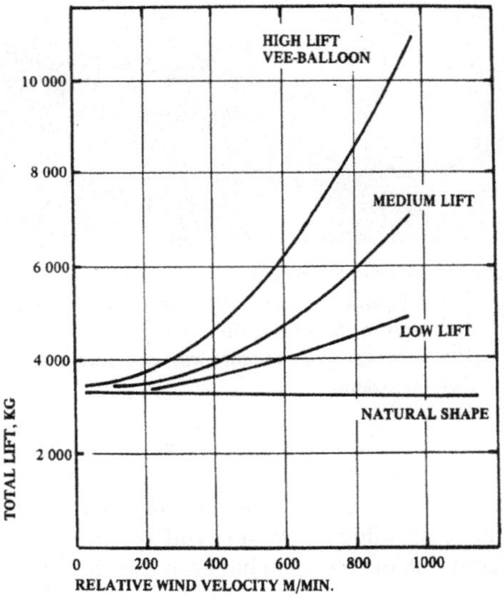

Figure 519. The aerodynamic lift varies with the speed and the shape of the balloon.

Figure 520. Raven 530 K logging balloon.

13.3. The gas and the balloon envelope

The gas has to be lighter than air in order to give aerostatic lift. Generally three gasses have been considered in balloon logging contexts.

Hydrogen	(H)	Specific gravity	0.068
Helium	(He)	Specific gravity	0.138
Ammonia	(NH_3)	Specific gravity	0.587

Hydrogen has the highest aerostatic lift. It is inflammable and it is explosive when mixed with 25% – 96% air (% by volume). This means that all metal parts of the balloon and harness must be grounded. Lightning or any static electricity around the balloon could conceivably cause an explosion if any hydrogen was leaking out. Logging is a production oriented business, so the equipment is not treated as carefully as it could be done in a laboratory environment. Therefore, even if hydrogen is considerably cheaper than helium, it must be considered a serious safety hazard and should not be used. Ammonia is also flammable, however, it is not so dangerous as hydrogen. The main problems with ammonia are its incompatibility with conventional balloon fabrics (LYSONS 1966) and its relatively high density. It has not been used in balloon logging applications so far.

Table 71. Density and aerostatic lift for some gasses (N/m^3)

Gas		Air	–	Gas	=	Aerostatic lift
Hydrogen (H)	:	12.68	–	0.88	=	11.80 kgs/m^3
Helium (He)	:	12.68	–	1.75	=	10.93 kgs/m^3
Ammonia (NH_3)	:	12.68	–	7.44	=	5.24 kgs/m^3

To give an aerostatic lift of 1 metric ton (9.8 kN) at sea level a volume of 831 m^3 hydrogen or 897 m^3 helium or 1,873 m^3 ammonia are needed. The aerostatic lift varies with the conditions in the air, the air pressure, the air temperature and the purity of the air (figure 521).

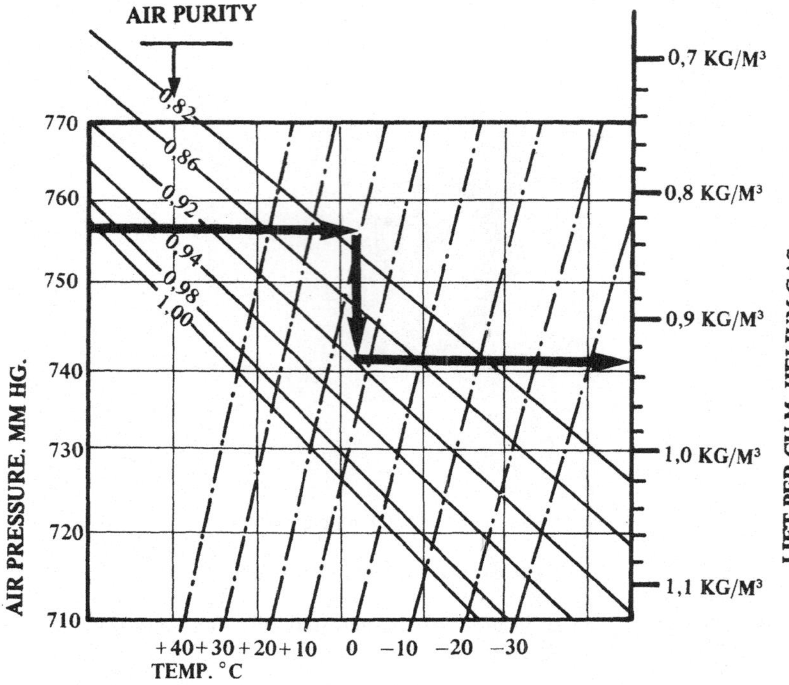

Figure 521. The lift of Helium gas in kg/m³ and its variation with air pressure, air purity, and air temperature.

Balloon envelopes are made of film-fabric laminates, two-ply laminates, or single films, depending upon the application of the particular balloon. The envelope of a non-rigid airship is usually made of 3 ply laminates. The exterior coating is usually aluminized polyurethane or aluminized hypalon to protect against the weather and the ultra-violet radiation. The envelope has two important functions, namly to support the structural loads by means of the cloth and to contain the helium gas by means of the elastomers.

The cloth can be made from a multitude of materials such as: cotton, rayon, acetate, nylon, dacron, orlon, acrilon, polyprophene, polyethylene, fibreglass (MYERS 1968).

The following elastomers which provide the gas barrier for the envelope have been used: natural rubber, butyl, neoprene, polyurethane, teflon, hypalon, viton, silicone.

The fabric construction is an important factor in determining service life, permeability to gasses, strength to weight ratio, and packaging characteristics. The most commonly used construction has been the elastomer coated fabrics. Multiple layers of cotton, nylon or dacron woven cloths have been used, and the elastomer coating was added. A non-rigid airship filled with 28,000 m³ helium gas uses a 1 mm thick envelope which weighs approximately 12 N/m². Figure 522 shows the film-fabric laminates.

The approximate weights of balloon or airship envelope materials can be obtained from figure 523.

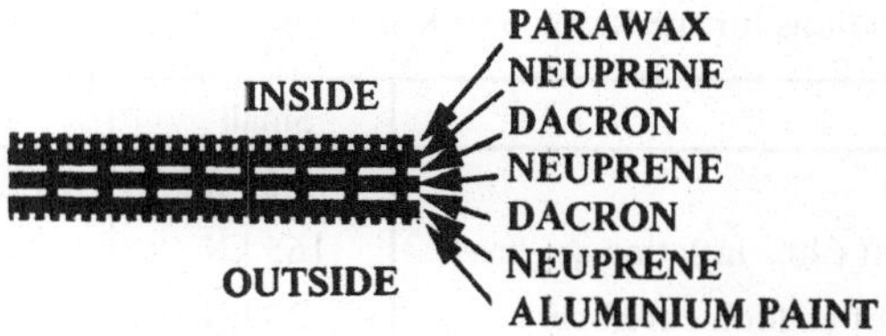

Figure 522. The construction of the balloon envelope.

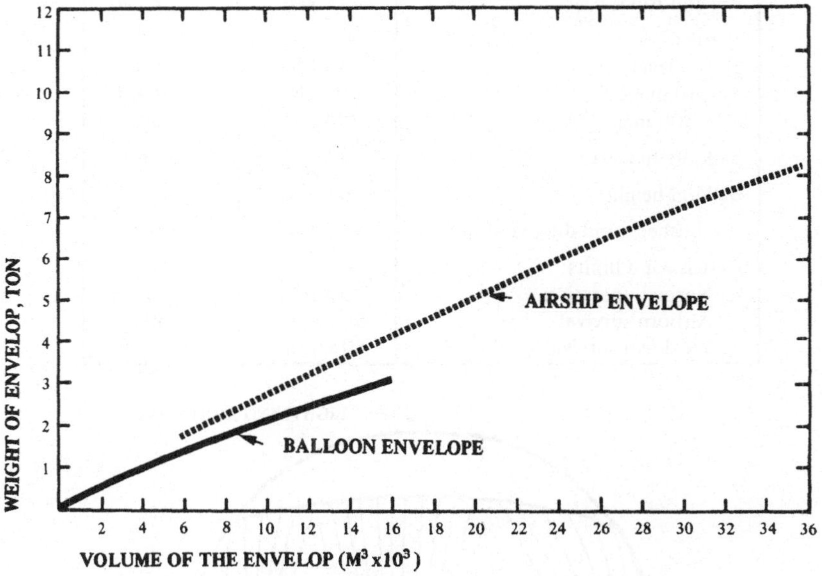

Figure 523. The weight of the balloon or non-rigid airship envelopes.

13.4. The equipment

Balloon logging is a winch system with yarder, winch drums and wire ropes. The balloon carries the load by means of aerostatic lift and replaces the skyline and the supports in a cable crane.

13.4.1. The natural shape balloon

The balloon which is used for balloon logging of big timber on the American West Coast by Bohemia Lumber Co. is constructed by Raven Industries Inc. It is a natural shape balloon which sometimes is called an onion-shaped balloon. It is made of a dacron-neoprene fabric envelope with reflective aluminium coloured outer surface (figure 520). The weight of the envelope material is 3.1 N/m^2. It has a tangent harness load suspension system and has a ram air pressurized skirt (ballonet). The balloon is attached to the butt rigging by means of a main swivel with a load divider and two tether lines. The balloon also has a lightning arrestor with six wires connected to earth via the tether lines. Figure 524 shows also the handling lines that are used when the balloon is bedded down. The balloons which have been used by Bohemia Lumber Co. have two dimensions, one 14,800 m^3 balloon with a net lift of 100 kN and a big model with 17,360 m^3 helium gas and a net lift of 130 kN (table 72).

Table 72. Specifications for the Raven 530 K Balloon.

Unit	Small	Big
Volume	14,800 m^3	17,360 m^3
Gross lift (90% inflation sea level)	165 kN	194 kN
Weight of balloon envelope	30 kN	34 kN
Balloon lift (sea level)	135 kN	160 kN
Rig and ropes	22 kN	22 kN
Net lift:		
a) Sea level	113 kN	138 kN
b) 610 m.a.s.	106 kN	131 kN
c) 1,500 m.a.s.	94 kN	116 kN
Balloon diameter	32 m	34 m
Balloon height	34 m	36 m
Established wind drag at 40 km/h	18.5 kN	19.0 kN
Design wind limits:		
Normal operation	40 km/h	40 km/h
Airborn survival	56 km/h	56 km/h
Tie down survival	80 km/h	80 km/h

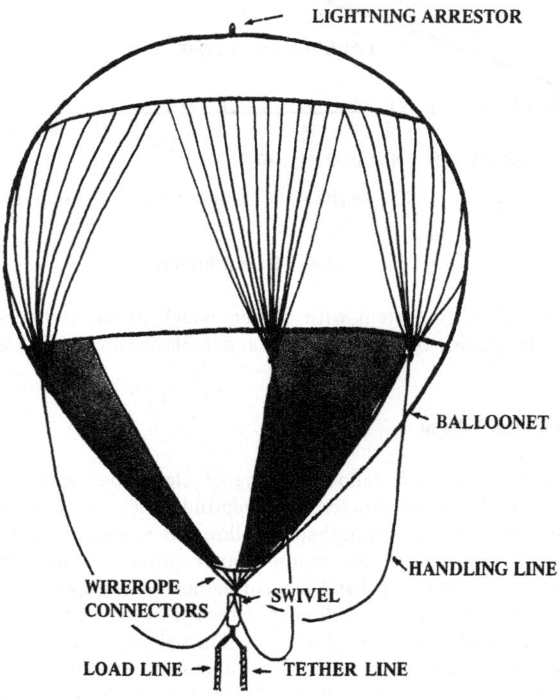

Figure 524. The balloon equipped for logging

Usually the balloon will not be completely filled to the design volume. If the balloon was fully inflated at standard atmospheric conditions, any change in the flight altitude or any increase in the temperature of the helium would cause the balloon to become pressurized, or the gas would be forced out of the balloon. Any initial fill of the balloon should be less than design volume in order to compensate for the changing atmospheric conditions. This is particularly easy to do with the natural shape balloon configuration, because the balloon will continue to keep its flight characteristics even when the actual gas volume is far below design volume.

13.4.2. The yarder

The yarder must be heavy in order to resist the lift of the balloon including possible wind forces (figure 525). Washington Iron Works developed a yarder with a double-drum interlocked winch for the balloon logging. The yarder is mounted on a crawler tractor chassis.The two drums are a main line drum and a haul-back drum. In some cases the haul-back line is used as an inverted skyline.

When a balloon running skyline system is used a 3-drum interlock winch is needed for the main line, the slack-pulling line and haul-back line.

Table 73. Some specifications of the Washington 608 A yarder (440 kW).

Drum type	Max. line speed		Max. line pull	
	Empty drum m/s	Full drum m/s	Empty kN	Full kN
Main drum	6.7	8.1	408	340
Haul-back drum	8.1	10.9	209	158
Straw line drum	6.6	18.5	165	56

Figure 525. The Washington single-drum yarder for balloon yarding. Power: 440 kW. Line speed: 10 m/s

Bohemia Lumber Co. has changed the system. Instead of one yarder with a double-drum winch, two yarders with single-drum winches are being used. One of them is pulling the main line and the other the haul-back line. This gives a certain freedom in the set up, the two yarders may be placed one at each end or both at the same end. The haul-back yarder may be placed anywhere on the yarding area where it is convenient for the set up. The two yarder operators communicate with each other by radio. These yarders are powered by 440 kW diesel engines and the line speed is up to 10 meters per second. The following lines are used:

Main line	L = 2,250 m	d = 25 mm
Haul-back line	L = 2,250 m	d = 25 mm
Straw line	L = 2,250 m	d = 11 mm
Load line	L = 200 m	d = 25 mm
Tether line	L = 200 m	d = 20 mm
Tagline	L = 30–150 m	d = 25 mm

13.4.3. Transfer vehicle

The Washington yarder is self-propelled, and may be used to move the balloon over shorter distances. Usually, however, the main and haul-back lines are fixed to the yarder and a transfer vehicle is used for moving the balloon to the bedding area and for tying it down to the ground by means of the handling lines. The transfer vehicle may also be used to do additional work on the landing (figure 526).

13.4.4. Landing machines

The balloon logging has a high daily production and it is necessary to remove the logs from the landing during the yarding operation. A grapple skidder is used to move the logs from where the yarder engineer dumps them on the landing to the log-pile. A loader continuously loads the logs on trucks.

Figure 526. Transport tractor to take down the balloon and move it from one place to another.

13.4.5. The bedding area

The bedding area is needed to park the balloon when the operation is stopped or when the balloon must be refilled. There is an anchoring point with a suckerdown block at the centre of the bedding area. The balloon is transferred from the transfer vehicle to the suckerdown block. The balloon is being pulled down to the suckerdown block and fastened. The handling lines are then fastened to anchoring points on the periphery of the bedding area.

A helium trailer is an important part of the equipment. It carries the containers with helium gas for refilling the balloon.

13.4.6. Balloon size for yarding 2 m^3/load

Most of the balloon logging so far, has been carried out in virgin forests with very large tree dimensions averaging 9 m^3 per tree. Other regions have smaller tree dimensions. If they are less than 3 m^3 per tree and cut into logs before the yarding, 2 m^3 per load would normally be sufficient. This corresponds to a net lift of 15 kN in coniferous forests. Sometimes it is interesting to carry out full tree yarding. In this case a net lift of 15 kN would correspond to 1.0–1.4 m^3 per load (see chapter 4.1., table 7 and 8).

Balloon logging is of interest under extremely difficult terrain conditions when the timber has to be yarded down from very rough mountainous terrain to a landing. If the height is 1,000 m.a.s. and the temperature is 25°C, the gross lift of the helium gas is 9 N/m^3 (figure 523). Under these conditions a balloon containing 3,700 m^3 helium gas would have a gross lift of 33 kN.

Table 74. The size of a logging balloon for 2 m^3 loads (SVENDSEN 1972).

Balloon volume	4,100 m^3
Volume helium gas (90% inflation)	3,700 m^3
Gross lift at 1,000 m.a.s.	33 kN
Weight of envelope	9 kN
Balloon lift (1,000 m.a.s.)	24 kN
Rig and ropes	9 kN
Net lift (1,000 m.a.s.)	15 kN
Design wind limits	
Normal operation	30 km/h
Tie down survival	90 km/h

The yarder should have a double-drum interlock winch powered by 100–150 kW engine. For a 1,000 meters balloon cable crane the following wire ropes may be needed:

Main line	L = 1,600 m	d = 12 mm
Haul-back line	L = 2,300 m	d = 12 mm
Straw line	L = 3,000 m	d = 6 mm

13.5. The operational method

The yarder with the 2- or 3-drum interlock winch may be placed at the top or at the bottom of a hillside, depending on the terrain conditions and the location of the access and feeder roads. The yarding may go uphill or downhill independent of where the yarder is located.

The balloon logging should only be used where the forest conditions allow a large logging volume for each set up. One has to build large landings and bedding areas. There must be one landing for each set up, but one bedding may serve several set ups.

The balloon logging is suitable for uphill yarding, downhill yarding or even in places where the timber has to be yarded over a ridge or a crest of a hill. The balloon replaces both the supports and the skyline of a cable crane.

When the yarder is equipped with a 2- or 3-drum winch for the yarding of big timber, the following crew is needed (SVENDSEN 1972):

1	foreman (hooktender)
1	yarder operator
1	rigging slinger
2	choker setters
1	chaser
1	operator on the grapple skidder
1	operator on the loader
8	men

13.5.1. The turn cycle

Each of the turn cycles during the balloon yarding is approximately the same for the various operational methods.

1. While the balloon is at the landing and the load is unhooked, the choker crew (one rigging slinger and 2 choker men) is pre-setting one set of chokers (usually 4 chokers per set).

2. The yarder engineer slacks off the main line and reels in the haul-back line. He applies the interlock, which causes the balloon to travel at a fixed height above the ground, until it is above the next load. When two single-drum yarders are used the two yarder engineers communicate by radio in order to return the balloon above the next load. At this time the rigging slinger radios the yarder engineers to stop the balloon.

3. The chokers are brought down to the ground by either tightening the main line or the haul-back line, depending on where the rigging slinger wants the chokers deposited. Once the chokers are so close to the ground that one of the crew members can catch them, the rigging slinger calls the yarder engineers to stop the lines. The set of empty chokers is then released from the tagline.

4. The rigging slinger radios the yarder engineer to tighten the main line or haul-back line whereby the tagline with the toggle is slacked off and the crew can attach the load where the chokers already have been set, to the tagline.

5. When the chokers are attached to the tagline and the crew has moved a safe distance away, the rigging slinger calls the engineer to slacken the haul-back line. In this way the balloon rises and lifts the load off the ground by pure static lift. When the turn is in the air the yarder engineer starts the winch and pulls the load to the landing.

6. When the load is close to the landing the engineer stops the haul-back line and tightens the main line whereby the logs are lowered to the ground. The chaser unhooks the logs and signals the engineer to slacken the main line and thus a new cycle is started.

13.5.2. Main and haul-back lines with double-drum yarder

During the planning of the operation the yarding area is marked on the maps and so is the landing, the bedding area and the location of the balloon - cable crane. When the landing and the bedding area has been prepared, the equipment can be transported to the site and the balloon filled with gas. The hooktender selects the cable corridor. The crew strings out the straw line. The haul-back line is being pulled around the corridor by means of the straw line. The balloon with the load line, the tether line and butt rigging is moved to the yarder by means of the transfer vehicle and the rigging is connected to the main and haul-back lines. The yarding can start.

The length of the tagline varies from 30 – 150 meters depending on the conditions. The head end of the tagline consists of 15 meters sections attached to each other by means of quick couplings. A suckerdown block, fastened to a stump in the centre of the area where the logs are to be collected, helps to pull the balloon down (figure 527). The haul-back line pulls out the rig with the balloon and pulls it down towards the suckerdown block. When the butt rig approaches the ground, the tagline is loose and it is easy for the crew to hook on the pre-set chokers. Then the load transport can start.

When the logs on the area around the suckerdown block are yarded the yarder engineer pulls in the balloon towards the yarder. When the balloon is near the yarder the engineer slackens the haul-back line so that the balloon only hangs in the main line. When the haul-back line lies slack on the ground, the crew move the suckerdown block from one stump in the area where the yarding is finished to a new stump where the yarding can continue (figure 531–14 A).

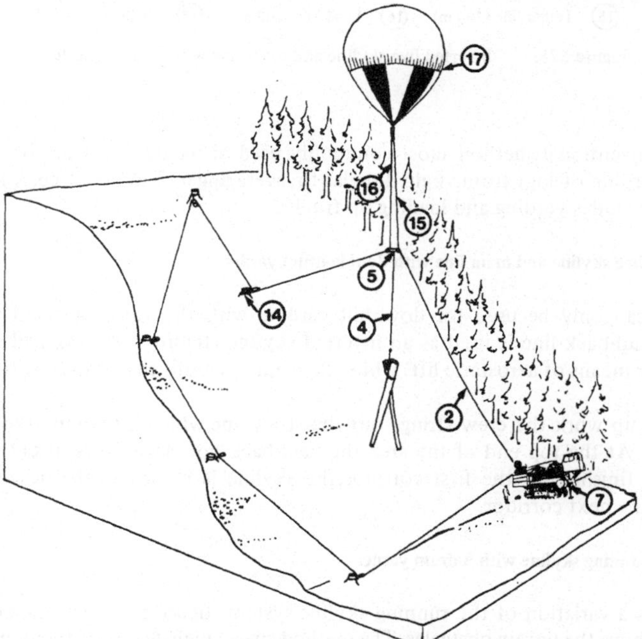

(2) Main line (4) Tagline (70 m) (5) Butt rigging (7) 2-drum yarder

(14) Suckerdown block (15) Tether line (200 m) (16) Load line (200 m) (17) Balloon

Figure 527. Main and haul-back line with 2-drum yarder.

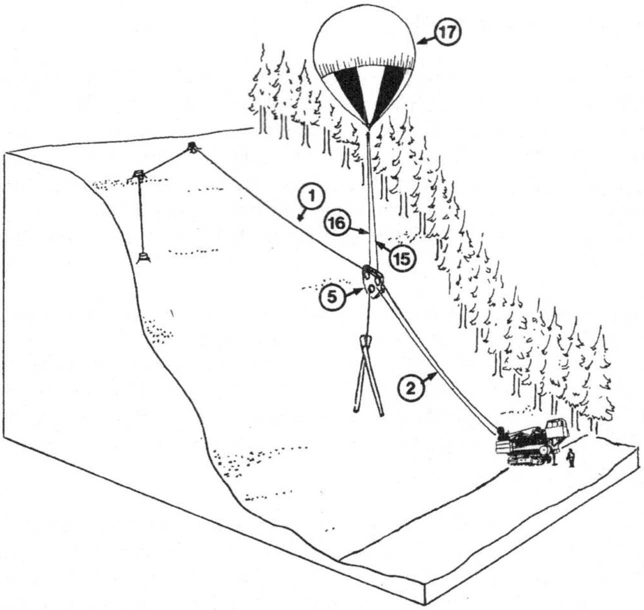

(1) Skyline (2) Main line (5) Carriage (7) 2-drum yarder

(15) Tether line (200 m) (16) Load line (200 m) (17) Balloon

Figure 528. Inverted live skyline and main line with 2-drum yarder.

This is the conventional method most commonly used in balloon yarding. An average production for yarding of logs from virgin douglas fir, averaging 6–7 m^3 per tree, is 160–165 m^3 per day. This includes yarding and loading on trucks.

13.5.3. Inverted live skyline and main line with double-drum yarder

This method can only be used for downhill yarding with the yarder located at the valley station. The haul-back line is used as an inverted skyline (figure 528). The balloon hauls the carriage out by means of aerostatic lift, while the main line pulls the carriage with the load in.

During the rig up work the crew strings out the straw line which is then used to pull out the haul-back line. At the top end of the area the haul-back line passes several tail blocks.When the yarding is finished on the first corridor, the skyline is moved to the next tail block in order to yard the next corridor.

13.5.4. Balloon running skyline with 3-drum yarder

This method is a variation of the running skyline system (figure 529). The balloon is used to lift the loads above the terrain obstacles. The method gives a high degree of freedom concerning the location of the yarder under different terrain conditions. Uphill or downhill yarding may be carried out whether the yarder is placed at the valley station or at the top of the yarding area. It is also possible to yard the logs out from behind a ridge. The rig up work is approximately the same as for a conventional running skyline cable crane.

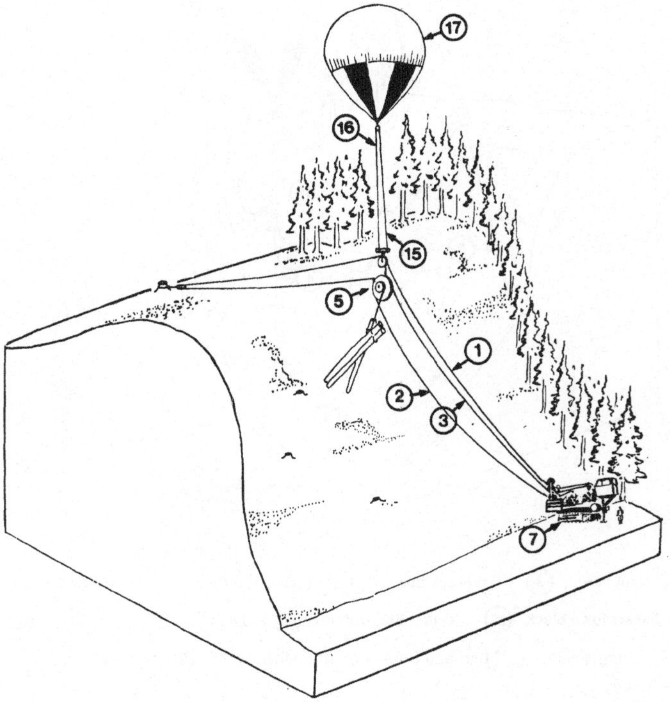

(1) Skyline (2)(3) Main and slack-pulling line (5) Carriage (7) 3-drum yarder

(15) Tether line (200 m) (16) Haul-back line (200 m) (17) Balloon

Figure 529. Balloon running skyline with 3-drum yarder.

13.5.5. Main and haul-back lines with two single-drum yarders

This is a similar method to the one described in chapter 13.5.2. Instead of one yarder with a double-drum winch, the main and haul-back lines are pulled by two single-drum yarders (figure 530). Bohemia Lumber Co. has changed to this method because the production is higher than with any of the other systems. The two yarders may be located at each end of the yarding area. The main line yarder should be located at the landing. The haul-back yarder may be placed anywhere on the yarding area where it is convenient to find a yarder station. The following crew is needed:

1	hooktender
2	yarder engineers
1	operator on the grapple skidder
1	operator on the loading machine
1	chaser
2	choker setters
1	rigging slinger
9	men

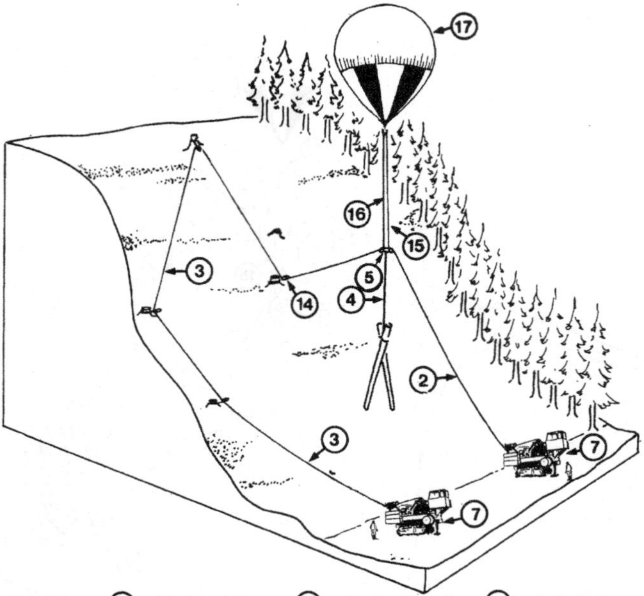

(2) Main line (3) Haul-back line (4) Tagline (70 m) (5) Butt rigging

(14) Suckerdown block (15) Tether line (200 m) (16) Load line (200 m) (17) Balloon

Figure 530. Main and haul-back line with two single-drum yarders.

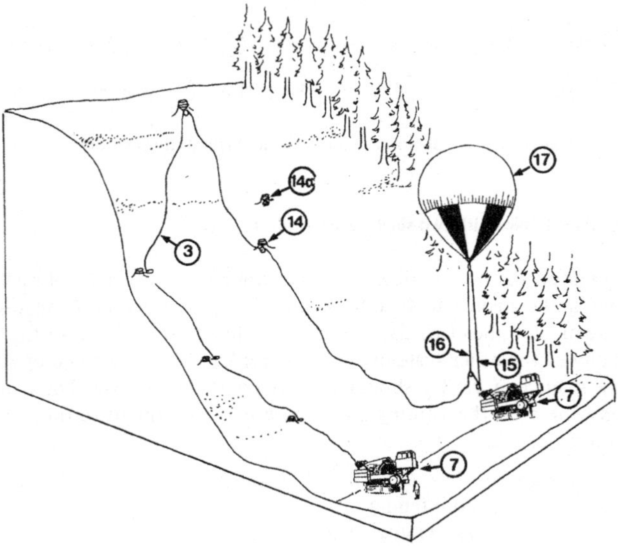

(2) Main line (3) Haul-back line (7) 1-drum yarder (14) Suckerdown block

(15) Tether line (200 m) (16) Load line (200 m) (17) Balloon

Figure 531. The method of changing place for choker setting on the yarding area.
The balloon is hauled in to the main line yarder. The haul-back line lies loose in the terrain. This simplifies the work when moving the suckerdown block from one logging place (14) to another (14a).

Figure 532. Balloon logging carried out by Bohemia Lumber Co. at Goose Bay, California, September 1983. The logging balloon has 15,000 m^3 Helium gas. There are two Washington 608 A single-drum yarders: one for the main line and the other for the haul-back line. The production was 280 m^3 per day.

(1) The logging balloon with 15,000 m^3 of Helium gas. (2) The two yarders are placed at the top of the very steep logging area: one yarder for the main line and one for the haul-back line. (3) The Washington 608 A yarder powered by a 440 kW diesel engine. The yarder is equipped with a single-drum winch. (4) The main line yarder has pulled in the rig.

The two yarder operators have radio contact with each other and can move the butt rigging and the load with high speeds, up to 10 meters per second. The tagline varies between 30 and 150 meters, depending on the conditions. The head end of the 30–150 m line is a 15 meters long tagline. The choker crew pre-sets the chokers, and hooks them on the 15 m long tagline. If one tagline is not enough two sections with a quick coupling between them are used. When the balloon arrives at the choking area, the choker crew simply replaces the empty tagline with a full tagline. While the yarder and the balloon transports the load to landing the choker crew pre-sets the chokers and prepare the load on a new tagline. Two different balloon sizes have been used shown in table 72. The average load per turn depends on the balloon:

The small balloon:	Average 7.0 m^3/load	Max. 12 m^3/load
The big balloon :	Average 8.5 m^3/load	Max. 14 m^3/load

The maximum length of the cable crane has been 2,000 meters but usually the length is between 900 – 1,200 meters. A balloon logging operation was carried out in an old stand of Douglas fir (figure 532). The average tree dimension was 9.5 m^3 per tree and the biggest tree was 28 m^3 per tree. The typical production was 280 m^3 per day.

13.5.6. One main line and two haul-back lines with 3-drum winch (Japanese system)

One system has been tested in Japan where frequent road changes are avoided by using two haul-back lines (figure 533). The yarder is equipped with a 3-drum winch, one main drum and two haul-back drums. The two haul-back lines run from the spar tree to two tail trees.The yarding area is located between the two haul-back line corridors. The entire area may be yarded using one set-up, since the main line and the two haul-back lines can move the balloon with the tagline to any point within the yarding area. A practical test of the system was carried out in the Itonose National Forest near Numata in Japan 1979. The natural shaped

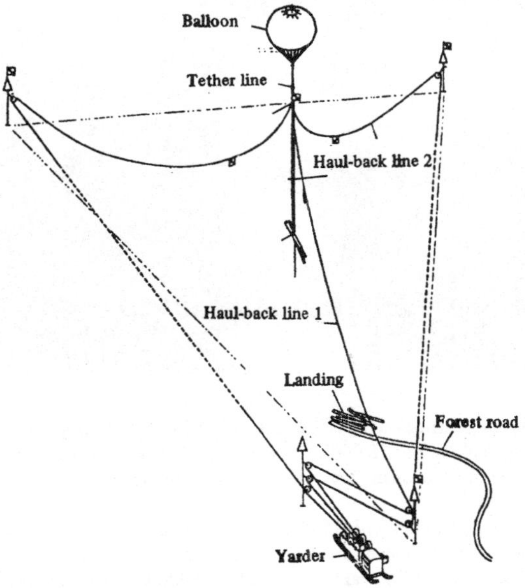

Figure 533. A Japanese method of balloon logging, covering the entire area between the two haul-back line corridors. One main line and two haul-back lines with a 3-drum yarder (KONUMA and SHIBATA 1976).

balloon was filled with 1,400 m^3 helium gas which gave a net lift of 7.3 kN. The balloon lifted the front end of the loads during the high skidding towards the landing. The 3-drum yarder was powered by a 77 kW diesel engine (Iwafuji Industrial, Modell Y-33 HD6). The yarder had four speeds, and the highest speed was 9 m/s. The main line dimension was d = 12 mm and the two haul-back lines d = 10 mm.The dimension of the tagline was d = 10 mm. The load line between the butt rigging and the balloon was 45 meters long and d = 12 mm. No production figures have been reported.

14.0. THE DIMENSIONS, PRODUCTIVITY AND SOME ECONOMIC ASPECTS OF CABLE CRANE YARDING

When comparing production and productivity of different operational methods used under various conditions and in different regions of the world, one has to take into account the variety in the lenght of the working days and the way in which the working day is being utilized. It is even more difficult to compare cost figures from one region to another. The cost of manual labour as well as the machine costs are different from one country to another and so is the relationship between manual and machine costs.

The costs also varies with the time. An example to illustrate the problem is given in table 75.

Table 75. Total working time, work place time, and costs of manual labour including social costs. Piece work in Norway.

Year	Total hours per week	Work place hours per 8 hrs.day	Increase in manual labour cost		Consumer price index
			Nkr./hr 8 hrs.day	%	%
1950	48	7.5	3.6	100	100
1955	48	7.3	5.8	161	136
1960	45	7.0	6.9	192	156
1965	45	6.8	11.0	306	190
1970	42.5	6.5	20.9	581	242
1975	42.5	6.3	37.6	1,044	361
1980	40	6.0	57.5	1,597	490

In Norway the cost of manual labour was 16 times higher in 1980 than in 1950. The consumer price index was only five times higher after the same period. The saleries increased more in forestry than in other industries in order to give the forest workers the same income as other workers. Another problem is the working time. The total working time was reduced from 48 hours per week to 40 hours per week. With the increasing social considerations in the country the work place hours per 8 hour day decreased from 7.5 to 6.0 work place hours per 8 hour day, according to investigations made by the Norwegian Forest Research Institute.

Trends of similar character but with different rates may be found in the various countries. It is therefore, difficult to compare production in volume per working day since small contractors (the Gypo-contractors) sometimes work 10 or 12 hours per day. Big forestry companies have to follow the rules and regulations of the country. A comparison between daily productions can be meaningless.

Instead, we have chosen to compare productivity figures per work place hour. This is the effective working time pluss the normal delay times when the worker or the working crew is occupied with the operation.

In the attempt to find out whether there exist some general trends of relationship between tree dimensions, the sizes of the cable cranes and the productivity, 85 different cable systems have been analysed. These are cable systems commonly used in different parts of the world, Japan, Northern Europe, Central Europe, Soviet Union, USA and British Columbia. These systems have been described in chapter 11. The average figures, grouped in five classes, are shown in table 76.

Table 76. Average figures for 85 investigated cable systems in Japan, Northern Europe, Central Europe, Soviet Union, USA and British Columbia.

Cable systems	Number of systems	Value	Mean value	Std. error of mean	Minimum value	Maximum value
Old systems	21	m^3/tree	**2.5**	0.8	0.2	9.0
		Yarding dist. m	**348**	79	100	1,250
		Crew size	**4.5**	0.5	3	8
		Manhours/m^3	**0.62**	0.09	0.12	1.80
		ˣ Crew hour/m^3	**0.138**			
		ˣ m^3/hr	**7.3**			
Running skyline	19	m^3/tree	**4.9**	0.9	0.3	9.0
		Yarding dist. m	**111**	8	50	200
		Crew size	**5.9**	0.5	2	9
		Manhours/m^3	**0.31**	0.04	0.06	0.69
		ˣ Crew hour/m^3	**0.053**			
		ˣ m^3/hr	**19.0**			
Grapple yarding	5	m^3/tree	**4.9**	1.7	0.3	9.0
		Yarding dist. m	**151**	25	110	250
		Crew size	**4.0**	0	4	4
		Manhours/m^3	**0.18**	0.06	0.06	0.43
		ˣ Crew hour/m^3	**0.045**			
		ˣ m^3/hr	**22.2**			
Shotgun systems (gravity outhaul)	15	m^3/tree	**2.7**	0.8	0.2	9.0
		Yarding dist. m	**128**	11	71	214
		Crew size	**4.3**	0.5	2	8
		Manhours/m^3	**0.39**	0.05	0.17	0.80
		ˣ Crew hour/m^3	**0.091**			
		ˣ m^3/hr	**11.0**			
Other new systems	25	m^3/tree	**3.7**	0.6	0.5	9.0
		Yarding dist. m	**168**	21	57	500
		Crew size	**4.9**	0.4	3	9
		Manhours/m^3	**0.29**	0.02	0.15	0.57
		ˣ Crew hours/m^3	**0.059**			
		ˣ m^3/hr	**16.9**			

ˣ calculated

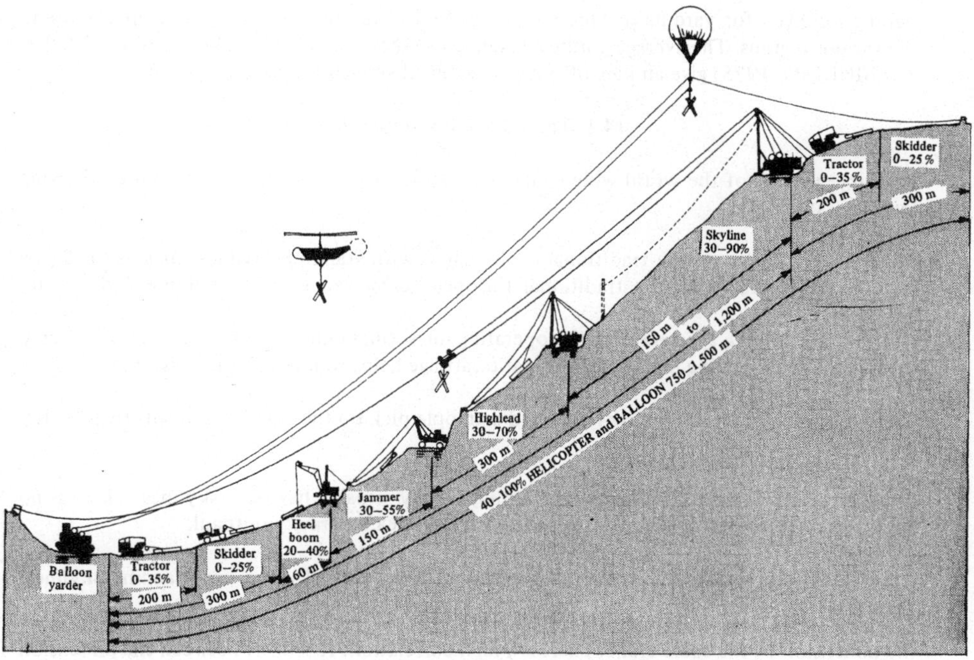

Figure 534. Optimum yarding distances and slope per cent of various logging systems, suggested by STUDIER and BINKLEY (1975).

The 85 cable systems have been described in scientific or practical reports from the various countries. The author has visited these countries several times and studied these systems in the field. Each of the 85 cable systems has been described with comparable figures based on the scientific or practical reports as well as the authors notes.

Tree dimensions. In most cases there is a positive correlation between the size of the trees and the volume to be cut per hectare. Cu.m. per tree has been used to describe the size of the timber.

The size of the cable crane. The weight of the yarder cannot be used to describe the size of a cable crane. A light top mounted yarder may have the same size of engine as a much heavier mobile yarder at the road side. They may both use the same dimensions on the wire ropes. The engine power is not a good description of a size of a cable crane. A 40 kW top mounted yarder may use gravity to pull down the loads and may have a higher production than a 100 kW yarder, where motor power is used to transport the loads uphill.

The carrying cross-sectional area of the metal in the wire ropes may be used as an indication of the size of the cable cranes. The cross-sectional area is given in mm^2. It is the sum of the cross-sectional area of each individual wire in the wire rope. We have calculated the carrying cross-sectional area of the main line, as well as of the skyline. For a running skyline system this is the sum of the carrying cross-sectional areas of the lines which are carrying the load.

Length of the cable crane. With better layout of the road network the yarding distance tends to be shorter for modern cable cranes than for the more traditional types. The productivity figures are given for yarding distances which is typical for the specific cable cranes in use in the various regions. The average yarding distances are shown in table 76. Figure 534 (STUDIER and BINKLEY 1975) give an idea of typical yarding distances for various systems.

14.1. The cable crane dimensions

The large variety of the world-wide cable crane systems has been grouped into the following five classes:

Old systems. These are the traditional cable cranes with standing skylines, such as the Tylor system, the North Bend, the traditional European cable cranes with standing skylines etc.

The running skyline systems uses the operating lines (main line and haul-back lines) to carry the load. A slack-pulling line or an automatic carriage helps to pull out the hoist line.

Grapple yarders uses running skylines and a grapple pick up the logs and unloads them at the landing.

Shotgun systems are used for uphill yarding. They usually have live skylines. There is no haul-back line and the carriage returns to the yarding area by gravity outhaul.

Other new systems are the modern cable cranes which usually use live standing skylines.

The sum of the cross-sectional areas of the main line and the skyline has been used to indicate the dimension of the cable crane. There is a high correlation, R = 0.91, between the dimension of the main line and the skyline. The standard error of mean is S = 41 %. The relationship is illustrated in figure 535. As expected there is a high degree of variations in the the world-wide range of cable systems. Clear relationships have been found, however, between the cross-sectional areas of the main line (Main mm^2) and the skyline (Sky mm^2):

Old systems	$\text{Sky} = \text{Main} \cdot 2.46$	(135)
Running skyline	$\text{Sky} = \text{Main} \cdot 1.48$	(136)
Grapple yarders	$\text{Sky} = \text{Main} \cdot 1.44$	(137)
Shotgun systems	$\text{Sky} = \text{Main} \cdot 2.50$	(138)
Other new systems	$\text{Sky} = \text{Main} \cdot 1.90$	(139)

All the 85 investigated systems are represented in figure 535. Each point represents one or more systems because many systems had the same dimensions of the skylines and the main line.

The formulae 135 to 139 show that modern cable cranes use smaller skylines than the older ones. One of the reasons may be that the cable cranes are shorter today than in the past, due to a denser road network. This has resulted in reduced skyline tensions and bigger deflection than what they used on the standing skylines in the past.

The relationship between the tree volume ($\overline{v}$ m^3/tree) and the cross-sectional area of the main line (Main mm^2) can be described by the following equations (R = 0.77, S = 63%).

Old systems	$\text{Main} = 32.8 \cdot \overline{v} + 30$	(140)
Running skyline	$\text{Main} = 27.6 \cdot \overline{v} + 30$	(141)
Grapple yarders	$\text{Main} = 28.6 \cdot \overline{v} + 30$	(142)
Shotgun systems	$\text{Main} = 20.7 \cdot \overline{v} + 30$	(143)
Other new systems	$\text{Main} = 19.7 \cdot \overline{v} + 30$	(144)

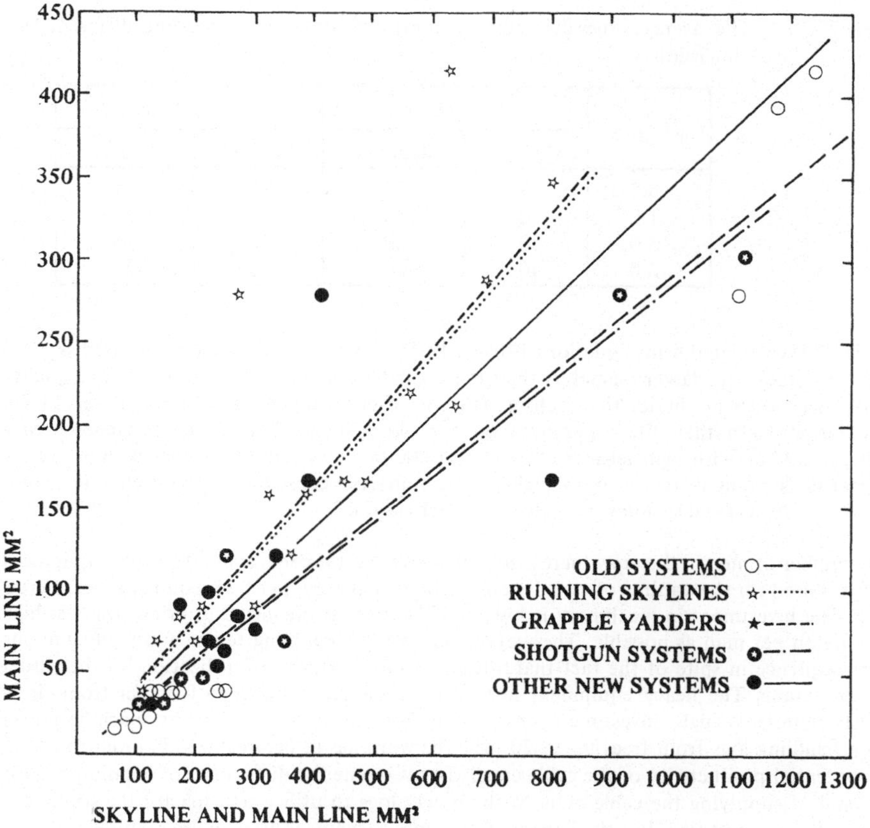

Figure 535. The relationship between the skyline and the main line dimensions for various yarding systems. (The dimensions are defined by means of the carrying cross-section in mm^2).

The relationship between the tree volume ($\overline{v}$ m^3/tree) and the cross-sectional area of the skyline (Sky mm^2) ($R = 0.77$, $S = 58\ \%$) is as follows:

Old systems	$Sky = 78.8 \cdot \overline{v} + 80$	(145)
Running skyline	$Sky = 33.2 \cdot \overline{v} + 80$	(146)
Grapple yarders	$Sky = 33.6 \cdot \overline{v} + 80$	(147)
Shotgun systems	$Sky = 49.6 \cdot \overline{v} + 80$	(148)
Other new systems	$Sky = 30.3 \cdot \overline{v} + 80$	(149)

The wire rope dimensions which are needed for a cable crane can be found by the calculations discussed in chapter 5 and 7. The equations 140 to 149 simply show what dimensions were normally used when yarding timber from stands with varying tree dimensions. The volume $\overline{v}$ is the average tree volume in a stand. There will always be a distribution of the volume per tree around the average and this is especially the case where the average tree volume is high because they represent virgin stands. If the average volume is 6 m^3 per tree, the largest trees to be yarded may have a volume as high as 20 m^3 per tree. The smaller average volumes originates usually from stands which have been managed by a good silvicultural programme on a sustained yield basis and the variation of the tree volumes around the average is less in such cases.

Table 77. The average line diameters (d mm) of cable cranes yarding different tree dimensions.

m^3/tree	Old systems		New systems	
	Skyline	Main line	Skyline	Main line
0.5	19	12	17	11
1.0	22	14	18	12
3.0	30	20	23	16
6.0	40	26	28	21

Table 77 is calculated using equations 140 – 149. It shows that the old systems used skylines and main lines with larger diameters than the diameters of the lines used today. The quality of the wire ropes are better than before. The wire ropes which are used today have a higher breaking strength than the ropes used on the old systems. The old cable cranes used a standing skyline with high tension. A modern cable crane system uses a live skyline and the shorter cable crane distances allow higher deflection. As a consequence the skyline tension is lower than for the old systems for a given tree dimension.

The number of men in the crew increases with the size of the cable crane. The light equipment which is used for the yarding of tree dimensions less than 1 m^3 per tree requires a 2-man crew only. The investment in machinery is low, and the work is organized in order to get as high production per man as possible. The workers are often combining felling and yarding in one work sequence in spite of the fact that it leads to a lower degree of utilization for the yarder and the crane. The heavy equipment needed for cable crane yarding of timber from virgin forests represents high investment costs. Manpower must be replaced by machine power when handling logs from trees up to 20 m^3. The work has to be organized in order to utilize the potential productivity of the cable crane. To overcome the difficulties of handling the big logs and of supplying the cable crane with enough logs to utilize its capacity the size of the crew has to be increased. For the largest cable cranes as many as 10 men are required.

The 85 investigated cable systems had the following relationships between the number of men in the crew and the size of the cable crane classified according to the carrying cross-sectional areas of the main line and skyline (figure 536).

Main line ($R = 0.78$, $S = 26\ \%$):

$$\text{Crew} = 3.79 \cdot \log A - 2.24 \tag{150}$$

Main line and skyline ($R = 0.75$, $S = 27\ \%$):

$$\text{Crew} = 4.59 \cdot \log A - 6.16 \tag{151}$$

The equations 150 and 151 can be used to calculate the number of men required to operate a given size of cable crane.

Table 78 shows the average sizes of the crews in the 85 investigated cable systems. Figure 536 shows, however, that there are big deviations around the average figures. The trend toward smaller wire rope dimensions in modern cable cranes has already been discussed. New cable yarding methods also reduces the size of the crew required. Grapple yarding is one example. Future development of cable yarding systems will probably reduce rather than increase the number of men in the yarding crew.

Table 78. Average number of men in the crew for various dimensions of the cable cranes.

Crew Average number of men	Main line d (mm)	Skyline d (mm)
2 – 3	7 – 8	12 – 15
4	8 – 12	15 – 21
5	12 – 16	21 – 26
6	16 – 21	26 – 30
> 7	> 21	> 30

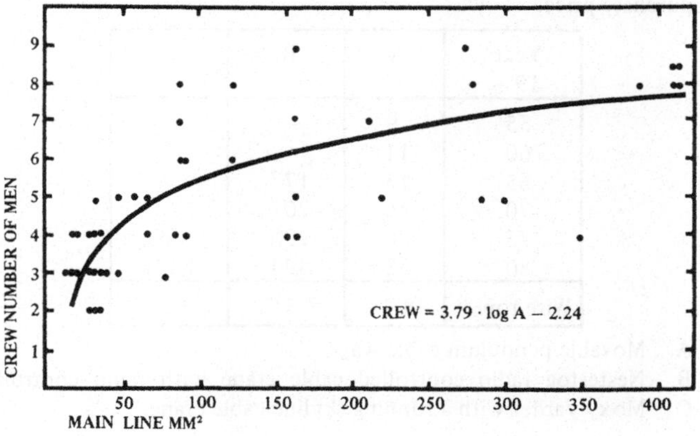

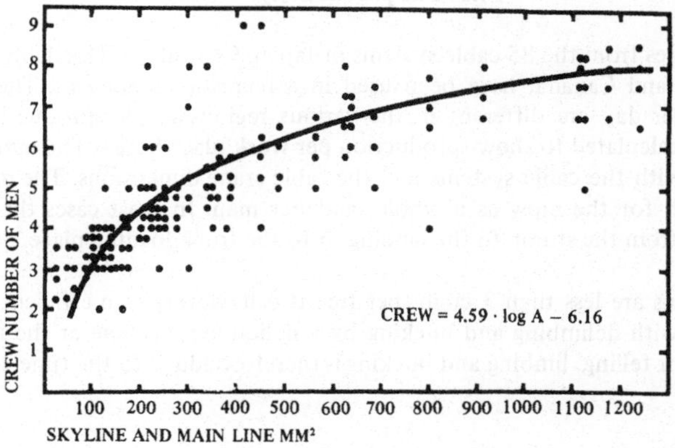

Figure 536. Relationship between the size of the cable crane (defined by the carrying cross-section of the cables) and the number of men in the yarding crew.

The technical development has led to more sophisticated yarders and cable cranes. Every effort has been made to design machinery which can help the yarding crew to achieve a high production and a low rig up time. Together with the inflation this has led to a considerable increase in the purchase prices for the cable cranes. As a result the investment in machinery has increased. An example from Norway is shown in table 79. The wire ropes represented approximately 50 % of the purchase price for a pendulum cable way powered by gravitational forces. They represent 7.5 % of the purchase price for a radio controlled cable crane and only 3 % for a running skyline cable crane with a mobile yarder.

The fact that the cost of the wire ropes represents a smaller part of the purchase price does not alter the fact that the sum of the cross-sectional areas of the carrying ropes is a good criterion for describing the cable crane size. The ropes must be dimensioned in accordance with the size of the logs or the loads which have to be transported.

Table 79. The purchase price (including taxes) in 1,000 Nkr. for some Norwegian 700 meters long cable cranes. The purchase price of wire ropes in percent of total purchase price.

Year 19–	A	B	C
55	9	–	–
60	11	–	–
65	13	177	–
70	16	201	–
75	26	290	890
80	32	494	1,110
Wire ropes	48.4%	7.5 %	3.2%

A. Movable pendulum cable way.
B. Nestestog radio controlled cable crane with radio controlled carriage.
C. Moxy yarder with a running skyline cable crane.

14.2. The productivity

The production figures from the 85 cable systems in Japan, Central and Northern Europe, the Soviet Union, USA and Canada, have been used in a regressional analysis. The number of work place hours per day are different in the various regions so the »production per day» figures have been recalculated to show »production per work place hour». The number of men in the crew, varies with the cable systems and the cable crane dimensions. The production is therefore given both for the crew as a whole, and per man. In most cases the production cover yarding only, from the stump to the landing or to the truck loading place.

When tree dimensions are less than 3 cu.m. per tree it is,however,often efficient to combine felling and yarding with delimbing and bucking by a delimbing machine at the landing. The time consumption for felling, limbing and bucking is therefore added to the time consumption for yarding.

In the cases when the cutting is being carried out by chainsaw, the time consumption is shown in figure 537. It covers the preparation of unbarked logs from coniferous trees on snow free ground and under normal cable crane conditions with smooth and steep terrain surface. The time consumption during cutting may be higher if there is extreme terrain difficulties.

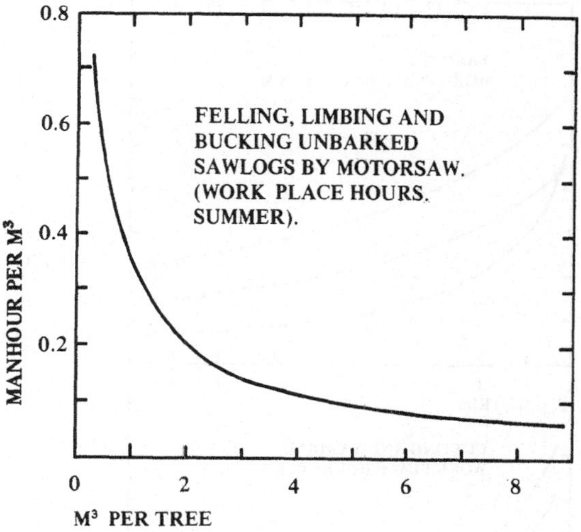

Figure 537. Time consumption for motor-manual felling, limbing, and bucking of unbarked logs of softwood (coniferous trees). The time consumption is higher under extreme steep and rough terrain conditions.

The average production for the 85 investigated cable systems are given in the equations 152 – 160. The equations show how the production varies with the tree dimensions ($\overline{v}$ m^3 per tree). The typical yarding distances for the various systems are shown in table 76.

Yarding production m^3/hour for the crew: ($R = 0.889$, $S = 40$ %).

Standing skyline systems	$y = 3.30 \cdot \overline{v} + 7.20$	(152)
Running skyline systems	$y = 3.23 \cdot \overline{v} + 9.60$	(153)
Grapple yarding	$y = 5.83 \cdot \overline{v} + 7.54$	(154)

Yarding production m^3/man/hour: ($R = 0.819$, $S = 46$ %).

Standing skyline systems	$y = 0.30 \cdot \overline{v} + 1.61$	(155)
Running skyline systems	$y = 0.52 \cdot \overline{v} + 2.11$	(156)
Grapple yarding	$y = 1.44 \cdot \overline{v} + 1.94$	(157)

Yarding and cutting production m^3/man/hour: ($R = 0.894$, $S = 30$ %).

Standing skyline systems	$y = 0.28 \cdot \overline{v} + 1.14$	(158)
Running skyline systems	$y = 0.32 \cdot \overline{v} + 1.31$	(159)
Grapple yarding	$y = 0.60 \cdot \overline{v} + 1.56$	(160)

There is a good correlation (R) and an acceptable standard error of mean (S) which shows that there exist clear tendencies in the world-wide range of cable crane systems that the production increases with the tree dimention. As must be expected from such a comparison there is a great deal of variation around the averages. There are systems with much lower production than the averages which are described by the equations. There are also systems with much higher production. This shows that there is still room for developments in the fields of cable yarding, felling, conversion and truck loading.

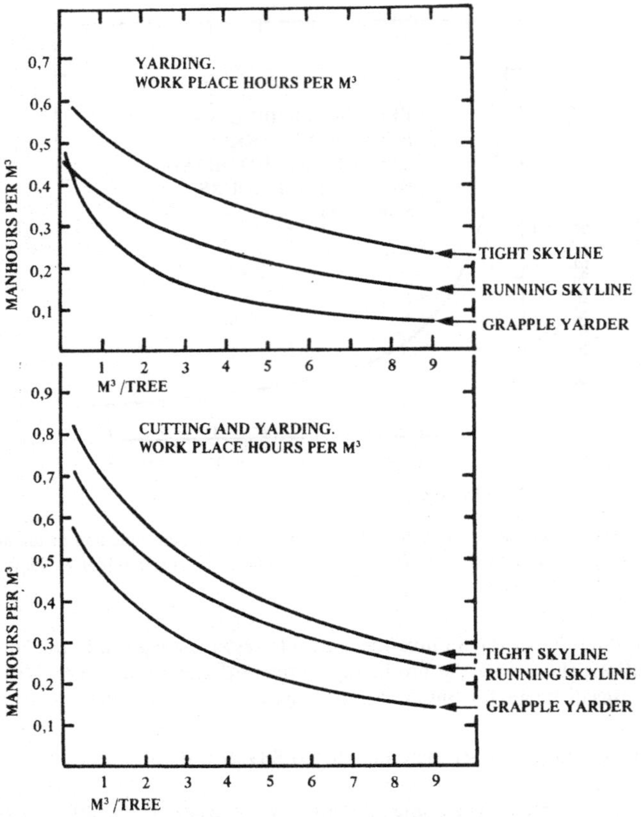

Figure 538. The productivity in manhours per m^3 for cable yarding and for cable yarding including cutting and conversion. Average figures for 85 cable systems in Japan, Soviet, Central Europe, Northern Europe, USA, and British Columbia.

The production based on the equations 152 – 160 are shown in table 42, chapter 11.4.3. and table 57, chapter 11.5.4.9. The corresponding time consumption in manhours per cu.m. are shown in figure 538.

14.3. Some economic aspects

The machine cost is the sum of all the capital costs and the running costs, such as depreciation, interest, fuel, oil, maintenance, repair, insucrance, garage etc. The costs of the operating lines such as main line, haul-back lines, as well as all auxiliary lines, may be treated as running costs while the cost of the wire rope in a standing skyline may be treated as another capital cost (the yarder, the carriage etc.). The life span of wire ropes has been discussed in chapter 2.

Figure 539 shows a comparison between the machine cost per kWh for some of the Norwegian cable cranes and the manual cost per hour. The manual cost has increased more than the machine cost. The cost level and the cost increase are different in the different countries, but most countries in the world show the same tendency namely that the manual cost is increasing more rapidly than the machine cost. This is the reason why it is necessary to develop better equipment in order to increase production and reduce the input of manual labour.

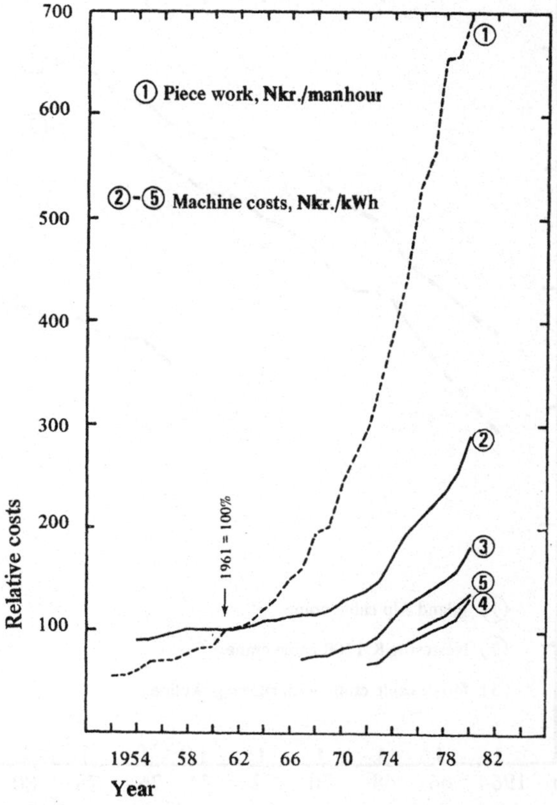

Figure 539. Manual costs per man-day and machine costs per kilowatt hour.
1. Manual costs, Nkr./man-day.
2. Bjønn cable crane. Barked timber.
3. Igland Alp winch and Nestestog K-1200 radio controlled cable crane. Motor-manual limbing. Tree length logs.
4. Nestestog K-1200 radio controlled cable crane and Logma T-300 limbing machine. Tree length logs.
5. Moxy cable crane and Logma T-300 delimbing machine. Tree length logs.

The total yarding and cutting cost is the sum of the machine cost and the labour cost. The distribution between them varies with the yarding method. The manual labour costs as a percentage of the total yarding costs for some cable crane systems are listed below:

Gravity powered pendulum cable way.	3-man crew	< 1 m^3/tree	80–90%
Light cable cranes with top mounted winch.	3-man crew	< 1 m^3/tree	70–80%
Radio controlled cable crane.	3-man crew	< 2 m^3/tree	60–70%
Running skyline with mobile yarder.	4-man crew	< 2 m^3/tree	40–50%
Mobile tower. Slackline system.	8–10-man crew	< 20 m^3/tree	25–35%

While the manual costs represented 70–90% of the yarding costs with the labour intensive methods in the past the costs of manual labour represents 25–40% of the yarding costs with todays efficient cable cranes.

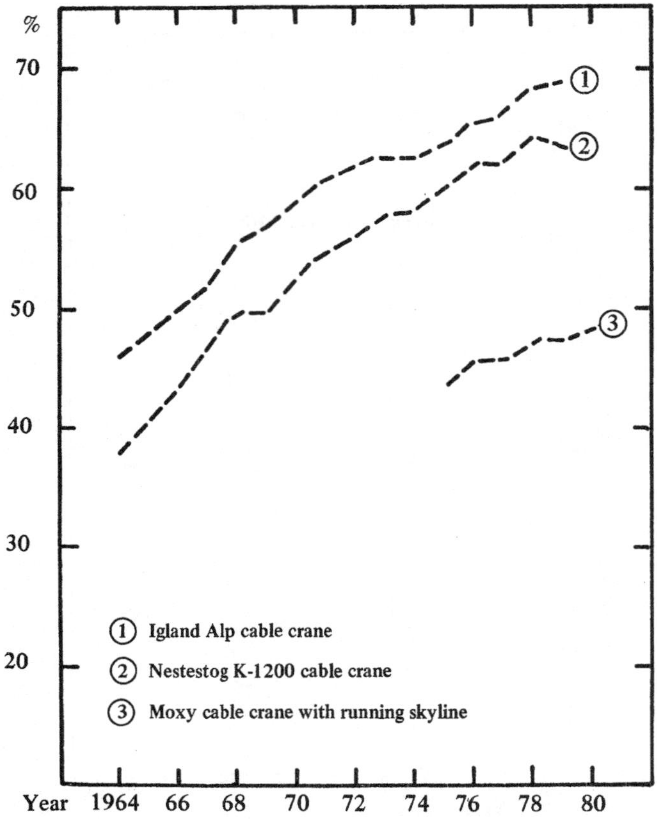

Figure 540. Cost of manual labour in per cent of the total yarding costs (nominal Nkr./m^3) for Igland, Nestestog, and Moxy cable cranes.

Since the costs of manual labour increases more rapidly than the machine costs, the relationship between them changes with time. This is illustrated in figure 540 for three different Norwegian cable cranes. The manual costs play a more important role today than a few years ago. The same tendency may be seen when studying the operating costs of cable crane yarding in developing countries where the labour costs are low compared to the industrialized countries.

14.4. A simple method for the calculation of machine costs

The Norwegian Forest Research Institute carried out an investigation of the machine costs for different types of work in forestry, such as tractor operations, skidding, harvesting, road building etc. From the studied material (143 machines operating during 491,759 work place hours) it was found that there is a linear relationship between the total machine cost per day during normal work in forestry and the purchase price of the machine (SAMSET 1972). As a result equation 161 may be used for approximate calculations of the machine costs:

$$y = 2\,A + 40 \qquad (161)$$

y = machine costs in Nkr./day
A = Purchase price in 1,000 Nkr.

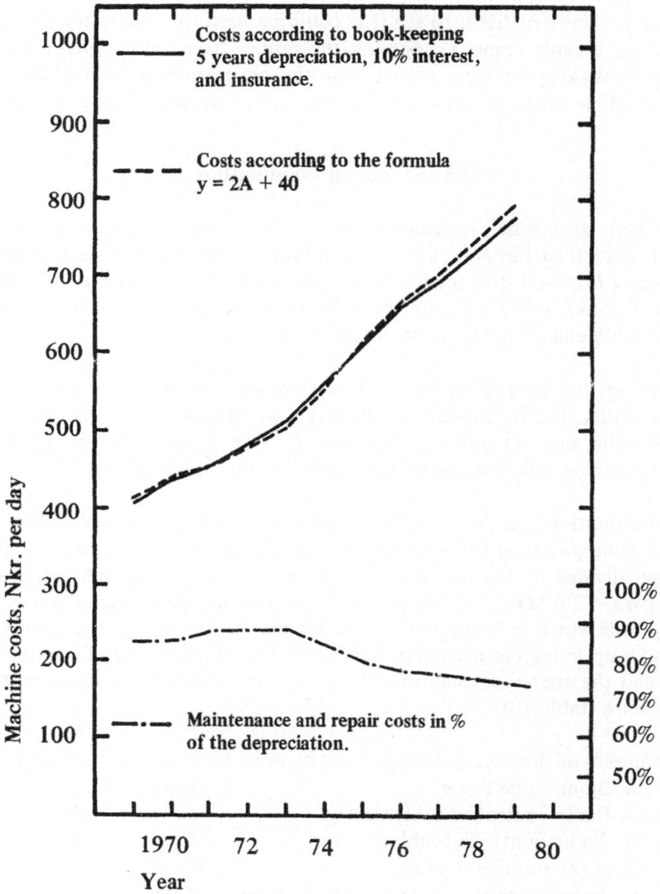

Figure 541. The machine costs in Nkr./day according to book-keeping compared with the costs calculated by the formula y = 2A + 40. The investigation was carried out for year-round yarding with Nestestog K-1200 cable crane during the period 1969–1979.

The factor 40 is related to the money rate of Nkr. The equation is applicable to any other currency if the constant 40 is changed according to the exchange rate. If one dollar costs 8 Nkr. the equation would read:

$$y = 2\,A + 5 \tag{162}$$

y = machine costs in \$/day
A = purchase price in 1,000 \$

If the total purchase price for a machine is 200,000 \$ equation 162 shows that the total machine costs are 405 \$/day. These costs includes both the capital costs and the running costs, but not the manual cost of operator or other workers in the crew. The machine costs according to this equation have been tested against more detailed calculations of the machine costs in many countries (Europe, Australia, America etc.) and the tests have shown that the equation gives a good approximation.

The equation has also been tried, to see if it could be used for calculating the total machine costs per day for a cable crane. Comparing the results from using equation 161 with the results from book-keeping have shown that the equation can be used to calculate the approximate machine costs provided all the wire ropes are included in the purchase price (figure 541).

14.5. The energy consumption

A cable crane may be powered by yarders with different engine capacities and operated by different sized crews. If cutting and yarding is combined in one work sequence using the Moxy cable crane, and a Logma T-300 telescope delimber converts the trees to logs at the landing, the total engine pöwer is 237 kW, and there are 6 men in the crew. A Bjønn cable crane is powered by a 18 kW engine and there are 3 men in the crew.

A large yarder does not necessarily have a higher energy consumption per m^3 yarded timber than a smaller yarder. In the long run it is an advantage to choose the right yarder for the job and to organize the work in order to minimize the energy consumption of the machinery as well as the labour consumption, in both cases per produced m^3 of timber.

The energy consumption may be given in kWh per m^3. It may be calculated by multiplying the optimum engine power in kW with the average number of work place hours per day.This product is then divided by the produced volume in m^3 per day. 1 kWh corresponds to 3.6 megajoule (1 kWh = 3.6 MJ). It is important to use the maximum engine power and not for example the power which is being used by the machine. If an over-sized yarder is being used, this should show up in the calculation of the energy consumption. As an example, the labour consumption and the energy consumption have been investigated for 5 different Norwegian cable crane systems (table 80).

Table 80. Examples of the energy consumption of some Norwegian cable cranes.
A. Bjønn cable crane (Chapter 11.4.2.2.)
B. Radio controlled cable crane (Chapter 11.5.2.1.)
C. Radio controlled cable crane (Chapter 11.5.2.1.)
D. Moxy running skyline (Chapter 11.5.4.4.)
E. Igland farm-tractor mounted cable crane (Chapter 11.5.4.6.)

Method	A	B	C	D	E
L_1 m	500	500	500	500	150
Crew	3	3	4	6	2
Limbing	chainsaw	chainsaw	Logma (x	Logma (x	chainsaw
Power kW	18	63	145	237	50
m^3/h (crew)	0.61	2.06	3.91	8.93	3.04
Manhours/m^3	5.0	1.45	1.02	0.67	0.67
kWh/m^3	29.5	30.6	37.1	26.5	16.4

(x Mechanized limbing and bucking by the Logma telescope delimbing machine.

This method of showing the energy consumption was first suggested by the Canadian logging engineer Alexander Koroleff (KOROLEFF 1961). It is rather interesting to see that the labour intensive method A, with the Bjønn cable crane had the highest labour consumption, but also a rather high energy consumption in spite of the fact that the engine power was 18 kW. The yarder and the limbing machine in the method D was powered by 237 kW, but the energy consumption was less than in the method A and the efficient cable crane operation gave a low labour consumption per m³. The light, skidder or farm-tractor mounted, cable crane with a small 2-man crew used in method E has a low energy consumption as well as a low labour consumption. For timber of less than 1 m³/tree this is an elegant method. Much emphasis should be laid on the construction of a dense network of feeder roads with 300 meters spacing in order to take advantage of such a cable crane method.

The calculation of the energy consumption and the labour consumption is a good way of finding out whether the yarder is well adapted to the forest and terrain conditions and whether the work is being organized in an appropriate manner.

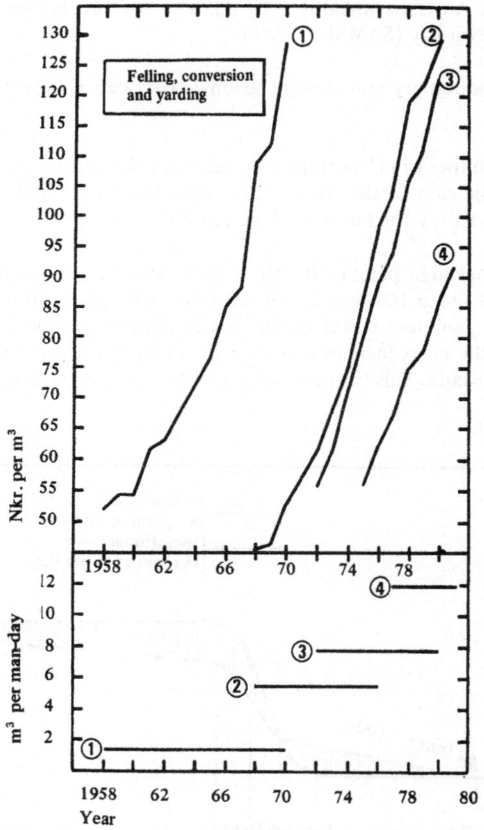

① Yarding with Bjønn cable crane. Felling and conversion with chainsaw (3 man crew)

② Yarding with Nestestog radio controlled cable crane. Felling and conversion with chainsaw (3-man crew)

③ Yarding with Nestestog radio controlled cable crane. Felling by chainsaw, limbing by Logma telescope delimber (3-man crew)

④ Moxy cable crane with running skyline. Felling by chainsaw. Limbing by Logma telescope delimber (6-man crew)

Figure 542. Operational costs (Nkr./m³) and productivity (m³/man-day) for felling, conversion and transport with some cable crane systems.

14.6. The law of discontinuous evolution

As long as the same yarding method is being used there is little variation in the productivity. The nominal costs of the yarding increases, however, and the only way to decrease the operational costs is to introduce a new yarding method either by using a new working method, a new organization of the work or by introducing new equipment. A study of the changes in operational methods showed that the evolution followed a certain pattern. We call it *the law of discontinuous evolution.*

The existence of this law has been shown for the evolution of operational methods in easy terrain (SAMSET 1966), for logging in *Pinus radiata* plantations and for logging big timber from *Eucalyptus* stands in Australia (SAMSET 1971). The law has also been shown to apply for cable crane yarding in Norway (SAMSET 1981).

Figure 542 shows the productivity and the operational costs per m^3 for various Norwegian cable crane methods.

The productivity is the number of m^3 produced per man per day during the work place time. *The operational costs* is the sum of the costs of the manual labour in the crew and the total machine costs per day, divided by the number of men in the crew.

The productivity (in m^3 per man per day and the costs in Nkr.per man per day) may be given in relative figures. Both are set to 100 % in a year where the total costs of men and machinery per man per day for the yarding method should not increase if one wishes to keep a good operational economy. If the costs increases too much in relation to the productivity of the method, a price pressure occurs. This happens according to a pattern illustrated in figure 543.

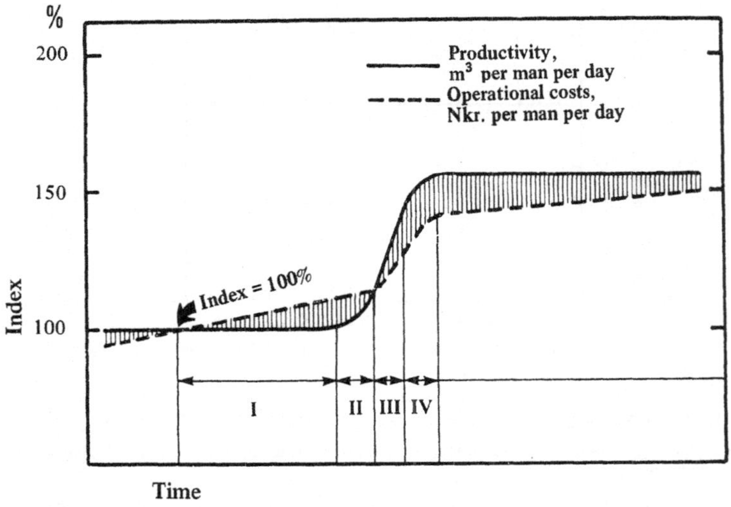

Figure 543. The stages of discontinous evolution.
I. The stage of price pressure.
II. The stage of developing new methods.
III. The stage of introducing the developed new method.
IV. The stage of stabilizing the developed new method in practical forestry.

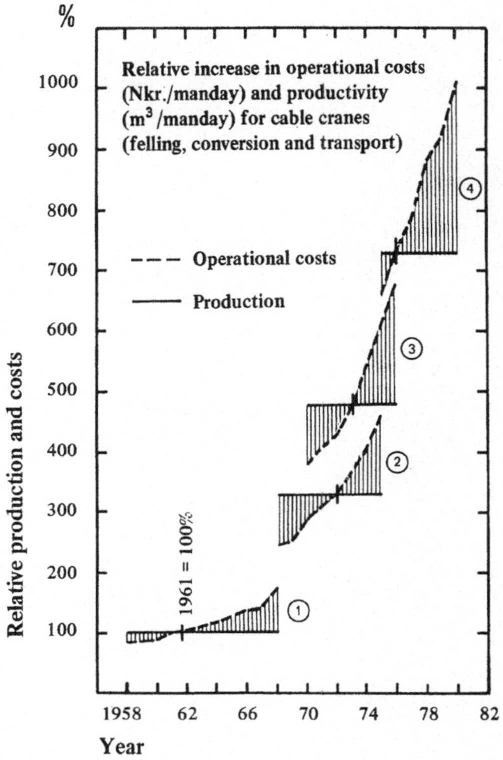

Figure 544. The law on the discontinous evolution applied for felling, conversion and cable transport with some cable systems:

1. Bjønn cable crane. Manual felling, conversion and transport of barked timber.
2. Nestestog radio controlled cable crane. 3 men. Motor-manual conversion and transport of tree length logs.
3. Nestestog radio controlled cable crane and Logma telescope delimber. 3 men. Felling, conversion, and transport tree length logs.
4. Moxy yarder, Logma delimbing machine. 6 men.

An example of the law of discontinuous evolution is shown for some Norwegian cable crane systems (figure 544). It is interesting to note that the relationship between the costs and productivity was approximately the same for a labour-intensive cable crane method in 1962 as for a capital-intensive cable crane method in 1975.

LITTERATURE

ADAMS, K. 1962. Common carriers in the woods. Forest History *6*(1 pg 2). 11 pp.

AGER, B. H., NILSSON, N.-E. och SEGEBADEN, G. v., 1964: Beskrivning av vissa skogstekniskt betydelsesfulla bestånds- och trädegenskaper samt terrengförhållanden. Studia Forestalia Suecica *20*: 1—159.

AMERICAN STEEL & WIRE COMPANY. 1943. Amerstrand steel strand. Catalog and construction handbook. New Jersey. 95 pp.

AMERICAN STEEL & WIRE COMPANY. 1943. Wire rope. Engineering hand book. New Jersey. 136 pp.

ANDERL, J. 1972. Praktische Möglichkeiten der übertrieblichen Zusammenarbeit. Allgemeine Forstzeitung, Folge 7. Juli. 5 pp.

ANDERSON, C. H., 1977: Engineering Calculations. C. H. Anderson and Associates Ltd. Vancouver. (stensil)

ANDREWS, W. 1954. This was logging! Superior Publishing Company, Seattle, Washington. 157 pp.

ANDREWS, W. 1956. Glory days of logging. Superior Publishing company, Seattle, Washington. 175 pp.

ANONYMOUS. 1892. Wire rope transportation. The Trenton Iron Co., New Jersey. 128 pp.

ANONYMOUS. 1900. Skogbruget paa verdensudstillingen i Paris. Tidsskr. Skogbr. *8*(8):259.

ANON, 1943: Wire Rope. Engineering Handbook. Columbia Steel Company. American Steel & Wire Company. New Jersey. 136 pp.

ANONYMOUS. 1945. Den nye jordbruksmaskinen «Den elektriske hesten» vinner mer og mer innpass. Morgenavisen *44*(9). 1 pp. [Morgenavisen 11. januar 1945.]

ANONYMOUS. 1950. «De flygande skottkärrorna». Några data ur linbanans 550-åriga historia. Svenska Dövstum-Tidningen nr. 7—9:6—8.

ANONYMOUS. 1952 a. Når tømmerstokkene går luftveien. Fædrelandsvennen *78*(278). 1 pp. [Fædrelandsvennen 29. november 1952.]

ANONYMOUS. 1952 b. Taubane for tømmertransport i Flå. Buskeruds Blad *71*(274). 1 pp. [Buskruds Blad 24. november 1952.]

ANONYMOUS. 1952 c. Taubane og winsj løser transportproblemer i våre skoger. Vårt Land *8*(249). 1 pp. [Vårt Land 25. oktober 1952.]

ANONYMOUS. 1953 a. Ideell taubane for kupert terreng. Fylket *27*(178). 1 pp. [Fylket 29. september 1953.]

ANONYMOUS. 1953 b. Mekaniseringen innen skogbruket krever standardisering av utstyret. Østlendingen *53*(171). 1 pp. [Østlendingen 28. juli 1953.]

ANONYMOUS. 1953 c. «Skogbruksutstillingen i Elverum peker inn i fremtiden» — Tømmer på luftfart i Vindheia. Østlendingen *53*(171). 1 pp. [Østlendingen 28. juli 1953.]

ANONYMOUS. 1954 a. Ei lunne- og transportbane som vil laga ein liten revolusjon i skogbruket. Hordaland *71*(41). 1 pp. [Hordaland 29. mai 1954.]

ANONYMOUS. 1954 b. Hugst også i avsidesliggende skoger. Aftenposten *95*(489). 1 pp. [Aftenposten 23. oktober 1954.]

ANONYMOUS. 1954 c. Ny tømmervinsj lansert av Skogforsøksvesenet. Nationen *37*(245). 1 pp. [Nationen 22. oktober 1954.]

ANONYMOUS. 1954 d. Vinsjen når fram til skogen i utilgjengelig terreng. Østlendingen *54*(25). 1 pp. [Østlendingen 30. januar 1954.]

ANONYMOUS. 1958 a. Angaben über Versuchstransport mit Langstreckenseilkranen in Rüti, Linie No. 4, Abteilungen 7 und 8, September/Oktober 1957. Eidg.Forstl. Versuchsanstalt Abt. III. Birmensdorf. 7 pp.

ANONYMOUS. 1958 b. Nürzung and Pflege des Bergwaldes. Die Blaue Nr. 6. 19 pp.

ANONYMOUS. 1963 a. Här kommer virke dansande på lina. Söderhamns-Kuriren *68*(34). 1 pp. [Söderhamns-Kuriren 9. februar 1963.]

ANONYMOUS. 1963 b. Tømmerdrift der alle vilde segja det er uframkomeleg lende. Hordaland *80*(19). 1 pp. [Hordaland 9. mars 1963.]

ANONYMOUS. 1966. 100 years after the birth of B.C.'s glory industry. Columbia Lumberm. *50*(8):32—41.

ANONYMOUS. 1975. Sikring av taubanedriften i Norge. Statens tilsyn med privatbaner og taubaner. Oslo. 65 pp.

ANONYMOUS. s.a. Internationale seilbahnterminologie. Industrie- und Fachverlag, Wien. Upag.

ANONUYMOUS. s.a. Loggers all. 60 years of progress in western logging. Caterpillar Tractor Co., Peoria, Illinois. 16 pp.

ANON, 1975: Norse Rope. Ståltaukatalog. Elkem Spigerverket A/S. Stål og Tau. Bergen. 67 pp.
ANON, 1978: Machinery for forestry — winches — specification, definitions and performance. The International Organization for Standardization. ISO/TC 23/SC 15N86E. 4 pp.
ANON. 1980. Oregon occupational and health code. Division 80. Logging. Salem, Oregon. 97310. 107 pp.
ANON. 1983. Production of various cable-systems on the US West Coast. Oregon State University, Corvallis. Stencile. 23 p.
ANON. 1983. Illustrated cable logging systems. The Forestry Mechanization Society, Japan. Report no. 65. 133 pp.
AROLA, R.A. 1981. A steep terrain feller - buncher:The Menzi Muck. Technical Release 81–R–15. Washington.6 pp.
ARVESEN, A., 1970: Snarekjøring av tømmer i hele lengder med terrenggående traktorer. Meddr. norske Skogfors. Ves. *27* (nr. 99): 313—474.
AUSTIGARD, B. 1980. Skriftlege opplysningar om taubanar. (Brev 8/4. Molde. 1 pp. [Arkiv NISK.]
BACH, C., 1922, 1924: Maschinenelemente. I—II. A. Kröner, Leipzig. 560, 552 pp.
BAKKE, E. 1956. Driftstekniske kurs i skogbruket. Forelesninger om skogbruksspill. A/s Anleggstransport, Oslo. 26 pp.
BAKKEN, A. 1972. Restbruddstyrke i slitte ståltau. *Residue breaking strength in worn steelwire ropes.* Driftsteknisk rapport nr. 11. Tidsskr. Skogbr. *80*: 495—500.
BENNETT, W. D. 1960. Cableway systems for pulpwood transport. Woodlds Res. Index No. 119. 71 pp.
BENNET, W. D., 1962: Forces Involved in Skidding Full Tree and Tree Length Loads of Pulpwood. Pulp and Paper Research Institute of Canada. Techn. Rep. 302—1962, 39 pp.
BERGEN v., L. 1963. Lintransport av rundvirke. SDA-Redogörelse Nr. 1. Stockholm. 110 pp.
BERGER, B. 1951. Landbruksvinsjen i skogbrukets tjeneste. Sogn og Fjordane Tidende *20*(20). 1 pp. [Sogn og Fjordane Tidende 9. mars 1951.]
BERGSTEN, G. 1978. Hovedtræk i NSRs historie. Pp. 9—16 *in*:Nordiska Skogsarbetsstudiernas Råd 25 års jubilæumskonference, maj. Skovteknisk Institut, København. 42 pp.
BERING, K. 1956. Driftstekniske kurs i skogbruket. Forelesninger i taubaneberegninger. Det norske SkogforsVes., Vollebekk. 15 pp.
BIELAYEV, L. M. & FRANCUZOV, I. L., 1964: Erection of aerial ropeways. Moscow.
BINKLEY, V. W. 1964. A comparison of high-lead yarding production rates in windthrown and standing timber. USDA For.Serv.Res. Note PNW—13. 9 pp.
BINKLEY, V. W. 1965. Economics and design of a radio-controlled skyline yarding system. USDA For.Serv.Res.Pap. PNW—25. 30 pp.
BINKLEY, V. W. & LYSONS, H. H. 1968. Planning single-span skylines. USDA For.Serv. Res.Pap. PNW—66. 10 pp.
BINKLEY, V. W. s.a. Factors to consider when selecting skyline yarding systems for site protection. USDA For.Serv., Oregon. 10 pp.
BOTNEN, O. 1980. Eventyrgarden Brakandalen. Varden *106*(117). 1 pp. [Varden 24. mai 1980.]
BRABECK, W. 1972. Erfahrungsbericht über den Einsatz des Kurzstreckenseilkrans »Urus» bei der Hespa-Domäne. Sonderdruck aus Allgemeine Forstzeitschrift Nr. 22/23. 4 pp.
BRABECK, W. 1972. Neues Zweimann-Rückeseilgerät für Durchforstung. Sonderdruck aus Holz-Kurier Nr. 32/33 10. August. 2 pp.
BRIEGLEB, Ph. A. 1964. Balloon logging systems. Phase I — Analytical study. USDA For. Serv. Contract 19—25. 110 pp.
BRIGGS, R. D. 1975. Small logs make a big point: Loggers must hone sharper, finer operations. For. Industries *102*(10):44—46.
BROWN, N. C. 1936. Logging — transportation. The principles and methods of log transortation in the United States and Canada. John Wiley & Sons Ltd. New York. 327 pp.
BROWN, N. C. 1949. Logging. The principles and methods of log transportation in the United States and Canada. John Wiley & Sons Ltd. New York. 418 pp.
BROWN, W. E. 1948. Use and maintenance of logging ropes in Eastern Canada. Canadian Pulp and Papers Association. Woodlands Section. 27 pp.
BROX, A. 1976. Folkeminne frå Ytre Senja, løypestrengen. Norsk Folkeminnelags Skrifter nr. 117, Oslo. 41 pp.
BYRKJELAND, J. 1974. Løypestrengen. Fra Hardanger Historielag, Hardanger. Pp. 487—494.

CALVERT, W. W. and GARLICKI, A. M., 1967: Choker Line Forces in Skidding Saw Logs. Bi-monthly Research Notes. Department of Forestry and Rural Development. Ottawa. *23* (4): 28—29.

CAMPBELL, Ch. 1970. Skyline tension and deflection handbook. Supplement. USDA For. Serv.Res.Pap. PNW—39 (Supp.). 25 pp.

CAPPELONI. 1955: Teleferiche — Lavgnolo Torino.

CARSON, W. W., 1975: Analysis of Running Skyline with Drag. Pacific Northwest Forset and Range Experiment Station. USDA Forest Service Paper PNW-193. 8 pp.

CARSON, W. W. & JØRGENSEN, J. E. 1974. Understanding interlock yarders. Portland, Oregon. 13 pp. USDA, Forest Service. Res. Note. PNW-221.

CARSON, W., JØRGENSEN, J. & LYSONS, H. 1979. Symposium on mountain logging. IUFRO Proceedings. 222 pp.

CARSON, W. W. & MANN, Ch. N. 1970. A technique for the solution of skyline catenary equations. USDA For.Serv.Res.Pap. PNW—110. 18 pp.

CARSON, W. W. and MANN, C. N., 1971: An Analysis of Running Skyline. Pacific Northwest Forest and Range Experiment Station. USDA Forest Service Paper PNW-120. 9 pp.

CARSON, W. W., STUDIER, D. D. & THOMAS, W. M. 1970. Digitizing topographic data for skyline design programs. USDA For.Serv.Res.Note PNW—132. 13 pp.

CATERPILLAR TRACTOR CO. 1939. 60 years of progress in western logging. Peoria, Illinois. 16 pp.

CENTRE TECHNIQUE DU BOIS. 1966. Mise en æuvre des téléphériques forestiers. Cah.Cent. tech. Bois 71, Sér. II. 58 pp.

CENTRE TECHNIQUE DU BOIS. s.a. Principes de calcul des téléphériques. Cah. Cent. tech. Bois 65 b, Sér. II. 31 pp.

COLLIER, R. D. 1950. Use of expendable load attachments and slings in the transport of wood by gravity over a suspended wire. Woodlds Res. Index No. 87 (B—8—g). 4 pp.

COLLIER, R. D. 1951. Skidding by wire. Pulp and Paper Research Institute of Canada, Montreal. 4 pp.

COLUMBUS, L. 1893. Going to the isle of dogs. Bullivant & Co. London. 30 pp.

CONWAY, S. 1977. How to cope with smaller logs and their higher harvest costs. For. Industries *104*(10):34—36.

CORNWALL, G. 1909. Pacific Coast Logging Congress. The Timberman. Aug.

COTTELL, P. L., MCMORLAND, B. A. & WELLBURN, G. V. 1976. Evaluation of cable logging systems in interior B.C. and Alberta. Technical Report No. TR-8, September. FERIC, Canada. 48 pp.

CUMMING, D. 1972. Remote control rockets grapple yarder's yield. Br. Columbia Lumberm. *56*:6—7.

CZITARY, E. 1931. Spannkraftermittlung in Seilen durch Schwingungsmessung. Wasserwirtschaft Nr. 15—16. Pp. 246—249.

CZITARY, E. 1951. Seilschwebebahnen. Springer-Verlag, Wien. 390 pp.

CZITARY, E. 1962. Seilschwebebahnen. Zweite Aufl. Springer-Verlag, Wien. 467 pp.

DAHL, O. S., 1973: Trekkraftbehov ved slepekjøring av fulltrær og stammer. Hovedoppgave i Skogbrukets driftsteknikk. Norges Landbrukshøgskole. 40 pp + bilag.

DE MEGILLE, X. 1954. Du choix de tracteurs pour le débardage. Collection FAO: Mise en valeur des forets No. 1. Rome. 189 pp.

DENNINGER, W. 1979. Die leichteren mobilen europäischen Kippmastseilkräne — Technische Möglichkeiten und Eignung. Kuratoum für Waldarbeit und Forsttechnik, Mechanisch-Technische Abteilung. Gross-Umstadt. 7 pp.

DET NORSKE VERITAS. 1980. Regler for ståltau til taubaner og kabelkraner. Oslo. 23 pp.

DICKIE, D. E. & ENG, P. 1975. Rigging manual. First edition, October. Construction Safety Association of Ontario, Canada. 188 pp.

DRESSLER, M. 1979. New progressive logging technologies in mountains of Czechoslovakia. Problems of logging in Czechoslovakia. Brno. 20 pp.

DRESSLER, M. s.a. New progressive logging technologies in mountains of Czechoslovakia. Brno. 16 pp.

DROTNINGHAUG, Ø. 1975. Mandal reberbane — 200 år. Lindesnes *87*(217). 4 pp. [Lindesnes 20. september 1975.]

DUKIELSKII. H. I., 1951: Rope ways and rope cranes. Moscow — Leningrad.

DYKSTRA, DENNIS P. 1975. Production rates and costs for cable, balloon and helicopter yarding systems in old Douglas-fir. Research bulletin 18. Oregon State University, Corvallis, Oregon. 57 pp.

DYKSTRA, DENNIS P. 1976. Production rates and costs for yarding by cable, balloon and helicopter compared for clearcuttings and partial cuttings. Research bulletin 22. Oregon State University, Corvallis, Oregon. 44 pp.

DYKSTRA, DENNIS P. 1978. Some observations on roundwood metrication. Forest products journal. Vol. 28 no. 2: 18–20.

EHMKE, K. 1962. Schlepper und Seilwinde. Forstarchiv *33*(10):218—220.

EIDE, S. 1971. Ståltau i skogbruket. Landbruksforlaget, Oslo. 78 pp.

EISBACHER, J. 1973. Überblick über den Stand der stationäre Holzaufarbeitungsplätze (Holzhöfe) in Österreich. Forstliche Bundesversuchsanstalt, Wien. 147. Folge. Juni. 2 pp.

ELKEM SPIGERVERKET A/S STÅL og TAU. 1980. Ståltaukurs. Tønsberg. 45 pp.

ERIKSSON, H. och LÖJDSTRÖM, R., 1962: Undersøkning av tyngd, tyngdpunkt och släpmotstånd hos enskilda träd och stammer. Mo och Domsjö AB, Örnsköldsvik. T: 12. 14 pp (stensil).

ERNST, H. 1953. Die Hebezeuge. Band I: Grundlagen und Bauteile. Friedr. Vieweg & Sohn, Braunschweig. 304 pp.

ERNST, H. 1953. Die Hebezeuge. Band II: Winden und Krane. Friedr. Vieweg & Sohn, Braunschweig. 302 pp.

ERNST, H. 1953. Die Hebezeuge. Band III: Sonderausführungen. Friedr. Vieweg & Sohn, Braunschweig. 287 pp.

ETCHU, T. s.a. Skyline crane systems used for logging in national forests of Japan. Forestry Mechanization Society of Japan . 95 pp.

FEDERATION EUROPENNE DE LA MANUTENTION. s.a. Terminologie illustrée des téléphériques, blondins et scrapers. Section 3. Krausskopf-Verlag, Wiesbaden. 64 pp.

FELT, M. E. 1978. The enterprising mister murray. The Caxton Printers Ltd., Caldwell, Idaho. 159 pp.

FJALESTAD, I. 1975. Iglands alpevinsj. Utstyr, prestasjoner og driftskostnader. *Igland Alp winch. Equipment, performance rate and working expenses.* Driftsteknisk rapport nr. 13. Tidsskr. Skogbr. *83*: 153—162.

FJONE, H. 1977. Transportplanlegging i Bø i Telemark på grunnlag av tilgjenge-analyse med driftsteknisk terrengklassifisering. *Transport planning at Bø in Telemark on the basis of accessability analysis with operational terrain classification.* Driftsteknisk rapport nr. 15. Tidsskr. Skogbr. *85*: 251—259.

FJONE, H. 1977. Drift til flytande dalstasjon II. *Cable operation to a floating lower landing II.* Driftsteknisk rapport nr. 15. Tidsskr.Skogbr. *85*: 127—144.

FJONE, H. & FRØNSDAL, J. 1980. Traktordrifter i bratt og vanskelig terreng. Medd. Nor. inst. skogforsk. *35*:387—467.

FJÅGESUND, T. 1980. Intervju om gamle tømmerskott. 1 pp. [Arkiv NISK.]

FLATLANDSMO, K. 1979. Sikkerhet av taubaneanlegg. Smøremidler, studie av optimal virkningsgrad med forsøk. NTNF-prosjekt 0601.7573. Foreløpig rapport, status pr. 31.12.78. Hordaland Menkaniske Verkstad, Voss. 14 pp.

FRAUENHOLZ, O. 1972. Arbeitsverfahren und Arbeitsabläufe bei der Holzernte in kleineren Forstbetrieben. Allgemeine Forstzeitung, Folge 7. Juli. 7 pp.

FRENKIEL, Z. 1966. History of ropeways. Areal ropeways and funicular railways, Warszawa. 16 pp.

FRIISVOLD, J. A. 1979. Isskjæring på Oravannet. Romsdal Sogelags Årsskrift, Molde. Pp. 128—132.

FRÖLICH, J. 1948. Problemene der Holzbringung im Gebirge. Int. Holzmarkt *39*(12):6—9.

FRØNSDAL, J. 1980. Skogsbilveier i bratt terreng. Tidsskr. Skogbr. *88*(1): 93—97.

FØLSTAD, Ø. 1969. Arbeidsdaglengde i skogen på bakgrunn av lys- og klimamålinger. Tidsskr. Skogbr. *77*(1): 161—167.

FØRELAND, J. M. 1971. Stammeterminaler, produksjon og kostnader. *Stem terminals, production and costs.* Driftsteknisk rapport nr. 10. Tidsskr. Skogbr. *79*: 107—128.

GARLICKI, A. M., 1967: A rectangular components transducer for measuring skidder forces. Pulp and Paper Magazine of Canada *68*(8): WR 346 — WR 350.

GARLICKI, A. M. and CALVERT, W. W., 1969: A Comparison of Power Requirements of Full-Tree Versus Tree-Length Skidding. Pulp and Paper Magazine of Canada *70*(14): 83—86.

GIORDANO, G. 1959. Logging cableways. ECE/FAO Joint Committee on Forest Working and Training of Forest Workers. FAO/ECE/LOG 60. Geneve. 145 pp.

GIORDANO, G. 1967. Le téléferiche nelle utilizzazioni forestali — Calcolo delle funi — Installazione e messa in funzione.Contrti scient.-prat. migl. Conosc. Util. Legno 9(16):1—141.

GLYNN, J. 1873. Construction of cranes and other hoisting machinery. Lockwood & Co., London. 118 pp.

GLØERSEN, H. A. TANDBERG. 1883. Indberetning fra Skovdirektøren for Tidsrummet fra Januar 1875 til 30. juni 1880. Kristiania. Pp. 64,76.

GLØERSEN, H. A. TANDBERG. 1886. Indberetning fra Skovdirektøren for Aarene 1880—1883. Kristiania. Pp. 18.

GRAN, K. 1891. Om Tømmerrender. Den norske forstforenings aarbog. Kristiania. Pp. 153—160.

GRØNNEVIK, P. 1980. Intervju om gamle pendelbaner og rundbaner. 1 pp. [Arkiv NISK.]

GUDIM, Å. 1954. Taubaner og taubanedrift i Norge og Mellom-Europa. Tidsskr. Skogbr. *61*:417—433, 469—482.

GULISASCHVILI, B. G., 1952: Rope Ways for the Forestry. Goslesbumizdat. Moscow — Leningrad.

HAFNER, F. 1941. Zur Geschichte der Beforderung von Holz auf Draht und Seilriesen. Fortswissenshchaftliches Zentralblatt *63*:100—109.

HAFNER, F. 1948. Erfahrungen mit dem Pohlig-Seilfördergerät P 500. Allg. forst- u.holzw. Ztg. *59*:90—93.

HAFNER, F. 1958 a. Seilkräne mit offenem Zugseil und mit geschlossenem Zugseil (blondins) österreichischer Herkunft. FAO/ECE/LOG/Conf. Room Doc. No. 14. 17. July. 11 pp.

HAFNER, F. 1958 b. Seilkrananlage und Hangweg. Int. Holzmarkt Nr. 22. 3 pp.

HAFNER, F. 1958 c. Wirtschaftlicher Abstand von Seilkränen. FAO/ECE/LOG/Conf. Room Doc. No. 15, 17. July. 7 pp.

HAFNER, F. 1964. Der Holztransport. Handbuch für Rückung, Lagerung, Ladeverfahren und Haupttransport. Institut für forstliches Ingenieurwesen und der Hochschule für Bobenkultur in Wien. Österreichischer Agrarverlag, Wien. 460 pp.

HAFNER, F. 1971. Forstlicher Strassen und Wägebau. Österreichischer Agrarverlag, Wien. 359 pp.

HAFNER, F. & HEDENIGG, W. 1959. Untersuchungen bezüglich der erforderlichen Abstände von Wasserkehren bei Forstwegen im Gebirge. Allg. Forstztg *70*(11/12):138—140.

HAKKILA, P., 1969: Weight and Composition of Branches of large Scots pine and Norway spurce trees. Communicationes Instituti Forestalis Fenniae *67.6*: 1—37.

HALLDIN, S. (Ed.) 1979. Comparison of forest water and energy exchange models. Proceedings of an IUFRO Workshop held at Uppsala, sept. 24—30, 1978. Elsevier Scientific Publishing Company, Amsterdam. 258 pp.

HARKNESS, W. D. 1962. Tree length cable yarding. A report on the comparative performance in loose and bunched wood. Woodlds Sect. Index Can. Pulp Pap.Ass.No. 2142 (B—8—9). WR—93—96.

HAUSKA, L. 1933. Riesanlagen und Seilbahnen. Bind I. Das forstliche Bauingenieurwesen. Druck und Verlag von Carl Gerold's Sohn, Wien. 289 pp.

HAWKES, E. G. 1979. Cable harvesting systems for small timber. The Vermont Department of Forests, Parks and Recreation. 69 pp.

HEINRICH, R. (Ed.) 1976. Technical report of FAO/Austria training course on forest roads and wood harvesting in mountainous forests. Held in Ossiach, Austria, 1.—29. June 1975. Rome. 320 pp.

HELLEFORS BRUKS AKTIEBOLAG. 1952. Stållinor. Hällefors. 255 pp.

HERRGEN, A. 1954. Wire skidding equipment. Woodlds Sect. Index Can. Pulp Pap.Ass. No. 1376 (B—8—g). 1 pp.

HILF, H. H. 1954. Zuggeräte zum Zufallbringen von Hängern. Forstarchiv *25*(3):80.

HOCEVAR, M. 1975. Stammholztransport mit Lasso-Seilbahn. Berichte Nr. 149. Eidgenössische Anstalt für das forstliche Versuchswesen, Birmensdorf. 43 pp.

HONEGGER. 1982. Powerphone system Honegger. Zollikerberg, Austria. Stencil. 5 p.

HOVLAND, T. 1962. Rettledning i kabelkrantransport med 1-tromlet vinsj. Vestlandske Taubaneråd, Voss. 22 pp.

HOWARD, P. L. 1962. The era of the Lombard Log Hauler. Forest History *6*(1 og 2):2—8.

HUDLER, S., 1937: Elastizitätsmodul der Drahtseile. Wasserwirtschaft und Technik. M. 28—30.

HAANSHUS, S. Ø. 1978. Transportplanlegging i bratt terreng — tiltak for økt avvirkning. *Transport planning in steep terrain — some ways to increase the annual cut.* Driftsteknisk rapport nr. 17. Tidsskr. Skogbr. *86*: 491—497.

IELSTRUP, H. 1908. Utkast til lov om tvungen avstaaelse av grund til anlæg av taugbaner og løipestrenger. Tidsskr. Skogbr. *16*(4):155—157.

ILO. 1973. Symposium on forest operations in mountainous regions. Technical report. 90 pp.

ILON, B. 1953. Winschdrift i skogen. Maskinteknikk i Jord och Skog *5*(1):18—19.

INTERNATIONAL UNION OF FORESTRY RESEARCH ORGANIZATIONS 1980. Report from 1979 IUFRO symposium on mountain logging. University of Washington, Seattle. 200 pp.

JACHWITZ, E. 1956. En kort beskrivelse av «Bjønn» kabelkrans tekniske konstruksjon og drift ved utferden for driftsteknisk kurs 13. sept. Det norske Skogforsøksvesen, Vollebekk. 4 pp.

JACOBSEN, Fr. & SOMDAL, J. A. 1942. Aluminium og stålaluminium kraftledninger. A/s Norsk Aluminium Company. Oslo. 175 pp.

JØRGENSEN, J. E., 1971: Steady State Analysis of Interlocking Yarders. A Preliminary Report. Pacific Northwest Forest and Range Experiment Station. Seattle, Washington. 52 pp.

JØRGENSEN, J. 1974. Skyline logging. Symposium proceedings. University of Washington, Seattle. 86 pp.

KAIERLE, L. 1979. Projektierungshilfe für forstliche Seilkrane mit Fixspannung und Einwagenbetrieb. Bericht der Bayerischen Forstlichen Versuchs- und Forschungsanstalt, München. 47 pp.

KARIGL, H. 1949. Eine Durchforstungsseilbahn der ob.-öst. Zentralforstgenossenchaft. Allg. forst- u. holzw. Ztg *60*(13/14):107.

KATO, S. 1960. «Free-cableway» — a new «curve-device» enables free allignment of a logging cableway. Inst. of Forestutilization, University of Tokyo, Tokyo. 7 pp.

KATO, S. 1961. Popular skyline cable systems used in the mountain forests of Japan. XIIIth IUFRO Congress. Section 32, Paper. Vienna. 11 pp.

KATO, S. et al. 1961. An experimental study on the fan-governor. Applied to a supplementary braking equipment of the logging cableways. J. Jap.For.Soc. *43*:101—106.

KAUFMANN v., G. 1954. Zubringen von Holz im Hochgebirge mit Drahtseilen. Bayerischer Landwirtschaftsverlag, München. 80 pp.

KAUFMANN v., G. 1958. Kleinseilanlagen bei der Holzbringung. Allg. Forstztg *69*(7/8):95—99.

KEEN, R. E., 1963: Weights and centres of gravity involved in handling pulpwood trees. Pulp and Paper Research Institute of Canada. Technical Reports Series 340: 1—93. (Woodlands Research Index No. 147.

KEEN, and WEETMAN, G. F., 1961: Two methods of weighing trees. Pulp and Paper Research Institute of Canada. Research Note 17: 1—9. (Woodlands Research Index No. 123).

KELLOGG, LOREN D. 1979. Thinning young timber stands in mountainous terrain. Symposium on mountain logging. University of Washington, College of Forest Resources, Seattle. p.136–145.

KISTENFEGER, J. 1963. Die maschinelle Entrindung von Kiefern-Grubenlangholz. Allg. Forstz. *18*(18):298—300.

KLEM, G. G.., 1934: Undersøkelser av granvirkets kvalitet. Meddr. norske Skogfors. Ves. 5 (nr. 17): 197—348.

KOBER, R., 1938: Die Landwirtschaftlichen Seilwege. Verlag von Carl Geralds Sohn. Wien. 559 pp.

KOBER, R. 1938. Seilwegebau. Verlag von Carl Gerold's Sohn, Wien und Leipzig. 550 pp.

KOCH, C. DAHLERUP. 1970. Berømte naturvidenskabsmænd. Politikens Forlag, København. 3 pp.

KOHLHASE, F., 1950: Hand- und Tabellenbuch für die Drahtindustrie, das Drahtgewerbe und dessen Nebenzweige. Emil Griebsch Verlag. Hamm (Westf.). Upaginert.

KONUMA, J. & SHIBATA, J. 1976. Cable logging systems in Japan. Bulletin of the Government Forest Experiment Station No. 283, July. Tokyo, Japan. p. 117–174.

KOROLEFF, A. 1944. Skidding wood by wire. Woodlds Sect. Index Can.Pulp Pap.Ass. No.742 (B—8—g). 2 pp.

KOROLEFF, A. 1950. Wood transport by gravity over a suspended wire. Woodlds Res. Index No. 81 (B—8—g). 14 pp.

KOROLEFF, A. 1961. New approach to collective work evaluation of machinery aggregations (exemplified in logging). Woodlds Sect. Index Can.Pulp Pap.Ass. No. 2051(B—1). 6 pp.

KOROLEFF, A. & SILVERSIDES, C. R. 1948. Transport of wood by gravity over suspended wire. Woodlds Res. Index No. 43 (B—8—g). 8 pp.

KOROLEFF, A. & COLLIER, R. D. 1954. Wire skidding. Wood transportation by gravity over a suspended wire. Pulp and Paper Research Institute of Canada, Montreal. 137 pp.

KOSTENAPFEL, A. 1975. Die Berechnungen der Seile der forstlichen und übrigen Seilbahnen. Ljubljana. 29 pp.

KROHN, B. 1977. Verbietet die neue Unfallverhütungsvorschrift «Winden-Hub- und Zuggeräte» (VBG 8) in Zukunft den Einsatz von Kleinseilwinden? Forstarchiv *48*(7):125—127.

KUBASÁK, E. 1958. Seilbahnholztransport und die Perspektiven in der Waldwirtschaft der ČSR. Lesnictvi *4*:593—604.

KYLLO, O. 1980 a. Fulltredrift med Iglands interlukket taubanevinsj og NISK-kvisteren. Tidsskr.Skogbr. *88*(1):61—67.

KYLLO, O. 1980 b. Fulltredrift med radiostyrt kabelkran, Koller/Nestetogs løpekatt og mekanisert kvisting på dalstasjon. Tidsskr.Skogbr. *88*(1):55—59.

KYLLO, N.O. & LISLAND, T. 1981. Østerrikske vinsj- og kabelkranutstyr. Sør-Trøndelag Fylkesskogkontor, Trondheim. Stencil. 32 p.

KÅSA, J. 1952. Skognytting. H. Aschehoug & Co, Oslo. 373 pp.

LARSEN, R. 1975. Compendium of major cable logging systems. Harvesting and Transportation, Interforest AB, Stockholm. 112 pp.

LENGER, A. & BERNHARD, A. 1972. Leistungsuntersuchungen bei der mechanisierten Nadelholzernte. Forstliche Bundesversuchsanstalt, Wien. 139. Folge. März. 5 pp.

LEONARD, G. M. & HOUSLEY, R.M. 1983. Helistat program plan. Forest Service, U.S.Department of Agriculture. 33 pp.

LIDGEWOOD MANUFACTURING COMPANY. 1905. Logging by steam. The Logging Machinery Department. Lidgewood Manufacturing Company, New York. Upag.

LILEY, W. B. 1983. Cable logging handbook. New Zealand Logging Research Ass. Inc., Rotorua. 158 pp.

LIND, C. J. 1978. Big timber, big men. A history of loggers in a new land. Hancook House Publishers Ltd., Seattle, Wa. and Saanichton, B.C. 153 pp.

LINDVIK, L. 1930. Om løypestrenger. Hardanger *18*(25). 1 pp. [Hardanger 27. februar 1930.]

LISLAND, T. 1967. Laboratorietest av vinsjer. *Laboratory test of winches*. Driftsteknisk rapport nr. 6. Tidsskr. Skogbr. *75*: 29—48.

LISLAND, T. 1967. Prøving av Kunstfibertau. *Testing of synthetic ropes*. Driftsteknisk rapport nr. 6. Tidsskr. Skogbr. *75*: 49—54.

LISLAND, T. 1969. Laboratorietest av vinsjer. *Laboratory test of winches*. Driftsteknisk rapport nr. 7. Tidsskr. Skogbr. *77*: 151—160.

LISLAND, T. 1975. Cable logging in Norway. A description of equipment and methods in present use. Norsk institutt for skogforskning, Ås—NLH. 52 pp.

LUND, D. H. 1972. Fra Norges fjell til fjerne kyster. H. Aschehoug & Co., Oslo. Pp. 33—54.

LYNG. 1902. Foredrag i Polyteknisk forening 11/3. Referat i Teknisk Ukeblad *20*(15):162.

LYNGSTAD, J. 1959. Ståltau. NTHs materialprøvningsanstalt, Trondheim. Publikasjon nr. 57. 143 pp.

LYSONS, H. H. 1971. Cable logging in tomorrow's forest environment. ECE/FAO/ILO Symposium on forest operations in mountainous regions, Krasnodar 30. aug.—11.sept. 25 pp.

LYSONS, H. H. 1976. Why log with skylines? Skyline logging symposium. Ass. of B.C. Professional Foresters, USDA Forest Service, Forest Engineering Research Institute of Canada, U.B.C. Vancouver B.C. Pp. 1—8.

LYSONS, H. H. 1979. Harvesting thinnings on steep ground. USDA Forest Service, Pacific Northwest Forest and Range Experiment Station, Washington. 8 pp.

LYSONS, H. H. s.a. Grapple yarding revolutionizes logging. Logging Issue, Western Conservation Journal. 2 pp.

LYSONS, H. H. & MANN, Ch. N. 1965. Skyline logging handbook on wire rope tensions and deflections. USDA Forest Service. 34 pp.

LYSONS, H. H. & MANN, Ch. N. 1967 a. Single-span running skylines. USDA For.Serv.Res. Note PNW—52. 7 pp.

LYSONS, H. H. & MANN, Ch. N. 1967 b. Skyline tension and deflection handbook. USDA For.Serv.Res.Pap. PNW—39. 41 pp.

LYSONS, H. H. & WELLBURN, G. V. 1976. Planning mountain logging on the North American west coast. XVI IUFRO World Congress, Oslo. Proceedings, Division III:100—117.

LØVDAL, Aa. 1976. Drift til flytende dalstasjon. *Cable operation to a floating lower landing*. Driftsteknisk Rapport nr. 14. Tidsskr. Skogbr. *84*: 203—214.

MACEMAN, G. 1971. Power for prairie plows. Western Producer Prairie Books, Saskatchewan. 115 pp.

MALMBERG, C. E. 1955. Vinschens användning inom svenskt skogsbruk. (Föredrag hållet vid NSR-sammanträde den 24. maj 1955 på Aas). SDA, Stockholm. 8 pp.

A/S MANDAL STÅL, A/S NORSK STAALTAUGFABRIK & A/S TØNSBERG REPERBANE. s.a. Ståltautabeller. Mandal, Trondheim, Tønsberg. 27 pp.

MANN, Ch. N. 1969. Mechanics of running skylines. USDA For.Serv.Res.Pap. PNW—75.

MARKS, E. C. R. 1892. Notes on the construction of cranes and lifting machinery. The Technical Publishing Company Ltd., John Heywood, London. 138 pp.

MAY, M. 1957. Le debardage des bois a la traine. FAO/ECE/LOG/14. Comité mixte des techniques de travail en foret et de la formation des ouvriers forestiers. Geneve. 70 pp.
MAYER, H. 1972. Waldbau und Technik im Gebirge. Schweiz. Z. Forstwes. *123*(1):12—40.
McCOLL, B. J. 1949. Cable yarding. Woodlds Sect. Index Can. Pulp Pap.Ass. No. 1056 (B—8—g). 8 pp.
McCRAW, W. E. 1962. Variable factors affecting skidding production in logging. USDA, Forest Products Research Branch Technical Note 28. 15 pp.
MCDONALD, M.J. 1981. Evaluation of the thinning swing hydraulic yarder. Technical Note No. TN-40, January. FERIC, Canada. 28 pp.
MCMORLAND, B. & WONG, A. 1982. Evaluation of the model 9 micro master yarder. Technical Note No. TN-54, March. FERIC, Canada. 20 pp.
McNALLY, J. A. s.a. Cable skidders. Woodlds Sect. Index Can. Pulp Pap. Ass. No. 573 (B—8—g). 3 pp.
MÉGILLE, X. de, 1954: Du choix de tracteurs pour le dèbardage. Organisation des National Unies pour l'Alimentation et l'Agriculture. Rome. 189 pp.
MEYR, R. 1954 a. Anmärkningar rörande kortsträckskabelkranen «Vossavinsjen» efter gjord besiktning av redskapet hos Tytbo. Stockholm. 5 pp.
MEYR, R. 1954 b. Forstliche Seilbahnen in der Schweiz und in Frankreich. Allg. Forstztg *65*:210—212.
MEYR, R. 1954 c. Rückung und Nahförderung des Holzes unter Verwendung von Forstseilwinden. Allg. Forstztg *65*(5/6). 3 pp.
MEYR, R. 1954 d. Seilbahnbau in der Schweiz und in Frankreich. Berichte und Ergebnisse zweier Studienreisen. Österreichisches Produktivitäts-zentrum. Wien. 105 pp.
MEYR, R. 1960. Unfallverhütung bei Seilbringungsarbeiten. Inf.-Dienst Geb. Nachr Tech. Nr. 29. 2 pp.
MEYR, R. 1971. Übersicht über derzeitige Angebot an Schlepper seilwinden für die Holzrückung. Forstliche Bundesversuchsanstalt, Wien. 133. Folge. August. 9 pp.
MEYR, R. 1972. Technische Daten, Preise und Betriebsstundenkosten einiger Universalschlepper und Zusatzgeräte für die Holzrückung. Forstliche Bundesversuchsanstalt, Wien. 142. Folge. Oktober. 6 pp.
MEYR, R. 1972. Schlepper und Zusatzgeräte für die Holzernte im kleineren Forstbetrieb. Allgemeine Forstzeitung. Folge 7. Juli. 6 pp.
MEYR, R. 1980. Geschichtliche Entwicklung des Seilbahnwesens in den Alpenländern. Promemoria. 5 pp.
MEYR, R. & EISBACHER, J. 1973. Ergebnisse einer Datenerhebung über Knickschlepper. Forstliche Bundesversuchsanstalt, Wien. 145. Folge. April. 2 pp.
MIFFLIN, R. W. & MANN, Ch. N. 1980. Skyline yarding forces with partial ground support. USDA Forest Service. Pacific Northwest Forest and Range Experiment Station. 5 pp.
MILLER, M. S. & RESCHETOW, A. W. 1950. Das Holzschleppen mit den Winden TL—3 vom Abstand bis zu 500 mt. Sibirischer wissenschaftlichen Forschungsinstitut der Forstwirtschaft und Exploitation, Krasnojarsk. 6 pp.
MOROTO, T. & FUJIWARA, N. 1979. Small area operations in steep terrain. Symposium on mountain logging. University of Washington, College of Forest Resources, Seattle. p. 146 – 150.
MUIR, M. B. 1984. Washington 88 interlock system. Auburn, WA, USA. Stencil. 10 pp.
MÅGE, K. L. 1980. Løypestrenger i Måge. (Brev 2/3). Måge. 2 pp. [Arkiv NISK.]
MAARTMANN, H. R. 1954. Wire og kabel. Skogbrukets og Skogindustrienes Forskningsforening, Transportutvalget og Vestlandske Taubaneråd, Vollebekk. 32 pp.
NESBØE, S. 1978. Industriutviklinga i Svelgen. Ei vaksemd og eit samfunn i vekst. Elkem Spigerverket A/S og Bremanger Smelteverk. Svelgen. Pp. 12—17.
NIETHAMMER, F. 1897. Motoren und Hülfsapparate für elektrisch betriebene Hebezeuge. Verlag von Julius Springer, Berlin. 68 pp.
NORDISKA SKOGSARBETSSTUDIERNAS RÅD (NSR). 1978. Skoglig arbeidsstudienomenklatur. Norsk institutt for skogforskning. Ås—NLH. 130 pp.
NORGES STATSBANER. 1965. Terminologi for taubaner. RTT's forslag. Oslo. Upag.
NORGES STATSBANER. 1975. Sikring av taubanedriften i Norge. Innstilling ang. bedring av sikringen ved norsk taubanedrift etter oppdrag fra Samferdselsdepartementet. Oslo. 65 pp.
NORGES STATSBANER. 1976. Tilsyn med privatbaner og taubaner. Trykk nr. 398. Oslo. Upag.

A/S NORSK STAALTAUGFABRIK & STÅL OG TAU. 1965. Ståltau-katalog. Mandal. 28 pp.
NYGAARD, J. 1909. Driftsforhold og trælasthandel i Bratsberg amt. Tidsskr.Skogbr. *17*(1):13—22.
O'LEARY, JOHN (1970). Production data on various cable systems. Oregon State University. School of Forestry. Corvallis. 23 pp. (Stencil).
OLSEN, O. H. 1980. Svekket infanteritrening. Aftenposten *121*(338). Oslo. 1 pp. [Aftenposten 26. juli 1980.]
OMNES, H. 1974. Prestasjoner og kostnader for aptering, kapping og sortering av tømmer ved taubanedrifter. *Performances and costs of marking for bucking, and sorting on cable crane operations.* Driftsteknisk rapport nr. 13. Tidsskr. Skogbr. *82*: 37—44.
OMNES, H. 1975. Fulltredrift med radiostyrt kabelkran og mekanisert kvisting på dalstasjonen. *Full tree yarding with radio controlled cable crane and mechanized delimbing at the lower landing.* Driftsteknisk rapport nr. 13. Tidsskr. Skogbr. *83*:99—106.
OMNES, H. 1977. Bruk av VOLVO BM 985 «TVIGG» kvistemaskin i forbindelse med taubanedrift. *VOLVO BM 985 «TVIGG» delimber used in cable crane operations.* Driftsteknisk rapport nr. 15. Tidsskr. Skogbr. *85*: 117—126.
OMNES, H. 1980. Prestasjoner og kostnader for radiostyrt kabelkran K-1200 og Iglands alpevinsj. Rapp. Nor.inst. skogforsk.1/80:1—98.
OPSAL STEEL LIMITED 1977. Catalogue No. 15. Vancouver, B. C. 17 pp.
PAPÁNEK, Fr. & JURGA, Fr. 1948. Oekonomische Aspekte der neuen Seilbahnen in der Forstwirtschaft. Veröffentlicht in der Monatschrift Lesnickà Prace *27*(1):9—24.
PEARCE, J. K. 1947. Logging hoists. Skagit Steel & Iron Works. Sedro-Volley, Washington. 16 pp.
PESTAL, E. 1948 a. Der Wyssen Seilkran. Allg.forst-u.holzw. Ztg *59*(11/12). 7 pp.
PESTAL, E. 1948 b. Die Rentabilität des Wyssen-Seilkranes. Int. Holzmarkt *39*(12):10—13.
PESTAL, E. 1950. Die Messung der Tragseilspannung nach dem «Seilschwingungsverfahren». Int. Holzmarkt *41*(17):93—95.
PESTAL, E. 1951. Der Seilzug «GREIF» und seine Verwendung in der Forstwirtschaft. Int. Holzmarkt *42*(20): 113—115.
PESTAL, E. 1953. The transportation of wood by long skylines. University of British Columbia, Faculty of Forestry. Translation No. 15. 9 pp.
PESTAL, E. 1961. Seilbahnen und Seilkrane für Holz- und Materialtransport. Institut für forstliches Ingenieurwesen der Hochschule für Bodenkultur in Wien. Verlag Georg Fromme & Co. Wien. 544 pp.
PESTAL, E. 1964. Transportprobleme der Faserholzaufbringung. Holzkurier Nr. 26. 2 pp.
PESTAL, E. 1965. Umstellung der Rückungsmethoden im Gebirgsforst. Int. Holzmarkt *56*(8):1—6.
PESTAL, E. s.a. Allgemeine technische Beschreibung der Materialseilbahn PH 1000. Internationale Seilbahn-Rundschau. Wien. 4 pp.
PESTAL, E. 1971. Entwicklung und Verfahren der kurzstrechenseilkrane in Österreich. Sonderdruck aus Allgemeine Forstzeitung. Folge 5. Mai. 5 pp.
PESTAL, E. 1972. Vorfürungen und Diskussionen beim Holzernteseminar »Mittel- und Kleinbetrieb» in Gmunden. Allgemeine Forstzeitung. Folge 7. Juli. 6 pp.
PICKETT, G. 1979. Harvesting clearcuts with Peewee yarder. Symposium on mountain logging. University of Washington, College of Forest Resources, Seattle. p. 151 – 152.
PLATZER, H. B. & SIKORA, H. 1963. Drahtseil, Drahtseilverbindungen und Zubehör. Forstarchiv *34*:285—294, 306—308.
POHLHAUSEN, A. 1894. Berechnung und Konstruktion der wichtigsten Lasthebemaschinen. Verlag von R. Schulze, Mittweida. 150 pp.
PROCKTER, C. E., 1977: Kempe's Engineers Year-Book for 1977. 82nd edition.I—II. Morgan-Grampian (Publishers) Ltd. London. 1360, 1408 pp.
REINTON, L. 1961. Sæterbruket i Noreg III. Instituttet for sammenlignende kulturforskning, Oslo. 599 pp.
ROLSTON, K. S. 1963. Acco doubl-bite log choker chain. Am. Pulpwood Ass. Tech. Release No. 62—R—48. 2 pp.
ROSKO, P. 1970. Die Seilrückung des Holzes im Gebirgsgelände. Lesnícke Studie 6. 112 pp.
ROSKO, P. & HABOVSTIAK, J. 1963. Agricultural transport in Montane Regions. Zemedelská technika *9*:461—474.
ROSSELAND, S. 1958. Rapport fra det andre internasjonale taubanekurs. FAO/ECE Italia og Sveitz, 14/9—4/10. 15 pp.
ROSSELAND, S. 1960. Lunning med 2-tromlede traktorvinsjer. *High-lead yarding by 2-drum tractor winch.* Driftsteknisk rapport nr. 1. Tidsskr. Skogbr. *68*: 271—275.

ROSSELAND, S. 1960. Vinsjlunning kombinert med lessing og framkjøring. *Winch skidding and transport in one operation.* Driftsteknisk rapport nr. 1. Tidsskr. Skogbr. *68*: 281—287.
ROSSELAND, S. 1961. Kabelkranlunning for vassdragslevering på høst- og vintertid. *Cable crane transport. Lower landing mounted on ice on the lake.* Driftsteknisk rapport nr. 2. Tidsskr. Skogbr. *69*: 290—300.
VON ROTEN. 1970. Holznutzung und Holzhof bei der Hespa-Domäne. Sonderdruck aus Allgemeine Forstzeitschrift, Nr. 48. 3 pp.
SAMSET, I., 1950: Hogstundersøkelser i norsk granskog. Meddr. norske Skogfors. Ves. 10 (nr. 37): 397—595.
SAMSET, I., 1950: Inntrykk fra studiereise i Nord-Amerika. I. Tømmertransport med taubane. Tiddsk. Skogbr. 58: 259—291.
SAMSET, I. 1953. Inndeling av taubanetypene. Det norske Skogforsøksvesen, Vollebekk. 4 pp.
SAMSET, I. 1955. Om lunning med to-tromlet skogbruksvinsj. Skogeieren *42*(2):36—48.
SAMSET, I. 1956 a. Arbeidsundersøkelse og hogsttariff. Norsk Skogbr. *2*(20):659—670.
SAMSET, I. 1956 b. Skogbrukets Arbeidslære II. Det norske Skogforsøksvesen, Vollebekk. 172 pp.
SAMSET, I. 1957. Taubaneberegninger. Det norske Skogforsøksvesen, Vollebekk. 60 pp.
SAMSET, I. 1958. Forsøkskjøring med unimogtraktor og Isachsenvinsj. Det norske Skogforsøksvesen, Vollebekk. 12 pp.
SAMSET, I. 1959 a. Eksempler på økonomiske kalkyler over lunne- og transportomkostninger for en del vinsj- og taubanetyper. Det norske Skogforsøksvesen. Vollebekk. 39 pp.
SAMSET, I. 1959 b. Examples of calculation of cost prices for skidding with two-drum winches. Det norske Skogforsøksvesen, Vollebekk. 21 pp.
SAMSET, I. 1959 c. Monteringsskjema for vinsjer og kabelkraner. Det norske Skogforsøksvesen, Vollebekk. 7 pp.
SAMSET, I. 1960. Tømmerutislag med vinsj. Tidsskr.Skogbr. *68*(4):231—240.
SAMSET, I. 1960. Skogforsøksvesenets motorhåndvinsj. *Motor hand-winch.* Driftsteknisk rapport nr. 1. Tidsskr. Skogbr. *69*: 241—249.
SAMSET, I. 1961. Skogbrukets Arbeidslære I. Det norske Skogforsøksvesen, Vollebekk. 130 pp.
SAMSET, I. 1962. Overflaterenner på skogsbilveger. Norsk Skogbr. *8*(18):577—579.
SAMSET, I. 1962. Skogbrukets interne transportnett. *Network in forestry.* Driftsteknisk rapport. nr. 3. Tidsskr. Skogbr. *70*: 286—298.
SAMSET, I., 1962: Vekten av hele grantrær (*Picea abies*). Driftsteknisk rapport nr. 3. Tidsskr. Skogbr. 70: 342—348.
SAMSET, I. 1963. Ground-skidding with single- and double-drum winches. (Short description of various methods). Det norske Skogforsøksvesen, Vollebekk. 9 pp.
SAMSET, I . 1964. Tømmertransport med Bell—204—B helikopter i fjellterreng. Meddr norske SkogforsVes. *19*:198—284.
SAMSET, I. 1966 a. A pilot study of the logging and transport problems in the forest areas of the Acheloos watershed. Det norske Skogforsøksvesen, Vollebekk. 151 pp.
SAMSET, I. 1966 b. Utviklingen av skogbrukets driftsmetoder II. Loven om den sprangvise utvikling. Norsk Skogbr. *20*:737—741.
SAMSET, I. 1967. Tømmertransport med MI—6 helikopter i fjellterreng. Et eksperiment med verdens største helikopter. Tidsskr.Skogbr. *75*(1):67—78.
SAMSET, I. 1969. Arbeidsmetoden ved montering og drift av radiostyrt kabelkran. *Erection and operation of the Norwegian radio-controlled cable crane.* Driftsteknisk rapport nr. 7. Tidsskr. Skogbr. *77*: 41—74.
SAMSET, I. 1969. Skogterrenget i Norge. *The forest terrain in Norway.* Driftsteknisk Rapport nr. 8. Særtr. av Taksering av Norges Skoger. Landsskogtakseringen 1919—1969. Pp. 7—54.
SAMSET, I. 1969. Kostnadskalkyler for skobruksmaskiner. Tidsskr. Skogbr. *77*(1):21—39.
SAMSET, I. 1970 a. Loven om den sprangvise utvikling — i ny utgave. Norsk Skogbr. *19*:399—400.
SAMSET, I. 1970 b. Skogsbilvegene våre. Oppbyggingen av skogbrukets transportnett. Skogeieren *57*(8):5—10.
SAMSET, I. 1970 c. Skogterrenget i Norge. Skogeieren *57*(5):4—10.
SAMSET, I. 1970 d. Tømmertransport med Sikorsky S61 N helikopter i fjellterreng. Tidsskr. Skogbr. *78*(1):117—127.
SAMSET, I. 1971. Some observations on forest operations and techniques in the Australian forestry. Det norske Skogforsøksvesen, Vollebekk. 43 pp.
SAMSET, I. 1973 a. Skogsdrifter i bratt terreng på den amerikanske vestkyst. Del I. Norsk Skogbr. *19*(19):325—329.

SAMSET, I. 1973 b. Skogsdrifter i bratt terreng på den amerikanske vestkyst. Del II. Norsk Skogbr. *19*(20):360—363.

SAMSET, I. 1973 c. Skogsdrifter i bratt terreng på den amerikanske vestkyst. Del III. Norsk Skogbr. *19*(21):383—387.

SAMSET, I. 1973 d. Skogsdrifter i bratt terreng på den amerikanske vestkyst. Del IV. Norsk Skogbr. *19*(22):417—420.

SAMSET, I. 1975 a. Proposal of a project on mechanical harvesting and transportation for coniferous forests in the Himalayas and reduction in logging wastes. Norsk institutt for skogforskning, Ås—NLH. 59 pp.

SAMSET, I. 1975 b. Skogterrengets tilgjengelighet og terrengforholdenes innflytelse på skogtilstanden i Norge. Medd. Nor. inst. skogforsk. *32*(1): 1—92.

SAMSET, I. 1978 a. Interlukket to-tromlet skogbruksvinsj. Tidsskr.Skogbr. *86*(4):391—405.

SAMSET, I. 1978 b. Signaler ved traktor-, vinsj- og taubanedrift i skogbruket. Tidsskr. Skogbr. *86*(4):267—270.

SAMSET, I. 1978 c. Utviklingsarbeidet med taubaneutstyr 1977/78. Tidsskr.Skogbr. *86*(4): 406—417.

SAMSET, I. 1979 a. Inntrykk mens man streifer omkring. Foredrag. Ås Rotary Klub's 20-års jubileum. Ås. 44 pp. [Arkiv NISK.]

SAMSET, I. 1979 b. Krefter og effekter i vinsj- og taubanesystemer. Medd. Nor. Inst. skogforsk. *35*:20—211.

SAMSET, I. 1980. Skogsdrifter i bratt terreng i det indre av British Columbia. Tidsskr.Skogbr. *88*(1):41—54.

SAMSET, I. & LISLAND, T. 1977 a. Østerrikske skogsdrifter i bratt terreng. Del I. Norsk Skogbr. *23*(10):10—13, 28.

SAMSET, I. & LISLAND, T. 1977 b. Østerrikske skogsdrifter i bratt terreng. Del II. Norsk Skogbr. *23*(9B): 4—10.

SAMSET, I., STRØMNES, R. og VIK, T., 1969: Hogstundersøkelser i norsk gran- og furuskog. Meddr. norske Skogfors. Ves. 26 (nr. 95): 293—607.

SAMSET, I. & SØRHAGEN, O. 1983. Skogsdrifter i bratt terreng i det indre av British Columbia. 10 pp.

SANKTJOHANSER, L. 1979. Logging in the Bavarian alpine area. Symposium on mountain logging. University of Washington, College of Forest Resourses, Seattle. p. 93–106.

SAUDER, E.A. 1980. An Evaluation of grapple-crane yarding in Costal B.C. Technical Note No. TN-36, November. FERIC, Canada. 35 pp.

SAUDER, E.A. 1981. Grapple yarding with a steel tower. Technical Note No. TN-53, December. FERIC, Canada. 36 pp.

SAUDER, E.A. 1982. An evaluation of highlead and slackline yarding with a Madill 046 yarder. Technical Note No. TN-56, March. FERIC, Canada. 32 pp.

SCHANTL, M. 1971. Die Holzbringung mit dem Kippmast-Seilkran Urus 250/500. Allgemeine Forstzeitung, Wien. 82 Jahrgang.Folge 5. p. 6–8.

SCHELDE-MØLLER, E. 1972. Videnskabens hvornår skete det. Politikens Forlag, Köbenhavn. 351 pp.

SCHINDLER, A. 1936. Die statische und dynamische Fahrbahnreibung und die Mittel zu deren Bestimmung. Eidgenössische technische Hochschule in Zürich. 80 pp.

SCHMITZ, H. 1973. The long road travelled. An account of forestry. University of Washington, Seattle, Washington. Pp. 129—143.

SCHNEIGERT, Z. 1966. Aerial ropeways and funicular railways. (Engl. translation by Z. Frenkiel). Pergamon Press, London. 554 pp.

SCHÖNAUER, H. 1972. Das Waldwegenetz als Voraussetzung einer zeitgemässen Mechanisierung. Allgemeine Forstzeitung. Folge 7. Juli. 3 pp.

SELLÆG, M. og GISLERUD, O., 1972: Grønn granflis fra tynninger til produksjon av trefiberplater. Driftsteknisk rapport nr. 11. Tidsskr. Skogbr. 80: 423—432.

SELVAAG, O. 1980. Synkende tempo på byggeplassen. Aftenposten *121*(249). 1 pp. [Aftenposten 4. juni 1980.]

SHAMES, I. H., 1970: Engineering Mechanics. Statics and Dynamics. Prentice/Hall International Inc. London. XVIII + 748 pp.

SILVERSIDES, C. R. 1946. Skidding of pulpwood over a suspended wire. Woodlds Sect. Index Can. Pulp Pap. Ass. No. 877 (B—8—g). 12 pp.

SILVERSIDES, C. R. 1947. Gravity skidding of pulpwood over suspended wire or cable. Woodlds Res. Index No. 32 (B—8—g). 5 pp.

SIMPSON, J.W. 1983. Rigging procedure for spars. N.Z. Logging Industry Research Assoc. Inc. Rotorua. 35 pp.

SKAGIT STEEL & IRON WORKS. 1947. Logging hoists. Washington. 8 pp.

SKATVEDT, A., EIDE, E. & JUUL, J. G. 1931. Norske Forstmenn fra 1893—1928. Tandberg & Jensens Boktrykkeri, Oslo. 384 pp.

SKINNEMOEN, K. A. 1979. Forstmester J. B. Barth. Mannen og livsverket. Norsk Skogbruksmuseums særpublikasjon nr. 4. Elverum. 480 pp.

SKRÅMO, G. 1969. Detaljplanlegging av taubanefelt på flyfoto — en beskrivelse av metodikken. Tidsskr.Skogbr. *77*(1):75—92.

SKRÅMO, G. 1971. Detaljplanlegging av taubanefelt på flyfoto. Måleresultater. Tidsskr. Skogbr. *79*(1):145—155.

SKRÅMO, G. & FJONE, H. 1983. Taubane- og vegplanlegging med bordkalkulator, flyfoto og kart. Rapp. Nor. inst. skogforsk. 12/83: 1–90.

SLÅTTEN, T. & AMLIE, E. 1965. Opptak av synketømmer ved hjelp av kabelkran. Tidsskr. Skogbr. *73*(4):412—422.

SMALLEY, C. H. 1949. Skidding pulpwood by gravity on a suspended wire. Woodlds Sect. Index Can.Pulp Pap.Ass. No. 1028 (B—8—g). 2 pp.

SODERSTON, H. R. & DAY, C. W. R. 1940. Pulpwood skidding with mechanical cable skidder. Woodlds Sect. Index Can.Pulp Pap.Ass. No. 570 (B—8—g). 3 pp.

SOLBRAA, O. & SKAAR, R. 1969. Overflatestabilisering av skogsbilveier. Tidsskr.Skogbr. *77*(1):131—150.

SPÅNGHOLMS BRUKS AB. 1955. SYA. Stållinor. Katalog nr. 33. Karlstad. 102 pp.

STALLELAND, K. 1966. Utfyllende kommentarer til taubaneforelesning av stud. Torjus Nordbø, fremført 18/1-1966. Vollebekk. 4 pp. [Arkiv NISK.]

STATENS ARBEIDSTILSYN. 1967. Ståltau. Veiledning ved valg, bruk og vedlikehold. Rundskriv nr. 264. 15 pp.

STAVSETH, R. 1948. Av repslagerens saga. Tønsberg Reperbane gjennom 150 år. Johan Grundt Tanum, Oslo. 207 pp.

STEINLIN, H. 1953. Seilförderung bei kleinen Holzmengen. Allg. Forstztg *64*:57—59.

STEINLIN, H. 1954. Einige Möglichkeiten zur Bringung des Holzes aus dem Bestand an den Abfuhrweg unter schwierigen Geländeverhältnissen. Holz Nr. 48. 7 pp.

STEINLIN, H. 1958. Der wirtschaftliche Abstand von Seillinien bei systematischer Erschliessung mit Langstreckenseilkranen. Zürich. 12 pp.

STEINLIN, H. 1959. Die Seilbahn im Dienste des Waldes. HESPA Mitt. *9*(Aug.):7—24.

STEINLIN, H. 1961. Effecient cable operations. Schweiz. Z. Forstwes. *112*(3):224—234.

STEINLIN, H. & ZEHNTNER, K. 1953. Das Aufseilen mit Tragseil und leichter Kabelkraneinrichtung. Mitt. Schweiz. Anst forstl. VersWes. *29*(1):140—168.

STEINLIN, H. & ZEHNTNER, K. 1954. Untersuchungen zur Verbesserung des Holztransportes im Gebirge. II. Mitteilung. Mitt. Schweiz. Anst forstl. VersWes. *31*(1):102—152.

STEPHAN, P. 1926. Die Drahtseilbahnen (Schwebebahnen). Einschliesslich der Kabelkrane und Elektrohängebahnen. 4. Ausg. Verlag von Julius Springer, Berlin. 572 pp.

STEVENS, P. M. & CLARKE, E. H. 1974. Helicopters for logging. Charateristics, operations and safety considerations. USDA For.Serv.Gen.Tech.Rep. PNW—20.

STEYR-DAIMLER-PUCH AG. 1979. Forsttechnik. Steyr KSK 16/20. Information. Wien. Upag.

STREHLKE, B. 1971. Problems of occupational safety and health related to forest operations in mountainous regions. FAO/ILO/ECE symposium on forest operations in mountainous regions. LOG/SYMP. 5/58. 14 pp.

STUDIER, D.D. & BINKLEY, V.W. 1975. Cable logging systems. Oregon State University, Book Stores Inc.,Corvallis, Oregon. 209 pp.

STÖHR, G. 1974. Undersuchungen über die Eignung mobiler Kurzstreckenseilkrane bei Durchforstungen am Steilhang. Dissertation zur Erlangung des Doktorgrades des Fachbereichs Forstwissenshaft der Ludvig-Maximilians-Universität München. 176 pp.

STÅL OG TAU 1971. Norse rope. Ståltaukatalog. Tønsberg. 31 pp.

SVENDSEN, O. 1953. Pendelbaner. Upubl.arb. 29 pp.

SVENDSEN, O. 1980. Pendelbaner. (Brev 31/3). Drammen. 1 pp. [Arkiv NISK.]

SVENDSEN, O. & UBISCH v., J. 1953. Arbeidshåndbok. Løvenskiold-Vækerø, Oslo. Upag.

SØLSNES, L., 1941: Forelesninger i Maskindeler. Rem- og Taudrift. NTH, Trondheim. 94 pp. (stensil).

SØRHAGEN, O. 1978 a. Måleenheter i SI-systemet. Tidsskr.Skogbr. *86*(4):263—267.

SØRHAGEN, O. 1978 b. Sammenligning av kjøring med totromlet vinsj som returbremset system og som interlukket system. Tidsskr.Skogbr. *86*(4):419—424.
TAKLE, J. 1977. Utviklings- og utformingsalternativer for vinsj- og taubaneutstyr. Norsk institutt for skogforskning, Ås—NLH. 88 pp.
TARALDRUD, H. & SAMSET, I. 1972. Kostnadskalkyler for skogbruksmaskiner (En forts. av Samset, I. 1969. Kostnadskalkyler for skogbruksmaskiner. Tidsskr.Skogbr. *77*(1):21—39). Tidsskr.Skogbr. *80*(4):357—374.
THIERY, E. & DEMONET, Ch. 1896. Les transports par cables aeriens. Imprimerie A. Nicolle, Nancy. 112 pp.
TRZESNIOWSKI, A. 1972. Fragen der Ausbildung von Kleinwaldbesitzern und Forstarbeitern. Allgemeine Forstzeitung. Folge 7. Juli. 2 pp.
TUFFIER, M. 1954. Du choix et de l'entretien des cables. Bois Forets Trop. *37*(Sept.—Oct.):31—66.
A/S TØNSBERG REPERBANE. 1960. Moderne ståltau for heiser og kraner. Tønsberg. 11 pp.
A/S TØNSBERG REPERBANE. s.a. Katalog over tauverk og ståltrådtau. Oslo. 102 pp.
UNION WIRE ROPE CORPORATION. 1958. Wire rope splicing. Kansas, Missouri. 32 pp.
USHER, I. A., and SPROULE, L. W., 1951: Wire Rope Research at the Ontario Research Foundation. I: Wire Ropes in Mines. Proceedings of a Conference held at Ashorne Hill. Leamington Spa, Warwickshire in September 1950, pp. 485—516. The Institute of Mining and Metallurgy. London.
VALLILLEE, J. E. s.a. Brake for pulpwood hauling operations. Woodlds Sect.Index Can.Pulp Pap.Ass. No. 516 (B—11). 2 pp.
VESTAD, O. 1971. Analyse av driftsstatistikk for Logma T-300 teleskopkvister. Tidsskr. Skogbr. *79*(1):71—82.
VIK, T. 1983. Arbeidstyngden ved drift av Moxy kabelkran. Driftsteknisk rapport nr. 29, Rapp. Nor. inst. skogforsk. 18/84.
VIK, T. 1983. Arbeidstyngden ved drift av Variovinsj. Driftsteknisk rapport nr. 29. Rapp. Nor. inst. skogforsk. 18/84.
VIK, T. 1983. Arbeidstyngden ved drift av Iglands Teleskopvinsj. Driftsteknisk rapport nr. 29. Rapp. Nor. inst. skogforsk. 18/84.
VIK, T., KRISTIANSEN, S. A. & NYLAND, T. H. 1980. Hogst av stammer og trær i meget bratt terreng. Prestasjoner og arbeidsbelastning. Tidsskr.Skogbr. *88*(1):17—41.
DE VRIES, J. 1967. The cost implications of owning and operating forest machinery in Australia. Forestry and Timber Bureau. Canberra. Leaflet No. 108. 21 pp.
VYPLEL, K. 1964. Die Holzbringung im Bergland. Holzkurier Nr. 8 (5 pp.).
VYPLEL, K. 1979. Mechanized harvesting in steep terrain. Symposium on mountain logging. University of Washington, College of Forest resources, Seattle. p. 117–121.
VYPLEL, K., SCHAFHUBER, F. & AUER, E. 1962. Bericht über einen Seilbringungsversuch im Schwachholz. Göss. 37 pp.
WELLBURN, G. V. 1976. Cable logging is effective but to costly. Pulp Pap.Can. *77*(10):37—40.
WESETH, G. 1952. Vinsjdrift på brattlendte bruk. Landbruksteknisk Institutt. Melding nr. 5. Vollebekk. 86 pp.
WESTERN, H., 1970: Tre- og kvistdata. Driftsteknisk rapport nr. 10. Tidsskr. Skogbr. 79: 83—93.
WETTSTEIN, R., 1971: Construction and Lifetime of Wire Ropes. FAO/ILO/ECE — LOG/SYMP. 5/79. Geneve. 5 p.
WILLAMETTE IRON & STEEL WORKS. 1925. Logging machinery. Portland, Oregon. 155 pp.
WRIGHT, G. H. von. 1967. Essay om naturen, mennesket og den vitenskapelig-tekniske revolusjon. Naturen *91*(3):155—180.
WYSS, Th., 1956: Stahldrahtseile der Transport- und Förderanlagen. Schweizer Druck und Verlagshaus A. G. Zürich. 118 pp.
YAMAWAKI, S. 1958. On a relation between the coefficient of rolling resistance of logging vehicles and the degree of sinking of tires into the ground. J. Jap.For.Soc. *40*:515—516.
YOUNG, H. E., 1967: Complete Tree Mensuration. The Forestry Chronicle 43: 360—364.
YOUNG, STRAND, L. and ALTENBERGER, R., 1964: Preliminary fresh and dry weight tables for seven tree species in Maine. Maine Agricultural Experiment Station. Technical bulletin 12: 1—76.
ZEEH, Fr. 1927. Die Zugspitzbahn. Siemens Zeitschrift *9*:618—628.
ØSTTVEDT, E. 1954. De norske jernbaners historie. Bind III. Norges Statsbaner i kommisjon hos Cappelens Forlag, Oslo. 587 pp.

INDEX

Acceleration distance 63
Access road 178, 415
Acoustic signal 172
Aerodynamic lift 478, 488
Aerostatic lift 478
A-framed support 113, 431
Airships 481
Ammonia 489
Anchoring strap 434
Anchors 97
Andrews 27, 39, 40, 41, 44
Andritz hydrocrane 350, 351, 353
Anleggstransport cable crane 282
Arm signals 173
Arola 223
Automatic carriage stop 259, 265, 338, 339
Automatic unhooking 189, 215
Auxiliary winch 418
Availability 22

Back-pack spool 420, 430
Baco cable crane 35, 280, 284
Bacon 25
Balloon envelope 491
Balloon logging 478, 505
Ballonet 492
Band brake 76, 77
Barth 46
Basin-shaped hillsides 417
Beartrap binder 194, 203
Bedding area 495
Belt variator 159, 386, 394
Bending of ropes 57
Berger yarder 346, 360
Bergkuli 33
Bielayev 140
Bjønn cable crane 282
Bleichert 32
Block in the bight system 280
Braking capstan 182
Braking line 182, 192, 200
Braking load 19
Braking pulley 192
Byrkeland 50

Cable length 140
Cable oscillations 60, 161
Cable shoe 139
Cable skidding 81, 176
Cable transport 81, 176
Cable yarding 176
Calendar time 44
Cappeloni 140
Capstan 66, 68, 69, 159
Carriage stop 251, 254, 256, 257, 302, 326, 333
Carriage with hoist winch 320, 322, 324, 325, 330
Carrying cross-section 507
Carson and Jørgensen 379
Catcher (steel egg) 396
Chain chokers 214
Chainsaw winch 425
Chaser 167
Chasing 229
Choker setter 167
Christy carriage 273
Clamp carriage with 3-way swivel 306
Clark 44
Climbing shoe 108
Closed valleys 417
Coefficient of friction 82, 83, 87
Coefficient of skidding 82, 83
Coefficient of transport 82, 83
Combi-Cat 321
Come-along 419
Communication radio 171
Concave profile 438
Cone-shaped hillsides 417
Consumer price index 503
Convex profile 438
Conway 227, 359, 477
Cornwall 44
Cost formulae 514, 515
Cottell, McMorland and Wellburn 476
Crew 508, 509
Critical site yarder 272
Crossline support 113, 114, 175, 431
Cross-sectional area 506
Culverts 415
Cyclo-Crane 485
Czitary 140, 162

Dead man 101
Deflection 17
Delay time 21
DePrelies 35
Dolbeer 27, 40
Donkey yarder 40
Double haul-back line 276
Downhill North Bend 299
Downhill yarding 250
Drake 38
Driving power 158
Drottninghaug 28
Drum capacity 64

Drum carriage 374, 378
Drum speed variation 159
Dunham system 366
Dukielskii 140
Dycker 31
Dykstra 359

Earth anchor 102
Eccentric grip 429
Eco logger 359
Effective time 21
Efficiency 152, 153
Elastic elongation 18, 119, 120
Endless braking line 200
Endless running skyline 153, 391
Endless Tyler 297
Energy consumption 516
English system 31
Ergonomic work load 168
Extensometer 161

Factor of safety 54
Fan brake 75, 76
Feeder roads 178, 415
Fibre rope chokers 214
Field crew 167
Field telephone 169
Finger support 108, 115
Fish-hook yarding 223, 268, 269
Fixed hoist line 397
Fixed skyline 176
Fleet angle 60
Foreman 167
Forest operations 13

Gantner 35
German system 30, 31, 33
Giordano 31, 36
Girak 35
Glynn 36
Goose-neck boom loader 237
Gosch 35
Grabinski 42
Grabinski system 280
Grapple carriage 376
Grapple skidder 238
Grapple yarder 280, 408, 413, 504
Gravity inhaul system 211, 250
Gravity outhaul system 211, 264, 275, 277, 294, 388
Gravity rolling 224
Gravity transport 26, 181
Ground skidding 80, 81, 84, 86
Grooved chain pulley 201, 203, 208
Groove wheel pulley 37, 65, 69, 70, 71, 334
Gulisashvili 140

Gypo contractors 503
Guyline 105
Guyline drum 376, 380, 381, 418
Gösser cable crane 35, 240

Hafner 31, 34, 415
Handling line 492
Harpoon gun 421, 422
Haul-back line 85, 92, 175
Head spar 175
Heel boom 505
Heelhold 175
Helicopter logging 479
Helistat 484
Helium 489, 490
Herringbone skidding 292
High climber 42, 435
Highlead 505
Highlead skidding 81, 88, 89, 90, 91
Hilf 44
Hinteregger 35
Hoist line 85, 93, 95, 175
Holt, Benjamin 43
Hooktender 167
Hordaland Mek. Verkstad 51
Horse team 38
Hot logging 230, 233, 238
HPC-yarder 382
Huber 35
Hybrid vehicles 483
Hydrogen 489
Hydrostatic drive 159
Hydrostatic interlocked winch 377
Håland 51

Igland Alp cable crane 51, 326
Igland highlead yarder 364
Inhaul direction 210
Interlocked winch 157
Interlocked winch with hydraulic speed variation 382
Interlocking torque 160
Intermediate support 112
Isachsen 36, 51
Iwafuji 503

Jammer 505

Kato 45
Kaierle and Löffler 382
Kaufmann 36
Knuckleboom loader 238, 241
Koller cable crane 35
Koller carriage 269
Koller/Nestestog carriage 337
Koller yarder 265, 267, 274

Konuma and Shibata 295, 298, 299, 324, 366, 367, 502
Koroleff 516
Kumamoto system 281, 299
Küpfer 35
König 32

Labour costs 503
Landing 229, 231
Lang's lay 28
Lasso mono cable 240, 242
Lateral skidding 93, 94, 95, 239
Law of discontinuous evolution 518
Lidgerwood 43
Life time of ropes 56
Line horse 39
Lines of transport 176
Link belt yarder 273
Live skyline 176, 211, 273, 388
Load chain 186
Load curve 117, 128
Load line 492
Load path 128, 137
Load resistance 78, 149
Loading platform 186, 199, 206
Log deck 232
Logging railway 41
Lund 48
Lysons 489
Løland cable crane 53, 324

Machine utilization 22
MacIntosh 486
Madill slack-pulling carriage 356
Madill Yarder 3-section drum carriage 345
Main line 85, 175
Mann 486
Mass 19
McMorland and Wong 359
Means of transport 176
Megahertz 20
Meyr 34
Micro Master 359
Milliwatt 20
Mobile yarder 35, 266, 315, 316, 344
Modulus of elasticity 19, 120, 121
Mono cable 240, 244
Moxy cable crane 374
Multiple sheave block 61
Multiple sheave block carriage 302, 304
Multispan skyline 136, 137

Nansei mono cable 245
Nansei yarder 285, 287
Natural shaped balloon 491, 492
Nessler yarder 253
Nestestog cable crane 51, 330
Nestestog carriage 337
Non-riggid airships 482
North Bend 280

Obihiro cable crane 367
Okawara system 311, 313
O'Leary 280
Omnes 412
Oriental civilization 25
Oscillating wave 164
Outhaul direction 210
Owren Vario Crane 384, 386
Owren Vario Winch 382, 383
Oxen team 38

Paternoster 246
Pearse 44
Pendulum cable way 51, 179, 181
Pestal 33, 36, 165
Planetary drive variation 372, 373
Planetary gear 159
Planning 417
Plateaus 417
Portal support 113
Power 20, 62, 157
Pre-bundling 222
Pre-rigging 436
Pre-tension 142
Production cycle 212, 496
Productivity 510, 511, 512
Pulley 159
Pulley brakes 73, 76
Pulley drives 66

Quick coupling 224
Quick drop system 371

Radio communication 171
Radio controlled cable crane 324, 329
Radio controlled carriage 337, 338, 339, 357
Raven logging balloon 489
Regeneration of power 157, 159
Reinton 49, 50
Relative costs 513
Relative height 18
Retardation distance 63
Rigging 213, 227
Rigging slinger 168
Rig up ladder 109
Rig up line 376, 381, 421, 423, 430
Rig up spool 421
Rig up time 473 – 478
Rig up tools 418

Rig up winch 423, 427
Road network 15, 414
Rock bolt anchor 103
Rope deflection 125
Rope deflection angle 138
Rope sag 17, 125, 143
Running skyline 146, 176, 211, 280, 362, 372, 388, 413, 504
Rørlien 51

Salvage winch 85
Sauder 477
Schmitz 44
Schneigert 23, 141
Seim 51
Self-propelled carriage 341
Self releasing choker 218, 220, 221
Shield Bantam yarder 405
Shoot gun system 280, 504
Signaller 167
Signal systems 169
Single cable with circulating cables 209
Skagit bullet carriage 358
Skagit cable crane 280, 345, 360
Skagit torpedo carriage 280, 358
Skidder mounted yarder 316
Skyline 175
Skyline drum with storage section 267
Skyline tension 117
Skråmo 417
Slackline system 280
Slack-pulling block 396
Slack-pulling carriage 376, 378
Slack-pulling line 356
Slipping clutch interlocked winch 159, 367
Smith Berger cable crane 280
Smith Berger slack-puller carriage 375
Solbraa and Skaar 415
South Bend 280
Southwell 30
Spar tree 104, 106, 434, 436
Stavseth 28
Steel spar 43, 108, 110, 111
Steel wire for gravity transport 48, 49, 50, 179
Steinlin 35
Stephan 23, 24, 26, 29, 32
Steyr 35
Steyr carriage 356
Steyr thinning-paternoster 246
Steyr yarder 351, 361
Storage drum 65
Straw line 419, 428
Stress 19
Studier and Binkley 506
Suckerdown block 500
Sundberg 486
Support 175
Support on running skyline 402
Surface culverts 415
Sveen cable crane 306
Svendsen 495
Swing boom 112
Swing boom yarder 406

Taccola 29
Tackle 429
Tagline 224, 226, 497
Tail anchor 432
Tail block swivel 434
Tailhold 175
Tail spar 175
Tail tackle 432
Talkie-tooter 172
Taylor's expansion 140
Telescope spar 108, 364, 365
Telescope tower 395
Tension measurement 59, 161
Tension measurement by stop watch 161
Tension weight 118, 199
Terminal functions 177
Terminal places 178, 416
Terminal time 21
Tether line 492
TDT 60 yarder (USSR) 363
Thinning-paternoster 246
Tightening device 419
Tightening line 391, 429
Tight skyline 119, 176, 211
Timber shoot 46, 47
Tip spar 108
Tip/telescope tower 318
Tip tower 318, 387
Top mounted cable crane 251
Top mounted yarder 249
Torque 20
Total working time 21, 503
Transfer vehicle 494
Tree-weight 79
Truck mounted yarder 316
Tröim cable crane 320
Tröim carriage 319
Turn cycle 212, 496
Twin cable with circulating carriages 31, 37, 179, 281
Tyler system 280, 295, 296

Under-carriers 317
Underway functions 177
Unhooking line 218
Uniform hillsides 417
Uphill skidding 150, 152

Uphill yarding 264
Urus cable crane 270
Urus yarder 271

Vossa yarder 282
V-shaped balloon 486, 488

Washington yarder 280, 370
Wedge grip 101
Weseth 33
West Coast Tower 347, 348, 360
Willamette 42
Winch operator 167
Winch trailer 427
Wire ropes 28
Wooden groove wheel 193
Work place hour 473
Work place time 21, 503
Wright 25, 26
Wyssen 34, 35
Wyssen cable crane 263, 280
Wyssen yarder 253

Yarder 176
Yarder for balloon logging 493
Yarder on crawler chassis 316
Yarding 176
Yarding crane 404
Yarding crew 166
Young cable crane 357, 404
Young carriage 357